T0211757

Texts in Theoretical Computer Science.
An EATCS Series

Series Editors

Juraj Hromkovič, Gebäude CAB, Informatik, ETH Zürich, Zürich, Switzerland

Mogens Nielsen, Department of Computer Science, Aarhus Universitet, Aarhus, Denmark

More information about this series at https://link.springer.com/bookseries/3214

Markus Roggenbach · Antonio Cerone ·
Bernd-Holger Schlingloff · Gerardo Schneider ·
Siraj Ahmed Shaikh

Formal Methods
for Software Engineering

Languages, Methods, Application Domains

With a foreword by *Manfred Broy*
and a contribution on the origins and
development of Formal Methods by *John V. Tucker*

 Springer

Markus Roggenbach ⓘ
Department of Computer Science
Swansea University
Swansea, UK

Antonio Cerone
Department of Computer Science
Nazarbayev University
Astana, Kazakhstan

Bernd-Holger Schlingloff
Institut für Informatik
Humboldt-Universität zu Berlin
Berlin, Germany

Gerardo Schneider
Department of Computer Science
and Engineering
University of Gothenburg
Gothenburg, Sweden

Siraj Ahmed Shaikh
Institute for Future Transport and Cities
Coventry University
Coventry, UK

ISSN 1862-4499   ISSN 1862-4502 (electronic)
Texts in Theoretical Computer Science. An EATCS Series
ISBN 978-3-030-38802-7   ISBN 978-3-030-38800-3 (eBook)
https://doi.org/10.1007/978-3-030-38800-3

This Springer imprint is published by the registered company Springer Nature Switzerland AG
The registered company address is: Gewerbestrasse 11, 6330 Cham, Switzerland

The title page shows detail from the cover of the book "Rechnung auff der Linihen und Federn" by Adam Ries, Sachsse, Erffurdt, 1529.[1] It depicts the following scene: "A table-abacus competes against longhand calculations using 'Arabic' numerals, which were still new in Europe. Either one could trounce calculating by hand with Roman numerals—but which was faster? Are the coins a wager on the outcome?"[2]

In his book Ries describes two practices: working with the calculation board (established practice) and numerical calculations with digits (new practice). Historically, as we all know, the new practice as the superior one took over.

In the same sense, the authors of this book hope that their advocated approach of utilising Formal Methods in software engineering will prove to be of advantage and become the new standard.

---

[1] Cover of the book "Rechnung auff der Linihen und Federn" by Adam Ries, Sachsse, Erffurdt, 1529. Digitized by SLUB Dresden.
   Link to the image: http://digital.slub-dresden.de/id267529368/9.
   Link to the rights notice: https://creativecommons.org/publicdomain/mark/1.0/.
[2] See: https://www.computerhistory.org/revolution/calculators/1/38/139.

Foreword by Manfred Broy

The development of programs and software engineering are fascinating technical challenges. If software runs on a piece of hardware, if the hardware is embedded into a cyber-physical system, and as soon as the system is started, a process is initiated and the system shows some behaviour and—if designed and programmed in a careful way—it performs a certain task and it generates a behaviour which fulfills specific expectations.

As we have painfully experienced, software systems show a complexity, especially if they are large and used in complicated applications that are often beyond the imagination of engineers. As a result, we all have learned to work and live with imperfect software systems that often show behaviours and properties which are different from what we expect and—in the worst case—do not perform the task software was written for. This is unacceptable—not only in safety critical applications.

As a result, there is a lot of research to find better ways to engineer software systems such that they become reliable and show high quality. High quality means that they provide adequate user interfaces, guarantee the expected functionality, or, even more, over-fulfill the purposes they are built for and that they behave never in an incorrect way. For the engineering of such systems, a large number of proposals have been published and also experimented with, in practice. Some of them being quite useful and successful, others did not deliver what they promised.

An important observation is that computer programs and software in general are formal objects. They are written in a formal language, they are executed on a machine with a formal operational semantics, and each statement of the programming language results in precisely defined behaviours of the machine (state changes, input, and output) exactly determined by the software. In the end, strictly speaking, software is just a huge formula—however, usually not written in the classical style of mathematical formulas, but in the style of algorithmic languages. But, after all, it is a formal object. This means that we are and should be able to provide a kind of a formal theory that describes the elements of the programming languages and the behaviour of programs that is expressed and generated by these elements.

This underlines that there is a difference between writing a text in a natural language and writing a program. Soon, we have learned that writing a program is error-prone. Too many things have to be kept in mind and thought of when writing a line of program text such that it is very likely that what we are writing is sometimes not what we want.

Here formal theories can help a lot to give a precise meaning and some deep understanding, not only for programs and the behaviours they generate, but also for specifications which formally describe certain aspects of program behaviour. The main contribution of formalization is precision, abstraction, and helpful redundancy. Redundancy means that we work out different—if possible formal—more adequate formulations of specific aspects that support the concentration onto specific properties. This way, relevant aspects are represented in isolation to be able to study them independently which may reduce complexity. This has led to a number of formal theories addressing quite different aspects of programs including their functional behaviour, quality issues, and questions of robustness.

This shows that theories providing formal foundations for formalisms, languages, and also for methods in software construction are indispensable artifacts to support software development.

In the academic community, having all this in mind, soon the term "Formal Methods" has been become popular. This term is carefully defined and explained in this book. It is illustrated both by examples and use cases as well as by careful discussion and proper definitions.

For Formal Methods, the challenge is to keep the balance between being formal and providing methods. In this book, numerous examples are given for such a line of attack, but we have to always keep in mind that it is dangerous to define a formalism and to believe that this formalism is already a development method. However, here is another challenge: in the details of the definitions of formalisms, we have to decide about concepts that are critical and difficult. A simple example would be the use of partial functions in specifications: as long as all functions are total, expressions written with these functions have well-defined values. For partial functions, it gets much trickier: what is the value of an expression when certain subexpressions are built of partial functions which happen not to provide a result for this particular application? What is the value of the overall expression then? Is it always undefined? What are the rules to deal with this and to reason about it? Of course, there are many different ways to provide a theory for expressions with partial functions, but obviously not all of them are equally well-behaving and well-suited for engineering. Therefore, when defining formal theories, a rich number of delicate questions have to be solved—many of them related to the goal to use the formalism as an element of a Formal Method.

Another example is how to represent concurrency. Concurrency is a fact of everyday life. We are living in a concurrent world. Our software systems are connected and run concurrently. There are a number of constructs that have been invented to describe concurrent structures and concurrent processes of software systems by formal theories, and again there are challenges—first of all, to come up with a sound theory and a formal model and second to deal with the question whether the theory is exactly addressing the structures and behaviours which are typical for practical systems on one side and are easy to deal with on the other side.

Therefore, it is a very valuable contribution of this book to present an interesting selection of formal theories and to explain how they can be used in the context of methods for software engineering. Certainly, this is a book highly relevant for people interested in formal theories for software engineering usable as elements of methods. It also addresses students in informatics who want to learn about this subject and, even more, scientists who work on formal theories and methods.

I hope this book will also find interest by practical engineers to give them some clue how formal foundations and rigorous methods could be combined to formal methods to help them in their everyday development tasks.

July 2021

Manfred Broy

The original version of the book was revised: The Author names have been updated on Springer Link for all chapters. The correction to the book is available at https://doi.org/10.1007/978-3-030-38800-3_10

Preface

This book is about Formal Methods in software engineering. Although software engineering is nowadays a largely empirical science, its foundations rely on mathematics and logic. Ultimately, the task of a software engineer is to transform ideas into programs. Ideas are by nature informal, and they are often vague and subjective. In contrast, a program is a formal entity with a precise meaning, and this meaning is independent of the programmer. Therefore, the transition from ideas to programs necessarily involves a formalisation at some point. An early formalisation has several benefits:

- It allows to formulate concepts on an abstract level;
- it is a means for unambiguous communication of ideas;
- it helps to resolve misunderstandings, thus preventing errors at a later stage; and
- it enables to gain insights by transformation, simulation, and proof.

Formal Methods are a way to realize these advantages in a rigorous process.

This book elaborates on several views of how to do this. In Chap. 1, we approach a definition of what actually constitutes a Formal Method. The rest of the book is structured into three parts: languages, methods, and application domains. These parts represent different dimensions of the views:

1. A *language* is a means to formally describe ideas;
2. a *method* is a set of procedures for manipulating such descriptions; and
3. an *application domain* represents a concrete way in which real-life problems drive the different views.

Each part consists of several chapters which are more or less independent.

In the *languages part*, we present "classical" views on elements of computation.

Chapter 2: Logics are formal languages to describe reasoning.
Chapter 3: The process algebra Csp is a formal language to describe behaviours.

In the *methods part*, we discuss a variety of procedures.

Chapter 4: Casl is a computer supported method for the specification of software, which is based on classical logic as discussed in the language part.
Chapter 5: Specification-based testing is a computer supported method for the validation of software.

Finally, the *application part* provides three contributions to apply Formal Methods to real-world problems.

Chapter 6: In the chapter on specification and verification of normative documents, we discuss a way to reason about legal contracts with logic.
Chapter 7: In the chapter on Formal Methods for human-computer interaction, we discuss how to capture cognitive theories with logic and CSP.
Chapter 8: In the chapter on formal verification of security protocols, we discuss how to verify authentication properties with CSP.

These three chapters have in common that they present solutions to general challenges in software engineering. These solutions are based on the application of one specific Formal Method. It should be noted, though, that other Formal Methods would be applicable to these challenges as well.

We conclude our book by providing a historical perspective on Formal Methods for software engineering:

Chapter 9: In the chapter on the history of Formal Methods, John V. Tucker surveys some of the problems and solution methods that have shaped and become the theoretical understanding and practical capability for making software.

This is followed by some summarizing and reflecting remarks from the book authors.

Audience, Prerequisites, and Chapter Dependencies

This book addresses final year B.Sc. students, M.Sc. students, and Ph.D. students in the early phases of their research. It is mainly intended as a underlying textbook for a university course. Formal Methods are one means in software engineering that can help ensure that a computer system meets its requirements. They can make descriptions precise and offer different possibilities for analysis. This improves software development processes, leading to better, more cost-effective, and less-error-prone systems.

Due to their ubiquity, software failures are overlooked by society as they tend to result in nothing more serious than delays and frustrations. We accept it as mere inconvenience when a software failure results in a delayed train or an out-of-order cash machine or a need to repeatedly enter details into a website. However, the problems of systems failures become more serious (costly, invasive, and even deadly) as automatic control systems find their way into virtually every aspect of our daily lives. This increasing reliance on computer systems makes it essential to develop and maintain software in which the possibility and probability of hazardous errors are minimised. Formal Methods offer cost-efficient means to achieve a high degree of software quality.

However, in computer science and software engineering education, Formal Methods usually play a minor role only.[3] Often, this is due to the lack of suitable textbooks. Typical questions an academic teacher faces when preparing such a course include the following: Which of the many Formal Methods shall be taught? Will the topics be relevant to mainstream students? Which examples and case studies should be used? This book offers constructive answers to such questions. It does not focus on one specific Formal Method, but rather provides a wider selection of them. For each method, material from basic to a more advanced level is presented. Thus, the teacher can choose to what depth a specific method shall be studied. All material is illustrated by examples accessible to the target audience.

Moreover, for individual students, this book can serve as a starting point for their own scientific work, e.g., in a thesis. Even if the reader does not plan to work directly in one of the addressed areas, the book offers solid background knowledge of Formal Methods as a whole.

We assume some basic knowledge of mathematical notation as taught in the first two years of typical B.Sc. curricula in computer science or software engineering. However, we will introduce all formal concepts from scratch, whenever they are used. For the casual reader, the book contains an index, where one can look up the defining page for each technical term.

The material in Part I is foundational for the subsequent parts, whereas the chapters in Parts II and III can be read in any order. General dependencies are depicted in Fig. 1.

More specifically, dependency on the introduction is only from a motivational, but not from a technical point of view. In reading the book, Part I can serve as a "reference", Part II and Part III depend on Part I only in some technical aspects. The reader interested just in specific topics of these parts can safely start there and refer to Part I only when needed.

Although the linear order of reading the chapters would be preferred, for readers who want to focus on specific aspects, the authors suggest two possible alternative paths through the book. Chapter 1 provides a common start to both.

The first path is for those who wish to stay with logic: Chapter 2 leads on to Chap. 4 to provide a grounding in logic and the use in algebraic specifications. Chapter 6 follows on as an area of application for modal logics.

---

[3] See, e.g., Cerone et al., *Rooting Formal Methods within Higher Education Curricula for Computer Science and Software Engineering*, 2020, https://arxiv.org/abs/2010.05708.

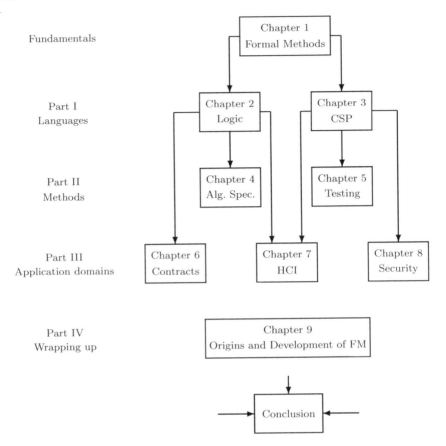

Fundamentals

Part I
Languages

Part II
Methods

Part III
Application domains

Part IV
Wrapping up

Fig. 1 Structure of the book

An alternative path starts with Chap. 3, thoroughly covering CSP both in theory and practice. Chapter 5 offers a formal perspective on testing. Chapters 7 and 8 provide case studies both using CSP to demonstrate how the process algebra is applied. Only the last part of Chap. 7 depends on logic, limited to temporal logic.

Chapter 9, written by our colleague John V. Tucker, puts the contents of the previous chapters into the historical context. It can be read at any time and it is independent of any of the other chapters.

The conclusion serves to summarise and remind the reader of the final message of the book. It is the natural ending to any reading path.

Book Use and Online Supporting Materials

This is not a typical software engineering book. Nor is it promoting a particular formal approach as many books on the subject do. Formal methods are increasingly *acknowledged* amongst the wider software community. However, there is no evidence to suggest that they are widely *adopted*. It is this gap that this book is designed to address. The use of tools is emphasised and supported; the expectation is that certain parts are to be *done* rather than just *read*. Therefore, the authors have set up a website for the book which contains exercises and links to tools. Currently, this website can be accessed at

https://sefm-book.github.io.

Book History

The inception of this book is due to the first International School on Software Engineering and Formal Methods held in Cape Town, South Africa, from late October to early November of 2008, organised by Antonio on behalf of the United Nations University International Institute for Software Engineering (UNU-IIST), which was located in Macau, SAR China. The two-week school consisted of five courses on the application of Formal Methods to software design and verification delivered to an audience of graduate and research tudents from several African countries, who were hosted by UNU-IIST. In line with the UNU-IIST mandate, the authors were encouraged to find young minds taking up the challenge of Formal Methods and demonstrating commitment to it. The book draws upon the topics of the school with a similar audience in mind and a strong desire to make the subject more widely accessible. Hence learning is promoted through examples running across the book. The pedagogic style is largely owed to the instructional setting offered by the school.

Two more schools followed, in Hanoi, Vietnam, in November 2009, and in Thessaloniki, Greece, in September 2012, are also hosted by UNU-IIST. These events provided additional opportunities for feedback and reflection from school participants. UNU-IIST hosted Markus for one week in 2009. During that meeting, Antonio and Markus sketched the first structure of the book. UNU-IIST organised a one-week workshop in August 2012 in Mezzana (Val di Sole), Italy. During this workshop, the authors decided the final structure and content of the book. After the closing of UNU-IIST in 2013, the authors continued the collaboration through regular virtual meetings and some physical meetings in Coventry and Swansea, UK. Since January 2020, Antonio Cerone, School of Engineering and Digital Sciences, Nazarbayev University, Nur-Sultan, Kazakhstan, has been partly funded to work on the book by the Project SEDS2020004 "Analysis of cognitive properties of interactive systems using model checking", Nazarbayev University, Kazakhstan (Award number: 240919FD3916).

During the years since the Mezzana workshop, the book content has been updated and widely tested in undergraduate and postgraduate courses by the authors and a number of their colleagues at various universities around the world. The intense cycle of collaborative writing, internal reviewing, and in-class testing was followed by an external reviewing process, in which the reviewers offered their reflections on individual chapters and then incorporated in the final revision by the authors.

Author Team

The book's content, organisation, and writing style were curated by the five book authors. The author team reached out to John V. Tucker, who kindly accepted our invitation to contribute a chapter on the origins and development of Formal Methods. For the writing of some individual chapters, the author team invited Liam O'Reilly for the chapter on algebraic specification in CASL and Hoang Nga Nguyen for the chapter on formal verification of security protocols. The book authors are grateful for their contributions, which made it possible for the book to appear in its current form.

Swansea, UK   Markus Roggenbach
Nur-Sultan, Kazakhstan   Antonio Cerone
Berlin, Germany   Bernd-Holger Schlingloff
Gothenburg, Sweden   Gerardo Schneider
Coventry, UK   Siraj Ahmed Shaikh
September 2021

Acknowledgments

The authors would like to express their gratitude to a number of people. The following colleagues, listed in alphabetical order, were so kind to contribute in different ways, including reading (parts of) the book and commenting on it: Ulrich Berger, Mihai Codescu, Alan Dix, Stephen Fenech, Marie-Claude Gaudel, Michael Harrison, Magne Haveraaen, Yoshinao Isobe, Alexander Knapp, Ranko Lazić, Antónia Lopes, Faron Moller, Till Mossakowski, Gordon Pace, Jan Peleska, Cristian Prisacariu, Jörn Müller-Quade, Fernando Schapachnik, Steve Schneider, and David Williams. We are grateful for their input and feedback that helped greatly shaping and improving our book. Manfred Broy provided a foreword to our book, which we very much appreciate. We also would like to thank Alfred Hoffman, Ronan Nugent, Wayne Wheeler, and Francesca Ferrari from Springer Verlag for their help, support, and patience.

Swansea, UK	Markus Roggenbach
Nur-Sultan, Kazakhstan	Antonio Cerone
Berlin, Germany	Bernd-Holger Schlingloff
Gothenburg, Gothenburg	Gerardo Schneider
Coventry, UK	Siraj Ahmed Shaikh
September 2021	

Contents

Contributors

Antonio Cerone Nazarbayev University, Nur-Sultan, Kazakhstan

Hoang Nga Nguyen Coventry University, Coventry, United Kingdom

Liam O'Reilly Swansea University, Wales, United Kingdom

Markus Roggenbach Swansea University, Wales, United Kingdom

Bernd-Holger Schlingloff Humboldt University and Fraunhofer FOKUS, Berlin, Germany

Gerardo Schneider University of Gothenburg, Gothenburg, Sweden

Siraj Ahmed Shaikh Coventry University, Coventry, United Kingdom

John V. Tucker Swansea University, Wales, United Kingdom

List of Examples

Chapter 1
Formal Methods

Markus Roggenbach, Bernd-Holger Schlingloff, and Gerardo Schneider

Abstract Formal Methods are one means in software engineering that can help to ensure that a computer system meets its requirements. Using examples from space industry and every programmer's daily life, we carefully develop an understanding of what constitutes a Formal Method. Formal Methods can play multiple roles in the software design process. Some software development standards actually require the use of Formal Methods for high integrity levels. Mostly, Formal Methods help to make system descriptions precise and to support system analysis. However, their application is feasible only when they are supported by tools. Consequently, tool qualification and certification play a significant role in standards. Formal Methods at work can be seen in the many (academic) surveys, but also in numerous published industrial success stories. Hints on how to study Formal Methods in academia and on how to apply Formal Methods in industry conclude the chapter.

1.1 What Is a Formal Method?

You have just bought a book on Formal Methods and are making holiday plans in the Caribbean to read it on the beach. In order to guarantee the reservation, your travel agent requires a deposit. You decide to pay electronically via credit card.

When performing such a transaction, obviously you have certain expectations on the electronic payment system. You don't want the agent to be

Markus Roggenbach
Swansea University, Wales, United Kingdom

Bernd-Holger Schlingloff
Humboldt University and Fraunhofer FOKUS, Berlin, Germany

Gerardo Schneider
University of Gothenburg, Sweden

© Springer Nature Switzerland AG 2022, corrected publication 2022
M. Roggenbach et al., *Formal Methods for Software Engineering*,
Texts in Theoretical Computer Science. An EATCS Series,
https://doi.org/10.1007/978-3-030-38800-3_1

able to withdraw more than required. The agent wants at least the amount which was asked for. Thus, both you and the agent expect that the payment system gets its numbers right. The payment should go through—as, clearly, your credit card is a valid one. Also, you don't want too much information to be disclosed, e.g., your PIN should stay secret. The transaction should solely concern the holiday reservation, no further contracts shall follow from this. Finally, you want to be able to use the system without the need to consult a user manual of hundreds of pages. All these points are typical requirements for an electronic payment system. Formal Methods are one way how software engineering can help to ensure that a computer system meets such requirements.

So, what is a Formal Method? Instead of trying to start with a comprehensive definition of the term, we give two motivating examples.

1.1.1 An Application in Space Technologies

Formal Methods are often used in safety-critical areas, where human life or health or a large sum of money depends on the correctness of software. We start with an example from the largest aerospace project mankind has endeavored so far.

Example 1: ISS Fault Tolerant Computer

The International Space Station (ISS) which was docked on November 2nd, 2000 (ISS-Expedition 1), has provided a platform to conduct scientific research that cannot be performed in any other way.

At the heart of the ISS is a fault tolerant computer (FTC) "to be used in the ISS to control space station assembly, reboost operations for flight control and data management for experiments carried out in the space station" [BKPS97].

In outer space, the probability of hardware faults due to radiation is much higher than on earth. Thus, in the ISS-FTC there are four identical interconnected hardware boards, which perform essentially the same computation. A software fault management layer is responsible for detecting, isolating, rebooting and reintegrating malfunctioning boards.

One problem in the design of this layer is the recognition of a faulty board, since it not only can generate wrong messages, but also modify messages of the other (correct) boards. To overcome this problem, a so-called Byzantine agreement protocol is used, which abstractly models the problem of distributed consensus in the presence of faults.

Lamport et al. use the following story to exemplify the distributed consensus problem [LSP82]:

We imagine that several divisions of the Byzantine army are camped outside an enemy city, each division commanded by its own general. The generals can communicate with one another only by messenger. After observing the enemy, they must decide upon a common plan of action. However, some of the generals may be traitors, trying to prevent the loyal generals from reaching agreement. The generals must have an algorithm to guarantee that

- **A.** All loyal generals decide upon the same plan of action.

The loyal generals will all do what the algorithm says they should, but the traitors may do anything they wish. The algorithm must guarantee condition **A** regardless of what the traitors do. The loyal generals should not only reach agreement, but should agree upon a reasonable plan. We therefore also want to ensure that

- **B.** A small number of traitors cannot cause the loyal generals to adopt a bad plan.

In Lamport's paper, various pseudocode algorithms for this problem are given and proven to be correct. For these proofs, certain assumptions about the possible actions of the generals are made, e.g., that a traitorous general may send different, contradicting messages (attack and retreat) to different other divisions.

Even though Lamport et al. prove their algorithms to be correct, the questions on whether the communication by messengers can block (deadlock) or the exchange of a message can lead to infinite internal chatter (livelock) in the communication system are not in the scope of his consideration.

Example 1.1: ISS Fault Tolerant Computer – Findings

For the implementation of the fault management layer in the FTC, one of the algorithms presented by Lamport et al. [LSP82] was coded in the programming language Occam.

As mentioned above, the algorithm is proven to be correct, and great care was taken to assure that the actual code matches the pseudocode as closely as possible. However, this still did not guarantee that the software worked as expected: in a series of publications [BKPS97, BPS98, PB99], Buth et al. report that using code abstraction into the process algebra Csp,
- "seven deadlock situations were uncovered", and
- "about five livelocks were detected"
in the software of the FTC fault management layer.

The language Occam uses *synchronous communication* between tasks: the sender of a message is blocked until the receiver is willing to pick up this message. With such a communication paradigm, a deadlock can occur if two actors send each other messages at the same time. Thus, even though the algorithm on which the code is based was proven to be correct on the conceptual layer, there still was the possibility of errors in the underlying communication layer.

In Chap. 3, which is devoted to the process algebra CSP, we will give precise definitions of deadlock and livelock. Furthermore, we will provide proof techniques to show their absence.

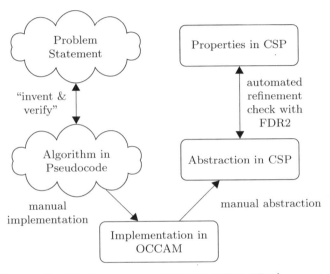

Fig. 1.1 The overall verification approach [BKPS97, BPS98, PB99]

Example 1.2: ISS Fault Tolerant Computer – Reflections

The programming language Occam has been designed as an implementation language for the process algebra CSP. Thus, it is rather easy to abstract an Occam program, e.g., the fault management layer of the FTC, into CSP. Compared to the original Occam program, the CSP abstraction has a significantly reduced state space, since all computations that have nothing to do with the Byzantine agreement protocol can be omitted. However, it still preserves the deadlocks and livelocks of the original program.

CSP has a formal semantics and proof methods to verify properties. In this example, the FDR tool was used to automatically analyse the CSP code and thus—indirectly—the Occam program. If a deadlock or livelock is found in the CSP abstraction, FDR generates a sequence of events which exhibits the problem. Figure 1.1 shows the overall verification approach. The sequence generated by FDR can be used to trace the problem in the original code, which then can be analysed and corrected.

Interestingly, most of the observed issues concerned the message exchange, i.e., the layer which was not in the scope of the formal correctness proof of the algorithm. This situation is rather typical in such a context. In Sect. 3.3, we will discuss in more detail how to avoid such problems.

In Example 1, Formal Methods were used for quality assurance rather than in the design and implementation phase. For the process algebra CSP, we will discuss a different approach in Sect. 3.3. There, the idea is to automatically translate a CSP model into a C++ program. Before translation, the model can first be analysed using a model checker such as FDR. The C++ program can then be enriched with additional functionality without compromising the properties established earlier for the CSP model.

1.1.2 An Everyday Application

Our second example is from the area of formal languages and text processing. It shows the importance of having a precise formal semantics even in common-day tools such as text processing.

Example 2: Text Processing

Assume that we want to replace all occurrences of certain patterns in a text, e.g., remove all comments from an HTML document. In HTML, comments are marked beginning with '<!--' and ending with '-->'. Most editors offer a facility for replacement based on *regular expressions*, that is, you may specify the symbol '*' as a wildcard in the search. With this, replacing '<!--*-->' by an empty string yields the desired result.

Regular replacements are a convenient tool for text processing. However, the semantics (meaning) is not always easy to understand.

The wildcard sign is explained in the documentation of Word 2007 as

matching any string of characters. Word does not limit the number of characters that the asterisk can match, and it does not require that characters or spaces reside between the literal characters that you use with the asterisk.

For GNU Emacs, it is defined by the following explanation:

The matcher processes a '*' construct by matching, immediately, as many repetitions as can be found. Then it continues with the rest of the pattern. If that fails, backtracking occurs, discarding some of the matches of the '*'-modified construct in case that makes it possible to match the rest of the pattern.

These descriptions might or might not be intelligible to the ordinary reader. However, if the text processing component is used as part of a safety-critical tool chain, it is important that it has a clear semantics. Imagine that the regular replacement is used for macro expansion as part of a compiler. In this case, it is essential that the result of any replacement command is unique and predictable.

Example 2.1: Tool Experiments for Text Processing

What happens if we replace the wildcard sign '*' by the single character 'x'? As original text we take the string 'abc'.
- In Word, the result of replacing '*' by 'x' in 'abc' is 'xxxx';
- in Word, the same result is returned when taking the wildcard symbol '?@';
- in Emacs, replacing '*' by 'x' in 'abc' gives 'abc';
- in Emacs, replacing '.*' by 'x' in 'abc' gives 'x'.

This might come as a surprise.

The problem is that both for Word and Emacs, there is no formal semantics of "replacement of regular expressions". Whereas the syntax of admissible regular expressions is (more or less) fixed in the documentation, the semantics is only informally explained.

This example allows us to show the key ingredients of a Formal Method: syntax, semantics and method.

Syntax

Syntactically, each Formal Method deals with objects from a formal language. A *formal language* is a well-defined set of words from a given alphabet. Usually it is defined by a *grammar*, which is a set of rules determining the membership of the language.

There are also Formal Methods dealing with graphical objects (e.g., Petri nets). In most of these cases, there is a textual description of these objects as well (e.g., in XML). For such models, the syntax is fixed by a metamodel, which determines membership of the class of graphical objects under consideration.

Example 2.2: Syntax of Regular Expressions

Given an alphabet $\mathcal{A}$, the language of regular expressions is given by the following grammar:
- every letter from the alphabet is a regular expression.
- $\emptyset$ is a regular expression.
- if φ and ψ are regular expressions, then $(\varphi\ \psi)$ and $(\varphi+\psi)$ are regular expressions.
- if φ is a regular expression, then φ^* is a regular expression.

The same definition can be written in so-called Backus–Naur-Form [Bac59]:

$$Regexp_{\mathcal{A}} \quad ::= \quad \mathcal{A} \quad | \quad \emptyset \quad | \quad (Regexp_{\mathcal{A}} \ Regexp_{\mathcal{A}}) \quad |$$
$$(Regexp_{\mathcal{A}} + Regexp_{\mathcal{A}}) \quad | \quad Regexp_{\mathcal{A}}^{*}$$

According to this definition, each regular expression is a string which contains only letters of the alphabet and the symbols '$\emptyset$', '(', ')', '+' and '*' (the so-called *Kleene-star*).[a] Of course, for such a definition to make sense, these symbols themselves must not be letters.[b] Backus–Naur-Form (BNF) notation will be used also for several other formal languages in this book.

---

[a] If the alphabet contains letters composed of several characters, there might be several ways to parse a given string into a regular expression. For example, if $\mathcal{A} = \{a, aa\}$, then (aaa) could be read as $(a\ aa)$ and $(aa\ a)$. A solution to this problem is to require appropriate white spaces in strings.

[b] To allow the use of special symbols in the alphabet, some tools use a 'quoting' mechanism: '\+' refers to the letter of the alphabet, whereas '+' denotes the symbol.

The benefit of having a 'minimal' syntax is that the definition of semantics and proofs are simplified. For practical applications, often the core of a formal language is extended by suitable definitions.

Example 2.3: Extended Syntax of Regular Expressions

Assume that the alphabet $\mathcal{A} = \{a_1, a_2, \ldots, a_n\}$ is finite.

Most text processing systems allow the following abbreviations, including the above mentioned '*' and '.*' notation of Word and GNU Emacs. In this book, the symbol "$\triangleq$" stands for "equal by definition" or "is defined as".

- $\varepsilon \triangleq \emptyset^*$ ('the empty word'),
- $\varphi^+ \triangleq (\varphi \ \varphi^*)$ ('one or more repetitions of φ')
- $. \triangleq (((a_1 + a_2) + \ldots) + a_n)$ ('any letter'),
- $* \triangleq .^*$ ('any word'),
- $\varphi? \triangleq (\varepsilon + \varphi)$ ('maybe one φ'),
- $\varphi^0 \triangleq \varepsilon$ and $\varphi^n \triangleq (\varphi \varphi^{n-1})$ for any $n > 0$ ('exactly n times φ'),
- $\varphi_m^n \triangleq (\varphi^m \varphi?^{n-m})$ for $0 \leq m \leq n$ ('at least m and at most n φ')
  (Here we assume that $\{\emptyset, +, \ ^*, (,), \varepsilon, \ ^+, \ ., *, ?, \ ^n, \ _m^n\} \cap \mathcal{A} = \emptyset$).

Semantics

In the context of Formal Methods, a formal language comes with a formal semantics which explains the meaning of the syntactical objects (words or graphs) under consideration by interpreting it in some domain.

The semantics identifies for each syntactical object a unique object in the chosen interpretation domain. Probably the fundamental question which can be answered by a semantics is: when can two different syntactical objects be considered equal? For our regular expression case study this means to determine when two different expressions are to be the same. As another example from computer science, we would like to know whether two different programs compute the same result.

Other questions include whether one object entails another one. For instance, we would like to know whether one regular expression includes another one, one program extends another one, or one specification refines another one.

In contrast to syntax, the semantics of Formal Methods is not always decidable. That is, membership of a word or model in the formal language defined by a grammar or metamodel is usually trivial to decide. Semantical equality, however, is often undecidable as can be seen by the example of program equivalence.

There are three main ways of defining a semantics for formal languages: denotational, operational, and axiomatic.

In denotational semantics the *denotation* of an object is defined. That is, a denotational semantics is a function defining for each syntactic object an object in some semantical domain. For example, a regular expression denotes a language (a set of words) over the alphabet. That is, the semantical domain for regular expressions is the set of all languages. As another example, a program in a functional language denotes a function (set of tuples) from the input parameters to the output type. In Chap. 3 we will discuss three different denotational semantics for the process algebra CSP.

Example 2.4: Denotational Semantics of Regular Expressions

For any regular expression φ, we define the denoted language $[\![\varphi]\!]$ by the following clauses:

- $[\![a]\!] \triangleq \{a\}$ for any $a \in \mathcal{A}$. That is, the regular expression 'a' defines the language consisting solely of the one-letter word 'a'.
- $[\![\emptyset]\!] \triangleq \{\}$. That is, $\emptyset$ denotes the empty language.
- $[\![(\varphi\,\psi)]\!] \triangleq \{xy \mid x \in [\![\varphi]\!], y \in [\![\psi]\!]\}$. That is, $(\varphi\,\psi)$ denotes the language of all words which can be split into two parts, such that the first part is in the denotation of φ and the second in the denotation of ψ.
- $[\![(\varphi + \psi)]\!] \triangleq [\![\varphi]\!] \cup [\![\psi]\!]$. That is, $(\varphi + \psi)$ denotes the union of the denotations of φ and ψ.
- $[\![\varphi^*]\!] \triangleq \{x_1 \ldots x_n \mid n \geq 0, \text{ and for all } i \leq n, x_i \in [\![\varphi]\!]\}$.
  That is, φ^* denotes the language of all words which can be split into a finite number of n parts, such that each part is in the denotation of φ.

> With the special case $n = 0$, this definitions entails that for any φ, the empty word (consisting of zero letters) is in $[\![\varphi^*]\!]$.

Operational semantics describes the execution of the syntactic object by some virtual machine. In our example, from each regular expression we can construct an automaton accepting its language. For an imperative or object-oriented programming language, the operational semantics defines, for instance, the change of the memory content induced by an assignment. In Chap. 3 we will discuss an operational semantics for the process algebra CSP.

Example 2.5: Operational Semantics of Regular Expressions

For any regular expression φ, we define an *automaton* $\mathbf{A}(\varphi)$, that is, a graph (N, E, s_0, S_F), where N is a nonempty set of *nodes*, $E \subseteq (N \times \mathcal{A} \times N) \cup (N \times N)$ is a set of (labelled) *edges*, $s_0 \in N$ is the *initial node* and $S_F \subseteq N$ is the set of *final nodes*.

- $\mathbf{A}(a) \triangleq (\{s_0, s_1\}, \{(s_0, a, s_1)\}, s_0, \{s_1\})$ for any letter $a \in \mathcal{A}$.
- $\mathbf{A}(\emptyset) \triangleq (\{s_0\}, \{\}, s_0, \{\})$.
- If $\mathbf{A}(\varphi) = (N_\varphi, E_\varphi, s_{0,\varphi}, S_{F,\varphi})$ and $\mathbf{A}(\psi) = (N_\psi, E_\psi, s_{0,\psi}, S_{F,\psi})$ (where we assume all elements to be disjoint), then $\mathbf{A}((\varphi\ \psi)) \triangleq (N_\varphi \cup N_\psi, E_\varphi \cup E_\psi \cup \{(s, s_{0,\psi}) \mid s \in S_{F,\varphi}\}), s_{0,\varphi}, S_{F,\psi})$.
- $\mathbf{A}((\varphi + \psi))$ is constructed from $\mathbf{A}(\varphi)$ and $\mathbf{A}(\psi)$ by $\mathbf{A}((\varphi + \psi)) \triangleq (N_\varphi \cup N_\psi \cup \{s_0\}, E_\varphi \cup E_\psi \cup \{(s_0, s_{0,\varphi}), (s_0, s_{0,\psi})\}, s_0, S_{F,\varphi} \cup S_{F,\psi})$, where s_0 is a new node not appearing in N_φ or N_ψ.
- If $\mathbf{A}(\varphi) = (N, E, s_{0,\varphi}, S_F)$, then $\mathbf{A}(\varphi^*) \triangleq (N \cup \{s_0\}, E \cup \{(s_0, s_{0,\varphi})\} \cup \{(s, s_0) \mid s \in S_F\}, s_0, \{s_0\})$, where again s_0 is a new node not appearing in N_φ.

A word w is *generated* or *accepted* by an automaton, if there is a path from the initial node to some final node which is labelled by w. It is not hard to see that for every regular expression φ the automaton $\mathbf{A}(\varphi)$ accepts exactly $[\![\varphi]\!]$. That is, denotational and operational semantics coincide.

An axiomatic semantics gives a set of proof rules from which certain properties of the syntactical object can be derived. For example, for regular expressions an axiomatic semantics might consist of a list of rules allowing to prove that two expressions are equal. For logic programming languages, the axiomatic semantics allows to check if a query is a consequence of the facts stated in the program. In Chap. 2 we will discuss a Hilbert-style proof system for propositional logic, which—thanks to its correctness and completeness— can also serve as axiomatic semantics of propositional logic.

Example 2.6: Axiomatic Semantics of Regular Expressions

Axiomatic systems for equality of regular expressions were given by various authors [Sal66, Koz94, KS12]. We call an equation $\alpha = \beta$ *derivable* and write $\vdash \alpha = \beta$, if it is either an instance of one of the axioms below or follows from a set of such instances by a finite number of applications of the below rules. Salomaa gives the following axioms

- $\vdash (\alpha + (\beta + \gamma)) = ((\alpha + \beta) + \gamma)$, $\quad \vdash (\alpha \, (\beta \, \gamma)) = ((\alpha \, \beta) \, \gamma)$
- $\vdash (\alpha \, (\beta + \gamma)) = ((\alpha \, \beta) + (\alpha \, \gamma))$, $\quad \vdash ((\alpha + \beta) \, \gamma = ((\alpha \, \gamma) + (\beta \, \gamma))$
- $\vdash (\alpha + \beta) = (\beta + \alpha)$, $\quad \vdash (\alpha + \alpha) = \alpha$, $\quad \vdash (\varepsilon \, \alpha) = \alpha$
- $\vdash (\emptyset \, \alpha) = \emptyset$, $\quad \vdash (\alpha + \emptyset) = \alpha$
- $\vdash \alpha^* = (\varepsilon + (\alpha^* \, \alpha))$, $\quad \vdash \alpha^* = (\varepsilon + \alpha)^*$

  and derivation rules

- If $\vdash \alpha = \beta$ and $\vdash \gamma = \delta$, then $\vdash \gamma[\alpha := \beta] = \delta$ and $\vdash \gamma[\alpha := \beta] = \gamma$
- If $\vdash \alpha = ((\alpha\beta) + \gamma)$ and not $\varepsilon \in \beta$, then $\vdash \alpha = (\gamma\beta^*)$

  Here, $\gamma[\alpha := \beta]$ means γ with one or more occurrences of α replaced by β. For the second rule, $\varepsilon \in \beta$ means that

1. β is of form ρ^* for some regular expression ρ, or
2. β is of form $(\rho_1 + \rho_2)$ where $\varepsilon \in \rho_1$ or $\varepsilon \in \rho_2$, or
3. β is of form $(\rho_1 \, \rho_2)$ where $\varepsilon \in \rho_1$ and $\varepsilon \in \rho_2$.

Without the restriction "not $\varepsilon \in \beta$" the rule would not be correct: $a^* = (a^*a^*) + \emptyset$, but not $a^* = (\emptyset \, a^{**})$, since $(\emptyset \, a^{**}) = \emptyset$

It can be easily proven that $[\![\alpha]\!] = [\![\beta]\!]$ if $\vdash \alpha = \beta$, that is, the system is *correct* with respect to the denotational semantics. The proof proceeds by showing that all axioms are correct, and that the rules allow only to derive correct equations from correct ones. In passing we mention that the system also can be proven to be *complete*, that is, if $[\![\alpha]\!] = [\![\beta]\!]$ then $\vdash \alpha = \beta$. Completeness usually is much harder to show than correctness.

Methods

A formal language is described by an unambiguous syntax and a mathematical semantics. For a Formal *Method* (as opposed to a formal language) it is essential that there are some *algorithms* or *procedures* which describe what can be done with the syntactic objects in practice. According to the Oxford dictionary, a method is a particular procedure for accomplishing or approaching something, especially a systematic or established one. A Formal Method describes how to 'work with the language', that is, perform some activity on its elements in order to achieve certain results. In general, this information processing is some form of transformation (metamorphosis, *Gestaltwandlung*), where the syntactic objects are modified from one form to another.

Usually, a formal language is designed for a specific purpose. For example, a logical language is supposed to formalise human reasoning. A specification language should allow to describe the functionality of a system. A program formulated in a programming language should be executable on a machine. A model expressed in some modelling language should help humans to understand a concept or design.

The methods associated with a formal language usually are constructed to support this purpose. For a logical language, the transformation can be a calculus with which to derive theorems from axioms. For a specification language, it can be a set of rules to transform a specification into an implementation. For programs written in any programming language, the execution on a virtual machine can be seen as a form of transformation. For modelling languages, model transformations allow to change between different levels of abstraction.

In a Formal Method, the transformation must be according to fixed rules; these rules operate on syntactical objects of the formal language under discussion, and result in some 'insight' about them. Such an insight might be the result of the transformation, or the realisation that the (repeated) transformation does not come to an end. Other insights we might want to achieve are whether a program is correct with respect to its specification, or whether one model refines another one.

Continuing our example, we show how regular expressions can be used in everyday text processing.

Example 2.7: Regular Replacements

A frequent task while writing scientific articles is to consistently replace certain text passages by others in the whole text. A replacement $[\alpha := \beta]$ consists of a regular expression α and a word β over $\mathcal{A}$. The word δ is the result of the replacement $[\alpha := \beta]$ on a word γ, denoted as $\delta = \gamma[\alpha := \beta]$ if one of the following holds:

1. either there exist γ_1, γ_2 and γ_3 such that

  1.1. $\gamma = \gamma_1 \gamma_2 \gamma_3$,
  1.2. $\gamma_2 \in [\![\alpha]\!] \setminus [\![\varepsilon]\!]$,
  1.3. γ_1 is of minimal length, that is, there are no γ_1', γ_2' and γ_3' such that $\gamma = \gamma_1' \gamma_2' \gamma_3'$, $\gamma_2' \in [\![\alpha]\!] \setminus [\![\varepsilon]\!]$ and $|\gamma_1'| < |\gamma_1|$,
  1.4. γ_2 is of maximal length, that is, there are no γ_2' and γ_3' such that $\gamma = \gamma_1 \gamma_2' \gamma_3'$, $\gamma_2' \in [\![\alpha]\!] \setminus [\![\varepsilon]\!]$ and $|\gamma_2'| > |\gamma_2|$,

  and $\delta = \gamma_1 \beta (\gamma_3[\alpha := \beta])$, or
2. there are no γ_1, γ_2 and γ_3 satisfying the above 1.1.–1.4., and $\delta = \gamma$.

The definition of the first case is recursive; it is well-defined because condition 2. requires that γ_2 is a nonempty string. Therefore, $|\gamma_3| < |\gamma|$, and the recursion must terminate.

As an application of regular replacement, we note that $(p \Rightarrow (q \Rightarrow p))[(p + q) := (p \Rightarrow q)] = ((p \Rightarrow q) \Rightarrow ((p \Rightarrow q) \Rightarrow (p \Rightarrow q)))$.

Coming back to our introductory tool experiments in Example 2.1, the above definition determines:

- if the wildcard sign '*' has been defined to express the iteration of the empty word, then $(abc)[\varepsilon^* := x] = abc$ as condition 1.2 of Example 2.7 can not be fulfilled;
- if the wildcard sign '*' stands for 'any word', then $(abc)[* := x] = x$, because the wildcard sign matches exactly 'abc' and therefore condition 1 of Example 2.7 is fulfilled with $\gamma_1 = \varepsilon$, $\gamma_2 =' abc'$, and $\gamma_3 = \varepsilon$.

Thus, the formal treatment allows to calculate a reliable result which is independent from the particular text editor being used, and against which the tools can be verified.

We now have discussed all ingredients of what constitutes a Formal Method, and thus are in a position to give a definition.

Definition 1 A *Formal Method* $\mathbb{M}$ consists of three components:

- syntax,
- semantics, and
- method.

The syntax gives a precise description of the form of objects (strings or graphs) belonging to $\mathbb{M}$. The semantics describes the 'meaning' of the syntactic objects of $\mathbb{M}$, in general by a mapping into some mathematical structure. The method describes algorithmic ways of transforming syntactic objects, in order to gain some insight about them.

1.2 Formal Methods in Software Development

Having developed an understanding of what Formal Methods are, we now consider their role in software development. To this end, we briefly recall the notion of the software life cycle, describe how this cycle is realised, and discuss where to use Formal Methods in the life cycles. While life cycle models describe development activities and their order, software development standards give a legal framework prescribing which activities have to be performed, including Formal Methods. This leads to a discussion of the main purposes for the use of Formal Methods in systems development.

1.2.1 The Software Life Cycle

A *software life cycle (SLC)* (also referred to as "software development life cycle", "software development process", and "software process") is a structure describing development, deployment, maintenance, and dismantling of a software product. There are several models for such processes, describing different approaches on how to develop software (waterfall, spiral, iterative, agile development, etc.).

Our objective is to discuss the use of Formal Methods in the software development process, independently of the model used, rather than to provide a survey on such different models. For that reason we concentrate on the *V-model*, and the general ideas behind *agile* methodologies.

The V-model

Models of software development often describe the development process as being composed of separate phases. For example, there usually are a project definition phase, an architectural and software design phase, a coding phase and a testing phase. Traditionally, these phases are ordered sequentially, which leads to the so-called waterfall model. In this model, the results of one phase are starting points for the subsequent phase, like water falls from one level to the next in a cascade. The waterfall model has several deficits and today is considered to be archaic. Mainly, it does not pay respect to the fact that quality assurance takes place on several levels. For instance, system testing is considered with the systems specification, whereas in unit testing individual units (methods, procedures, functions etc.) rather than the overall system are checked.

Traditionally, the V-model usually has been depicted like the waterfall model, however in the shape of a big V. The V-model has been developed over many years in various versions. A newer version is the V-model XT (for "eXtreme Tailoring") [dBfl12], see Fig. 1.2. In particular, the German federate office for information security (BSI) was a driving force in its elaboration. In Germany, the V-model is mandatory for safety-critical governmental projects. Instead of describing phases, the V-model XT describes states in the development process. It refrains from prescribing a specific order to these states.

Figure 1.3 hints at where and how validation and verification could be used in the software development process according to the V-model XT. It shows four design levels (from top to bottom: requirements, design specification, architecture, and implementation). At each level, the realisation and integration artefacts (on the right) should comply with the corresponding specification and subdivision artefacts (on the left).

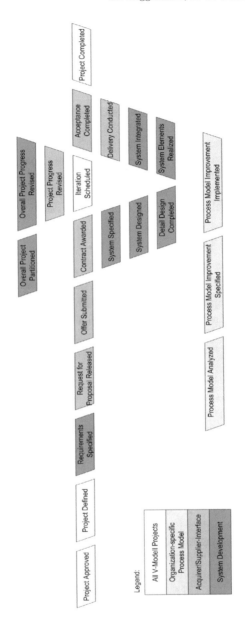

Fig. 1.2 V-model XT [dBfI12]

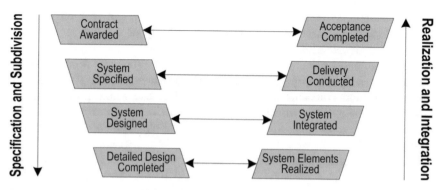

Verification and Validation

Fig. 1.3 Verification and validation in the V-model XT [dBfl12]

Many researchers and practitioners in software engineering differentiate between 'validation' and 'verification' in the following way: *validation* tries to answer the question

"Are we building the right product?",

whereas *verification* is concerned with the question

"Are we building the product right?"

Thus, validation refers to the user's needs according to the requests, while verification checks that the specification is correctly implemented. In Fig. 1.3 this means that validation could be associated with the compliance between 'contract awarded' and 'acceptance completed', whereas verification concerns the other compliances.

Note, however, that many researchers use a slightly different definition, where *validation* is a general term covering all activities for checking that the product is correct, while *verification* is used for the process of formally proving that a program meets its specification. Since such a formal proof requires formal languages, verification is only applicable at the three lower levels of Fig. 1.3. This can be for establishing the horizontal compliances, for refinement between different levels, or for proving properties about formal artefacts.

Agile Development

Phase-based models of software development, such as the waterfall model, have been criticised for a number of reasons. A main point is that each new phase has to wait until the previous one is completed. This can lead to delays in the project. Moreover, if an error is detected it might be necessary to go back to an earlier phase, causing further delays. Finally, the waterfall model assumes that all system requirements are known from the very beginning of the development. There is no provision to modify or extend requirements during the process. This can be a severe restriction.

Therefore, many other process models have been proposed. A current trend is to develop software in a manner where teams are small, the phases are not clearly identified, and the user is represented in the whole process of software development. This procedure has been called *agile development* [Coc00]. Design, development and testing are done almost simultaneously and in short iterations.

In an agile development process, a natural way to work is to follow a test-driven development approach [Bec02]. That is, before starting to write code, tests for the system are produced. These tests represent user scenarios and requirements for the system. As long as the code is non-existent or erroneous, the tests will fail. Then the code is written in order to make the tests pass. When all tests pass, one system development cycle is completed.

The Scrum management methodology identifies roles and responsibilities in an agile development process [SB01]. It also defines activities like daily and weekly meetings, where the basic unit of development is organised in project time slots (*sprints*). Each time slot should produce a potentially deliverable result (e.g., a piece of software).

In an agile process, roles (manager, analyst, programmer, tester, verifier etc.) are frequently swapped amongst group members. Therefore, each developer should in principle have knowledge of all development activities in the project. In particular, if Formal Methods are used, the group members should know the capabilities and limitations of the available formal development tools.

1.2.2 Formal Methods: When and Where

While process models describe development phases and their order, software development standards give a legal framework prescribing which activities have to be performed. For example, a standard might prescribe that "for each phase of the overall ...system and software safety lifecycles, a plan for the verification shall be established concurrently with the development for the phase. The verification plan shall document or refer to the criteria, techniques, tools to be used in the verification activities." [IEC10].

There are various standards on the development of software, e.g., EN-50128, IEC-61508 and DO-178 B/C. Some of these standards prescribe that Formal Methods are being used for high integrity levels. For example, IEC 61508, the international standard on functional safety of electrical/electronic/programmable electronic safety-related systems, is a 'meta-standard', from which several other domain-specific standards are derived. In Part 7 (2010), Sect. C.2.4.1 it defines the aim of Formal Methods as "the development of software in a way that is based on mathematics. This includes formal design and formal coding techniques. ... A Formal Method will generally offer a notation (generally some form of discrete mathematics being used), a technique for deriving a description in that notation, and various forms of analysis for checking a description for different correctness properties." In Sect. B.2.2. it states that "Formal Methods ... increase the completeness, consistency or correctness of a specification or implementation".

However, the IEC 61508 standard also states that there can be disadvantages of Formal Methods, namely: "fixed level of abstraction; limitations to capture all functionality that is relevant at the given stage; difficulty that implementation engineers have to understand the model; high efforts to develop, analyse and maintain model over the lifecycle of system; availability of efficient tools which support the building and analysis of model; availability of staff capable to develop and analyse model."

Several Formal Methods are described in the standard (CCS, CSP, HOL, LOTOS, OBJ, temporal logic, VDM and Z). The use of Formal Methods is recommended for achieving the highest safety integrity level (SIL 4), where the average frequency of a dangerous failure of the safety function must be provably less than 10^{-8}/h, i.e., a failure may occur on average at most once in 10,000 years of operation.

DO-333 is the Formal Methods supplement to DO-178C and DO-278A for safety-critical avionics software. It defines Formal Methods as "mathematically based techniques for the specification, development and verification of software aspects of digital systems". Formal Methods can be used to "improve requirements, reduce error introduction, improve error detection, and reduce effort". The supplement further states that "the extent to which Formal Methods are used [in the software development] can vary according to aspects such as preferences of the program management, choice of technologies, and availability of specialised resources".

Use of Formal Methods

The use of Formal Methods in software development is not constrained to a specific process and life cycle model followed by a company. That is, Formal Methods can be used with traditional as well as agile models. Moreover, Formal Methods should not constitute separate phases or sprints, but should rather be integrated as part of the general verification activities.

Mishra and Schlingloff [MS08] evaluate the compliance of Formal Methods with process areas identified in CMMI-DEV, the capability maturity model integration for development. Their result is that out of 22 process areas from CMMI, six can be satisfied fully or largely with a formal specification-based development approach. Notably, the process areas requirements management,

product integration, requirements development, technical solutions, validation and verification are supported to a large extent. They also show the
possibility of automation in process compliance, which reduces the effort for
the implementation of a process model.

Formal Methods are used in system development for two main purposes:

1. as a means to make descriptions precise, and
2. to help in different kinds of analysis.

Concerning the first purpose, descriptions of interest include requirements,
specifications, and models, which appear at different levels and moments in
the life cycle. It is common practice to write software descriptions in natural
language. In spite of the apparent advantage of being written in a language
understandable to everybody, its inherent ambiguity and lack of precision
makes the realisation of such descriptions problematic.

The literature distinguishes between linguistic and domain-specific ambiguities. Kamsties et al. [KBP+01] provide the following examples: the 500
most used words in English have on average 23 meanings; the sentence "The
product shall show the weather for the next 24 h" exhibits the linguistic
ambiguity if the phrase 'for the next twenty-four hours' is attached to the
verb 'show' or to the noun 'weather'; the sentence "Shut off the pumps if
the water level remains above 100 m for more than 4 s" is ambiguous as in
the given domain the term 'water level' can refer to the mean, the median,
the root mean square, or the minimum water level. An attempt to address
such problems is the use of controlled natural language. Here, the grammar
and vocabulary of natural language is restricted in order to avoid or reduce
ambiguity of sentences. Present day controlled languages, however, are often
felt to be either too restrictive or too informal to be practical. In this book we
advocate formal languages, which have a well-defined syntax and semantics.
They may be used to resolve such ambiguities and to achieve the required
level of precision.

To illustrate such a process of removing ambiguities, consider the regular
replacements discussed in Example 2. We showed that the informal description is ambiguous and can lead to unexpected results. In contrast, in Example
2.7 we formally defined $\gamma[\alpha := \beta]$ to be the word resulting from the word γ
by the replacement of a regular expression α with the word β. The formal
definition cares for all special cases; there is no need for explaining what, e.g.,
the replacement of the empty language by the empty string in a one-letter
word is. $a[\emptyset := \varepsilon]$ has a well-defined meaning which can be derived from the
definition.

This book focuses largely on using Formal Methods in the second way,
i.e., to assist with analysis. The use of Formal Methods in Verification and
Validation (often abbreviated as V&V) is wide and includes techniques such
as static analysis, formal testing, model checking, runtime verification, and
theorem proving. All the above are complementary techniques to standard

methods such as code review, testing and debugging. They increase the confidence in the correctness of the software under consideration.

This is shown in the software development for the ISS as described in Example 1. It illustrates how Formal Methods can help to analyse a system. Given a system model, the model checker FDR could prove the presence of several deadlocks and livelocks. This helped to improve the quality of the safety-critical system.

Aligned with current practices in software development, Formal Methods may be used from the very beginning (when a system is initially conceived) up to the end (when the final product is shipped). Model checking, for instance, does not require that a single line of code has been written: it may already be used when the first formal specifications and high-level models are available. As another example, runtime verification can be used in a pre-deployment phase, when (part of) the code is ready to run, and even after deployment to control/enforce desirable properties at runtime.

1.2.3 A Classification Scheme for Formal Methods

In Definition 1, we said that a Formal Method consists of syntax, semantics and specific methods or algorithms. Thus, e.g., "CTL model checking" or "Z theorem proving with HOL" are particular Formal Methods.

Although there has been quite a debate in the Formal Methods community on the 'right' syntax, the 'best' semantics and the 'most effective' algorithms, these aspects can be subsumed within other categories in a taxonomy of Formal Methods.

In order to give an orientation, we provide a classification scheme which allows to categorise each Formal Method along the following dimensions.

- **Method definition**—syntax, semantics and procedures as described above.

  - Syntactic aspects, e.g., whether the language allows user-defined mixfix operators, linear or non-linear visibility, graphical or textual notation, etc., are related to the usability-aspect of a Formal Method.
  - Semantic aspects—which semantic domains are employed and how they are characterised (denotational, operational, axiomatic semantics)— determine the application range and underlying technology.
  - Algorithmic aspects (describing what can be done with the method) dominate the underlying technology and properties of concern. Typical procedures include simulation and symbolic execution, model checking, automated or interactive theorem proving, static analysis, refinement checking, etc.

- **Application range**—this dimension determines the application domain (e.g., avionics, railway, finance) and the specific needs of this domain (whether the systems are mainly reactive, interactive, real time, spatial, mobile, service oriented, etc.)
- **Underlying technology**—this dimension notes how the method can be realised. Technologies are, for example, SAT solving, logical resolution, term rewriting, symbolic representation, etc.
- **Properties of concern**—This dimension categorises properties of the systems which are the subject of the Formal Method and which are supported by the method (safety, liveness, fairness, security, consistency, functional correctness, etc.)
- **Maturity and applicability**—this dimension describes how fit the method is for actual use (universality, expressivity, usability, learning-curve, intuitive level, tool support, etc.)

Each particular application of Formal Methods can be located within the space that these dimensions span. For illustration, consider Example 1: the language used in this example is the process algebra CSP, with its failures semantics, and automated refinement checking as a procedure (see Fig. 1.1). The application domain is that of fault-tolerant algorithms in aerospace. The technology used in the FDR tool is the hierarchical compression of the state space, a technique specific for this tool. Properties of concern are livelock and deadlock. The case study was conducted in a collaboration between industry and academia, since the abstraction process from Occam to CSP and the use of FDR was outside the standard routine of the aeronautic engineers.

The second example from this chapter, regular replacement, can be classified as follows: the syntax is the language of regular expressions, with the usual denotational (set-theoretic) semantics, and text transformation as a procedure. Application domain are text editors or macro processors. There is no specific technology involved with this example, as we refrain from giving an implementation; one possibility would be to use list processing in a functional programming language. The property to be achieved is to give a well-defined transformation, open to formal argument about the correctness of any implementation. The Formal Method of regular expressions belongs to the standard knowledge of computer science and is accessible at an undergraduate level.

Of course, there are other dimensions which could be added to this classification scheme. These include specification focussed versus analysis focussed, correctness-by-construction versus design-and-proof-methodology, lightweight versus heavyweight Formal Methods, etc.

1.2.4 Tool Support for Formal Methods

Formal Methods usually start on the 'blackboard': toy examples are treated in an exemplary way. With paper and pen one checks if a method works out. In the long run, however, Formal Methods need tool support in order to become applicable. This is the case as software systems are fundamentally different compared to mathematical theories:

Numbers of axioms involved.   In Chap. 4, we will formalise and verify control programs written in Ladder Logic. Here, each line of code is represented by one logic axiom in the specification language CASL. The toy example presented, a traffic light controller, cf. Example 44, has about 10 lines of code, i.e., the model consists of about 10 CASL axioms. Our approach scales up to actual railway interlockings. Such interlockings describe for a railway station how to position the points and how to light the signals in such a way that trains can move safely. A typical interlocking program written in Ladder Logic consist out of 500–1500 lines of code, i.e., its model has 500–1500 CASL axioms in its formalisation.
In contrast, the whole realm of group theory is based on three axioms only, namely that the operation $+$ is associative (axiom 1), that $+$ has the unit 0 (axiom 2), and, finally, that every element has an inverse (axiom 3).
These example indicate that the number of axioms when applying Formal Methods is by magnitudes larger than the number of axioms involved in mathematical theories. Consequently, tool support is needed in Formal Methods for sheer book keeping.

Ownership and interest.   The interlocking program for a railway station is commissioned by a rail operator. Intellectual property rights ensure that, besides the rail operator, only the company writing the code and the railway authorities have access to design documents, code, and verification documentation, etc. These artefacts are studied only when the software lifecycle dictates it:

- during production by the company programming it,
- for acceptance by the company running the train station,
- for approval by the railway authorities, and
- when maintaining the code by a possibly different company hired for the task.

Thus, any verification of a ladder logic program, say in CASL, will be studied only at few occasions.
In contrast, group theory is public, its theorems and their proofs are published in books and journals, everyone has access to them. The proofs of group theory are taught for educational purposes at universities. Every year, the fundamental theorems of group theory are proven and checked in lecture halls all over the world.

Many software systems are the intellectual property of a company. This restricts access to the actual code; interest in their design is limited. Mathematical theories are part of the scientific process and publicly available. There is scientific interest in them. Therefore, proofs related to a specific software system are studied by few people only, and only when necessary—while mathematical proofs are studied by many, over and over again. Consequently, tools play the role of 'proof checkers' for quality control in Formal Methods.

Change of axiomatic basis. Every ten to fifteen years, the design of a railway station changes. New safety regulations have to be implemented, the station shall deal with more trains, new technology shall be introduced such as the European Train Control System (ETCS). This requires changes to the interlocking program and, consequently, to the proofs on it.

In contrast, mathematical theories are stable. Already in the 1830s Galois worked with the axioms of group theory, which have not changed ever since.

Requirements of software systems are bound to change in small time intervals. This means that design steps involving Formal Methods need to be repeated several times, sometimes already during the design phase, certainly when maintaining the system. Mathematical theories, however, are stable over centuries. Consequently, tools are needed to help with management of change in Formal Methods.

The technology underlying tools for Formal Methods is generic. The Heterogeneous Tool Set HeTS—to be discussed in Chap. 4 "Algebraic Specification in CASL"—for example is a 'broker' which offers, amongst other functionalities, translations from the language CASL to various tools. Yet another example is the Process Analysis Toolkit PAT, which supports reasoning about concurrent and real-time systems. Other tools have been built specifically for one Formal Language. An example is the model checker FDR which has been designed specifically for the process algebra CSP—see Chap. 3 "The process algebra CSP". The current trend in Formal Methods is to offer (integrated) work environments for different Formal Languages and Methods, e.g., HeTS.

Tool Qualification

When software tools are used to validate software, the questions is, who is validating the tools? In other words, for highly safety-critical systems there needs to be evidence why the tools which are used in their development should be trusted. There are two kinds of tools: for artefact generation and for artefact validation. This holds for all artefacts occurring in the software design cycle, e.g., binary code, program text, formal model, specification, or even the user manual. In industry, there are contradicting views concerning the importance of these tool classes. In some areas, generating tools are considered to be more critical than validating tools, since they directly affect the behaviour of the system. In other areas it is argued that faulty behaviour of a generator can anyway be found by the validation tools, which therefore are more critical.

If a generating tool is faulty, then the generated artefact will not correspond to its source. In the case of executable code, e.g., this may mean that the runtime behaviour is not as expected. In the case of a model transformation, the generated model might miss out on properties already established for the source model.

Example 3: Public-Domain C Compilers

Yang et al. [YCER11] found more than 325 errors in public-domain C compilers using a specialised compiler testing tool. They report:

"Every compiler we tested was found to crash and also to silently generate wrong code when presented with valid input."

"A version of GCC miscompiled this function:

```
1    int x = 4;
2    int y;
3
4    void foo (void) {
5        for (y = 1; y < 8; y += 7) {
6            int *p = &y;
7            *p = x;
8        }
9    }
```

When foo returns, y should be 11. A loop-optimisation pass determined that a temporary variable representing *p was invariant with value x+7 and hoisted it in front of the loop, while retaining a dataflow fact indicating that x+7==y+7, a relationship that no longer held after code motion. This incorrect fact led GCC to generate code leaving 8 in y, instead of 11."

If a validation tool is inaccurate or faulty, there are two cases: the tool might report an error where there is none (false positive), or the tool might miss to report an error where there is one (false negative). For example, an erroneous program verifier might fail to verify a correct program (false positive), or it might claim to have found a proof for an incorrect program (false negative). Often, false negatives are more critical than false positives since they convey a deceptive certainty. False positives are a hassle, because they need to be dealt with manually.

To make this more concrete, consider the tool Cppcheck for static analysis of C programs:

Example 4: Static Analysis of C Programs

When Cppcheck (Version 1.59) checks the following program, it issues for line 5 the error message

```
Array 'x[7]' accessed at index 13, which is out of bounds.
1    int main() {
2        int x[7];
3        int i = 13;
4        int flag; if (i<7) flag = 1; else flag = 0;
5        if (flag) x[i] = 0;
6        x[3] = 33; x[x[3]] = 0;
7    }
```

Since the assignment in line 5 is never executed, this is a false positive. Surprisingly, this false positive disappears when we replace line 4 by the equivalent

```
4        int flag = (i<7)?1:0;
```

Cppcheck does not issue an out-of-bounds warning for line 6. This is a false negative, since the assignment x[33] = 0; clearly might cause problems.

What are now the possibilities for the validation of tools? The usual approach is to resort on tools which are *proven-in-use*, i.e., where experience from many previous projects suggests that the tool is 'correct'. This is especially the case for certain public-domain tools which have been applied by many users for a long period of time. In order to claim that a tool is proven-in-use, it is necessary to provide evidence in which comparable projects it was used. As the above example of the GCC compiler error shows, proven-in-use is no guarantee against subtle, hidden errors.

For new or newly introduced methods and tools, the proven-in-use principle poses the problem of how to begin such a chain of trust. So, what to do when proven-is-use is not applicable? In order to be allowed to use tools which are not proven-in-use in a safety-oriented development, at least one has to perform a *tool qualification*. That is, the tool has to be applied to a number of selected examples under controlled conditions, where the tool's behaviour must be analysed and documented in detail. *Tool certification* is the process of confirming the qualification by a designated authority. Usually this is done only for tools to be applied in several different projects. The software development standard DO-333, e.g., prescribes in detail how to qualify and certify tools for different safety integrity levels in aerospace.

Of course there is still a possibility that even certified tools might contain errors. There are further methods that can improve the reliability of Formal Methods tools.

In the case of generating tools, one possibility is to verify the generator itself. For instance, there are a number of research projects and results dealing with compiler verification. One challenge here is that the correctness argument needs to deal with several different languages: the language the generator is written in, the source language, and the target language. Due to the sheer size of the problem, often compiler verification is supported by automated tools. Yet another possibility is to generate, besides the code, also certain proof conditions from program annotations, which can be checked automatically in the generated code with a suitable verification tool. This way, if the compiler is faulty in its code generation part, this will be detected by the following verification tool.

Both these suggestions to improve the dependability of generating tools rely on the existence of correct verification tools. In order to increase the trust in the correctness of verification tools themselves, one can run several different provers on the same problem and compare their results. If at least two of them agree, then, under the assumption that different tools do not make the same mistake, the result is as given. If one prover claims to have found a proof, while another one claims that the property is not valid, one of them must be faulty.

Another approach to increase the trust in theorem provers is to augment them with a proof checking component. For this, the prover must not only give a boolean result but also produce some term which allows to check whether the proof is a valid one.

1.3 Formal Methods in Practice

We present various case studies on the application of Formal Methods. These come in two flavours: comparative case studies compiled by academics, and the application of Formal Methods in industry.

1.3.1 Comparative Surveys and Case Studies

The Formal Methods community has compiled several surveys with the aim of comparing different approaches for the application of Formal Methods in various areas. The characteristic of these surveys is to discuss *one* coherent example in the context of *several* Formal Methods.

- Lewerentz and Lindner [LL95] discuss a production cell. This cell is considered to be safety critical, i.e., a number of properties must be enforced in order to avoid injuries of people. The system is reactive, i.e., it has to react permanently to changes of the environment. In principle this is a real-time problem, as the reaction of the control software must be guaranteed within a certain interval of time.

Example 5: Production Cell

"The production cell is composed of two conveyor belts, a positioning table, a two-armed robot, a press, and a travelling crane. Metal plates inserted in the cell via the feed belt are moved to the press. There, they are forged and then brought out of the cell via the other belt and the crane." [LL95].

This case study reflects a typical scenario as it arises in industrial automation. For this case study, various safety and liveness requirements are to be established. Efficiency (w.r.t. production time) and flexibility (w.r.t. the effort it takes to adapt a solution to changed requirements) should also be taken into account. Besides presenting 18 contributions, the book includes a summary and evaluation of the different solutions.

- Broy et al. [BMS96] study a memory cell which can be accessed by remote procedure calls. Such a call is an indivisible action. A return is an atomic action issued in response to a call. There are two kind of returns, *normal* and *exceptional*. A return is issued only in response to a call.

Example 6: RPC Memory Cell

"The component to be specified is a memory that maintains the contents of a set **MemLocs** of locations. The content of a location is an element of a set **MemVals**. This component has two procedures... [Procedure] **Read** returns the value stored in address **loc**. [Procedure] **Write** stores the value **val** in address **loc.** The memory must eventually issue a return for every **Read** and **Write** call." [BMS96].

Broy et al. [BMS96] collect fifteen solutions in various Formal Methods, including Petri nets, temporal and higher-order logics, various forms of transition systems or automata, and stream-based approaches.

- Abrial et al. [ABL96] study a classical control problem:

> **Example 7: Steam-Boiler Controller**
>
> "[The steam-boiler control program] serves to control the level of water
> in a steam-boiler. The program communicates with the physical units
> through messages which are transmitted over a number of dedicated
> lines connecting each physical unit with the control unit. ... The pro-
> gram follows a cycle and a priori does not terminate. This cycle takes
> place each five seconds and consists of the following actions:
> • Reception of message coming from the physical units.
> • Analysis of informations which have been received.
> • Transmission of messages to the physical units." [Abr94].

Abrial et al. [ABL96], use Formal Methods for various purposes: formal
requirement specifications, intermediate refined models, analysis of system
properties, proofs, automated synthesis of conditions implying safety for
parameters of the controller, design or generation of executable code. The
overall twenty-one contributions used algebraic, logical, and operational languages similar to those treated in the subsequent chapters of this book.

• Frappier and Habrias [FH01] discuss a classical commercial software application:

> **Example 8: Invoicing Software**
>
> "To invoice is to change the state of an order (to change it from the
> state "pending" to "invoiced"). On an order, we have one and one only
> reference to an ordered product of a certain quantity. The quantity
> can be different to other orders. The same reference can be ordered
> on several different orders. The state of the order will be changed into
> "invoiced" if the ordered quantity is either less or equal to the quantity
> which is in stock according to the reference of the ordered product."
> [FH01].

The volume collects specifications in the state-based methods Z and B, in the
event-based methods Action Systems, UML with a behaviour-driven method,
VHDL, Estelle, SDL, and E-Lotos, and in other formal approaches as CASL,
Coq, and Petri Nets.

• Jones and Woodcock [JW08] collect approaches to mechanise the proof of
  correctness of the Mondex smart-card for electronic finance. This was one
  of the first comparative case studies dealing with security issues.

> **Example 9: Electronic Purse Mondex**
>
> "The system consists of a number of electronic purses that carry financial value, each hosted on a Smartcard. The purses interact with each other via a communication device to exchange value. Once released into the field, each purse is on its own: it has to ensure the security of all its transactions without recourse to a central controller. All security measures have to be implemented on the card, with no real-time external audit logging or monitoring." [SCW00].

The methods applied to the Mondex case study are the Alloy model-finding method, the KIV system, Event-B, UML and OCL, RAISE, and Z/Eves.

- Rausch et al. [RRMP08] document a competition on the so called Common Component Modelling Example (CoCoME). Given a prescribed architecture, the challenge lies in using a specific formalism for modelling and analysing the CoCoME according to this architecture.

> **Example 10: CoCoME Trading System**
>
> The CoCoMe case study concerns a trading system as it can be observed in a supermarket handling sales. At a Cash Desk the Cashier scans the goods the Customer wants to buy and the paying (either by credit card or cash) is executed. The central unit of each Cash Desk is the Cash Desk PC which wires all other components with each other. Also the software which is responsible for handling the sale process and amongst others for the communication with the Bank is running on that machine. A Store itself consists of several Cash Desks organised in Cash Desk Lines. A Cash Desk Line is connected to a Store Server which itself is also connected to a Store Client. A set of Stores is organised as an Enterprise where an Enterprise Server exists to which all Stores are connected. (Formulated closely following Herold et al. [HKW+07].)

Rausch et al. [RRMP08] documents more than ten formal component models and their use in verification and quality prediction. A jury evaluation concludes the volume. We discuss similar examples in Chap. 6 on "Specification and Verification of Electronic Contracts".

- Cortier and Kremer [CK11] collect several popular symbolic approaches to formal security. Different research groups demonstrate their techniques, taking a flawed public key protocol (and its correction) as a common example:

> **Example 11: Handshake Security Protocol**
>
> "The aim of the protocol is that A and B share a secret key s at the end. Participant A generates a fresh session key k, signs it with his secret key $sk(A)$ and encrypts it using B's public key $pk(B)$. Upon receipt B decrypts this message using the private secret key, verifies the digital signature and extracts the session key k. B uses this key to symmetrically encrypt the secret s." [CK11].

This protocol shall provide secrecy: the secret s shall only be known to A and B. The above handshake protocol, however, is vulnerable to a 'man in the middle' attack. Various techniques are used to (automatically) find this flaw and to prove that adding the identities of the intended participant changes the protocol into a correct one. These techniques include rewrite rules, Horn clauses, strand spaces, constraint systems, process algebra, and Floyd–Hoare style logics. Further protocols are consider in order to demonstrate the strengths of individual techniques. In Chap. 8 on the "Formal Verification of Security Protocols" we verify a protocol for authentication.

For good reason, the above compilations refrain from giving a concrete recommendation which method is 'the best'. Similar to the problem of selecting a suitable programming language for a particular project, the suitability of a method strongly depends on the context. For example, a method for analysing functional correctness might not be suitable for deadlock analysis of protocols.

1.3.2 Industrial Practice

Formal Methods play an increasing role in industrial practice: "yesterday's Formal Methods are today's best practice". For example, the theory of static program analysis and abstract interpretation has been developed since the mid-1970s. A first tool, Lint, for checking source code of C programs has been released in 1979. Subsequently, more specialised tools for safety-critical applications based on this theory were developed. Today, more than one hundred tools for static code analysis exist, with varying strengths and application ranges. However, static analysis is also performed in ordinary compilers: the Java specification (for version 7 in Sect. 14.21.) requires that "it is a compile-time error if a statement cannot be executed because it is unreachable" [GJBB13]. There are detailed instructions in the language definition on how to figure out whether a statement is reachable; in general, this is a static analysis task which is performed by the Java compiler.

In hardware design, modelling a chip lay out and checking it with model checking and theorem proving (both techniques were developed during the

1990s) is an established practice today. Graph grammars originated in the late 1960s; the theory of graph transformation provides the mathematical foundation for code generators in current model-based development environments. Existing standards, such as DO-333 (see Sect. 1.2.2), which was released in 2012, allow to replace informal validation steps such as code inspections, code reviews, and testing by Formal Methods.

Knowledge transfer from academia, however, is a slow process in general. The Formal Methods community itself reflects on this topic. We document this reflection within the last twenty years. In 1990, Hall [Hal90] identifies "Seven Myths of Formal Methods": he argues that unrealistic expectations can lead to the rejection of Formal Methods in industry, and presents a more realistic picture of what Formal Methods can achieve. Five years later, Bowen and Hinchey [BH95a] formulate "Seven more myths of Formal Methods", identify them as misconceptions, and conclude that "Formal Methods are not a panacea, but one approach among many that can help to improve system reliability". Complementing this work, Bowen et al. [BH95b] formulate "Ten Commandments of Formal Methods" which give guidelines of how to make good use of Formal Methods. Ten years later, Bowen and Hinchey [BH06] observe that the "application of formal methods has been slower than hoped by many in the Formal Methods community" and conclude that, at least for highly safety-critical systems, Formal Methods have found a niche. Over the years, the perception of Formal Methods has become more positive. In 2011, e.g., Barnes states that "the application of Formal Methods is a cost effective route to the development of high integrity software" [Bar11]. The 2020 white paper "Rooting Formal Methods within Higher Education Curricula for Computer Science and Software Engineering" [CRD+20] argues:

- Current software engineering practices fail to deliver dependable software.
- Formal Methods are capable of improving this situation, and are beneficial and cost-effective for mainstream software development.
- Education in Formal Methods is key to progress things.
- Education in Formal Methods needs to be transformed.

The "2020 Expert Survey on Formal Methods" [GBP20] compiles a collective vision on the past, present, and future of FMs with respect to research, industry, and education. They report: "A huge majority of 90% thinks the use of Formal Methods will likely become more widespread in industry, while only nine experts [out of 130] doubt this and four have no opinion."

Success Stories

In order to support positive views on Formal Methods, we report on a number of industrial experiments and experiences.

Example 12: Model Checking at Intel

In 1993, Intel released the first Pentium® processor. Shortly afterwards, a bug in the floating point arithmetic was detected, which caused wrong computation results with certain division operations. As a consequence, Intel had to exchange more than one million faulty processors, with cost of more than 475 million dollars. Subsequently, Intel initiated major changes in its validation technology and methodology. "Since 1995 Intel engineers have been using formal verification tools to verify properties of hardware designs" [Fix08]. In hardware design, bugs have traditionally been detected by extensive testing, including pre-silicon simulation. However, this procedure is rather slow, and there are too many input combinations for an exhaustive testing. Therefore, Intel now employs temporal logic model checking for this task. Here, a model of the system (i.e., the hardware design) is compared with a formal specification of system properties (see Chap. 2 on Logics). For describing hardware properties, Intel developed the specification language ForSpec, which was later made into the IEEE 1850 standard PSL (property specification language). In this language, properties of floating point arithmetic as required by the relevant IEEE 754 standard were formulated. As a model, the register transfer level description of the design is used. Thus, the verification is done with the same gate-level design that is used for traditional dynamic validation. Given suitable model checking tools, the verification is fast and can be easily done within the development timeframe. Therefore, such a full formal verification of floating-point processing units is now standard practice at Intel, see also the work of Harrison [Har03b].

Regarding this case study, L. Fix [Fix08] of Intel remarks:

The barrier to moving from a limited deployment to wide spread deployment of formal property verification in Intel was crossed mainly due to two developments: the first was the introduction of ForSpec assertions inside the Verilog code, thus allowing the designers to easily code and maintain the properties (assertions). The second was the integration of the formal verification activity with other validation efforts. In particular, the RTL designer had two reasons to annotate his/her code with assertions. The assertions were always checked during simulation and in addition the assertions served as assumptions and properties for formal verification. In case an assertion was too complex to be verified formally it was still very useful as a checker in simulation.

Example 13: Microsoft's Protocol Documentation Program

Due to legal negotiation with the U.S. Department of Justice and the EU, Microsoft decided to make available to competitors the interfaces of certain client-server and server-server communication protocols used in the Windows operating system. In order not to disclose the source code of the implementation, a series of technical documents were written describing the relevant protocols. This documentation was quite extensive, consisting of more than 250 documents with approximately 30,000 pages in all. The actual implementation of the protocols had previously been released on millions of servers, as part of the Windows operating system. To ensure that the informal specifications conform to the actual code, in the Winterop project a formal model of the specification was produced [GKB11]. For this, the specification language Spec# was used, which is based on the notion of abstract state machine, with C# syntax. The effort took more than 50 person-years to complete. From these specifications, test cases were automatically generated by the tool Spec-Explorer. These test cases could be executed with the existing implementation, exposing over 10,000 "Technical Document Issues" in the specification [GKSB10]. The endeavour was such a big success, that SpecExplorer was turned into a product which is now distributed as a 'power-tool' add-on to the software development environment Visual Studio.

Several other formal specification and verification project within Microsoft have been done. Hackett et al. [HLQB] use the modular checker HAVOC to check properties about the synchronisation protocol of a core Microsoft Windows component in the NT file system with more than 300,000 lines of code and 1500 procedures. The effort found 45 serious bugs (out of 125 warnings) in the component, with modest annotation effort.

Das [Das06] writes on Formal Methods at Microsoft:

"Today, formal specifications are a mandated part of the software development process in the largest Microsoft product groups. Millions of specifications have been added, and tens of thousands of bugs have been exposed and fixed in future versions of products under development. In addition, Windows public interfaces are formally specified and the Visual Studio compiler understands and enforces these specifications, meaning that programmers anywhere can now use formal specifications to make their software more robust."

Example 14: Electronic Voting

Secure communications have become fundamental to modern life for the purposes of electronic commerce, online banking and privacy over the Internet to name but a few applications. As a design problem, security protocols have inspired the use of Formal Methods for well over two decades. The distributed and parallel nature of communications facilitated by protocols, along with various assurances desired, means that designing secure message exchange is not straightforward. A good example of this problem is electronic voting, which has a complex set of security and privacy requirements all of which must be guaranteed if digital democracy is to be truly realised in the modern world.

An electronic voting system subject to formal scrutiny is the Prêt à Voter system [RBH+09], which is essentially a multi-party cryptographic protocol offering privacy, voter verifiability, coercion-resistance, and receipt-freeness. Some of these properties have been subject to formal examination, using various methods including process algebra and refinement checks [HS12], and zero-knowledge proofs [KRT13], with the ultimate goal to providing a formal proof of the relevant property.

An implementation of the Prêt à Voter system has been demonstrated for the state of Victoria in Australia [BCH+12].

Undoubtedly some legal [DHR+12] and usability [SSC+12, SLC+11] challenges exist for such electronic voting systems. However, the above case study demonstrates considerable progress for providing assurances to the government and public to ensure confidence and trust in the election system.

Example 15: The Operating Systems seL4 and PikeOS

Formal verification of operating systems remains a difficult task to achieve given the scale and complexity of the software involved. One such attempt stands out to provide a benchmark of how Formal Methods have been effectively applied towards achieving such a goal. The L4 family of microkernels [Lie96] for embedded systems serves as an operating system with typical features of concurrency in terms of threading and inter-process communication, virtual memory, interrupts and process authorisation features.

A secured version of such an operating system, known as seL4 [KAE+10], has been established through formal specification and verification. Formal Methods have been applied at various levels of the development of seL4. Starting with an abstract specification a prototype is generated in Haskell [KDE09], which is a functional and executable language. This has the advantage of translating all data structure and implementation details desired for the final implementation. The Haskell prototype is formalised using Isabelle/HOL, an interactive-theorem prover allowing for machine-checking of proofs, and functional correctness is demonstrated using refinement. A C implementation is manually achieved from Haskell with a view to optimising the code for better performance. The implementation is then translated into Isabelle (using a formal semantics defined for a subset of C) for checking.

The methods used for seL4 have influenced the verification of PikeOS [BBBB09], which is a commercial microkernel operating system based on L4. Core parts of the embedded hypervisor, and, in particular, the memory separation functionalities, have been formally verified using the VCC verification tool. PikeOS is certified according to various safety standards and is used in several critical real-time applications, e.g., in the integrated modular avionics modules of Airbus planes.

The German Verisoft project [Ver07] demonstrates that with present Formal Methods it is not only possible to verify an operating system, but that the systematic use of computer-aided verification tools is possible throughout all layers of abstractions.

Example 16: Model-Based Design with Certified Code Generation

Lustre is a synchronous data-flow programming language which evolved in the 1980s from academic concepts similar to the ones existing in algebraic specification languages (see Chap. 3) [Hal12]. Its main focus was programming reactive real-time systems such as automatic control and monitoring devices. From the beginning, it had a strict denotational and operational semantics. The formalism was very similar to temporal logics (see Chap. 2) which allowed the language to be used for both writing programs and for expressing program properties. In the mid-1980s, the company Merlin Gerin (now Schneider Electric) in collaboration with researchers from VERIMAG developed an industrial version of Lustre for the development of control command software in nuclear power plants. This version was called SAGA and provided a mixed textual/-graphical syntax for the language as well as a simple code generator. In order to further industrialise the tool, the company Verilog took over SAGA, renamed it SCADE (for "Safety Critical Application Environment Development") and adapted it to the needs of Aerospatiale (now part of Airbus). In the aerospace domain, any tool used for the development of a critical equipment must have at least the same quality as the equipment itself. Therefore, the SCADE code generator KCG was qualified according to the highest criticality level A. (In this qualification, it was shown that the development processes for KCG conform to the requirements of the standard; note that this does not amount to a full compiler verification!) Verilog itself was acquired in 1999 by Telelogic, a Swedish telecommunications tool provider (now IBM). In 2001, Esterel Technologies bought SCADE from Telelogic for 1.4 million Euro. It extended SCADE by various additional components, e.g., the tool IMAGE by Thales for the design of the cockpit of the A380 aircraft, as well as formal verification technology, SysML support, and software lifecycle management. In 2012, Esterel Technologies was taken over by Ansys Inc. for the sum of 42 million Euro. Ansys plans to integrate SCADE with its own tool Simplorer for modelling and simulating physical systems.

Today, more than 230 companies in 27 countries use SCADE to develop safety-critical control components. Success stories include the use in the primary flight control system of the Airbus A380, the autopilot for several Eurocopter models, several nuclear power plants as well as the interlocking and train control system of the Eurostar trains between London and Paris.

Example 16 demonstrates that tools for Formal Methods not only significantly contribute to system safety, but also can have a considerable market.

Example 17: Transportation Systems in France

The Paris Métro line 14 (Est-Ouest Rapide) was opened in 1998. It is the fastest and most modern line in the Paris subway network, being operated driverless, with a high train speed and frequency. For ensuring the correctness of the control and signalling software, it was decided to use the B method and the associated Atelier B programming tool.

The B method is based on the idea of refinement of abstract machines. Mathematical specifications written in first-order logic are stepwise refined, and it is proven that the refinement is coherent and includes all the properties of the abstract machine. Throughout all of the development steps the same specification language (B notation) is used. The process of refinement is repeated until a deterministic, fully concrete implementation is reached, which is then automatically translated into Ada code.

In the above mentioned Météor project, over 110,000 lines of B specifications were written, generating 86,000 lines of safety-critical Ada code. With this model, 29,000 proofs were conducted. No bugs were detected after the proofs, neither during the functional validation of the software, during its integration in the train, during the on-site tests, nor since the metro lines operate. The software is still operated in version 1.0 today, without any bug detected so far [LSGP07].

Other uses of the B method include the automatic train protection system for the French railway company SNCF, which was installed on 6,000 trains since 1993. For the verification, 60,000 lines of B specifications and approximately 10,000 proofs have been written. In the Roissy VAL project, an automatic pilot for a driverless shuttle in the Paris-Roissy airport has been developed and verified with 180,000 lines of B specification and 43,000 proofs.

In the report [BA05] of the Roissy VAL project mentioned in Example 17 the authors conclude:

The process described here is suitable for any industrial domains, not only for railways command/control software. Actually this process deals with designing procedural software based on logical treatments, not based on real or floating-point numbers. It is all the more suitable that software specification can be easily formalised into set-theoretical expressions.

From the management point of view, the project went off according to the initial schedule, although the software produced is quite large, thanks to a straightforward process and efficient tools.

Every verification stage throughout the process was useful and led to early error detection: analysis of software document specification, type checking, inspections, proof of abstract model safety properties, refinement proof of correct implementation.

Section 1.3.4 of Garavel's report [GG13] provides a collection of further success stories.

1.3.3 How to Get Started

The previous sections give the right impression that the variety of Formal Methods is overwhelming. This might leave the beginner or potential user to be lost in the field. Which method shall be selected in a given context? We discuss this question in two different scenarios. One possible scenario is that of a research student learning a Formal Method. The other scenario is that of a Formal Method to be selected in a specific industrial project.

Learning a Formal Method

For the first scenario, this book provides a good starting point. It offers a non-representative selection of methods, where each chapter provides a solid introduction to one method. Specialisation in one method is unproblematic, as the foundations of Formal Methods are well connected. Concepts studied, say, in the context of process algebra, are also to be found in temporal logics, which again are closely connected to automata theory, and are applied, e.g., in testing. Within a discipline, there are often attempts to unify structural insights. In logics, for example, the theory of institutions provides a general framework in which logical properties can be studied in a uniform way (see Chap. 2). The methodological approach to different Formal Methods often is comparable. Consequently, one should not be afraid of intensively studying one specific method, even if it is not in the direct line of one's own research.

The best approach of studying a specific method is by example. One should select a suitable case study of medium complexity (this book is full of these). The first step is to formalise the case study, i.e., to transfer it into the language of the chosen method. Already in this step one might find limitations or restrictions that one would like to study further. The next step is to check if the formalisation is an adequate representation of the case study. The modelling of systems and of proof obligations needs to be faithful.

Now, it is time for reflection: what insight can be gained into the formal representation with the chosen Formal Method? Here, one can try to derive properties manually—using a calculus, or even directly applying the semantics. Only in the next step one should reproduce the manual results in tools, if available. This order of first thinking and then using tools is important for keeping one's mind open to the possibilities of a Formal Method. Tools usually cover only certain aspects, namely those most relevant to their developers. Experience suggests that such a case study-driven approach creates good research topics.

Choosing a Formal Method in an Industrial Project

In the industrial scenario, it is often a challenge to choose an appropriate Formal Method for a particular industrial problem. Factors to be considered include

- the qualification and availability of staff,
- the degree of formalisation of existing documents,
- the development processes and capability maturities within the company, and
- the available budget in relation with the expected benefits.

Moreover, for each Formal Method to be considered, the availability of industrial strength tools is a decisive factor. In order to be usable for an industrial project, a tool has to satisfy certain criteria.

- It needs to be supported: that is, during a certain amount of time (usually, a time period well beyond the lifespan of the product, which can be several years) there must be a reliable partner offering maintenance, error correction, adaptation to evolving platforms, further development, and advice to users.
- It needs to be documented: that is, there must exist user manuals, online help, training material, and coaching resources for the engineers who shall use the tool. To this end, competences and skills profiles need to be established.
- It needs to integrate smoothly into the existing development processes. That is, exchange formats need to be available and translations between different representations should exist or be easily implementable.
- Its use should be predictable: there need to be good arguments that the intended task can be accomplished with the help of the tool, within a time frame which is allocated in advance.
- In some cases, it even needs to be qualified: for the development of safety-critical systems, it is not permitted to use an arbitrary tool; at least, an analysis of capabilities and alternatives must be conducted.

One risk in selecting a Formal Method is that most practitioners tend to favour their own area of expertise. Other approaches, which actually might be better suited, are easily overlooked. Thus, it is a good idea to consult several experts covering different areas.

Having identified a suitable Formal Method, the next step is to carry out a pilot project. Here, a small but relevant part of the problem is solved in an exemplary way using the selected Formal Method. This involves thorough time measurement and documentation for all activities that the Formal Method incurs: modelling, installing the tools, using the tools, integrating the tools, interpreting the results, etc. Reflecting upon this allows to check if the chosen approach is indeed feasible. On the management level, the pilot project then has to be evaluated as to whether it is an improvement with respect to current practice. It should be both more effective, in that it allows to achieve better results than previous methods, and more efficient, i.e., in the long run it should offer a better cost/result ratio.

1.4 Closing Remarks

In this chapter we developed an understanding of the key ingredients of Formal Methods: syntax, semantics and method. The syntax is usually given in Backus–Naur-Form; the semantics is mostly presented in either operational, denotational, or axiomatic style; the method says how to work with the language. Formal Methods are useful in classical as well as in agile software development processes. They are used to achieve precision in design documents and to support various forms of system analysis. International standards recognise and recommend the use of various Formal Methods. In practise, Formal Methods require tool support. As several academic and industrial success stories demonstrate, Formal Methods play an increasing role in industrial practice.

1.4.1 Current Research Directions

In this section we point out several challenges and current research topics for Formal Methods.

Advancement. An account of the historical development of Formal Methods is given in Chap. 9. In this context, the question is whether there still is a need for Formal Methods to evolve further. Considering computational systems, we see that their size and complexity is ever increasing. Also, computers for executing Formal Methods tools become more powerful. However, the increase of the problem size often outgrows what tools can handle. This is due to the fact that most algorithms in Formal Methods are complex. As a consequence, there is a constant need to improve methods and tools. Therefore, the questions of how to develop 'good' Formal Methods, i.e., Formal Methods which are efficient and usable, will stay.

Integration. As was shown above, various formal software modelling techniques have been developed. In systems' design, these can be used to describe different aspects of the very same system. In Example 1 (see Sect. 1.1.1) concerning the ISS, for instance, the correctness of the fault management layer was analysed using the process algebra CSP. In order to guarantee a minimal throughput on the station's MIL bus, Schlingloff performed a stochastic analysis using Timed Petri-Nets [THSS98]. Generally, in such circumstances the question arises whether different models provide a consistent view of the system, and whether analysis results for one aspect can be re-used and integrated into the analysis of other aspects. Here, UML provides an integration of various modelling frameworks. However, this integration is on the syntactical level only. Semantical and methodological integrations are still being researched [KM18].

Industrial Practice. The long standing question of how to turn Formal Methods into good industrial practice still remains a challenge. For example, the aerospace standard DO-333, published in 2012, allows Formal Methods to replace traditional engineering practice, e.g., in testing, code inspection, and code review. However, there are not yet sufficiently many qualified tools available. Moreover, it is not always clear where Formal Methods offer better results than the established processes.

Parallelisation. Another current research trend is that the impending multi/many core revolution poses the question of how to develop efficient parallel algorithms. In Formal Methods, e.g., for model checking, SAT and SMT solving, and automated theorem proving there are first proposals of algorithms tailored towards the execution on multi/many core machines.

Re-use. Nowadays, systems are rarely constructed from scratch. New, functionally increased and more complex software products are built on top of existing ones. Systems are rather improved than newly developed, i.e., there is a constant software evolution. Like other industrial products, also software is designed in product lines. Formal Methods have not yet come up with adequate techniques to reflect these development processes by evolutionary modelling and verifying of systems. The main challenge is how to re-use verification artefacts.

Compositionality. As systems become more and more complex and spatially distributed, there is an increasing need to verify large, parallel systems. For example, there are Formal Methods being developed to deal with service-oriented architectures, where autonomous software agents in the Internet cooperate in order to achieve a certain task. Questions include the interaction with an unknown, non-predictable environment, functional correctness, quantitative analysis, verification of service level agreements, and security (see Chap. 8).

Cyber-physical agents. Yet another challenge concerns the application of Formal Methods in cyber-physical systems. These are 'agent-based systems in the physical world', i.e., intelligent sensor networks, swarms of robots, networks of small devices, etc. Part of the problem is the combined physical and logical modelling. For modelling systems which have both discrete and continuous state changes, hybrid automata have been suggested as a formal framework. However, current methods are not yet sophisticated enough to allow the verification of industrial strength applications. Additionally, cyber-physical systems have to deal with unreliable communication, imprecise sensors and actors, faulty hardware etc. For each of these problems, initial individual approaches are being developed, however, it is an open research topic to develop convincing Formal Methods tackling them in combination.

Artificial intelligence and machine learning. Techniques based on artificial intelligence (AI) in general, and machine learning (ML) in particular, are massively being used in deployed software. Many applications using AI/ML are safety-critical (e.g., autonomous cars), so correctness is paramount. But the interaction between Formal Methods and AI/ML goes beyond the standard 'let us use a Formal Methods technique to prove the correctness of this algorithm—which happens to use AI/ML.' Indeed, the use of AI/ML introduces new challenges for formal verification, in particular in the area of deep neural networks where sometimes an algorithm has been learned without a clear understanding of the process of its creation. This makes it difficult to assert the correctness of the outcomes of the algorithm, which might require *transparency* in the underlying models and the used techniques and methods for learning algorithms, that is to get 'explainable' AI [Mol19]. Other interesting research directions are the use of machine learning to improve Formal Methods [ALB18], and the application of Formal Methods to AI/ML [HKWW17, SKS19, WPW+18].

Finally, we briefly mention further applications of Formal Methods beyond software engineering, such as biological systems and, more recently, ecology, economics and social sciences.

Biological systems. Formal Methods started to be used to model *biological systems* following 1998 Gheorghe Păun's definition of P systems [Pău98, Pău00], a computational model inspired from the way the alive cells process chemical compounds in their compartmental structure. Variations of this model led to the birth of a research field known as *membrane computing* [PRS09]. Although P systems were originally intended as a biologically-inspired computational model, it was soon understood that they could provide a modelling language to formally describe biological systems and on which to base tools to reason about their evolution. The Grand Challenge for computing that David Harel proposed in 2002 [Har03a] to model a full multicellular animal, specifically the *C. elegans* nematode worm, as a reactive system led to the extension of various Formal Methods, traditionally used in computer science, to make them suitable to the modelling of biological systems. For example, the Performance Evaluation Process Algebra (PEPA) [GH94] was extended in order to handle some features of biochemical networks, such as stoichiometry and different kinds of kinetic laws, thus resulting in the Bio-PEPA language [CH09a], whose models can be fed to PRISM [KNP10] for (stochastic) model checking. Another development in this application area has been to move from the modelling of a single organism to the modelling of population dynamics. Some formal notations, such as the process algebra-based BlendX language [DPR08], have been developed to model *ecological systems* consisting of various populations (or *ecosystems*, in a wider context), aiming at overcoming the technical and cultural gap between life scientists and computer scientists [CS15]. The final objective of this modelling approach is not only to understand the functioning of the ecosystem but also to test possible control interventions on some of the system components aiming at performing adjustments to the system behaviour and evaluate the impact of such intervention on the entire ecological system [CS15]. Examples are: pest eradication [BCB+10], preservation/reintroduction of species [CCM+10], disease control [CH09b] and even tumour control (a tumour can be seen as an 'ecosystem' consisting of various populations of normal and mutant cells) [SBMC15].

Economics. The most successful application of Formal Methods to economics is in the area of business process management. Will van der Aalst has been using variants of Petri nets to model enterprise resource planning systems, cooperative work, resource allocation and inter-organisational business processes [VDAS11]. It is in this application area that the two analytical philosophies of the Formal Methods community and the data mining/big data community are getting closer and closer. Rozinat and Van der Aalst developed methodologies to perform *conformance checking*, also called *conformance analysis*, that is, the detection of inconsistencies between an a priori process model and an a posteriori model produced by applying process mining to the corresponding execution log [Aal11, RA08]. The future of this approach goes well beyond the specific application to business process management, in particular in humanities.

Social Sciences. In fact, conformance checking seems appropriate for the analysis of social networks and peer-production systems, and the first attempts in this direction have being done in the areas of collaborative learning and OSS (Open Source Software) development [MCT15]. More in general, data mining, text mining and process mining, through conformance checking, can provide appropriate and effective validation tools for formal models of social systems, opening the application of Formal Methods to the vast area of social sciences.

Another promising use of Formal Methods in social sciences is the modelling of privacy. For example, privacy is an issue in sociology, politics, and legislation. Formalising privacy policies and realising enforcing mechanisms is not easy. The challenges of privacy for Formal Methods have been discussed for instance in [TW09]. Also, there is an increasing need for technology-based solutions to help lawyers to draft and analyse contractual documents, and citizens to understand the huge amount of different kinds of agreements and terms of services on paper and digital devices. Formal Methods can play a crucial role in providing solutions to help handling such complex documents (see Chap. 6).

The above items present opportunities for research on topics which are both scientifically exciting and have a large impact on society. In order to start such research, one can build upon the material presented in the subsequent chapters.

References

[Aal11]   Wil M. P. Van Der Aalst. *Process Mining: Discovery, Conformance and Enhancement of Business Processes.* Springer, 2011.

[ABL96]   Jean-Raymond Abrial, Egon Börger, and Hans Langmaack, editors. *Formal Methods for Industrial Applications, Specifying and Programming the Steam Boiler Control (the book grew out of a Dagstuhl Seminar, June 1995)*, LNCS 1165. Springer, 1996.

[Abr94]   Jean-Raymond Abrial. Steam-boiler control specification problem. https://www.informatik.uni-kiel.de/~procos/dag9523/dag9523.html, 1994.

[ALB18]   Moussa Amrani, Levi Lúcio, and Adrien Bibal. ML + FV = ♡? A survey on the application of machine learning to formal verification, 2018. http://arxiv.org/abs/1806.03600.

[BA05]   F. Badeau and A. Amelot. Using B as a high level programming language in an industrial project: Roissy VAL. In *ZB 2005*, LNCS 3455, pages 334–354. Springer, 2005.

[Bac59]   J. W. Backus. The syntax and semantics of the proposed international algebraic language of the zurich acm-gamm conference. In *Proceedings of the International Conference on Information Processing.* UNESCO, 1959. Available via the web site of the Computer History Museum's Software Preservation Group, http://www.softwarepreservation.org.

[Bar11]   Janet Elizabeth Barnes. Experiences in the industrial use of formal methods. In *AVoCS'11.* Electronic Communications of the EASST, 2011.

[BBBB09]   Christoph Baumann, Bernhard Beckert, Holger Blasum, and Thorsten Bormer. Formal verification of a microkernel used in dependable software systems. In *SAFECOMP 2009*, LNCS 5775, pages 187–200. Springer, 2009.

[BCB+10]   Thomas Anung Basuki, Antonio Cerone, Roberto Barbuti, Andrea Maggiolo-Schettini, Paolo Milazzo, and Elisabetta Rossi. Modelling the dynamics of an aedes albopictus population. In *AMCA-POP 2010*, volume 33 of *Electronic Proceedings in Theoretical Computer Science*, pages 18–36. Open Publishing Association, 2010.

[BCH+12]   Craig Burton, Chris Culnane, James Heather, Thea Peacock, Peter Y. A. Ryan, Steve Schneider, Sriramkrishnan Srinivasan, Vanessa Teague, Roland Wen, and Zhe Xia. A supervised verifiable voting protocol for the Victorian electoral commission. In *EVOTE 2012*, volume 205 of *LNI*, pages 81–94. GI, 2012.

[Bec02]   Kent Beck. *Test Driven Development: By Example.* Addison-Wesley, 2002.

[BH95a]   Jonathan P. Bowen and Michael G. Hinchey. Seven more myths of formal methods. *IEEE Software*, 12(4):34–41, 1995.

[BH95b]   Jonathan P. Bowen and Michael G. Hinchey. Ten commandments of formal methods. *IEEE Computer*, 28(4):56–63, 1995.

[BH06]   Jonathan P. Bowen and Michael G. Hinchey. Ten commandments of formal
              methods ... Ten years later. *IEEE Computer*, 39(1):40–48, 2006.
[BKPS97]   Bettina Buth, Michel Kouvaras, Jan Peleska, and Hui Shi. Deadlock analysis
              for a fault-tolerant system. In *AMAST*, LNCS 1349. Springer, 1997.
[BMS96]   Manfred Broy, Stephan Merz, and Katharina Spies, editors. *Formal Sys-
              tems Specification, The RPC-Memory Specification Case Study*, LNCS 1169.
              Springer, 1996.
[BPS98]   Bettina Buth, Jan Peleska, and Hui Shi. Combining methods for the livelock
              analysis of a fault-tolerant system. In *AMAST*, LNCS 1548. Springer, 1998.
[CCM+10]   Mónica Cardona, M. Angels Colomer, Antoni Margalida, Ignacio Pérez-
              Hurtado, Mario J. Pérez-Jiménez, and Delfí Sanuy. A P system based model
              of an ecosystem of some scavenger birds. In *WMC 2009*, LNCS 5957, pages
              182–195. Springer, 2010.
[CH09a]   Federica Ciocchetta and Jane Hillston. Bio-PEPA: a framework for the mod-
              elling and analysis of biochemical networks. *Theoretical Computer Science*,
              410:3065–3084, 2009.
[CH09b]   Federica Ciocchetta and Jane Hillston. Bio-PEPA for epidemiological models.
              In *PASM 2009*, volume 261 of *Electronic Notes in Theoretical Computer
              Science*, pages 43–69. Open Publishing Association, 2009.
[CK11]   Véronique Cortier and Steve Kremer, editors. *Formal Models and Techniques
              for Analyzing Security Protocols*. IOS Press, 2011.
[Coc00]   Alistair Cockburn. *Agile Software Development*. Addison-Wesley, 2000.
[CRD+20]   Antonio Cerone, Markus Roggenbach, James Davenport, Casey Denner,
              Marie Farrell, Magne Haveraaen, Faron Moller, Philipp Koerner, Sebastian
              Krings, Peter Ölveczky, Bernd-Holger Schlingloff, Nikolay Shilov, and Rustam
              Zhumagambetov. Rooting formal methods within higher education curricula
              for computer science and software engineering – A White Paper, 2020.
              https://arxiv.org/abs/2010.05708.
[CS15]   Antonio Cerone and Marco Scotti. Research challenges in modelling ecosys-
              tems. In *SEFM 2014 Collocated Workshops*, LNCS 8938, pages 276–293.
              Springer, 2015.
[Das06]   Manuvir Das. Formal specifications on industrial-strength code – from myth
              to reality. In *Computer Aided Verification*, LNCS 4144. Springer, 2006.
[dBfI12]   Die Beauftragte der Bundesregierung für Informationstechnik. Das V-Modell
              XT. http://www.v-modell-xt.de, 2012.
[DHR+12]   Denise Demirel, Maria Henning, Peter Y. A. Ryan, Steve Schneider, and
              Melanie Volkamer. Feasibility analysis of prêt à voter for german federal elec-
              tions. In *VoteID 2011*, LNCS 7187, pages 158–173. Springer, 2012.
[DPR08]   Lorenzo Dematté, Corrado Priami, and Alessandro Romanel. The BlenX lan-
              guage: a tutorial. In *Formal Methods for Computational Systems Biology*,
              LNCS 5016, pages 313–365. Springer, 2008.
[FH01]   Marc Frappier and Henri Habrias, editors. *Software Specification Methods*.
              Springer, 2001.
[Fix08]   Limor Fix. Fifteen years of formal property verification in Intel. In *25 Years
              of Model Checking*, LNCS 5000, pages 139–144. Springer, 2008.
[GBP20]   Hubert Garavel, Maurice H. ter Beek, and Jaco van de Pol. The 2020 expert
              survey on formal methods. In *Formal Methods for Industrial Critical Systems*,
              pages 3–69. Springer, 2020.
[GG13]   Hubert Garavel and Susanne Graf. *Formal Methods for Safe and Secure Com-
              puters Systems*. Federal Office for Information Security, 2013. Available, e.g.,
              via the book webpage https://sefm-book.github.io.
[GH94]   Stephen Gilmore and Jane Hillston. The PEPA workbench: A tool to support
              a process algebra-based approach to performance modelling. In *International
              Conference on Modelling Techniques and Tools for Computer Performance
              Evaluation*, LNCS 794, pages 353–368. Springer, 1994.

[GJBB13]   James Gosling, Bill Joy, Guy Steele Gilad Bracha, and Alex Buckley. The Java language specification, 2013.

[GKB11]   Wolfgang Grieskamp, Nico Kicillof, and Bob Binder. Microsoft's protocol documentation program: Interoperability testing at scale. *Communications of the ACM*, 2011.

[GKSB10] Wolfgang Grieskamp, Nicolas Kicillof, Keith Stobie, and Victor Braberman. Model-based quality assurance of protocol documentation: tools and methodology. *Softw. Test. Verif. Reliab.*, 2010.

[Hal90]   Anthony Hall. Seven myths of formal methods. *IEEE Software*, 7(5):11–19, 1990.

[Hal12]   Nicolas Halbwachs. *A Synchronous Language at Work: The Story of Lustre*, pages 15–31. Wiley, 2012.

[Har03a] David Harel. A grand challenge for computing: Towards full reactive modeling of a multi-cellular animal. *Bull. EATCS*, 81:226–235, 2003.

[Har03b] J. Harrison. Formal verification at Intel. In *18th Annual IEEE Symposium of Logic in Computer Science*, pages 45–54, 2003.

[HKW+07] Sebastian Herold, Holger Klus, Yannick Welsch, Constanze Deiters, Andreas Rausch, Ralf Reussner, Klaus Krogmann, Heiko Koziolek, Raffaela Mirandola, Benjamin Hummel, Michael Meisinger, and Christian Pfaller. CoCoME - the common component modeling example. In *CoCoME*, LNCS 5153, pages 16–53. Springer, 2007.

[HKWW17] Xiaowei Huang, Marta Kwiatkowska, Sen Wang, and Min Wu. Safety verification of deep neural networks. In *CAV'17*, LNCS 10426, pages 3–29. Springer, 2017.

[HLQB]   Brian Hackett, Shuvendu K. Lahiri, Shaz Qadeer, and Thomas Ball. Scalable modular checking of system-specific properties: Myth or reality?

[HS12]   James Heather and Steve Schneider. A formal framework for modelling coercion resistance and receipt freeness. In *FM 2012*, LNCS 7436, pages 217–231. Springer, 2012.

[IEC10]   International Electrotechnical Commission. *Functional Safety of Electrical/Electronic/Programmable Electronic Safety-related Systems*, IEC 61508:2010, 2010.

[JW08]   Cliff B. Jones and Jim Woodcock, editors. *Formal Aspects of Computing*, volume 20, No 1. Springer, 2008.

[KAE+10] Gerwin Klein, June Andronick, Kevin Elphinstone, Gernot Heiser, David Cock, Philip Derrin, Dhammika Elkaduwe, Kai Engelhardt, Rafal Kolanski, Michael Norrish, Thomas Sewell, Harvey Tuch, and Simon Winwood. seL4: formal verification of an operating-system kernel. *Communications of the ACM*, 53(6):107–115, 2010.

[KBP+01] Erik Kamsties, Daniel M. Berry, Barbara Paech, E. Kamsties, D. M. Berry, and B. Paech. Detecting ambiguities in requirements documents using inspections. In *First Workshop on Inspection in Software Engineering*, 2001.

[KDE09]   Gerwin Klein, Philip Derrin, and Kevin Elphinstone. Experience report: seL4: formally verifying a high-performance microkernel. In *ICFP 2009*, pages 91–96. ACM, 2009.

[KM18]   Alexander Knapp and Till Mossakowski. Multi-view consistency in UML: A survey. In *Graph Transformation, Specifications, and Nets - In Memory of Hartmut Ehrig*, LNCS 10800, pages 37–60. Springer, 2018.

[KNP10]   Marta Kwiatkowska, Gethin Norman, and David Parker. Probabilistic model checking for systems biology. In *Symbolic Systems Biology*, pages 31–59. Jones and Bartlett, May 2010.

[Koz94]   Dexter Kozen. A completeness theorem for Kleene algebras and the algebra of regular events. *Information and Computation*, 110:366–390, 1994.

[KRT13]   Dalia Khader, Peter Y. A. Ryan, and Qiang Tang. Proving prêt à voter receipt
            free using computational security models. *USENIX Journal of Election Tech-
            nology and Systems (JETS)*, 1(1):62–81, 2013.
[KS12]   Dexter Kozen and Alexandra Silva. Left-handed completeness. In *Relational
            and Algebraic Methods in Computer Science*, LNCS 7560, pages 162–178.
            Springer, 2012.
[Lie96]   Jochen Liedtke. Toward real microkernels. *Communications of the ACM*,
            39(9):70–77, September 1996.
[LL95]   Claus Lewerentz and Thomas Lindner, editors. *Formal Development of Reac-
            tive Systems – Case Study Production Cell*, LNCS 891. Springer, 1995.
[LSGP07]   T Lecomte, T Servat, and G G Pouzancre. Formal methods in safety-critical
            railway systems. In *Brazilian Symposium on Formal Methods: SMBF*, 2007.
[LSP82]   Leslie Lamport, Robert E. Shostak, and Marshall C. Pease. The Byzantine
            generals problem. *ACM Trans. Program. Lang. Syst.*, 4(3):382–401, 1982.
[MCT15]   Patrick Mukala, Antonio Cerone, and Franco Turini. Process mining event
            logs from floss data: State of the art and perspectives. In *SEFM 2014 Collo-
            cated Workshops*, LNCS 8938, pages 182–198. Springer, 2015.
[Mol19]   Christoph Molnar. Interpretable machine learning, 2019. `https://`
            `christophm.github.io/interpretable-ml-book/`.
[MS08]   Satish Mishra and Bernd-Holger Schlingloff. Compliance of CMMI process
            area with specification based development. In *Conference on Software Engi-
            neering Research, Management and Applications*, SERA '08, pages 77–84.
            IEEE Computer Society, 2008.
[Păun98]   Gheorghe Păun. Computing with membranes. Technical Report 208, Turku
            Centre for Computer Science, November 1998.
[Păun00]   Gheorghe Păun. Computing with membranes. *Journal of Computer and Sys-
            tem Science*, 61(1):108–143, 2000.
[PB99]   Jan Peleska and Bettina Buth. Formal methods for the international space
            station ISS. In *Correct System Design*, LNCS 1710. Springer, 1999.
[PRS09]   Gheorghe Păun, Grzegorz Rozemberg, and Arto Salomaa, editors. *The Oxford
            Handbook of Membrane Computing*. Oxford Handbooks in Mathematics.
            Oxford University Press, December 2009.
[RA08]   Anne Rozinat and Wil M. P. Van Der Aalst. Conformance checking of pro-
            cesses based on monitoring real behavior. *Information Systems*, 33(1):64–95,
            2008.
[RBH+09]   Peter Y. A. Ryan, David Bismark, James Heather, Steve Schneider, and Zhe
            Xia. Prêt à voter: a voter-verifiable voting system. *IEEE Transactions on
            Information Forensics and Security*, 4(4):662–673, 2009.
[RRMP08]   Andreas Rausch, Ralf Reussner, Raffaela Mirandola, and Frantisek Plasil,
            editors. *The Common Component Modeling Example: Comparing Software
            Component Models [result from the Dagstuhl research seminar for CoCoME,
            August 1-3, 2007]*, LNCS 5153. Springer, 2008.
[Sal66]   Arto Salomaa. Two complete axiom systems for the algebra of regular events.
            J. ACM, 13(1):158–169, January 1966.
[SB01]   Ken Schwaber and Mike Beedle. *Agile Software Development with Scrum*.
            Prentice Hall, 2001.
[SBMC15]   Sheema Sameen, Roberto Barbuti, Paolo Milazzo, and Antonio Cerone. A
            mathematical model for assessing KRAS mutation effect on monoclonal anti-
            body treatment of colorectal cancer. In *SEFM 2014 Collocated Workshops*,
            LNCS 8938, pages 243–258. Springer, 2015.
[SCW00]   Susan Stepney, David Cooper, and Jim Woodcock. An electronic purse: Spec-
            ification, refinement, and proof. Technical monograph PRG-126, Oxford Uni-
            versity Computing Laboratory, July 2000.

[SKS19]   Xiaowu Sun, Haitham Khedr, and Yasser Shoukry. Formal verification of neural network controlled autonomous systems. In *HSCC 2019*, pages 147–156. ACM, 2019.

[SLC+11]   Steve Schneider, Morgan Llewellyn, Chris Culnane, James Heather, Sriramkrishnan Srinivasan, and Zhe Xia. Focus group views on prêt à voter 1.0. In *REVOTE 2011*, pages 56–65. IEEE, 2011.

[SSC+12]   Steve Schneider, Sriramkrishnan Srinivasan, Chris Culnane, James Heather, and Zhe Xia. Prêt á voter with write-ins. In *VoteID 2011*, LNCS 7187, pages 174–189. Springer, 2012.

[THSS98]   L. Twele, B-H. H. Schlingloff, and H. Szczerbicka. Performability analysis of an avionics-interface. In *Proc. IEEE Conf. on Systems, Man and Cybernetics*, 1998.

[TW09]   Michael Carl Tschantz and Jeannette M. Wing. Formal methods for privacy. In *FM'09*, volume 5850 of *LNCS*, pages 1–15. Springer, 2009.

[VDAS11]   Wil M. P. Van Der Aalst, and Christian Stahl. *Modeling Business Processes: A Petri Net-Oriented Approach*. MIT Press, May 2011.

[Ver07]   Eyad Alkassar, Mark A. Hillebrand, Dirk Leinenbach, Norbert Schirmer and Artem Starostin. The verisoft approach to systems verification. In *Verified Software: Theories, Tools, Experiments, Second International Conference, VSTTE 2008, Toronto, Canada, October 6–9, 2008. Proceedings*, volume 5295 of *LNCS*, pages 209–224. Springer, 2008. https://doi.org/10.1007/978-3-540-87873-5_18.

[WPW+18]   Shiqi Wang, Kexin Pei, Justin Whitehouse, Junfeng Yang, and Suman Jana. Formal security analysis of neural networks using symbolic intervals. In *USENIX'18*, pages 1599–1614. USENIX Association, 2018.

[YCER11]   Xuejun Yang, Yang Chen, Eric Eide, and John Regehr. Finding and understanding bugs in C compilers. *SIGPLAN Not.*, 46(6):283–294, June 2011.

Part I
Languages

Chapter 2
Logics for Software Engineering

Bernd-Holger Schlingloff, Markus Roggenbach, Gerardo Schneider, and Antonio Cerone

Abstract Logic is the basis for almost all formal methods in computer science. In this chapter, we introduce some of the most commonly used logics by examples. It serves as a reference for subsequent chapters. We start with propositional logic, introduce its syntax, semantics, and calculus. Then we extend our view of propositional logic as a so-called institution, discuss model transformations and modular specifications. Subsequently, we turn to first- and second-order logic and show how these can be obtained as natural extensions of propositional logic. Finally, we discuss non-classical logics: multimodal and deontic logics to deal with alternative and subjective viewpoints, respectively, and dynamic and temporal logics for reasoning about time. The chapter concludes with an elaborate example of how to transform an informal natural language description into a formal specification in linear temporal logic.

2.1 Logic in Computer Science

Today's your lucky day! You are the candidate in a big TV quiz show, and so far you have mastered all challenges. In the final round, there are three

Bernd-Holger Schlingloff
Humboldt University and Fraunhofer FOKUS, Berlin, Germany

Markus Roggenbach
Swansea University, Wales, United Kingdom

Gerardo Schneider
University of Gothenburg, Sweden

Antonio Cerone
Nazarbayev University, Nur-Sultan, Kazakhstan

© Springer Nature Switzerland AG 2022, corrected publication 2022
M. Roggenbach et al., *Formal Methods for Software Engineering*,
Texts in Theoretical Computer Science. An EATCS Series,
https://doi.org/10.1007/978-3-030-38800-3_2

baskets, one of which contains the big prize, the other two being empty. The quiz master gives you the following facts.

1. Either the prize is in the middle basket, or the right basket is empty.
2. If the prize is not in the left basket, then it is not in the middle.

Which basket do you choose?

This example is typical for a number of 'logic puzzles' which have been popular since the middle of the 19th century. In this chapter we will show how to formalize and solve such puzzles—not just for the sake of winning in TV quiz shows, but also to construct software solutions for all sorts of problems.

Seen from a hardware perspective, nowadays every 'classical' computer is a binary device, built from switching elements which can be in one of two states each—on or off, high or low, zero or one etc. Therefore, hardware can be described and analysed with logical means.

From a software perspective, every programming language incorporates some logic, e.g., propositional logic for Boolean conditions in control flow decisions such as branches or loops. Thus, the study of logic is fundamental for programming. There even are special programming languages like, e.g., PROLOG, in which logic is turned into a programming paradigm. For many programming languages, there are special logics to express certain properties of program phrases, e.g., pre- and postconditions. Examples include ACSL for C, JML for Java, and Spark Ada.

Moreover, logic can be used as a means for specification of systems, independent of the particular programming language. Examples are CASL (see Chap. 4) and the Z, B, and Event-B specification languages. Given a system specification, logic can be used to verify that a particular program is correct with respect to it. Specialised logics have been developed for this task, including temporal and dynamic logics which are treated in this chapter.

From these examples it should be clear that there is not 'one' logic in computer science, but many different logics are being used. Each of these can be considered to be a Formal Method:

- The syntax of a logic usually is given by a small grammar which defines the well-formed formulae.
- For the semantics, notions like "model", "interpretation", "satisfaction" and "validity" are defined.
- The method usually includes ways to show which formulae are satisfied in a given model, or valid in all models of a certain class.

Subsequently, we will show how logic as a formal method can be used to solve some of the verification tasks mentioned above.

2.2 Propositional Logic—An Archetypical Modelling Language

Propositional logic can be considered to be the oldest language of formalised reasoning, dating back to Aristotle and the medieval scholastics. Its basic philosophy can best be described by Wittgenstein's proverb "Die Welt ist alles, was der Fall ist" (the world consists of all which is the case), i.e., the world can be considered as the set of all true propositions about it. Propositions can be, e.g., "the sun is shining", "John loves Mary", or "one plus two equals three". Simple propositions can be combined by *operators* or *junctors* to form more complex ones, e.g., "if the sun is shining, then there are no clouds". In software engineering, propositional logic can be used to describe not only the actual world as it is, but also some artifacts which are yet to be built. For example, we can specify the behaviour of software with propositional formulae.

2.2.1 Example: Car Configuration

Before we go into details, we give a motivating example.

Example 18: Car Configuration

This example is inspired by certain car manufacturing software, see, e.g., Volkswagen's "My Configurations" [VW].

Most modern car models come in many variants. Choices include different colours, motors, gear shifts, tyres, audio equipment, etc. Advanced models have dozens of features which lead to thousands of different combinations. Many car manufacturers provide a website where the potential customers can individually configure their cars. Of course, during such a configuration, certain constraints have to be met. For example, certain combinations of motor and transmission are disallowed. So, if the customer decides on a certain motor, some options for the gears are no longer available. Once the configuration is complete, the customer can order the car; the description then is sent to the manufacturing plant where the individual car is built according to the chosen specification.

What is behind the scene of such a scenario? We now develop the example towards modelling and analysis with propositional logic.

Example 18.1: Fixing the Language for Car Configurations

All possible attributes (colour, engine and transmission, tyres, etc.) of a car can have one of a finite number of values. For example, the colour can be red, blue, or green, the engine can use petrol or diesel fuel, the gear shift can be automatic or manual, with four to six gears, etc.

Thus, for a specific car the 'presence' of a particular feature can be 'true' or 'false'. This idea allows us to describe a car configuration as a map from the set of attributes $\mathcal{P}$ into the set $\{true, false\}$. A possible configuration would be: (colour_red=$true$, colour_blue=$false$, colour_green=$false$, motor_diesel=$true$, gearshift_automatic = $false$, ...).

After the language for reasoning about the domain of discourse has been fixed, it is possible to formulate properties in this language.

Example 18.2: Constraints on Car Configurations

Not all combinations of attributes are feasible.

- For example, a car cannot have two colours at the same time, or both automatic and manual gear shift.
- Some low-power motors can be used with manual gear change only, while other are built for manual or automatic transmission. For example, an automatic gear shift is not available with the 59 kW Diesel engine.
- A navigational system is only available with certain audio equipment, at least the radio must have a CD or DVD drive to update the map data.

There are many more of such constraints, typically in the order of several hundreds. Given a set of proposition symbols for the various attributes, we can formalise the constraints with propositional formulae:

- $\bigoplus$(colour_red, colour_blue, colour_green)
  (This formula is short for the following one:
  (colour_red $\wedge$ ¬colour_blue $\wedge$ ¬colour_green) $\vee$ (¬colour_red $\wedge$ colour_blue $\wedge$ ¬colour_green) $\vee$ (¬colour_red $\wedge$ ¬colour_blue $\wedge$ colour_green))
- (gearshift_automatic $\oplus$ gearshift_manual)
- (motor_59kW $\Rightarrow$ (motor_diesel $\wedge$ ¬gearshift_automatic))
- (navi $\Rightarrow$ (audio_cd $\vee$ audio_dvd))

Given a formal specification, Formal Methods provide algorithms and tools how to put it into practical use.

Example 18.3: Satisfiability of Car Configurations

Many car manufacturers have web sites where customers can enter wishes for their new cars. The configurator running in the background of such a web site has built in a set of specification formulae such as above, according to the manufacturer's constraints. For each customer request, it checks whether the chosen combination of attribute instances can be realised such that all given constraints are respected, and displays the remaining choices.

For example, a customer might enter `colour_red`, `audio_dvd`, `gearshift_automatic`, `motor_diesel` and `motor_59kW`. In this case, the configurator would detect that this set of wishes is inconsistent with the constraint "no automatic gearshift with 59 kW motor". It will report this finding and allow the customer to revise the choices.

Technically, this check amounts to determining whether a given set of propositional formulae is *satisfiable*, i.e., whether there exists a *model* (car configuration) in which all constraints are met.

In the last few years, much effort has been put in efficient algorithms for satisfiability checking; we will come back to this later.

2.2.2 Syntax and Semantics of Propositional Logic

Now we are ready to introduce the syntax and semantics of propositional logic.

Definition 1 (*Syntax of propositional logic*) Assume we are given a finite set of proposition symbols $\mathcal{P}$.

We define the following set of formulae:

- Every $p \in \mathcal{P}$ is a formula.
- $\bot$ is a formula.
- If φ and ψ are formulae, then $(\varphi \Rightarrow \psi)$ is a formula.

In a computer science notation (*Backus–Naur-form*, see Example 2), this definition can be written as

$$\mathrm{PL}_{\mathcal{P}} ::= \quad \mathcal{P} \quad | \quad \bot \quad | \quad (\mathrm{PL}_{\mathcal{P}} \Rightarrow \mathrm{PL}_{\mathcal{P}})$$

Note that the set of proposition symbols may be empty; in this case, the only formulae which can be built according to the definition are $\bot$, $(\bot \Rightarrow \bot)$, $((\bot \Rightarrow \bot) \Rightarrow \bot)$, $(\bot \Rightarrow (\bot \Rightarrow \bot))$, etc.

This syntax is 'minimalistic' in the sense that it only contains the constant $\bot$ and the junctor $\Rightarrow$. This allows easy proofs by induction on the structure of a formula. Usually, some more propositional junctors (or boolean operators) are being used, which can be introduced as abbreviations:

- $\neg\varphi \triangleq (\varphi \Rightarrow \bot)$
- $\top \triangleq \neg\bot$
- $(\varphi \vee \psi) \triangleq (\neg\varphi \Rightarrow \psi)$
- $(\varphi \wedge \psi) \triangleq \neg(\neg\varphi \vee \neg\psi)$
- $(\varphi \Leftrightarrow \psi) \triangleq ((\varphi \Rightarrow \psi) \wedge (\psi \Rightarrow \varphi))$
- $(\varphi \oplus \psi) \triangleq (\varphi \Leftrightarrow \neg\psi)$
- $\bigoplus(\varphi_1, \ldots, \varphi_n) \triangleq \bigvee_{i \leq n}(\varphi_i \wedge \bigwedge_{j \leq n, j \neq i} \neg\varphi_j)$
- $\bigvee_{i \leq n} \varphi_i \triangleq (\bigvee_{i \leq n-1} \varphi_i \vee \varphi_n)$, if $n > 0$, and $\bigvee_{i \leq 0} \varphi_i \triangleq \bot$
- etc.

It should be mentioned that different formalisms use different symbols for these junctors. For example, the CASL syntax (given in Chap. 4) uses `false` and `true` instead of $\bot$ and $\top$, respectively. Implication is often written as $(\varphi \rightarrow \psi)$, or $(\varphi \succ \psi)$. For $\neg$, the symbols ! and $-$ are being used. For $\vee$ and $\wedge$, various representations are common, including `or`, |, +, \/ and `and`, &, *, /\, and others.

The syntax requires parenthesis around binary junctors in order to make each formula uniquely parseable. Usually, one fixes an order of precedence between junctors by $\neg < \wedge < \oplus < \vee < \Rightarrow < \Leftrightarrow$, declares binary junctors to be left-associative and omits parentheses whenever appropriate.

In order to give a semantics (a meaning) to propositional formulae, we observe that each proposition symbol describes some statement which may or may not hold in the (actual or imagined) world. In a sense, the determination which propositions are true and which are false models an actual or intended state of affairs.

Definition 2 (*Propositional model*) A *propositional model* $\mathcal{M}$ is a function $\mathcal{P} \rightarrow \{true, false\}$ assigning a truth value to every proposition symbol.

The restriction to two truth values $\{true, false\}$ is often attributed to George Boole [Boo47]; thus, they are called *boolean values*. Note that if $\mathcal{P}$ contains n different proposition symbols, then there are 2^n different models. Given a model $\mathcal{M}$, the truth assignment can be extended to arbitrary formulae. We define the *validation relation* $\models$ between a model $\mathcal{M}$ and a formula φ by the following clauses.

- $\mathcal{M}\models p$ if and only if $\mathcal{M}(p) = true$,
- $\mathcal{M}\not\models\bot$, and
- $\mathcal{M}\models(\varphi \Rightarrow \psi)$ if and only if $\mathcal{M} \models \varphi$ implies $\mathcal{M}\models\psi$.

That is, $\mathcal{M} \models (\varphi \Rightarrow \psi)$ if and only if $\mathcal{M} \not\models \varphi$ or $\mathcal{M} \models \psi$. If $\mathcal{M} \models \varphi$, then we say that $\mathcal{M}$ *satisfies* φ, or, equivalently, φ is *valid in* $\mathcal{M}$.

For the junctors introduced as abbreviations above, we get

- $\mathcal{M} \models \neg\varphi$ if and only if $\mathcal{M} \not\models \varphi$,

- $\mathcal{M} \models \top$,
- $\mathcal{M} \models (\varphi \vee \psi)$ if and only if $\mathcal{M} \models \varphi$ or $\mathcal{M} \models \psi$ (or both),
- $\mathcal{M} \models (\varphi \wedge \psi)$ if and only if   both $\mathcal{M} \models \varphi$ and $\mathcal{M} \models \psi$,
- $\mathcal{M} \models (\varphi \oplus \psi)$ if and only if $\mathcal{M} \models \varphi$ or $\mathcal{M} \models \psi$, but not both,
- $\mathcal{M} \models \bigoplus(\varphi_1, \ldots, \varphi_n)$ if and only if for exactly one $i \leq n$ it holds that $\mathcal{M} \models \varphi_i$.

The *model checking problem* is defined as follows: given a model $\mathcal{M}$ and a formula φ, determine whether $\mathcal{M} \models \varphi$. For propositional logic, this problem can be solved with a complexity which is linear in the length of the formula and the access time of the truth value of a proposition symbol in a model (which is, depending on the representation of the model, typically either constant or linear in the size of the model). Model checking for propositional formulae involves parsing the formula according to its syntax, looking up the values of the proposition symbols in the model, and then assigning the truth value to compound formulae according to the above clauses. Detailed pseudocode-algorithms for this task are given below.

Definition 3 (*Satisfiability, Validity*) Given any formula φ, we say that φ is *satisfiable* if there exists a model $\mathcal{M}$ such that $\mathcal{M} \models \varphi$. A formula φ which is valid in every model is called *universally valid* or *logically true* or a *tautology*, denoted by $\models \varphi$.

Clearly, each universally valid formula is satisfiable. Moreover, if φ is unsatisfiable, then $\neg\varphi$ must be universally valid. The *satisfiability problem* SAT problem is to find out whether any given formula is satisfiable; likewise, the *validity problem* is to determine whether a given formula is universally valid. Any algorithm which solves the satisfiability problem also is a solution to the validity problem: in order determine whether $\models \varphi$ it is sufficient to find out whether $\neg\varphi$ is satisfiable or not: If $\neg\varphi$ is satisfiable, then there is a model $\mathcal{M}$ such that $\mathcal{M} \models \neg\varphi$; hence $\mathcal{M} \not\models \varphi$ and thus φ cannot be universally valid. If $\neg\varphi$ is unsatisfiable, then for all models $\mathcal{M}$ we have $\mathcal{M} \not\models \neg\varphi$; hence for all $\mathcal{M}$ we have $\mathcal{M} \models \varphi$ and thus φ is universally valid. In summary, $\neg\varphi$ is satisfiable if and only if φ is not universally valid; i.e., φ is universally valid if and only if $\neg\varphi$ is not satisfiable. Thus, given an algorithm for SAT, we can find out whether a formula φ is universally valid by checking whether $\neg\varphi$ is unsatisfiable. Likewise, any algorithm for the validity problem can be used as a solver for the satisfiability problem.

Satisfiability can be generalized and relativized to sets of formulae: let Γ be any set of formulae and φ be a formula. We say that Γ is *satisfiable*, if there is a model $\mathcal{M}$ satisfying all $\psi \in \Gamma$. Similarly, φ *follows from* Γ (written as $\Gamma \Vdash \varphi$), if φ holds in every model $\mathcal{M}$ which satisfies all $\psi \in \Gamma$.[1] Similar to above, it holds that $\Gamma \Vdash \varphi$ if and only if $\Gamma \cup \{\neg\varphi\}$ is unsatisfiable.

---

[1] Here, we use the symbol $\Vdash$ for the semantic consequence relation between formulae, which is different from the symbol $\models$ for the satisfaction relation between a model and a formula. For universal validity, the notions coincide: $\models \varphi$ if and only if $\{\} \Vdash \varphi$.

Example 18.4: Validation and Validity with Car Configurations

To illustrate these notions with the example, let's assume that the set of proposition symbols is $\mathcal{P} = \{$colour_red, colour_blue, colour_green, motor_diesel, motor_59kW, gearshift_automatic, gearshift_manual, navi, audio_cd, audio_dvd$\}$, with $\mathcal{M}($colour_red$) = \mathcal{M}($motor_diesel$)$ = *true*, and $\mathcal{M}(p)$ = *false* for all other $p \in \mathcal{P}$. Then

- $\mathcal{M} \models \bigoplus($colour_red, colour_blue, colour_green$)$,
- $\mathcal{M} \not\models ($gearshift_automatic $\oplus$ gearshift_manual$)$,
- $\mathcal{M} \models ($motor_59kW $\Rightarrow \neg$gearshift_automatic$)$,
- $($gearshift_automatic $\oplus$ gearshift_manual$)$ is satisfiable,
- $(($motor_59kW $\Rightarrow \neg$gearshift_automatic$) \wedge$ motor_59kW$\wedge$ gearshift_automatic$)$ is unsatisfiable,
- $\models ($audio_cd $\vee$ audio_dvd $\Rightarrow$ audio_dvd $\vee$ audio_cd$)$,
- $\{($audio_cd $\oplus$ audio_dvd$)$, audio_dvd$\}$ is satisfiable, and
- $\{($audio_cd $\oplus$ audio_dvd$)$, audio_dvd$\} \models \neg$audio_cd.

In model theory, a set of formulae often is identified with the set of its models. We write $Mod(\Gamma) \triangleq \{\mathcal{M} \mid \mathcal{M} \models \psi$ for all $\psi \in \Gamma\}$ Clearly, Γ is satisfiable if and only if $Mod(\Gamma) \neq \emptyset$, and Γ is universally valid if and only if $Mod(\Gamma)$ is the set of all models for the given proposition symbols.

2.2.3 Propositional Methods

Various methods have been developed for propositional logic. In the following, we discuss model checking, checking for satisfiability, and so-called deduction systems.

Model Checking in Propositional Logic

Model checking is the task to determine the truth value of a formula in a model: Given $\mathcal{M}$ and φ, does $\mathcal{M} \models \varphi$ hold? Often, it is possible to directly follow the definition of $\models$ for this task. However, depending how $\mathcal{M}$ and φ are represented as data structures, there may be algorithmically different ways to do so.

A propositional model is a mapping from proposition symbols to truth values $\{true, false\}$. Usually, this mapping is given in a lookup data structure, e.g., a hash table, such that access to the value of any proposition symbol is in constant time. The formula may be given as a formula tree or in a linear (string) representation. In the first case, we can employ a recursive descent algorithm for model checking:

Algorithm 1: Recursive Propositional Model Checking

input : Propositional model $\mathcal{M}$, formula φ as a tree.
output: Is formula φ valid in model $\mathcal{M}$, i.e., does $\mathcal{M} \models \varphi$ hold?

function modelcheckRec($\mathcal{M}$, φ)
  if $top(\varphi) = $ "p" **then return** $\mathcal{M}(p)$
  else if $top(\varphi) = $ "$\bot$" **then return false**
  else // $top(\varphi) = $ " $\Rightarrow$ "
      if modelcheckRec ($\mathcal{M}$, $left(\varphi)$) **then return** modelcheckRec ($\mathcal{M}$, $right(\varphi)$)
      else return true

In the second case, we need a pushdown store of values {**true**, **false**} for parsing the formula.

Algorithm 2: Iterative Propositional Model Checking.

input : Propositional model $\mathcal{M}$, formula φ as a string of symbols.
output: Is formula φ valid in model $\mathcal{M}$, i.e., does $\mathcal{M} \models \varphi$ hold?

function modelcheckIt($\mathcal{M}$, φ)
  result := **false**;
  while $nonempty(\varphi)$ **do**
    if $first(\varphi) = $ "p" **then** result := $\mathcal{M}(p)$
    else if $first(\varphi) = $ "$\bot$" **then** result := **false**
    else if $first(\varphi) = $ " $\Rightarrow$ " **then** push(result, stack)
    else if $first(\varphi) = $ ")" **then**
      if $top(stack) = $ **false then** result := **true**;
      pop(stack);
    $\varphi := rest(\varphi)$;
  return *(result)*

The iterative algorithm assumes a syntactically correct (i.e., fully parenthesized) formula as input. "result" is a boolean variable for recording intermediate results. An implication "$(\varphi \Rightarrow \psi)$" is processed as follows. No action is required for the opening bracket "(", checking just proceeds with subformula "φ". If the implication sign "$\Rightarrow$" is found, the variable result contains the value of φ, which is pushed onto the stack. Then, the subformula "ψ" is checked. If the closing bracket ")" is found, result contains the value of ψ, and the result of φ is on top of the stack. If φ is false, then $(\varphi \Rightarrow \psi)$ is true, hence *result* must be set accordingly. Otherwise, $(\varphi \Rightarrow \psi)$ has the same value as ψ, and nothing is to be done.

Propositional Satisfiability

In contrast to the model checking problem, it is computationally much harder
to check whether a given formula is satisfiable: whereas in model checking,
the model is already given, for satisfiability we have to find out whether *any*
model satisfies the given formula. For a formula φ with n different proposition
symbols, there are 2^n possible models. A nondeterministic algorithm could
'guess' a model $\mathcal{M}$ and then check that it indeed satisfies φ. In the literature,
SAT therefore is known as the generic NP-complete problem. On a present-
day deterministic machine, in the worst case all 2^n different models have to
be checked, which yields an exponential complexity.

Algorithm 3: Propositional satisfiability.

---
input : Formula φ.
output: Is formula φ satisfiable, i.e., does $\mathcal{M} \models \varphi$ hold for some model $\mathcal{M}$?

function SAT(φ)
  forall *models* $\mathcal{M}$ **do**
    ⌊   **if** modelcheck$(\mathcal{M}, \varphi)$ **then return true**
  else return false

Much work has been invested in the design of efficient SAT-solvers [BBH+09].
Examples include Chaff, MiniSAT, PicoSAT and others. There are annual
competitions to compare the performance of these tools on large formulae.

Deduction Systems for Propositional Logic

If the validity problem is too complex for fully automated tools, sometimes it
is possible to 'manually' construct a proof. For propositional logic, there exist
quite a few proof methods: axiomatic systems, tableaux systems, systems of
natural deduction, resolution systems, and others. Each such system describes
how to obtain a proof for a given propositional formula. There are two basic
ways of constructing a proof for formula φ. Either, one starts with a set
of valid formulae, and derives further valid formulae from these, until φ is
obtained. Or, one starts with φ, and iteratively decomposes it until only valid
constituents remain.

    We give an example for the first of these methods. Our *Hilbert-style proof
system* for propositional logic is given by the following axioms and derivation
rule:

(weak) $\vdash (\varphi \Rightarrow (\psi \Rightarrow \varphi))$
(dist) $\vdash ((\varphi \Rightarrow (\psi \Rightarrow \xi)) \Rightarrow ((\varphi \Rightarrow \psi) \Rightarrow (\varphi \Rightarrow \xi)))$
(tnd) $\vdash (\neg\neg\varphi \Rightarrow \varphi)$
(mp) $\varphi, \ (\varphi \Rightarrow \psi) \ \vdash \ \psi$

2 Logics for Software Engineering

Here, the symbol $\vdash$ denotes *derivability* in the following sense. The first three lines are *axioms*; they are known as *weakening, distributivity,* and *tertium non datur* (law of the excluded middle), respectively. A propositional formula is a *substitution instance* (or simply *instance*) of an axiom, if it can be obtained by consistently replacing the symbols φ, ψ and ξ by propositional formulae.

Example 19: Instances of Axioms

$\vdash ((p \Rightarrow q) \Rightarrow (q \Rightarrow (p \Rightarrow q)))$ is an instance of axiom **(weak)** (with substitution $[\varphi := (p \Rightarrow q), \quad \psi := q]$)
$\vdash ((p \Rightarrow ((p \Rightarrow p) \Rightarrow p)) \Rightarrow ((p \Rightarrow (p \Rightarrow p)) \Rightarrow (p \Rightarrow p)))$ is an instance of axiom **(dist)**.

The fourth line of the Hilbert-style proof system is a *derivation rule*. It is known as *modus ponens* and allows to derive new formulae from given ones. This derivation rule consists of two formula schemes—called the *antecedents*—on the left side of the derivation sign $\vdash$, and one formula scheme—the *consequent*—to its right. The notion of substitution instance for derivation rules is the same as that for axioms; in fact, an axiom can be seen as a derivation rule which has no antecedents.

Example 20: Instance of a Derivation Rule

$p, (p \Rightarrow (p \Rightarrow p)) \vdash (p \Rightarrow p)$ is an instance of **(mp)**.

The notion of derivability is defined recursively. We say that a formula is *derivable*, if it is either

- an instance of an axiom, or
- an instance of the consequent of the derivation rule, where all antecedents are derivable.

If φ is derivable, we write $\vdash \varphi$. In order to show that φ is derivable, we have to give a *derivation*, that is, a finite sequence of formulae where each formula in this sequence is either an instance of an axiom, or an instance of the consequent of modus ponens, where all antecedents are already contained in the list.

Example 21: Derivation of a Formula

We show that $(p \Rightarrow p)$ is derivable, i.e., $\vdash (p \Rightarrow p)$:
(1) $\vdash (p \Rightarrow ((p \Rightarrow p) \Rightarrow p)) \Rightarrow ((p \Rightarrow (p \Rightarrow p)) \Rightarrow (p \Rightarrow p))$ **(dist)**
(2) $\vdash (p \Rightarrow ((p \Rightarrow p) \Rightarrow p))$ **(weak)**
(3) $\vdash ((p \Rightarrow (p \Rightarrow p)) \Rightarrow (p \Rightarrow p))$ (1,2,**mp**)
(4) $\vdash (p \Rightarrow (p \Rightarrow p))$ **(weak)**
(5) $\vdash (p \Rightarrow p)$ (3,4,**mp**)

Often, it is not an easy task to find such a derivation; we challenge the reader to prove the following:

$\vdash (p \Rightarrow ((p \Rightarrow q) \Rightarrow q))$
$\vdash ((p \Rightarrow q) \Rightarrow ((q \Rightarrow r) \Rightarrow (p \Rightarrow r)))$
$\vdash (p \Rightarrow \neg\neg p))$
$\vdash (p \Rightarrow (p \vee q))$
$\vdash ((p \vee q) \Rightarrow (q \vee p))$

Hilbert-style deduction systems have been criticised for being unintuitive and hard to use. In 1934, Gentzen proposed sequent calculi of *natural deduction*. Here, a *sequent* is a string $\Gamma \vdash \varphi$, where Γ is a set of formulae and φ is a formula. The idea is that $\Gamma \vdash \varphi$ is a syntactical counterpart of $\Gamma \Vdash \varphi$, i.e., a formula is the (syntactic) consequence of a set of formulae if and only if it (semantically) follows from these formulae. There is only one axiom: $\Gamma \vdash \varphi$, where $\varphi \in \Gamma$. Derivation rules are formulated as *operator-introduction* and *operator-elimination* rules, and are usually written with a horizontal line between the antecedences and consequence of the rule. There are four rules:

$$\frac{\Gamma \vdash (\varphi \Rightarrow \psi) \quad \Gamma \vdash \varphi}{\Gamma \vdash \psi} \text{ (}\Rightarrow\text{-elim)} \qquad \frac{\Gamma \cup \{\varphi\} \vdash \psi}{\Gamma \vdash (\varphi \Rightarrow \psi)} \text{ (}\Rightarrow\text{-intro)} \qquad \frac{\Gamma \vdash \bot}{\Gamma \vdash \varphi} \text{ ("ex falso quodlibet")} \qquad \frac{\Gamma \cup \{\varphi\} \vdash \psi \quad \Gamma \cup \{(\varphi \Rightarrow \bot)\} \vdash \psi}{\Gamma \vdash \psi} \text{ ("tertium non datur")}$$

These rules allow to depict a derivation as a tree of sequents. As an example, we give a derivation of the formula $(p \Rightarrow ((p \Rightarrow q) \Rightarrow q))$:

$$\frac{\dfrac{\dfrac{\{p, (p \Rightarrow q)\} \vdash (p \Rightarrow q) \quad \{p, (p \Rightarrow q)\} \vdash p}{\{p, (p \Rightarrow q)\} \vdash q} \text{ (}\Rightarrow\text{-elim)}}{\{p\} \vdash ((p \Rightarrow q) \Rightarrow q)} \text{ (}\Rightarrow\text{-intro)}}{\{\} \vdash (p \Rightarrow ((p \Rightarrow q) \Rightarrow q))} \text{ (}\Rightarrow\text{-intro)}$$

Both Hilbert-style proof systems and natural deduction may not be optimal for automated theorem proving. In 1960, Davis and Putnam proposed theorem proving by *resolution*, which is based on the single rule

$$(\varphi \vee \xi_1 \vee \ldots \vee \xi_n), (\neg\varphi \vee \psi_1 \vee \ldots \vee \psi_m) \vdash (\xi_1 \vee \ldots \vee \xi_n \vee \psi_1 \vee \ldots \vee \psi_m)$$

Since $(\neg\varphi \vee \psi)$ is equivalent to $(\varphi \Rightarrow \psi)$, modus ponens can be seen as the special case of this rule where $n = 0$ and $m = 1$. Resolution theorem proving has turned out to be very successful and still is at the heart of most modern theorem provers.

Given any proof system, logicians are usually interested in two questions:

Correctness: Are all provable formulae valid?
Completeness: Are all valid formulae provable?

For our Hilbert-style proof system, this amounts to showing that only valid formulae are derivable, and that for any valid formula there exists a derivation. The first of these questions usually is easy to affirm (one wouldn't want to work with an incorrect proof system!), whereas the second one often is more intricate. For reference purposes, we sketch the respective proofs; they are not necessary for the rest of the book and can be safely skipped by the casual reader.

Theorem 1 (Correctness of Hilbert-style proof system) *Any derivable formula is valid: if $\vdash \varphi$, then $\models \varphi$.*

Proof The proof is by induction on the length of the derivation. We show that all instances of axioms are valid, and that the consequent of (mp) is valid if both antecedents are. Consider the axiom **(weak)**. Assume any model $\mathcal{M}$, and any substitution of propositional formulae φ^σ and ψ^σ for φ and ψ, respectively. Then, $\mathcal{M} \models \varphi^\sigma$ or $\mathcal{M} \not\models \varphi^\sigma$, and likewise for ψ^σ. Thus, there are four possibilities for the truth values of φ^σ and ψ^σ in $\mathcal{M}$. We give a truth table for these possibilities:

φ^σ	ψ^σ	$(\psi^\sigma \Rightarrow \varphi^\sigma)$	$(\varphi^\sigma \Rightarrow (\psi^\sigma \Rightarrow \varphi^\sigma))$
false	*false*	*true*	*true*
false	*true*	*false*	*true*
true	*false*	*true*	*true*
true	*true*	*true*	*true*

Since in any case the value of $(\varphi^\sigma \Rightarrow (\psi^\sigma \Rightarrow \varphi^\sigma))$ is *true*, this instance is valid. For axioms **(dist)** and **(tnd)**, the argument is similar. For **(mp)**, assume that φ^σ and $(\varphi^\sigma \Rightarrow \psi^\sigma)$ are derivable. According to the induction hypothesis, φ^σ and $(\varphi^\sigma \Rightarrow \psi^\sigma)$ are also valid. Therefore, $\mathcal{M} \models \varphi^\sigma$ and $\mathcal{M} \models (\varphi^\sigma \Rightarrow \psi^\sigma)$ for any model $\mathcal{M}$. From the definition, $\mathcal{M} \models (\varphi^\sigma \Rightarrow \psi^\sigma)$ if and only if $\mathcal{M} \not\models \varphi^\sigma$ or $\mathcal{M} \models \psi^\sigma$. Therefore, $\mathcal{M} \models \psi^\sigma$. Since we did not put any constraint on $\mathcal{M}$, it holds that $\models \psi^\sigma$. ∎

Theorem 2 (Completeness of Hilbert-style proof system) *Any valid formula is derivable: if $\models \varphi$, then $\vdash \varphi$.*

Proof We show that any consistent formula is satisfiable. A formula φ is said to be consistent, if it is not the case that $\neg\varphi$ is derivable. More generally, a finite set of formulae Φ is consistent, if it is not the case that $\neg \bigwedge_{\varphi \in \Phi} \varphi$ is derivable. In other word, Φ is inconsistent, if $\vdash \neg \bigwedge_{\varphi \in \Phi} \varphi$.

For the completeness proof, we first show that for any finite consistent set Φ of formulae and any formula ψ it holds that $\Phi \cup \{\psi\}$ or $\Phi \cup \{\neg\psi\}$ is consistent. This fact is called the *extension lemma*. Assume for contradiction, that both $\vdash \neg(\bigwedge_{\varphi \in \Phi \cup \{\psi\}} \varphi)$ and $\vdash \neg(\bigwedge_{\varphi \in \Phi \cup \{\neg\psi\}} \varphi)$. According to the definition of $\bigwedge$

and $\wedge$, this is the same as $\vdash \neg\neg(\bigwedge_{\varphi \in \Phi} \varphi \Rightarrow \neg\psi)$ and $\vdash \neg\neg(\bigwedge_{\varphi \in \Phi} \varphi \Rightarrow \neg\neg\psi)$. With **(tnd)** and **(mp)**, we get $\vdash (\bigwedge_{\varphi \in \Phi} \varphi \Rightarrow \neg\psi)$ and $\vdash (\bigwedge_{\varphi \in \Phi} \varphi \Rightarrow \neg\neg\psi)$. Using also **(dist)** and **(weak)**, we can show that $\vdash (\bigwedge_{\varphi \in \Phi} \varphi \Rightarrow \psi)$. The proof system also allows to show $\vdash ((\varphi \Rightarrow \psi) \Rightarrow ((\varphi \Rightarrow \neg\psi) \Rightarrow \neg\varphi))$. Therefore, $\vdash \neg \bigwedge_{\varphi \in \Phi}$, which is a contradiction to the consistency of Φ.

To continue with the completeness proof, let $SF(\varphi)$ denote the set of all *subformulae* of a given formula φ. That is, $SF(\varphi)$ contains all constituents according to the inductive definition of φ, including φ itself. For example, if $\varphi = ((p \Rightarrow q) \Rightarrow p)$, then $SF(\varphi) = \{\varphi, (p \Rightarrow q), p, q\}$. For any consistent formula φ, let $\varphi^{\#}$ be a *maximal consistent extension* of φ (i.e., $\varphi \in \varphi^{\#}$, and for every $\psi \in SF(\varphi)$, either $\psi \in \varphi^{\#}$ or $\neg\psi \in \varphi^{\#}$). The existence of such a maximal consistent extension is guaranteed by the above extension lemma. However, for any given consistent formula, there might be many different maximal consistent extensions. Any maximal consistent extension defines a unique model: for a given $\varphi^{\#}$, the *canonical model* $\mathcal{M}_{\varphi^{\#}}$ is defined by

$$\mathcal{M}_{\varphi^{\#}}(p) = true \text{ if and only if } p \in \varphi^{\#}.$$

The *truth lemma* states that for *any* subformula $\psi \in SF(\varphi)$ it holds that $\mathcal{M}_{\varphi^{\#}} \models \psi$ if and only if $\psi \in \varphi^{\#}$. The proof of this lemma is by induction on the structure of ψ:

- Case ψ is a proposition: by construction of $\mathcal{I}$.
- Case $\psi = \bot$: in this case, $\Phi \cup \{\psi\}$ cannot be consistent ($\vdash \neg(\bigwedge_{\varphi \in \Phi} \varphi \wedge \bot)$).
- Case $\psi = (\psi_1 \Rightarrow \psi_2)$: by induction hypothesis, $\mathcal{M}_{\varphi^{\#}} \models \psi_i$ if and only if $\psi_i \in \varphi^{\#}$, for $i \in \{1, 2\}$. Since $\varphi^{\#}$ is maximal, either $\psi \in \varphi^{\#}$ or $\neg\psi \in \varphi^{\#}$, and the same holds for ψ_1 and ψ_2. There are four cases: (i) $\neg\psi_1 \in \varphi^{\#}$ and $\neg\psi_2 \in \varphi^{\#}$, (ii) $\neg\psi_1 \in \varphi^{\#}$ and $\psi_2 \in \varphi^{\#}$, (iii) $\psi_1 \in \varphi^{\#}$ and $\neg\psi_2 \in \varphi^{\#}$, and (iv) $\psi_1 \in \varphi^{\#}$ and $\psi_2 \in \varphi^{\#}$. In case (i) and (ii), we have $\mathcal{M}_{\varphi^{\#}} \not\models \psi_1$, therefore $\mathcal{M}_{\varphi^{\#}} \models \psi$; and since $\neg\psi_1 \in \varphi^{\#}$ and $\vdash (\neg\psi_1 \Rightarrow (\psi_1 \Rightarrow \psi_2))$, we must have $\psi \in \varphi^{\#}$ (or otherwise $\varphi^{\#}$ would be inconsistent). Likewise, in case (ii) and (iv), we have $\mathcal{M}_{\varphi^{\#}} \models \psi_2$, therefore $\mathcal{M}_{\varphi^{\#}} \models \psi$; and since $\psi_2 \in \varphi^{\#}$ and $\vdash (\psi_2 \Rightarrow (\psi_1 \Rightarrow \psi_2))$, we must have $\psi \in \varphi^{\#}$. In case (iii), if $\mathcal{M}_{\varphi^{\#}} \models \psi_1$ and $\mathcal{M}_{\varphi^{\#}} \not\models \psi_2$, then $\mathcal{M}_{\varphi^{\#}} \not\models \psi$; and from $\vdash (\psi_1 \Rightarrow (\neg\psi_2 \Rightarrow \neg(\psi_1 \Rightarrow \psi_2)))$ we conclude that $\neg\psi \in \varphi^{\#}$.

Let φ be any consistent formula. By definition, $\varphi \in \varphi^{\#}$ for any maximal consistent extension $\varphi^{\#}$ of φ. Therefore, $\mathcal{M}_{\varphi^{\#}} \models \varphi$. Hence we have shown that any consistent formula is satisfiable. In other words, any unsatisfiable formula is inconsistent. Let φ be any valid formula ($\models \varphi$). Then $\neg\varphi$ is unsatisfiable, and therefore inconsistent. This means that $\vdash \neg\neg\varphi$. From this, with **(tnd)** and **(mp)** we get $\vdash \varphi$. ∎

Similar completeness proofs can be given for natural deduction and resolution calculi.

Correctness and completeness add up to the statement

$$\Vdash \varphi \text{ if and only if } \vdash \varphi.$$

In computer science, we are usually interested in consequences of a set Γ of specification formulae. Therefore, we want to make sure that

$$\Gamma \Vdash \varphi \text{ if and only if } \Gamma \vdash \varphi.$$

It is not hard to see that the above proofs can be adapted to this case.

2.3 A Framework for Logics

The above discussion of propositional logic showed the typical elements that constitute a logic. The syntax definition started with a set of proposition *symbols* $\mathcal{P}$. Often, this collection of symbols is called a *signature*. Relatively to the signature $\mathcal{P}$, the set of propositional *formulae* $\text{PL}_{\mathcal{P}}$ was introduced. Symbols need interpretation. This is where *models* come into play. Models are the semantical side of a logic: each symbol gets an interpretation. In propositional logic, models are functions $\mathcal{P} \to \{true, false\}$ assigning a truth value to every proposition symbol. Syntax and semantic level of a logic is finally connected via a *validation relation* $\models$. Some people also speak of a *satisfaction relation*. $M \models \varphi$ holds if the formula φ is valid in the model M.

The notion of an *institution* by Goguen and Burstall [GB92] captures the general idea of what a constitutes a logic. An institution consists of

- a signature Σ which collects the symbols available,
- a set of formulae $For(\Sigma)$ defined relatively to the signature,
- a class of models $Mod(\Sigma)$ which interpret the symbols, and
- a satisfaction relation $\models \subseteq Mod(\Sigma) \times For(\Sigma)$,

where furthermore the so-called *satisfaction condition* needs to be fulfilled— which we will discuss in Sect. 2.3.1 below.

Given a logic with these four elements, one can state several *algorithmic questions* on the relations between models and formulae:

Model checking problem.   Given a model M and a formula φ, is it the case that $M \models \varphi$?

Satisfiability problem.   Given a formula φ, is there any model M such that $M \models \varphi$ holds?

Validity problem.   Given a formula φ, does $M \models \varphi$ hold for any model M? This problem is dual to the satisfiability problem.

Uniqueness problem.   Given a formula φ, is there one unique model M satisfying this formula, i.e., are any two models for φ isomorphic (identical up to renaming)? A specification formula with this property sometimes is called *monomorphic* or *categorical*.

Model checking and satisfiability for propositional logic have been considered above. Subsequently, we study possible relation between different models.

2.3.1 Specification

A typical way of using logic for specification is to say: all models that satisfy a given set of formulae are considered to be 'interesting'. Consider for instance Example 18 on Car Configuration. When a customer has still some choices left, one might be interested in the question: what car configurations can the customer still choose? The customer might be interested in a minimal choice, as it probably is the cheapest. The car seller, however, might be interested in a maximal choice. This leads to several questions on logics:

1. How do models relate to each other? Is the relation between the models compatible with the one defined by the logic?
2. Are we interested in all models, or just a selected 'minimal' or 'maximal' model?

These questions can be studied for logics in general. In this section we will show how to answer these questions for Propositional Logic.

To address the first question, we relate models by the notion of a model morphism:

Definition 4 (*Model morphism for Propositional Logic*) Let $\mathcal{M}_1$ and $\mathcal{M}_2$ be models over the same signature $\mathcal{P}$. There is a model morphism from $\mathcal{M}_1$ to $\mathcal{M}_2$ if for all $p \in \mathcal{P}$ holds: $\mathcal{M}_1(p) = true$ implies $\mathcal{M}_2(p) = true$.

The set of variables which are set to *true* is growing along a model morphism. In other words: truth is preserved along a morphism. We write $\mathcal{M}_1 \leq \mathcal{M}_2$ if there is a model morphism from $\mathcal{M}_1$ to $\mathcal{M}_2$.

Example 18.5: Incomparable Models

Models are not necessarily comparable. Take the above signature $\mathcal{P} = \{\texttt{colour\_red}, \texttt{colour\_blue}, \texttt{colour\_green}, \texttt{motor\_diesel}, \texttt{motor\_59kW},$ $\texttt{gearshift\_automatic}, \texttt{gearshift\_manual}, \texttt{navi}, \texttt{audio\_cd}, \texttt{audio\_dvd}\}$ from the Car Configuration Example. Define $\mathcal{M}_1(\texttt{colour\_red}) = true$, $\mathcal{M}_1(p) = false$ for all other $p \in \mathcal{P}$; define $\mathcal{M}_2(\texttt{motor\_diesel}) = true$ and $\mathcal{M}_2(p) = false$ for all other $p \in \mathcal{P}$. Then there is neither a morphism from $\mathcal{M}_1$ to $\mathcal{M}_2$ nor a morphism from $\mathcal{M}_2$ to $\mathcal{M}_1$.

The relation $\leq$ is reflexive, transitive and antisymmetric:

Theorem 3 (Partial ordering of models by morphisms) *Let $\mathcal{M}$, $\mathcal{M}_1$, $\mathcal{M}_2$ and $\mathcal{M}_3$ be models over the same signature $\mathcal{P}$. Then*

1. $\mathcal{M} \leq \mathcal{M}$.
2. If $\mathcal{M}_1 \leq \mathcal{M}_2$ and $\mathcal{M}_2 \leq \mathcal{M}_3$, then $\mathcal{M}_1 \leq \mathcal{M}_3$.
3. If $\mathcal{M}_1 \leq \mathcal{M}_2$ and $\mathcal{M}_2 \leq \mathcal{M}_1$, then $\mathcal{M}_1 = \mathcal{M}_2$.

Proof 1. holds trivially. For 2. let $\mathcal{M}_1(p) = true$ for some $p \in \mathcal{P}$. Then, by $\mathcal{M}_1 \leq \mathcal{M}_2$, we know that $\mathcal{M}_2(p) = true$. Moreover, by $\mathcal{M}_2 \leq \mathcal{M}_3$, we know that $\mathcal{M}_3(p) = true$. Thus, $\mathcal{M}_1 \leq \mathcal{M}_3$. For 3. we argue as follows. If $\mathcal{M}_1 \leq \mathcal{M}_2$, then it holds that for all p with $\mathcal{M}_1(p) = true$, also $\mathcal{M}_2(p) = true$. Similarly, if $\mathcal{M}_2 \leq \mathcal{M}_1$ then for all p with $\mathcal{M}_2(p) = true$, also $\mathcal{M}_1(p) = true$. Thus, $\mathcal{M}_1(p) = true$ if and only if $\mathcal{M}_2(p) = true$. As a propositional model is a total function from proposition symbols to truth values, this has as a consequence that $\mathcal{M}_1 = \mathcal{M}_2$.   ∎

The result in Theorem 3 (3.) is unusual for a logic. In general, models mutually related by model morphisms need not be identical.

With the notion of a morphism available, there are now two ways to compare models. The first one uses the logical formulae. Two models are defined to be 'equivalent' if they validate the same set of formulae. The second way is to use the notion of a morphism: two models are considered 'equivalent' if they cannot be distinguished by morphisms, i.e., in our context, if they are identical. The following proposition shows that for PL these two notions coincide:

Theorem 4 (Isomorphism property for PL) *Let $\mathcal{M}_1$ and $\mathcal{M}_2$ be models over the same signature $\mathcal{P}$. Then the following holds:*

$$\{\varphi \mid \mathcal{M}_1 \models \varphi\} = \{\varphi \mid \mathcal{M}_2 \models \varphi\} \text{ if and only if } \mathcal{M}_1 = \mathcal{M}_2$$

Proof "$\Leftarrow$" holds trivially. For "$\Rightarrow$" we argue as follows. If $\mathcal{M}_1$ and $\mathcal{M}_2$ satisfy the same formulae, then in particular for all $p \in Prop$ it holds that $\mathcal{M}_1(p) = \mathcal{M}_2(p)$. Thus, $\mathcal{M}_1 = \mathcal{M}_2$.   ∎

Model morphisms can be used to determine the semantics of specifications, i.e., descriptions of objects we are interested in. Consider a subsignature of our above example $\mathcal{P}' = \{\, navi, \text{audio\_cd}, \text{audio\_dvd} \,\}$. The customer has chosen that she wants an audio_cd, the company says that a car either plays cd or dvd for audio, i.e., $\text{audio\_cd} \oplus \text{audio\_dvd}$. In the specification language CASL, which we will formally meet in Chap. 4, this specification can be written as follows:

spec CAR_CONFIG =
    preds *navi, audio_cd, audio_dvd* : ()
    • *audio_cd*   %(customer_request)%
    • $(audio\_cd \wedge \neg\ audio\_dvd) \vee (audio\_dvd \wedge \neg\ audio\_cd)$
                                       %(company_setting)%
end

The specification has the name CAR_CONFIG. After the keyword **preds** we declare the propositional signature. The requirements for the car are listed after the bullet points. Within "%(" and ")%" we provide a name for the axiom.

Of the eight potential models for $\mathcal{P}'$, the following two satisfy both formulae:

	navi	audio_cd	audio_dvd
$\mathcal{M}_1$	false	true	false
$\mathcal{M}_2$	true	true	false

With regards to morphisms, we have $\mathcal{M}_1 \leq \mathcal{M}_2$. We now discuss the question what the 'intended meaning' of the specification CAR_CONFIG is.

In CASL, the meaning of a specification is the collection of all models which satisfy all specification formulae:

$$Mod_{Casl}(\text{CAR\_CONFIG})$$
$$= \{\mathcal{M} \mid \mathcal{M} \models \text{audio\_cd}, \mathcal{M} \models \text{audio\_cd} \oplus \text{audio\_dvd}\}$$
$$= \{\mathcal{M}_1, \mathcal{M}_2\}.$$

As the meaning of a specification may contain more than one model, this kind of semantics for the specification language is called *loose*. There are two other possible choices.

- The specification language can be defined such that it uses *initial* semantics; that is, the meaning of a specification is the smallest model with respect to the relation $\leq$. In Propositional Logic, this is the model with a minimal number of propositional symbols set to true, which validates all formulae in the specification. According to Theorem 3, there is at most one such model. For the car configuration example, this is the car with a minimal number of features:

$$Mod_{initial}(\text{CAR\_CONFIG})$$
$$= \{\mathcal{M} \in Mod_{Casl}(\text{CAR\_CONFIG}) \mid$$
$$\quad \text{for all models } \mathcal{M}' \in Mod_{Casl}(\text{CAR\_CONFIG}) \text{ holds: } \mathcal{M} \leq \mathcal{M}'\}$$
$$= \{\mathcal{M}_1\}.$$

- Alternatively, the semantics of a specification language can be defined to be *final*; that is, the meaning of a specification is the largest possible model (with respect to model morphisms) validating the specification. Again, from Theorem 3 it follows that there is at most one such model. This is the car where a maximal number of features is incorporated:

$$Mod_{final}(\text{CAR\_CONFIG})$$
$$= \{\mathcal{M} \in Mod_{Casl}(\text{CAR\_CONFIG}) \mid$$
$$\quad \text{for all models } \mathcal{M}' \in Mod_{Casl}(\text{CAR\_CONFIG}) \text{ holds: } \mathcal{M}' \leq \mathcal{M}\}$$
$$= \{\mathcal{M}_2\}.$$

Tucker & Bergstra have studied systematically which kind of data can be described with initial, loose, and final semantics [BT87]. Algebraic specification languages usually employ initial or loose semantics. Final semantics is studied in the field of Coalgebra. Co-Casl [MSRR06] provides a specification language which offers initial, loose, and final semantics. It is an open field for research to study the interplay between these three approaches.

Modular Specification

Modularisation is important in specification practice. In the above Example 18 on car configuration, for instance, one would like to separate concerns: car colours, car engines and gear shifts, and equipment can be treated independent of each other. The overall car configuration then imports these three specifications and deals with properties where colours, engines, and equipment are combined:

spec COLORS =
    preds $color\_red$, $color\_blue$, $color\_green$: ()
    • $(color\_red \land \neg color\_blue \land \neg color\_green)$
      $\lor (color\_blue \land \neg color\_red \land \neg color\_green)$
      $\lor (color\_green \land \neg color\_blue \land \neg color\_red)$ %(only_one_color)%
end

spec ENGINES_AND_GEARSHIFTS =
    preds $motor\_diesel$, $motor\_59kW$: ()
    preds $gearshift\_automatic$, $gearshift\_manual$: ()
    • $(gearshift\_automatic \land \neg gearshift\_manual)$
      $\lor (gearshift\_manual \land \neg gearshift\_automatic)$
                            %(automatic_or_manual)%
    • $motor\_59kW \Rightarrow \neg gearshift\_automatic$
                     %(59kW_engine_has_no_automatic)%
end

spec EQUIPMENT =
    preds $navi$, $audio\_cd$, $audio\_dvd$: ()
    • $(audio\_cd \land \neg audio\_dvd) \lor (audio\_dvd \land \neg audio\_cd)$
                            %(cd_or_dvd)%
end

spec CONFIGURATION =
    COLORS **and** ENGINES_AND_GEARSHIFTS **and** EQUIPMENT
then • $motor\_diesel \Rightarrow \neg audio\_dvd$ %(dvd_not_available_with_diesel_engine)%
end

All the above specifications have different signatures. In order to support such modular specification approach, one needs to relate models over different signatures which each other. The theory of institutions with its *satisfaction condition* provides the foundations for this.

Definition 5 (*Signature morphism for propositional logic*) Let $\mathcal{P}$ and $\mathcal{P}'$ be (not necessarily disjoint) sets of propositional symbols. A signature morphism $\sigma : \mathcal{P} \to \mathcal{P}'$ is a function between these two sets.

We can lift the translation of the propositional symbols to formulae:

Definition 6 (*Formula translation for propositional logic*) Let $\sigma_0 : \mathcal{P} \to \mathcal{P}'$ be a signature morphism. We inductively define a map $\sigma : \mathrm{PL}_\mathcal{P} \to \mathrm{PL}_{\mathcal{P}'}$:

- If $\varphi = p$ for some $p \in \mathcal{P}$: $\sigma(\varphi) = \sigma_0(p)$.
- $\sigma(\bot) = \bot$.
- $\sigma(\varphi \Rightarrow \psi) = \sigma(\varphi) \Rightarrow \sigma(\psi)$.

Signature morphisms allow us to rename symbols. Let

- $\mathcal{P} = \{\texttt{colour\_red}, \texttt{colour\_blue}, \texttt{colour\_green}\}$,
- $\mathcal{P}' = \{\texttt{red}, \texttt{blue}, \texttt{green}\}$,
- $\sigma(\texttt{colour\_red}) = \texttt{red}$, $\sigma(\texttt{colour\_blue}) = \texttt{blue}$, $\sigma(\texttt{colour\_green}) = \texttt{green}$, and
- $\varphi = \bigoplus(\texttt{colour\_red}, \texttt{colour\_blue}, \texttt{colour\_green})$.

Then $\sigma(\varphi) = \bigoplus(\texttt{red}, \texttt{blue}, \texttt{green})$.

How does now the satisfiability of φ relate to the satisfiability of $\sigma(\varphi)$? Concerning $\sigma(\varphi)$, for instance $\mathcal{M}'$ with $\mathcal{M}'(\texttt{red}) = true$, $\mathcal{M}'(\texttt{blue}) = false$ and $\mathcal{M}'(\texttt{green}) = false$ is a model. From this, we can obtain a model $\mathcal{M}$ of φ with $\mathcal{M}(\texttt{colour\_red}) = true$, $\mathcal{M}(\texttt{colour\_blue}) = false$ and $\mathcal{M}(\texttt{colour\_green}) = false$. There is a 'construction principle' behind this: for every variable $p \in \mathcal{P}$, we look up how $\mathcal{M}'$ interprets its image $\sigma(p)$ under the signature morphism σ and define this value as the interpretation that model $\mathcal{M}$ gives to p.

Definition 7 (*Reducts in propositional logic*) Let $\sigma : \mathcal{P} \to \mathcal{P}'$ be a signature morphism, and let $\mathcal{M}'$ be a model over $\mathcal{P}'$. Its reduct $\mathcal{M}$ along σ is the model over $\mathcal{P}$, which is defined as

$$\mathcal{M}(p) = \mathcal{M}'(\sigma(p)).$$

We often write $\mathcal{M}'|_\sigma$ for $\mathcal{M}$.

With these definitions we can prove the main theorem of this section:

Theorem 5 (The satisfaction condition holds in propositional logic) *Let $\mathcal{P}$ and $\mathcal{P}'$ be sets of propositional symbols. Let $\sigma : \mathcal{P} \to \mathcal{P}'$ be a signature* *morphism. Then the following holds for all formulae $\varphi \in \mathrm{PL}_\mathcal{P}$ and for all model $\mathcal{M}'$ over $\mathcal{P}'$:*

$$\mathcal{M}'|_\sigma \models \varphi \quad \text{if and only if} \quad \mathcal{M}' \models \sigma(\varphi) \quad (*)$$

This property can be illustrated by the following diagram:

$$
\begin{array}{ccc}
\mathcal{M}'|_\sigma & \xleftarrow{\quad -|_\sigma \quad} & \mathcal{M}' \\[2mm]
\models & & \models \\[2mm]
\varphi & \xrightarrow{\quad \sigma \quad} & \sigma(\varphi)
\end{array}
$$

Proof The proof is by induction on the structure of a formula φ. Base cases:

- Let $\varphi = p \in \mathcal{P}$. By construction we have $\mathcal{M}'|_\sigma(p) = \mathcal{M}'(\sigma(p))$. Thus,

$$\mathcal{M}'|_\sigma \models p \text{ if and only if } \mathcal{M}'|_\sigma(p) = true = \mathcal{M}'(\sigma(p)) \text{ if and only if}$$
$$\mathcal{M} \models \sigma(p).$$

- Let $\varphi = \bot$. For all models $\mathcal{M}$ over $\mathcal{P}$ and $\mathcal{M}'$ over $\mathcal{P}'$ holds: $\mathcal{M} \not\models \bot$ and $\mathcal{M}' \not\models \bot$.

Induction step: Let $\varphi = (\psi \to \xi)$. $\mathcal{M}'|_\sigma \models (\psi \to \xi)$ holds by definition if and only if

1. $\mathcal{M}'|_\sigma \not\models \psi$ or
2. $\mathcal{M}'|_\sigma \models \psi$ and $\mathcal{M}'|_\sigma \models \xi$.

Applying the induction hypothesis to both cases, this is the case if and only if

1. $\mathcal{M}' \not\models \sigma(\psi)$ or
2. $\mathcal{M}' \models \sigma(\psi)$ and $\mathcal{M} \models \sigma(\xi)$.

This is equivalent to $\mathcal{M}' \models \sigma(\psi \to \xi)$. ∎

In logic, the equivalence $(*)$ is called *satisfaction condition*. It formalizes the slogan

Truth is invariant under change of notation. [GB92]

Generally, the satisfaction condition is seen as an important check if a logic is well designed. The satisfaction condition provides the theoretical foundation for modular specification as we have seen in the example above. We will discuss modular specification further in Chap. 4 on Algebraic Specification. Note that our discussion on institution theory followed a light weight approach, where we illustrate the main ideas in terms of Propositional logic. In general, institution theory is formulated in the language of category theory, see, e.g., [Fia05].

2.4 First- and Second-Order Logic

Subsequently, we give a short introduction to classical predicate logic, as a natural extension of propositional logic. Readers who are familiar with first-order logic can safely skip this section.

2.4.1 FOL

First-order logic, or FOL for short, has been 'the' language for the formalization of mathematics. Let us start with an example.

Example 22: Strict Partial Order

A *strict partial order* is defined as a mathematical structure consisting of a set and an relation on this set which is asymmetric and transitive. A binary relation is *asymmetric* if no two element are mutually related; and it is *transitive* if whenever you have any three elements of the set where the first is related to the second and the second is related to the third, then also the first is related to the third. If you have problems understanding this sentence, look at the following elegant formulation in FOL:

$$\text{asymmetry:} \quad \forall x \forall y (x \prec y \Rightarrow \neg y \prec x)$$
$$\text{transitivity:} \quad \forall x \forall y \forall z ((x \prec y \wedge y \prec z) \Rightarrow x \prec z)$$

As a more practical example, we consider the task of building a house.

Example 23: Project Planning

When building a house, there are several activities which necessarily must be completed in a certain order. E.g., the walls cannot be built as long as the cellar is not completed, and the door and window frames can not be built in as long as there are no walls. Other activities are independent from each other. e.g., door and window frames can be mounted concurrently, and electricity and plumbing are independent.

In a project planning software, the user can specify which activities are dependent and which are independent. The software can construct a schedule of the overall project, if there is a strict partial order of all activities which is consistent with these two relations.

Assume we are given a finite set of activities $Act = \{a_1, \ldots, a_n\}$ and two binary relations $dep, ind \subseteq Act \times Act$. The question is whether there exists a strict partial order $\prec$ consistent with these two relations, i.e., such that for all activities a and b it holds that

$$\text{if } (a,b) \in dep \quad \text{then} \quad a \prec b \qquad (*)$$
$$\text{if } (a,b) \in ind \quad \text{then} \quad \neg\, a \prec b \quad (**)$$

In order to construct a project schedule, planning software includes a check for this. We now show that a solution exists if and only if the following conditions are satisfied.

1. the transitive closure dep^+ of dep ist asymmetric, i.e., for all $a, b \in Act$ it holds that $(a,b) \in dep^+$ implies $(b,a) \notin dep^+$, and
2. ind is a subset of the complement of dep^+, i.e., if $(a,b) \in ind$, then $(a,b) \notin dep^+$.

For one direction, assume that 1. and 2. above hold. We show that the transitive closure dep^+ of dep is a partial order satisfying $(*)$ and $(**)$. By definition, dep^+ is transitive, and by 1. it is asymmetric, hence it is a partial order. Since $dep \subseteq dep^+$, condition $(*)$ holds. $(**)$ follows directly from 2.

For the other direction, assume that $\prec$ is a partial order satisfying $(*)$ and $(**)$. We show that $dep^+ \subseteq \prec$. If $(a,b) \in dep^+$, then there are $a_1, \ldots, a_n$ such that $a = a_1$, $b = a_n$, and for all $i < n$ it holds that $(a_i, a_{i+1}) \in dep$. From $(*)$ it follows that $a_i \prec a_{i+1}$ for all i. Since $\prec$ is transitive, $a_1 \prec a_n$, i.e., $a \prec b$. Hence if both $(a,b) \in dep^+$ and $(b,a) \in dep^+$, it would follow that both $a \prec b$ and $b \prec a$, contradicting the asymmetry of $\prec$. Thus, dep^+ is asymmetric, proving claim 1. If $(a,b) \in ind$, then $\neg\, a \prec b$ from $(**)$. The assumption $(a,b) \in dep^+$ would lead to the contradiction $a \prec b$. Thus, claim 2. holds. ■

Properties 1. and 2. can be checked algorithmically by constructing the transitive closure of dep and checking whether it contains a loop or a nonempty intersection with ind.

Subsequently, we formalize the concepts used above. The formulae defining strict partial orders are made of various symbols: $\prec$, x, $\forall$, etc.

Definition 8 (*First-order signature*) A *first-order signature* Σ is a structure $\Sigma = (\mathcal{F}, \mathcal{R}, \mathcal{V})$ consisting of

- a finite set $\mathcal{F}$ of *function symbols*, where each function symbol consists of the *name* and the *arity* of the function. The arity is any cardinal number designating the number of arguments of the function. Usually, the arity is considered to be self-evident and omitted; if necessary, it is marked as a superscript to the function name, e.g., f^2. A 0-ary function is called a *constant*;

- a finite set $\mathcal{R}$ of *relation symbols*, where again each $R \in \mathcal{R}$ contains a name and an arity. Relations with arity 2, 3, etc. are called *binary, ternary*, etc.; a unary relation is called a *predicate*. Proposition symbols can be considered to be 0-ary relation symbols; and
- a set $\mathcal{V} = \{x, y, z, v, x_1, \ldots\}$ of *individual variables*.

In the above mentioned strict partial ordering formulae, there are no function symbols involved. The formulae contain only one binary relation symbol $\prec$, and individual variables from the set $\mathcal{V} = \{x, y, z\}$.

Definition 9 (*First-order term*) A *first-order term* over Σ is defined by the following clauses.

- Each variable $v \in \mathcal{V}$ is a term.
- If t_1, ..., t_n are terms and f is an n-ary function symbol from $\mathcal{F}$, then $f(t_1, \ldots, t_n)$ is a term.

In particular, this definition declares $f()$ to be a term if f is a constant (a function symbol without arguments). In this case, parentheses are usually omitted. Binary function symbols are often written in infix-notation, i.e., $(x + y)$ instead of $+(x, y)$. In mathematical logic, this possibility is silently assumed, whereas in formal specification languages such as CASL (see Chap. 4) it must be explicitly stated.

Since the definition is recursive, a term may contain another term, which in turn contain yet another term, etc. In a mathematical context, terms are, e.g., $\cos(x)$, $(x + 5)$, or $f(f(x + 1))$. In programming languages, terms are often called *expressions*; popular examples are (foo(x)+bar(y,z)) and sort(myList).

Definition 10 (*Atomic formula*) A *first-order atomic formula* over Σ is defined as follows.

- If t_1, ..., t_n are terms and R is an n-ary relation symbol from $\mathcal{R}$, then $R(t_1, \ldots, t_n)$ is an atomic formula.

Again, if $p^0 \in \mathcal{R}$ is a proposition symbol, in the atomic formula $p()$ parentheses are omitted; similar to binary functions, also binary predicates are mostly written in infix notation, e.g., $(x \prec y)$ instead of $\prec (x, y)$.

Typical atomic formulae are $(x \leq 5)$, $even(abs(x))$, or isSorted(myList).

Definition 11 (*First-order formula*) Given a signature Σ, a *first-order formula* over Σ is defined by the following clauses.

- Each atomic formula over Σ is a formula.
- $\bot$ is a formula, and $(\varphi \Rightarrow \psi)$ is a formula, if φ and ψ are formulae.
- If x is an individual variable and φ is a formula, then $\exists x\, \varphi$ is a formula.

As for propositional logic, we can write this definition in Backus–Naur-form. $\mathcal{T}_\Sigma$ denotes the set of terms over the signature $\Sigma = (\mathcal{F}, \mathcal{R}, \mathcal{V})$.

$$\mathcal{T}_\Sigma ::= \quad \mathcal{V} \mid \mathcal{F}(\mathcal{T}_\Sigma, \ldots, \mathcal{T}_\Sigma)$$
$$FOL_\Sigma ::= \quad \mathcal{R}(\mathcal{T}_\Sigma, \ldots, \mathcal{T}_\Sigma) \mid \bot \mid (FOL_\Sigma \Rightarrow FOL_\Sigma) \mid \exists \mathcal{V} \, FOL_\Sigma$$

$\forall x \, \varphi$ is an abbreviation for $\neg \exists x \, \neg \varphi$. Of course, all propositional abbreviations defined in previous sections can be used in FOL as well. In the formula $\exists x (x < 5 \wedge y < x)$, the two occurrences of variable x are said to be *bound* by the quantification $\exists x$. In contrast, variable y is *free*, i.e., not in the scope of any quantification $\exists y$. A formula is called *closed*, if it contains no free variables, i.e., if every occurrence of a variable x appears within the scope of a quantification ($\exists x$ or $\forall x$).

Examples for first-order formulae are asymmetry and transitivity given above.

Definition 12 (*First-order model*) A *first-order model* $\mathcal{M}$ for the signature $\Sigma = (\mathcal{F}, \mathcal{R}, \mathcal{V})$ is a structure $\mathcal{M} = (U, \mathcal{I}, \mathbf{v})$ consisting of

- a nonempty set U which is the *universe of discourse*,
- an *interpretation function* $\mathcal{I}$, where

  - $\mathcal{I}$ assigns an n-ary function $\mathcal{I}(f) : U^n \to U$ to every n-ary function symbol $f \in \mathcal{F}$, and
  - $\mathcal{I}$ assigns an n-ary relation $\mathcal{I}(R) \subseteq U^n$ to every n-ary relation symbol $R \in \mathcal{R}$,

  and

- a *variable valuation* $\mathbf{v} : \mathcal{V} \to U$ assigning a value to each variable in the signature.

Example 24: First-Order Model

As an example, consider the signature $\Sigma = (\mathcal{F}, \mathcal{R}, \mathcal{V})$ with $\mathcal{F} = \{+\}$, $\mathcal{R} = \{\prec\}$, and $\mathcal{V} = \{x, y, z\}$. A model for this signature is, e.g., $\mathcal{M} = (U, \mathcal{I}, \mathbf{v})$, with $U = \{Hugo, Erna\}$, $\mathcal{I}(+)$ is defined by $Hugo + Hugo = Erna + Erna = Hugo$ and $Hugo + Erna = Erna + Hugo = Erna$, $\mathcal{I}(\prec) = \{(Hugo, Erna)\}$, and $\mathbf{v}(x) = \mathbf{v}(y) = Hugo$, and $\mathbf{v}(z) = Erna$.

If $\mathcal{M} = (U, \mathcal{I}, \mathbf{v})$, we often write $\mathcal{M}(f)$, $\mathcal{M}(R)$ and $\mathcal{M}(x)$ instead of $\mathcal{I}(f)$, $\mathcal{I}(R)$, or $\mathbf{v}(x)$, respectively, to denote the 'meaning' of a function, relation or variable in the model.

Definition 13 (*Semantics of first-order terms*) In order to define a semantics of first-order formulae, we first have to declare what the 'meaning' of a term is. Generally, each term denotes a value; that is, it is *evaluated* to an element of the universe. The *term valuation* induced by a model $\mathcal{M} = (U, \mathcal{I}, \mathbf{v})$ is defined by

$$\mathbf{v}(f(t_1, \ldots, t_n)) = (\mathcal{I}(f))(\mathbf{v}(t_1), \ldots, \mathbf{v}(t_n)).$$

That is, in order to evaluate $f(t_1, \ldots, t_n)$, one has to apply the interpretation of f to the valuations of $t_1, \ldots, t_n$. For a term which is recursively built from other terms, the evaluation has to follow this recursion.

Example 24.1: Evaluation

In our model, the term $((x + y) + z)$ is evaluated as $\mathbf{v}((x + y) + z) = \mathcal{I}(+)(\mathbf{v}(x + y), \mathbf{v}(z))$, where $\mathbf{v}(x + y) = \mathcal{I}(+)(\mathbf{v}(x), \mathbf{v}(y)) = \mathcal{I}(+)(Hugo, Hugo) = Hugo$. Therefore $\mathbf{v}((x+y)+z) = \mathcal{I}(+)(Hugo, Erna) = Erna$.

Definition 14 (*Semantics of* FOL) Similar to the propositional case, the *validation relation* $\models$ between a model $\mathcal{M}$ and a formula φ is defined by the following clauses.

- $\mathcal{M} \models R(t_1, \ldots, t_n)$ if and only if $(\mathbf{v}(t_1), \ldots, \mathbf{v}(t_n)) \in \mathcal{I}(R)$,
- $\mathcal{M} \not\models \bot$,   and
- $\mathcal{M} \models (\varphi \Rightarrow \psi)$ if and only if $\mathcal{M} \models \varphi$ implies $\mathcal{M} \models \psi$.
- $\mathcal{M} \models (\exists x \varphi)$ if and only if $\mathcal{M}' \models \varphi$ for some $\mathcal{M}' = (U, \mathcal{I}, \mathbf{v}')$ with $\mathbf{v}'(y) = \mathbf{v}(y)$ for all $y \neq x$.

That is, $\mathcal{M} \models (\forall x \varphi)$ if and only if $\mathcal{M}' \models \varphi$ for all $\mathcal{M}'$ which differ from $\mathcal{M}$ at most in the valuation of x.

Example 24.2: Validity

Our example model with $\mathcal{I}(\prec) = \{(Hugo, Erna)\}$ satisfies all requirements for a strict partial order: it is asymmetric (if $x \prec y$ then $\mathbf{v}(x)$ must be *Hugo* and $\mathbf{v}(y)$ must be *Erna*, and therefore $y \prec x$ does not hold) and transitive (there is no valuation for y such that both $x \prec y$ and $y \prec z$: if $x \prec y$ then $\mathbf{v}(y)$ must be *Erna*, and if $y \prec z$ then $\mathbf{v}(y)$ must be *Hugo*).

Having defined the semantics of formulae, as discussed above in Sect. 2.2.2, we obtain a semantic consequence relation $\Gamma \Vdash \varphi$ which allows one to carry out proofs using models.

Example 25: A Semantic Proof

As an example, we show that every asymmetric relation is irreflexive. That is, we show that $\mathcal{M} \models \forall x \forall y (x \prec y \Rightarrow \neg y \prec x)$ implies $\mathcal{M} \models \neg \exists x (x \prec x)$. Assume to the contrary that $\mathcal{M} = (U, \mathcal{I}, \mathbf{v})$ is a model such that $\mathcal{I}(\prec)$ is an asymmetric relation, which is not irreflexive. If

$\mathcal{M} \not\models \neg \exists x (x \prec x)$, then $\mathcal{M} \models \exists x (x \prec x)$. That is, there is a valuation $\mathbf{v}'$ such that $\mathcal{M}' = (U, \mathcal{I}, \mathbf{v}')$ and $\mathcal{M}' \models (x \prec x)$. Let $a \in U$ be $\mathbf{v}'(x)$. Since $\mathcal{M}' \models (x \prec x)$, we have $(a, a) \in \mathcal{I}(\prec)$. Let $\mathcal{M}'' = (U, \mathcal{I}, \mathbf{v}'')$ be such that $\mathbf{v}''(x) = \mathbf{v}''(y) = a$. By asymmetry we have $\mathcal{M}'' \models (x \prec y \Rightarrow \neg y \prec x)$. Therefore $(a, a) \in \mathcal{I}(\prec)$ implies $(a, a) \notin \mathcal{I}(\prec)$. Since $(a, a) \in \mathcal{I}(\prec)$, we have $(a, a) \notin \mathcal{I}(\prec)$. This is a contradiction.

In order to extend our propositional Hilbert-style proof system for first-order logic, we use the following axioms and derivation rules:

(prop) all substitution instances of axioms of propositional logic
(mp) $\varphi,\ (\varphi \Rightarrow \psi)\ \vdash\ \psi$
(ex) $\vdash (\varphi[x := t] \Rightarrow \exists x \varphi)$
(part) $(\varphi \Rightarrow \psi) \vdash (\exists x \varphi \Rightarrow \psi)$, provided that x does not occur free in ψ

Axiom **(prop)** declares every formula to be derivable which can be obtained from a propositional tautology by consistently replacing proposition symbols by first-order formulae. That is, in first-order proofs we do not want to be bothered with having to prove propositional truths. **(mp)** is the *modus ponens* rule which we introduced in the propositional calculus. **(ex)** is the *exemplification*, which allows to deduct the existence of an object x with a certain property φ from a specific example of an object with that property. Here, $\varphi[x := t]$ is the formula which is obtained from φ by replacing every free occurrence of the variable x by the term t. **(par)** is the *particularization* rule: if some statement follows from a particular instance of an existential property, it also follows from the property itself, provided that it does not refer to the particularities of the instantiation.

Axiom **(ex)** and rule **(part)** are often written in a universal fashion, where they become *instantiation* axiom and *generalization* rule:

(in) $\vdash (\forall x \varphi \Rightarrow \varphi[x := t])$
(gen) $(\psi \Rightarrow \varphi) \vdash (\psi \Rightarrow \forall x \varphi)$, provided that x does not occur free in ψ

Intuitively, the generalization rule can be read as follows: if from ψ it follows that a particular x has property φ, and ψ does not mention this x, then from ψ it follows that all x must have property φ.

Example 26: A First-Order Derivation Proof

Using this calculus, we can derive the example property "every asymmetric relation is irreflexive" formally as follows:

(1) $\forall x \forall y (x \prec y \Rightarrow \neg\, y \prec x)$	(assumption)	
(2) $\forall y (x \prec y \Rightarrow \neg\, y \prec x)$	(1, in[x:=x])	
(3) $(x \prec x \Rightarrow \neg\, x \prec x)$	(2, in[y:=x])	
(4) $((x \prec x \Rightarrow \neg\, x \prec x) \Rightarrow (x \prec x \Rightarrow \bot))$	(prop)	
(5) $(x \prec x \Rightarrow \bot)$	(3, 4, mp)	
(6) $(\exists x\ x \prec x \Rightarrow \bot)$	(5, part)	
(7) $\neg\, \exists x\ x \prec x$	(6)	

It can be shown that the above Hilbert-style calculus is correct and complete, i.e., all derivable formulae are valid, and all valid formulae can be derived. The completeness proof proceeds as in the propositional case by constructing the canonical model from a maximal consistent extension of a satisfiable formula. For space reasons, it is omitted; the reader is referred to textbooks on logic.

As in the case of propositional logic, Hilbert-style axiom systems can be hard to use. Finding a proof by natural deduction can be much easier. Together with the propositional rules from above, the following sequent rules yield a correct and complete natural deduction system for first-order logic.

$$\frac{\Gamma \vdash \varphi[x := t]}{\Gamma \vdash \exists x \varphi} \qquad \frac{\Gamma \cup \{\varphi\} \vdash \psi}{\Gamma \cup \{\exists x \varphi\} \vdash \psi} \qquad \frac{\Gamma \vdash \forall x \varphi}{\Gamma \vdash \varphi[x := t]} \qquad \frac{\Gamma \vdash \varphi}{\Gamma \vdash \forall x \varphi}$$
$$\text{(ex)} \qquad\qquad \text{(part)} \qquad\qquad \text{(in)} \qquad\qquad \text{(gen)}$$

In rule **(part)**, we require that x does not occur freely in Γ or ψ; in rule **(gen)**, x must not occur freely in Γ.

Similarly to Propositional Logic, one can extend the First-Order Logic as presented here towards a framework that allows for modular specification. To this end, we need to define

- how to translate first-order signatures, terms and formulae,
- how to relate first-order models, and
- how to form the reduct of a first-order model along a signature translation.

With such notations in place, one can prove a satisfaction condition for first-order logic similarly to the one stated in Theorem 5.

2.4.2 Second-Order Logic

First-order logic is well-suited to formalize most aspects of mathematical reasoning. However, certain semantic concepts cannot be expressed with FOL formulae.

Example 27: Total Orders

A *strict linear* or *total order* is a mathematical structure consisting of a set and a binary relation, usually written $<$, which is asymmetric, transitive and total (or trichotomous).

$$
\begin{aligned}
\text{asymmetry:} \quad & \forall x \forall y (x < y \Rightarrow \neg\, y < x) \\
\text{transitivity:} \quad & \forall x \forall y \forall z\, ((x < y \wedge y < z) \Rightarrow x < z) \\
\text{totality:} \quad & \forall x \forall y ((x \neq y \Rightarrow (x < y \vee y < x))
\end{aligned}
$$

Asymmetry and transitivity clearly are first-order principles. In totality, however, the special relation symbol "$\neq$" is being used. For this relation, we usually assume that $x = y$ means that x and y denote the same object (element of the universe of discourse) and $x \neq y$, which is short for $\neg\, x = y$, means that x and y are different objects. According to the semantics of FOL, however, in any model the relation "$=$" is assigned *an arbitrary binary relation*. Consequently, there will be many models satisfying asymmetry, transitivity and totality, which do not match our intuition about a linear order. Take, for example, the model with $U = \{1, 2, 3\}$, $\mathcal{I}(<) = \{(1, 2), (2, 3), (1, 3)\}$ and $\mathcal{I}(=) = \{(1, 1), (2, 2), (3, 3), (1, 2)\}$ (and hence $\mathcal{I}(\neq) = \{(1, 3), (2, 1), (2, 3), (3, 1), (3, 2)\}$). In this model, equality is not even symmetric; yet all of the above axioms are satisfied.

With first-order formulae, we could force "$=$" to be an equivalence relation (reflexive, symmetric, and transitive). Even with these additional axioms, however, there will be strange interpretations, e.g., $\mathcal{I}(=) = U \times U$.

It can be shown that no finite set of FOL formulae can characterize equality on the universe in general. Therefore, in the logic $\mathrm{FOL}_\Sigma^=$ (FOL with identity) we add equalities as new atomic formulae.

$$FOL_\Sigma^= ::= \mathcal{R}(\mathcal{T}_\Sigma, \ldots, \mathcal{T}_\Sigma) \mid (\mathcal{T}_\Sigma = \mathcal{T}_\Sigma) \mid \bot \mid (FOL_\Sigma^= \Rightarrow FOL_\Sigma^=) \mid \exists x FOL_\Sigma^=$$

Semantically, for $\mathrm{FOL}_\Sigma^=$ we add the clause

- $\mathcal{M} \models (t_1 = t_2)$ if and only if $\mathbf{v}(t_1) = \mathbf{v}(t_2)$

That is, $(t_1 = t_2)$ holds in a model if t_1 and t_2 are evaluated to the same element of the universe.

In $\mathrm{FOL}^=$, we can define properties which are beyond the expressiveness of FOL. For example, the following formula states that the universe contains exactly two elements:

$$\exists x \exists y (\neg(x = y) \land \forall z(z = x \lor z = y))$$

$\mathrm{FOL}^=$ is very popular, especially in computer science; almost all automated theorem provers allow to use equality as a built-in relation. However, adding equality to the syntax and semantics of the logic might appear like a 'cheap trick'. Logicians do not wish to extend the logic for every non-expressible property with a new concept. What they would like is to be able to write a formula which 'defines' equality. Attributed to Leibniz is the following *extensionality principle*:

> Two objects are equal if they have all properties in common.

This principle involves 'all properties' an object can possibly have. Logically, this is a quantification going beyond FOL. Second-order logic provides the possibility to quantify also on predicates.

Definition 15 (*Syntax of* MSO) A *monadic second-order signature* Σ is a first-order signature $\Sigma = (\mathcal{F}, \mathcal{R}, \mathcal{V})$, where the set $\mathcal{V}$ is partitioned into a set $\mathcal{V}^0 = \{x, y, z, \ldots\}$ of individual variables and a set $\mathcal{V}^1 = \{X, Y, P, \ldots\}$ of (unary) *predicate variables*. Terms are built with relation symbols or predicate variables, and atomic formulae are built from terms as above.

In second-order logic, quantification can be both on individual and predicate variables:

$$\mathcal{T}_\Sigma ::= \quad \mathcal{V}^0 \mid \mathcal{F}(\mathcal{T}_\Sigma, \ldots, \mathcal{T}_\Sigma)$$
$$MSO_\Sigma ::= \quad \mathcal{R}(\mathcal{T}_\Sigma, \ldots, \mathcal{T}_\Sigma) \mid \mathcal{V}^1(\mathcal{T}_\Sigma) \mid \bot \mid (MSO_\Sigma \Rightarrow MSO_\Sigma)$$
$$\mid \exists \mathcal{V}^0 \; MSO_\Sigma \mid \exists \mathcal{V}^1 \; MSO_\Sigma$$

In the syntax, we allowed unary predicate variables only; hence the logic is called *monadic* second-order logic. In full second-order logic, quantification on arbitrary n-ary relations is allowed. As an example for a syntactically correct formula of MSO, consider the following *principle of transfinite induction*.

$$\textbf{(TFI)} \quad \forall P(\forall x(\forall y(y < x \Rightarrow P(y)) \Rightarrow P(x)) \Rightarrow \forall x P(x))$$

To understand the meaning of this somewhat complex formula, we need to define the semantics of the logic.

Definition 16 (*Semantics of* MSO) In a *second-order model*, the variable valuation $\mathbf{v} : \mathcal{V} \to U \cup 2^U$ provides

- a value $\mathbf{v}(x) \in U$ to each individual variable x in the signature, and
- a value $\mathbf{v}(X) \subseteq U$ to each predicate variable X in the signature.

Now, the semantics of second-order quantification is straightforward:

- $\mathcal{M} \models (\exists X \varphi)$ if and only if $\mathcal{M}' \models \varphi$ for some $\mathcal{M}' = (U, \mathcal{I}^\mathcal{F}, \mathcal{I}^\mathcal{R}, \mathbf{v}')$ with $\mathbf{v}'(Y) = \mathbf{v}(Y)$ for all $Y \neq X$.

That is, $\mathcal{M} \models (\forall X \varphi)$ if and only if $\mathcal{M}' \models \varphi$ for all $\mathcal{M}'$ which differ from $\mathcal{M}$ at most in the valuation of X.

With second-order logic, we can formulate Leibniz' principle within the logic:

$$\forall x \forall y (x = y \Leftrightarrow \forall P(P(x) \Leftrightarrow P(y)))$$

For further discussion of the Leibniz' principle see, e.g., [For16].

Other properties which are undefinable in FOL but definable in second-order logic include finiteness of the universe and mathematical induction.

For finiteness, there are several equivalent formulations. A popular one is that a set is finite if and only if there exists a bijection onto some initial segment of the natural numbers. Dedekind defined a set S to be finite if and only if every injective function from S to S is also surjective. In second-order logic, Dedekind's definition could be formulated as follows:

$$fun_1(f) \triangleq \forall x(S(x) \Rightarrow \exists y(S(y) \land f(x,y)))$$
$$fun_2(f) \triangleq \forall x \forall y \forall z(f(x,y) \land f(x,z) \Rightarrow y = z)$$
$$fun_3(f) \triangleq \forall x \forall y(f(x,y) \Rightarrow S(x))$$
$$inj(f) \quad \triangleq \forall x \forall y \forall z(f(x,z) \land f(y,z) \Rightarrow x = y)$$
$$surj(f) \triangleq \forall y(S(y) \Rightarrow \exists x(S(x) \land f(x,y)))$$
$$fin(S) \quad \triangleq \forall f(fun_1(f) \land fun_2(f) \land fun_3(f) \land inj(f) \Rightarrow surj(f))$$

Note that in this definition, second-order quantification is applied to the binary relation symbol f. Hence this characterization is outside of MSO.

Mathematical induction is one of the most widely used proof principles for properties of natural numbers. If one can show that property P holds for the number 0, and that whenever P holds for any number i it also holds for the successor $i+1$, then P holds for all natural numbers. This can be formulated in monadic second-order logic as follows:

$$\textbf{(MI)} \quad \forall P((P(0) \land \forall i(P(i) \Rightarrow P(i+1))) \Rightarrow \forall i P(i))$$

Sometimes it is easier to read such a formula 'in the opposite direction': If P holds for 0, but fails for some $i > 0$, then there must be some i where P starts to fail, i.e., $P(i)$ holds but $P(i+1)$ does not.

$$\textbf{(MI')} \quad \forall P(P(0) \land \exists i \, \neg P(i) \Rightarrow \exists i(P(i) \land \neg P(i+1)))$$

The above principle **(TFI)** of transfinite induction is a generalisation of this mathematical induction principle. If one can show that whenever property P holds for all numbers smaller than x it also holds for x, then P holds for all natural numbers. In other words, if P fails for some x, then there must be a smallest such x where it fails:

$$\textbf{(TFI')} \quad \forall P(\exists x \neg P(x) \Rightarrow \exists x(\neg P(x) \land \forall y(y < x \Rightarrow P(y))))$$

As an example which is more related to computer science, we can define the reflexive transitive closure R^* of a binary relation R in monadic second-order logic. Call a set P of elements *closed under* R if whenever x in P and xRy, then also y in P. As a formula this reads $\forall x \forall y((P(x) \land xRy) \Rightarrow P(y))$. Now mR^*n if every set P which contains m and is closed under R also contains n.

$$mR^*n \Leftrightarrow \forall P(P(m) \land \forall x \forall y((P(x) \land yRx) \Rightarrow P(y)) \Rightarrow P(n))$$

This defines the reflexive-transitive closure R^* of R as the *minimal* relation which comprises identity and is closed under R. On natural numbers, the "'less or equal" relation $\leq$ is the reflexive transitive closure of the successor relation $+1$.

$$m \leq n \Leftrightarrow \forall P(P(m) \land \forall x(P(x) \Rightarrow P(x+1)) \Rightarrow P(n))$$

Therefore, the principle of mathematical induction is equivalent to the claim that each natural number n is a reflexive transitive successor of 0:

$$\forall n(0 \leq n) \Leftrightarrow \forall P(P(0) \wedge \forall x(P(x) \Rightarrow P(x+1)) \Rightarrow \forall n P(n))$$

Another interesting second-order property related to induction is *term-generatedness*. Informally, it states that every object can be described by a ground term, where a *ground term* is a term without variables. That is, a ground term is a term which is built from constants and function applications to other ground terms.

For example, in the signature where the set of functions consists of the constant (0-ary function) c and the binary function symbol f the ground terms include $c, f(c, c), f(f(c, c), c)$, and $f(c, f(c, c))$.

Definition 17 (*Term-generated model*) A model is called *term-generated*, if for any element k in the universe there exists a ground term t such that $\mathcal{I}(t) = k$.

This property can be formulated by a second-order formula. For sake of simplicity, we give this formula only for the above signature which contains exactly one constant c and one binary function f. The following property holds in a model $\mathcal{M}$ if and only if it is term-generated.

$$\forall P(P(c) \wedge \forall x \forall y(P(x) \wedge P(y) \Rightarrow P(f(x, y)))) \Rightarrow \forall x P(x)) \qquad (*)$$

Note the similarity of this formula to the above induction principle: $P(c) \wedge \forall x(P(x) \Rightarrow P(f(x)))$ means that $\mathcal{M}(P)$ contains $\mathcal{M}(c)$ and with every object also all objects which can be obtained from $\mathcal{M}(c)$ by applying function $\mathcal{M}(f)$. That is, $\mathcal{M}(P)$ contains the interpretations of all ground terms in the signature. The formula $(*)$ states that any such P must contain all elements of the universe. Stated negatively, $\mathcal{M}$ is not term-generated if and only if there exists a set containing the interpretations of all ground terms, but not the whole universe.

Definition 18 (*Freely term-generated model*) A model is called *freely term-generated*, if for any element k in the universe there exists a unique ground term t such that $\mathcal{I}(t) = k$.

A model is freely term-generated, if it is term-generated and additionally it holds that functions do not overlap and function applications to different arguments yield different results. This can be expressed in FOL$^=$, here again for the above signature:

$$\forall x, y(c \neq f(x, y))$$
$$\wedge \forall x_1, x_2, y_1, y_2(x_1 \neq x_2 \vee y_1 \neq y_2 \Rightarrow f(x_1, y_1) \neq f(x_2, y_2))$$

We will use the notions of term-generatedness and freely term-generatedness in the next section on the formulation of the logic of CASL.

2.4.3 The Logic of CASL

Whereas the above first- and second-order logics were designed to be as small as possible to ease theoretical investigations, they are inconvenient for system specification and verification. Therefore, various dialects and extensions of these logics have been suggested, among them

- the program development languages "Vienna Development Method" VDM, Z, B and Event-B,
- the OBJ family of languages, including CafeOBJ and Maude,
- the Meta-Environment ASF+SDF, which combines the "Algebraic Specification Formalism" ASF and the "Syntax Definition Formalism" SDF,
- the logics for program verification "ANSI/ISO C Specification Language" ACSL and "Java Modelling Language" JML,

to name just a few. Here, we discuss the extension to first- and second-order logics that were introduced with the algebraic specification language CASL (to be discussed in Chap. 4). The material presented can be safely skipped by the casual reader.

The "Common Algebraic Specification Language" CASL [Mos04, BM04] is a specification formalism developed by the CoFI initiative [Mos97] throughout the late 1990s and early 2000s. The aim of the CoFI initiative was to design a *Common Framework for Algebraic Specification and Development* in an attempt to create a de facto platform for algebraic specification. The main motivation for the CoFI initiative came from the existence of a number of competing algebraic specification languages with varying levels of tool support and industrial uptake.

The logic of CASL basically consists of many-sorted first-order logic with equality, which allows for partial functions and also includes sort generation constraints. In the following we introduce these concepts step by step, and discuss finally syntax and semantics of CASL specifications.

Note that CASL also includes the concept of subsorting. Here, we refrain of introducing this concept and refer to the literature [Mos04].

FOL with Sorts

One issue with FOL is that the universe of discourse is not structured. Whereas this is mostly unproblematic for mathematical theories (which deal, e.g., solely with natural or real numbers), computer science specifications often refer to different sorts of data.

Example 28: Lexicographic Order

As a practical application, let us consider a telephone book containing a set of different names. If you want to write a program which lists these names, usually you have to order them *lexicographically*. The lexicographic order is a total order on the set of character strings over some alphabet. The Latin alphabet {A, ..., Z} is ordered by A < B < $\cdots$ < Z. Building on this, the lexicographic order on strings can be defined as follows: Let $x \triangleq x_1x_2 \ldots x_m$ and $y \triangleq y_1y_2 \ldots y_n$ be two strings. Then $x < y$ if and only if

- There exists a position i such that x and y are equal up to i, and $x_i < y_i$ (e.g., "bell"<"bet"), or
- x is an initial segment of y (e.g., "bet"<"better").

Mathematically, this can be written as follows.

$x_1x_2 \ldots x_m < y_1y_2 \ldots y_n$ if and only if
$$(\exists i(i \leq m \wedge i \leq n \wedge x_i < y_i \wedge \forall j(j < i \Rightarrow x_j = y_j))$$
$$\vee (m < n \wedge \forall j(j \leq m \Rightarrow x_j = y_j)))$$

Note that the above is not a FOL formula, as, e.g., it makes use of indexed variables in combination with the "..." notation—to be filled with meaning by the human reader.

In this mathematical formulation, i and j range over natural numbers, x and y denote strings, and x_i and y_j are characters. The symbols "<" and "≤" denote the ordering between integers, characters and strings, respectively, i.e., these symbols are 'overloaded'. However, each variable is used to refer to objects of one type only.

In order to deal with these issues, one can introduce types into FOL. Each variable ranges over a certain type. Functions and predicates have argument types and result types. Using the sorts *String* and *Nat*, we can formulate lexicographic ordering in many-sorted FOL$^=$:

$\forall x : String \ \forall y : String.(x < y \iff$
$\quad (\exists i : Nat.(i \leq min(length(x), length(y)) \wedge elem(x, i) < elem(y, i)$
$\quad\quad \wedge \forall j : Nat.(j < i \Rightarrow elem(x, j) = elem(y, j)))$
$\quad \vee (length(x) < length(y)$
$\quad\quad \wedge \forall j : Nat.(j \leq length(x) \Rightarrow elem(x, j) = elem(y, j)))))$

In the logic many-sorted FOL$^=$, sort symbols are added to the signature and relations, functions and variables are typed.

Definition 19 (*Many-sorted signature*) A *first-order signature with sorts* or *many-sorted signature* $\Sigma = (S, \mathcal{F}, \mathcal{R}, \mathcal{V})$ consist of a nonempty finite set S

of *sort symbols* (or *type symbols*[2]) and a first-order signature $(\mathcal{F}, \mathcal{R}, \mathcal{V})$. In contrast to plain FOL, where each function just has an arity, with each $f \in \mathcal{F}$ is associated a list of *argument types*, and a *result type*. A function symbol named f with argument types $(T_1, \ldots, T_n)$ and result type T is written as $f : T_1 \times \ldots \times T_n \to T$. Again, if the list of argument types is empty, then f is called a *constant* symbol of type T. Similarly as with functions, in a many-sorted signature each $R \in \mathcal{R}$ consists of a relation name and a list of argument types, but no result type. A relation symbol sometimes is also called a *predicate* symbol.[3] Finally, each variable from $\mathcal{V}$ consists of a name and a type. We write $x : T$ for the variable with name x and type T.

The definitions of terms, atomic formulae and formulae of many-sorted first-order logic are similar to the ones for normal first-order logic, with the exception that each term must respect *typing*.

Definition 20 (*Syntax of many-sorted* FOL$^=$) Terms and formulae of many-sorted first-order logic are defined as follows.

- If x is the name of a variable of type T, then x is a term of type T.
- If $t_1, \ldots, t_n$ are terms of types $T_1, \ldots, T_n$, respectively, and f is the name of an n-ary function symbol with argument types $(T_1, \ldots, T_n)$ and result type T, then $f(t_1, \ldots, t_n)$ is a term of type T.
- If t_1 and t_2 are terms of the same type, then $t_1 = t_2$ is an atomic formula.
- If $t_1, \ldots, t_n$ are terms of type $T_1, \ldots, T_n$, respectively, and R is the name of an n-ary relation with argument types $(T_1, \ldots, T_n)$, then $R(t_1, \ldots, t_n)$ is an atomic formula.
- An atomic formula is a formula.
- $\perp$ is a formula, and $(\varphi \Rightarrow \psi)$ is a formula, if φ and ψ are formulae.
- If x is the name of a variable of type T, and φ is a formula, then $\exists x : T . \varphi$ is a formula.

Example 28.1: Lexicographic Order in CASL

Within this logic, we can precisely express the above example. In CASL pretty printing it reads as follows (note that "." is typeset as "•"):

spec LEXICOGRAPHICORDER =
    sorts *String, Nat, Char*
    ops   *min* : *Nat* × *Nat* → *Nat*;
           length : *String* → *Nat*;
           elem : *String* × *Nat* → *Char*
    preds __<__ : *Nat* × *Nat*;

---

[2] In this book, the words "sort" and "type" are mutually exchangeable. Historically, logicians prefer the word "type", whereas computer scientists prefer to use the word "sort".

[3] In CASL, the word "predicate" is used instead of the word "relation". Thus, in algebraic specifications we will use the words "relation" and "predicate" interchangeably.

$\_\_ \leq \_\_ : Nat \times Nat;$

$\_\_ < \_\_ : String \times String;$

$\_\_ < \_\_ : Char \times Char$

$\forall \ x, \ y : String$

- $x < y$

    $\Leftrightarrow (\exists \ i : Nat$

    - $i \leq min(length(x), length(y)) \land elem(x, i) < elem(y, i)$

        $\land \ \forall \ j : Nat \bullet j < i \Rightarrow elem(x, j) = elem(y, j))$

        $\lor \ (length(x) < length(y)$

        $\land \ \forall \ j : Nat \bullet j \leq length(x) \Rightarrow elem(x, j) = elem(y, j))$

end

Definition 21 (*Semantics of many-sorted* FOL$^=$) In a many-sorted model, the universe of discourse is divided into sorts. That is, a model for many-sorted FOL$^=$ contains in addition to its universe of discourse U, interpretation function $\mathcal{I}$ and variable valuation $\mathbf{v}$, a mapping $\mathcal{T}$ from sort symbols to subsets of the universe.

Formally, $\mathcal{M} = (U, \mathcal{I}, \mathbf{v}, \mathcal{T})$, where $(U, \mathcal{I}, \mathbf{v})$ is a first-order model, and $\mathcal{T} : S \to 2^U$. For each sort T, the set $\mathcal{T}(T) \subseteq U$ is a non-empty set called the *carrier set* of T in $\mathcal{M}$.

Interpretation and valuation must respect the sorts declared in the signature:

- For $f : T_1 \times \ldots \times T_n \to T$, the interpretation $\mathcal{I}(f)$ of f must be a function from $\mathcal{T}(T_1) \times \cdots \times \mathcal{T}(T_n)$ to $\mathcal{T}(T)$.
- If $R \in \mathcal{R}$ has argument types $(T_1, \ldots, T_n)$, then the interpretation $\mathcal{I}(R)$ of R must be a subset of $\mathcal{T}(T_1) \times \cdots \times \mathcal{T}(T_n)$.
- If the variable x has type T, then the valuation $\mathbf{v}(x)$ must be an element of $\mathcal{T}(T)$.

With these restrictions, each term of sort T evaluates to an element of the carrier set of T. The validation relation $\mathcal{M} \models \varphi$ between model $\mathcal{M}$ and formula φ is defined exactly as in first-order logic.

Considering just the expressiveness, multi-sorted FOL$^=$ is no more expressive than pure FOL$^=$: For any sort symbol T we could introduce a monadic predicate isT and replace all formulae $\exists x : T. \ \varphi$ by $\exists x (isT(x) \land \varphi)$. However, for practical applications, using sorts greatly increases the usability for system specification and the readability of formulae. The situation is similar as with programming languages: whereas in the early days each program variable referred to 'a machine word', many modern programming languages rely on some type system.

Partial Functions

In mathematics and computer science, there are functions which are defined only for certain arguments. A prominent example is the division function *div* on naturals, which is undefined if the divisor is 0. Strictly speaking, division is not a function on pairs of naturals, as in a function each input is related to exactly one output. Another example would be the function which returns the first element of a string, which is undefined for the empty string. In order to deal with such a situation, there are several possibilities:

- One can restrict the domain to those arguments, where the function returns a value; in our example, $div : Nat \times Nat \setminus \{0\} \to Nat$.
- One can extend the range by a value ω for 'undefined'; in our example, $div : Nat \times Nat \to Nat \cup \{\omega\}$
- One can introduce the new syntactic category of *partial functions*, where in a partial function each input is related to at most one output; in our example, $div : Nat \times Nat \to? Nat$

What the 'best' solution would be, depends on personal taste, the application domain, and also the availability of tools. CASL offers total and partial functions.

Definition 22 (*Signature with partiality*) A *signature with partiality* $\Sigma = (S, \mathcal{F}_t, \mathcal{F}_p, \mathcal{R}, \mathcal{V})$ is a signature where $(S, \mathcal{F}_t, \mathcal{R}, \mathcal{V})$ forms a many-sorted signature as above, and $\mathcal{F}_p$ is a set of *partial function symbols*. We require that function symbols are either total or partial, i.e., there is no f with argument types $(T_1, \ldots, T_n)$ and result type T which is both in $\mathcal{F}_t$ and $\mathcal{F}_p$. A partial function $f \in \mathcal{F}_p$ with argument types $(T_1, \ldots, T_n)$ and result type T is written as $f : T_1 \times \ldots \times T_n \to? T$.

Example 29: Datatype of Strings in CASL

The datatype of Strings speaks about two kinds of data, namely data can be a character *Char* or a string *String*. We denote the empty string with the constant *eps*. It is possible to concatenate a character with a string, in order to obtain a new string. We write concatenation as an infix operation $\_\_ :: \_\_$, which includes two placeholders: the first for a character, the second for a string. We have two 'decomposition' methods for strings. The operation *first* returns the first character of a non-empty string. *first* is a partial operation. Given a string s as its parameter, the operation *rest* returns s without its first character. *rest* is a total operation. Finally, there is a predicate symbol *isEmpty*, which holds just for the empty string *eps*.

spec LOOSESTRING =
      sorts *Char*;
             String

ops   *eps* : *String*;
      __::__ : *Char* × *String* → *String*;
      first : *String* →? *Char*;
      rest : *String* → *String*
pred *isEmpty* : *String*
$\forall$ *c* : *Char*; *s* : *String*
- *isEmpty*(*s*) $\Leftrightarrow$ *s* = *eps*
- *def first*(*s*) $\Leftrightarrow$ ¬ *isEmpty*(*s*)
- *first*(*c* :: *s*) = *c*
- *rest*(*eps*) = *eps*
- *rest*(*c* :: *s*) = *s*
end

CASL has a definedness predicate *def* on terms, which holds if a term is defined. Using this predicate we state that *first* is defined for all non-empty strings.

A model $\mathcal{M} = (U, \mathcal{I}, \mathbf{v}, \mathcal{T})$ with partiality is a many-sorted model, where each n-ary partial function symbol $f : D_1 \times \cdots D_n \to? D_{n+1}$ is interpreted as a *partial function* $\mathcal{I}(f) : D_1 \times \cdots D_n \times D_{n+1}$. This is a $n + 1$-ary relation where from $(x_1, \ldots, x_n, y_1) \in \mathcal{I}(f)$ and $(x_1, \ldots, x_n, y_2) \in \mathcal{I}(f)$ it follows that $y_1 = y_2$. If for some $(x_1, \ldots, x_n)$ there is a y such that $(x_1, \ldots, x_n, y) \in \mathcal{I}(f)$, we say that f is *defined* for $(x_1, \ldots, x_n)$, else *undefined*.

Terms with partiality are formed in the same way as many-sorted terms, with function symbols from $\mathcal{F}_p \cup \mathcal{F}_p$. Formulae are built in the same way as many-sorted formulae.

Note that constants, which are 0-ary function symbols, can be defined or undefined. Semantically, we will treat a term to be undefined in a model if any of its arguments is undefined in this model. Sometimes this property is called *strictness*. For a predicate to be true, we will require that all of its arguments are defined. That is, $P(t_1, \ldots, t_n)$ is false if any of the t_i is undefined. An exception is equality: $(t_1 = t_2)$ is true if and only if both terms are undefined or both terms are defined and equal. In order to express that a term t is defined, we introduce a special unary predicate *def* which is true if and only if applied to a defined term. Note that in the presence of partial function symbols, $\mathbf{v}$ itself is a partial function from terms to values. In this case, variables can also be undefined.

Formally, the semantics is defined as follows:

Definition 23 (*Semantics of many-sorted* FOL$^=$ *with partiality*) Given a model $\mathcal{M} = (U, \mathcal{I}, \mathbf{v}, \mathcal{T})$ with partiality, we define:

- $\mathbf{v}(f(t_1, \ldots, t_n)) = \begin{cases} y, & \text{if all } \mathbf{v}(t_1), \ldots, \mathbf{v}(t_n) \text{ are defined, and} \\ & \quad (\mathbf{v}(t_1), \ldots, \mathbf{v}(t_n), y) \in \mathcal{I}(f) \\ \text{undefined, otherwise} \end{cases}$
- $\mathcal{M} \models R(t_1, \ldots, t_n)$ if and only if all $\mathbf{v}(t_1), \ldots, \mathbf{v}(t_n)$ are defined and $(\mathbf{v}(t_1), \ldots, \mathbf{v}(t_n)) \in \mathcal{I}(R)$.

- $\mathcal{M} \models (t_1 = t_2)$ if and only if $\mathbf{v}(t_1)$ and $\mathbf{v}(t_2)$ are both undefined, or both defined and $\mathbf{v}(t_1) = \mathbf{v}(t_2)$.
- $\mathcal{M} \models def\, t$ if and only if $\mathbf{v}(t)$ is defined.
- $\mathcal{M} \not\models \bot$,   and
- $\mathcal{M} \models (\varphi \Rightarrow \psi)$ if and only if $\mathcal{M} \models \varphi$ implies $\mathcal{M} \models \psi$.
- $\mathcal{M} \models \exists x : T.\varphi$ if and only if $\mathcal{M}' \models \varphi$ for some $\mathcal{M}' = (U, \mathcal{I}, \mathbf{v}', \mathcal{T})$ such that for all $y \neq x$ it holds that $\mathbf{v}'(y)$ and $\mathbf{v}(y)$ are both undefined, or both defined and $\mathbf{v}'(y) = \mathbf{v}(y)$.

We illustrate this definition by continuing our example of the *first* function on strings, namely the interplay of partiality with predicates and equality.

Example 29.1: Properties of Loose Strings

spec PROPERTYOFLOOSESTRINGS =
      LOOSESTRINGS
then pred __<__ : *Char* × *Char*   %(any relation on Char)%
then %implies
    ∀ *s, t* : *String*
    • *first*(*s*) < *first*(*t*) ⇒ ¬ *isEmpty*(*s*)   %(*)%
end

In CASL, a set of axioms is separated from its semantic consequences through the annotation **implies**. (For a further explanation of this construct, see Chap. 4 on CASL.) The formula (*) is a logical consequence of the specification LOOSESTRINGS extended by the operation _ < _. The predicate *first*(*s*) < *first*(*t*) is only true if both terms *first*(*s*) and *first*(*t*) are defined. Therefore, *first*(*s*) < *first*(*t*) implies that *s* (and also *t*) is not empty. If we would replace formula (*) by the following

    • *first*(*s*) = *first*(*t*) ⇒ ¬ *isEmpty*(*s*)   %(**)%

then the specification would become inconsistent: if both *s* and *t* are empty, then *first*(*s*) and *first*(*t*) are both undefined. Therefore, *first*(*s*) = *first*(*t*) is true. However, ¬*isEmpty*(*s*) is false, hence (**) is also false.

Sort Generation Constraints

CASL is intended to be a practical specification language. As such, it offers simple syntactic means to express complex second-order properties such as term-generatedness. To this end, CASL allows to declare generated and free data types. We present the key ideas by examples; for technical details the reader is referred to Mosses et al. [Mos04].

Example 28 on lexicographic order uses the data type of strings. In contrast to many programming languages, in CASL there are no predefined basic data types that are part of the language definition. There is a library of such data type definitions available ready for use, however, the specifiers can decide if they want to import these or not.

Example 29.2: CASL **Data Types**

We first recapitulate how Strings have been defined in Example 29. Here, we concentrate on part of the signature only:

spec STRING1 =
    sort   *Char*
    sort   *String*
    ops   *eps* : *String*;
           _::_ : *Char* × *String* → *String*
end

This signature declaration can be equivalently written using the CASL type construct.

spec STRING2 =
    sort   *Char*
    type *String* ::= *eps* | _::_(*Char*; *String*)
end

The type construct declares the sort *String*, and operations *eps* and _ :: _, which have the sort *String* as their result type.
    The type construct can be qualified by the keyword **generated**:

spec STRING3 =
    sort   *Char*
    generated type *String* ::= *eps* | _::_(*Char*; *String*)
end

This means that we are considering only models in which the sort STRING3 is term-generated by *eps*, _ :: _, and variables of sort *Char*.
    Note that it is possible to declare further operations with result type STRING3. However, these further operations don't contribute to the terms needed for generatedness.

Alternatively, it is possible to qualify the type by the keyword **free**:

spec STRING4 =
    sort   *Char*
    free type *String* ::= *eps* | __::__(*Char*; *String*)
end

This means that we are considering only models in which the sort STRING4 is freely term-generated by *eps*, __ :: __, and variables of sort *Char*.

As these examples demonstrate, CASL offers handy syntax for the second-order formulae discussed above in Sect. 2.4.

Example 28.3: Recursive Lexicographic Order in CASL

Utilizing strings as a freely generated type, and the partial function *first* on strings, we can give a recursive definition of lexicographic order in CASL, which resembles a recursive, functional program.

spec LEXICOGRAPHICORDER2 =
    STRING4
then ops   *first* : *String* →? *Char*;
           rest : *String* → *String*
    ∀ *c* : *Char*; *s* : *String*
    • ¬ *def first*(*eps*)
    • *first*(*c* :: *s*) = *c*
    • *rest*(*eps*) = *eps*
    • *rest*(*c* :: *s*) = *s*
    preds __<__ : *String* × *String*;
           __<__ : *Char* × *Char*;
           isEmpty : *String*
    ∀ *s* : *String* • *isEmpty*(*s*) ⇔ *s* = *eps*
    ∀ *x*, *y* : *String*
    • *isEmpty*(*x*) ∧ *isEmpty*(*y*) ⇒ ¬ *x* < *y*
    • *isEmpty*(*x*) ∧ ¬ *isEmpty*(*y*) ⇒ *x* < *y*
    • ¬ *isEmpty*(*x*) ∧ ¬ *isEmpty*(*y*) ⇒
          (*first*(*x*) < *first*(*y*) ⇒ *x* < *y*)
    • ¬ *isEmpty*(*x*) ∧ ¬ *isEmpty*(*y*) ⇒
          (*first*(*x*) = *first*(*y*) ⇒ *rest*(*x*) < *rest*(*y*))
end

For the operations *first* and *rest* we write the axioms using 'pattern matching'. For *first*, we consider two cases: for *eps*, the result is undefined; in the non-empty case we know that the string argument is composed from a character and a string—and we return the character. Analogously, we define *rest* for *eps* and the situation *c::s*.

The lexicographic order $<$ is also defined using 'pattern matching'. For empty strings x and y we do not have that $x < y$. For the empty string it holds that $eps < y$, provided y is not empty. If both x and y are non empty, there are two cases. In the first case, the first character of x is smaller than the first character of y. Then we know that $x < y$. In the second case, the first characters of x and y are identical. In this situation, we have to consider what holds for the rest, i.e., $rest(x) < rest(y)$.

Thanks to term-generatedness, it is possible to give inductive proofs of properties for such specifications.

The Semantics of CASL Specifications

The algebraic specification language CASL provides an intuitive syntax to ease the writing of system specifications in logic. This makes it possible to find potential mistakes in specifications already on the syntactic level. An example of this would be a message such as "*** Error: unknown sort". Here, the static semantics of CASL allows to check whether a signature is coherent and formulae are written using only the declared signature.

The result of the static analysis of a CASL specification is

- a first-order signature $\Sigma = (S, \mathcal{F}_t, \mathcal{F}_p, \mathcal{R}, \mathcal{V})$ (see Definition 22) with $\mathcal{V} = \emptyset$. In the CASL context, such a signature is referred to as a tuple (S, TF, PF, P) with total function symbols $TF = \mathcal{F}_t$, partial function symbols $PF = \mathcal{F}_p$, and predicate symbols $P = \mathcal{R}$, and
- a set of many-sorted closed formulae Φ over Σ, possibly including the sort generation constraints **generated** and **free**.

The semantics of a CASL specification Sp with (Σ, Φ) is then given as the model class of Φ with non-empty carrier sets:

$$Mod(Sp) \triangleq \{\mathcal{M} \text{ is a model over } \Sigma \mid \mathcal{M} \models \Phi \text{ and } M(s) \neq \emptyset \text{ for all } s \in S\}$$

There are many possibilities of how to define morphisms between models of many-sorted FOL with partiality. In the context of CASL, the following choice has been made:

Definition 24 (*Model homomorphism*) Let $\Sigma = (S, TF, PF, P)$ be a signature and $\mathcal{M}$ and $\mathcal{N}$ models over Σ. A *many-sorted Σ-homomorphism* $h : \mathcal{M} \to \mathcal{N}$ is a family of functions $h = (h_s : \mathcal{M}_s \to \mathcal{N}_s)_{s \in S}$ with the following three properties:

- Let $f : T_1 \times \cdots \times T_n \to T \in TF$, and $(a_1, \ldots, a_n) \in \mathcal{M}(T_1) \times \cdots \times \mathcal{M}(T_n)$. Then

$$h_T(\mathcal{M}(f)(a_1, \ldots, a_n)) = \mathcal{N}(f)(h_{T_1}(a_1), \ldots, h_{T_n}(a_n)).$$

- Let $f : T_1 \times \cdots \times T_n \to? T \in PF$, $(a_1, \ldots, a_n) \in \mathcal{M}(T_1) \times \cdots \times \mathcal{M}(T_n)$, and $\mathcal{M}(f)(a_1, \ldots, a_n)$ defined. Then

$$h_T(\mathcal{M}(f)(a_1, \ldots, a_n)) = \mathcal{N}(f)(h_{T_1}(a_1), \ldots, h_{T_n}(a_n)).$$

- Let $p \in P$ have argument types $T_1, \ldots, T_n$, and $(a_1, \ldots, a_n) \in \mathcal{M}(T_1) \times \cdots \times \mathcal{M}(T_n)$. Then

$$(a_1, \ldots, a_n) \in \mathcal{M}(p) \text{ implies } (h_{T_1}(a_1), \ldots, h_{T_n}(a_n)) \in \mathcal{N}(p).$$

In the CASL context, a model homomorphism preserves definedness of partial functions and truth of predicates. A *model isomorphism* is a model homomorphism in which all functions h_s are bijective. The model class of a specification is said to be *monomorphic* if and only if all models of the specification are isomorphic to each other.

In this section, we described *basic* CASL specifications. CASL also includes structured specifications and architectural specifications. For the semantics of these constructs we refer to the literature [Mos04].

2.5 Non-Classical Logics

Classical logics (i.e., PL, FOL, MSO, and extensions thereof) have been very successful as a basis for the formalization of mathematical reasoning. Hence these logics are often referred to as "mathematical logic". However, for the formalization of other areas of interest, many competing approaches have been proposed. The reason is that classical logics have a number of shortcomings. Basically, they support a "static" view onto things only, lacking (among other things)

- *alternative aspects:* Classical logic can describe how things are, but not how they could be;
- *subjective viewpoints:* Personal and common knowledge, beliefs, obligations, and ambitions are not easily described in classical logics;
- *dynamic aspects:* To model a change of state in classical logics, time is treated like any other relation parameter. This does not reflect the ubiquity of time in, e.g., computer science applications;
- *spacial aspects:* Likewise, for reasoning about robot positions or distributed computing, the location of objects plays a dominant role;
- *resource awareness aspects:* Dealing with objects which cease to exist or come into existence is not easy in classical logics, since the universe of discourse is fixed. However, it is essential when arguing about limited resources.

For each of these aspects, special logics have been proposed which allow a convenient modelling and arguing about systems in which the respective aspect is important.

2.5.1 Modal and Multimodal Logics

Modal logic is the logic of possibility and necessity. It started out as a syntactical exercise when philosophers were asking questions such as

> If it is necessarily the case that something is possibly true, does it follow that it is possible that this thing is necessarily true?
> E.g., if it necessarily is possible that tomorrow will be cloudy, is it possible that it tomorrow will be necessarily cloudy?

In the beginning of the 20th century this was formalized with the operators $\Diamond$ (for "possibly") and $\Box$ (for "necessarily").

Definition 25 (*Syntax of modal logic*) Given a propositional signature $\mathcal{P}$, the syntax of modal logic can be defined as follows.

$$ML_{\mathcal{P}} ::= \quad \mathcal{P} \quad | \quad \bot \quad | \quad (ML_{\mathcal{P}} \Rightarrow ML_{\mathcal{P}}) \quad | \quad \Diamond ML_{\mathcal{P}}$$

Of course, we use all abbreviations from propositional logic ($\wedge, \bigvee, \Leftrightarrow$, etc.). The $\Box$ operator can be defined by $\Box \varphi \iff \neg \Diamond \neg \varphi$. That is, a sentence is necessarily true if it is not the case that it could be possibly false.

With this syntax, the above question could be written as

$$(\Box \Diamond \ cloudy \Rightarrow \Diamond \Box \ cloudy)$$

Various deductions systems were proposed on top of propositional logic for the derivation of such formulae. For example, the most basic modal proof system uses the following axioms and rules:

Definition 26 (*Axiomatic system for modal logic*)

$$
\begin{array}{ll}
(\textbf{PL}) & \text{all propositional tautologies} \\
(\textbf{K}) & (\Box(\varphi \Rightarrow \psi) \Rightarrow (\Box\varphi \Rightarrow \Box\psi)) \\
(\textbf{N}) & \varphi \vdash \Box\varphi \\
(\textbf{MP}) & \varphi, (\varphi \Rightarrow \psi) \vdash \psi
\end{array}
$$

Here, a formula is a *propositional tautology* if it can be proven by propositional reasoning. Axiom (**K**) is named in honour of Saul Kripke. It can be read as "if it is necessarily the case that φ implies ψ, and φ is necessarily true, then also ψ must necessarily hold."[4]

---

[4] Due to this axiom, the logic sometimes is called "modal logic **K**".

Example 30: Derivation of a Modal Formula in K

With this proof system, we can, e.g., derive $(\Box(p \land q) \Rightarrow \Box p)$:
- (1) $\vdash ((p \land q) \Rightarrow p)$ (PL)
- (2) $\vdash \Box((p \land q) \Rightarrow p)$ (1, N)
- (3) $\vdash (\Box((p \land q) \Rightarrow p) \Rightarrow (\Box(p \land q) \Rightarrow \Box p))$ (K)
- (4) $\vdash (\Box(p \land q) \Rightarrow \Box p)$ (2,3,mp)

A number of additional axioms have been suggested. Originally, these axioms have been motivated by questions of philosophical nature. Consider, for example, the following ones.

(T) $(\Box \varphi \Rightarrow \varphi)$
(D) $(\Box \varphi \Rightarrow \Diamond \varphi)$
(4) $(\Box \varphi \Rightarrow \Box \Box \varphi)$
(B) $(\Diamond \Box \varphi \Rightarrow \varphi)$
(5) $(\Diamond \Box \varphi \Rightarrow \Box \varphi)$

One of these philosophical question was **(T)**: what are the logical consequences of the statement "if p is necessarily true, then p is true"? Each combination of such and other axioms has been studied intensely with respect to the question what can be derived.

Example 31: Derivation of a Modal Formula in T

Using just the **(K)** and **(T)** axioms, we can, e.g., derive $(\Box \Box p \Rightarrow p)$:
- (1) $\vdash (\Box p \Rightarrow p)$ (T)
- (2) $\vdash \Box(\Box p \Rightarrow p)$ (1, N)
- (3) $\vdash (\Box \Box p \Rightarrow \Box p)$ (2,K)
- (4) $\vdash ((\Box \Box p \Rightarrow \Box p) \Rightarrow ((\Box p \Rightarrow p) \Rightarrow (\Box \Box p \Rightarrow p)))$ (PL)
- (5) $\vdash ((\Box p \Rightarrow p) \Rightarrow (\Box \Box p \Rightarrow p))$ (3,4,MP)
- (6) $\vdash (\Box \Box p \Rightarrow p)$ (1,5,MP)

In the beginning of modal logic, researchers were busy to find out which formulae could be syntactically derived from other axioms in this way. A major step was the definition of a semantic foundation for modal logics by Saul Kripke [Kri59]. He proposed that a model for modal logic consists of a universe of *possible worlds*, a binary accessibility relation between these worlds, and an interpretation which assigns a set of possible worlds to each proposition symbol. With this semantics, $\Box \varphi$ is true in some possible world w, if φ is true in all possible worlds w' which are accessible from w. Thus, modal models basically are graphs, consisting of nodes (possible worlds) and edges (the accessibility relation between possible worlds). Modal logic therefore is well-suited to reason about all sorts of graph structures. Before giving the formal definitions, we discuss a widely known graph structure.

Example 32: Modelling the World Wide Web in Modal Logic

The world wide web consists of a large number of web pages, which are connected via hyperlinks. Assume that you have a number of homepages, which link to some pages with your hobbies and to pages with your work projects. All pages with your hobbies link back to the homepage, as well as to outside pages belonging to a club of which you are a member. Your work pages contain links to themselves and to the home pages of your lab.

The following set of modal formulae could describe this situation.

$(isHome \Rightarrow (\Diamond isHobby \wedge \Diamond isWork))$

$(isHobby \Rightarrow (\Diamond isHome \wedge \Diamond isClub))$

$(isWork \Rightarrow (\Diamond isWork \wedge \Diamond isLab))$

Questions you might be interested in asking about this structure include:

- Are there no "dangling references", i.e., does every page link to some other page?
- Is it possible to come back from any page to a homepage?

Often, in practical applications, there is more than one accessibility relation. For example, in our model of the world wide web (see Example 32), we might want distinguish between internal links (on the same server) and external links (on a server outside of our control). This could be done by having two different modal operators $\langle int \rangle$ and $\langle ext \rangle$, and by replacing, e.g., the second of the above formulae by

$(isHobby \Rightarrow (\langle int \rangle isHome \wedge \langle ext \rangle isClub))$

The resulting logic is called *multimodal logic* (MML). Formally, it is defined as follows.

Definition 27 (*Syntax of multimodal logic*) Given a signature $\Sigma = (\mathcal{P}, \mathcal{R})$ of modal proposition symbols $p \in \mathcal{P}$ and relation symbols $R \in \mathcal{R}$, the set of formulae of MML is defined as follows.

$$\text{MML}_\Sigma ::= \quad \mathcal{P} \quad | \quad \bot \quad | \quad (\text{MML}_\Sigma \Rightarrow \text{MML}_\Sigma) \quad | \quad \langle \mathcal{R} \rangle \text{MML}_\Sigma$$

The operator $[R]$ is the *dual* of the operator $\langle R \rangle$, much the same as $\forall$ is the dual of $\exists$. Formally, $[R]\varphi$ is defined to be an abbreviation for $\neg \langle R \rangle \neg \varphi$. The intended reading of $\langle R \rangle \varphi$ is "there is a node accessible via R in which φ holds", and $[R]\varphi$ reads "for all nodes accessible via R it holds that φ is true".

This is made precise by the semantics of MML. A (modal) *frame* for the modal signature Σ is a tuple $(U, \mathcal{I}^\mathcal{R})$, where

- U is a nonempty set of *possible worlds* or *evaluation points*, and
- $\mathcal{I}^\mathcal{R}$ is a mapping $\mathcal{R} \rightarrow 2^{U \times U}$ assigning a binary *accessibility relation* over U to every relation symbol.

A *model* $\mathcal{M}$ for MML is a structure $\mathcal{M} = (U, \mathcal{I}^\mathcal{R}, \mathcal{I}^\mathcal{P})$, where

- $(U, \mathcal{I}^{\mathcal{R}})$ is a modal frame, and
- $\mathcal{I}^P$ is a mapping $\mathcal{P} \to 2^{\mathcal{U}}$ assigning a set of possible worlds to each modal proposition symbol. The intention is that $\mathcal{I}^P(p)$ denotes those worlds where the modal proposition p is true.

If there is only one accessibility relation, then these models are also called *Kripke-structures*.

Given a multimodal formula φ, a model $\mathcal{M}$, and an evaluation point $w \in U$, the *validation relation* $\mathcal{M}, w \models \varphi$ can be defined.

Definition 28 (*Validation relation for multimodal logic*)

- $\mathcal{M}, w \models p$ if and only if $p \in w$ for $p \in \mathcal{P}$,
- $\mathcal{M}, w \not\models \bot$, and $\mathcal{M}, w \models (\varphi \Rightarrow \psi)$ if and only if $\mathcal{M}, w \models \varphi$ implies $\mathcal{M}, w \models \psi$, and
- $\mathcal{M}, w \models \langle R \rangle \varphi$ if and only if there exists $w' \in U$ such that $(w, w') \in \mathcal{I}^{\mathcal{R}}(R)$ and $\mathcal{M}, w' \models \varphi$.

From the definition of $[R]\varphi$ it follows that

- $\mathcal{M}, w \models [R]\varphi$ if and only if for all $w' \in U$ such that $(w, w') \in \mathcal{I}^{\mathcal{R}}(R)$ it holds that $\mathcal{M}, w' \models \varphi$.

A formula is *universally valid* in a model, if it holds at every point:

- $\mathcal{M} \models \varphi$ if and only if $\mathcal{M}, w \models \varphi$ for every $w \in U$.

Example 32.1: Checking Links in the WWW Model

Assume the following MML-model $\mathcal{M} = (U, \mathcal{I}^{\mathcal{R}}, \mathcal{I}^P, w_0)$ for our fragment of the world wide web:

- $U = \{w_0, w_1, w_2, w_3, w_4 \, w_5\}$
- $\mathcal{I}^{\mathcal{R}}(int) = \{(w_0, w_2), (w_0, w_3), (w_1, w_2), (w_1, w_3), (w_2, w_0)\}$
- $\mathcal{I}^{\mathcal{R}}(ext) = \{(w_2, w_4), (w_3, w_5), (w_5, w_5)\}$
- $\mathcal{I}^P(isHome) = \{w_0, w_1\}$, $\mathcal{I}^P(isHobby) = \{w_2\}$, $\mathcal{I}^P(isWork) = \{w_3, w_5\}$, $\mathcal{I}^P(isClub) = \{w_2, w_4\}$, $\mathcal{I}^P(isLab) = \{w_5\}$

It is easy to see that all of the specification formulae given above are universally valid in this model. A graphical description is as follows:

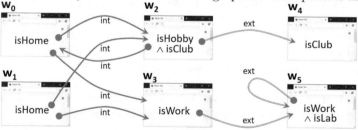

We can check whether there are no "dangling references" by evaluating the formula $(\langle int \rangle \top \vee \langle ext \rangle \top)$ for any $w \in U$. It turns out that this formula does not hold at w_4, since there is no link from this node. Likewise, the formula $[int]\langle int \rangle isHome$ expresses that every page reachable by an internal link has an internal link to the homepage. This formula is not satisfied at w_0 and w_1, since (w_0, w_3) and $(w_1, w_3) \in \mathcal{I}^{\mathcal{R}}(int)$, but w_3 does neither link back to w_0 nor to w_1. If we consider an alternative model, where

- $\mathcal{I}^{\mathcal{R}}(int) = \{(w_0, w_2), (w_0, w_3), (w_1, w_2), (w_1, w_3), (w_2, w_0), (w_3, w_1)\}$
  and
- $\mathcal{I}^{\mathcal{R}}(ext) = \{(w_2, w_4), (w_3, w_5), (w_5, w_5), (w_4, w_5)\}$,

then all specification formulae and both properties are satisfied.

The semantics of MML described above suggests that formulae of multimodal logic can be translated into first-order logic. In fact, there is a *standard translation* $ST_w :$ MML $\to$ FOL, where modal propositions $p \in \mathcal{P}$ are unary first-order predicates, and accessibility relation $R \in \mathcal{R}$ are binary first-order relations. The translation yields a first-order formula with exactly one free variable w.

- $ST_w(p) = p(w)$
- $ST_w(\bot) = \bot$
- $ST_w((\varphi \Rightarrow \psi)) = (ST_w(\varphi) \Rightarrow ST_w(\psi))$
- $ST_w(\langle R \rangle \varphi) = \exists v\ (wRv \land ST_v(\varphi))$,
  where v is a new variable not appearing in $ST_w(\varphi)$

In principle, this translation uses a new variable for each modal operator. By "re-using" bound variables, it suffices to use only the two variables w and v. Thus, modal logic is embedded in the *two-variable-fragment* of FOL, cf. [BdRV01].

2.5.2 Deontic Logic

Deontic logic is a branch of modal logic concerned, among other things, with moral and normative notions like *permission, prohibition, obligations, optionality, power, indifference, immunity*, etc. As any other logics, deontic logic cares about the logical consistency of (some of) the above notions, but also about the faithful representation of their intuitive meaning in different real-life context, like law, moral systems, business organizations and security systems.

Let us consider a first example on a workflow description of an airport ground crew describing what to do during the check-in process.

Example 33: Prescriptions for an Airport Ground Crew

1. The ground crew is obliged to open the check-in desk and request the passenger manifest from the airline two hours before the flight leaves.
2. The airline is obliged to provide the passenger manifest to the ground crew when the check-in desk is opened.

The complete work description has several more clauses similar to the above; the example will be continued in Chap. 6.

Below, we will show how the clauses may be formalised in deontic logic.

Our second example is part of a famous example in deontic logic on a moral system.

Example 34: John's Obligations for Partying

Mary is offering a party and has invited many friends, John among them. As John is not very reliable, Mary has asked him to respect the following agreement:
1. John ought to go to the party;
2. If John goes to the party, then he ought to tell them he is coming;
3. If John doesn't go to the party, then he ought not to tell he is coming.
   The above may be formalised in deontic logic.

There are many formal systems for deontic logic. In what follows we will introduce *Standard Deontic Logic (SDL)*. The starting point when defining SDL was to take different modal logics and to make analogies between "necessity" and "possibility", with "obligation" and "permission". Thus, the modal operators $\Diamond$ and $\Box$ became "$\mathbb{P}$" and "$\mathbb{O}$". However, this turned out to be difficult as many of the rules in modal logic did not transfer to deontic logic (as seen in the example below), though it was useful to make a start in understanding these normative concepts.

In modal logic with axiom **(T)** the following holds:

- If p then $\Diamond p$ (if p is true, then it is possible).
- If $\Box p$ then p (if it is necessary that p, then p is true).

While this makes perfect sense for the interpretation of $\Diamond$ and $\Box$ as possibility and necessity, it might not do so when considering possibility as permission and necessity as obligation:

- If p then $\mathbb{P}\, p$ (if p is true, then it is permissible).
- If $\mathbb{O}\, p$ then p (if it is obligatory that p, then p is true).

This is clearly counter-intuitive as any obligation then *must* be satisfied, and given any fact or action, it would be permissible.

We introduce now the syntax of the logic.

Definition 29 (*Syntax of SDL*) Assume we are given a propositional signature $\mathcal{P}$. The formulae of standard deontic logic are defined by

$$\text{SDL}_{\mathcal{P}}::= \quad \text{PL}_{\mathcal{P}} \mid (\text{SDL}_{\mathcal{P}} \Rightarrow \text{SDL}_{\mathcal{P}}) \mid \mathbb{P}\,\text{PL}_{\mathcal{P}}$$

That is,

- if φ is a propositional formula, then $\mathbb{P}\,\varphi$ is an SDL formula, and
- every propositional combination of SDL formulae is an SDL formula.

In contrast to multimodal logic, in SDL the modalities cannot be nested, since it appears pointless to state that "it is obligatory that it is obligatory that something must be done." Besides the usual derived operators inherited from propositional logic (conjunction, disjunction, etc), the following additional two modalities are useful:

$$\mathbb{O}\,p \triangleq \neg\mathbb{P}\,\neg p \quad \text{(obligation)}$$

$$\mathbb{F}\,p \triangleq \neg\mathbb{P}\,p \;(=\mathbb{O}\,\neg p) \quad \text{(prohibition)}$$

From the above we also get the intuitive relation that being permitted is the same as not being forbidden and vice-versa.

Example 33.1: Formalisation of the Ground Crew Procedure

The two clauses of Example 33 may be formalised in SDL as follows:
1. $(\textit{flight\_leaves} \Rightarrow (\mathbb{O}\;\textit{desk\_opens} \land \mathbb{O}\;\textit{request\_man}))$
2. $(\textit{desk\_opens} \Rightarrow \mathbb{O}\;\textit{provide\_man})$
where *flight_leaves* represents the fact that the flight leaves in at least two hours, *desk_opens* represents that "the ground crew opens the check-in desk", *request_man* that "the ground crew requests the passenger manifest from the airline", and *provide_man* "the airline provides the passenger manifest to the ground crew".

Note that going from a natural language description to a formal language is a modelling task, so it usually involves abstraction and certain subjectivity. It is worth noting the following in the example above: (i) As the logic is untimed we cannot represent the temporal aspect of the flight leaving in *at least two hours*; (ii) There is no notion of causality, thus both sentence should be interpreted in a certain context (e.g., we know that *desk_opens* should happen only if *flight_leaves* happens).

The following example shows a formalisation of an extension of Example 34.

Example 34.1: Formalisation of John's Obligations

Let us assume now that we add the following fact to the agreement between Mary and John:

1. John does not go to the party

   The above may be formalised in SDL as follows (including the agreement):

1. $\mathbb{O}\,party$
2. $\mathbb{O}(party \Rightarrow tell)$
3. $(\neg party \Rightarrow \mathbb{O}\,\neg tell)$
4. $\neg party$

SDL has a *Kripke-like modal semantics* based on: (i) a set of possible worlds (with a truth assignment function of propositions per possible world), and (ii) an accessibility relation associated with the O-modality. The accessibility relation points to *ideal* or *perfect deontic alternatives* of the current world, and to handle *violations* of obligations and prohibitions the semantics needs to be extended. Here, we will not present such a semantics; instead, we will instead see in more detail a proof system for SDL.

Definition 30 The axiomatic system of SDL consists of the following axioms and rules.

(PL) all propositionally valid formulae
($\mathbf{K}_O$) $(\mathbb{O}(\varphi \Rightarrow \psi) \Rightarrow (\mathbb{O}\,\varphi \Rightarrow \mathbb{O}\,\psi)))$
($\mathbf{D}_O$) $\neg\mathbb{O}\,\bot$
($\mathbf{N}_O$) $\varphi \vdash \mathbb{O}\,\varphi$
(MP) $\varphi, (\varphi \Rightarrow \psi) \vdash \psi$

In fact, this axiomatic system is the same as for the classic modal logic **K** (see Definition 26), with the additional axiom **($\mathbf{D}_O$)**. This axiom states that it cannot be obligatory to do something impossible. E.g., it cannot be obligatory to go and not go to the party.

Example 33.2: Deduction on the Ground Crew Procedure

We now present a small example for reasoning in deontic logic. Here, we outline only the major steps, the complete proof could be written as a Hilbert deduction as in Example 21.

Let us assume that the ground crew respects the two obligations on their procedure, namely $\mathbb{O}\,desk\_opens$ and $\mathbb{O}\,request\_man$. By adding the fact that the departure time is two hours from now, we can assume *flight_leaves*. From **(MP)** and the fact *flight_leaves* and the clause

$flight\_leaves \Rightarrow (\mathbb{O}\,desk\_opens \wedge \mathbb{O}\,request\_man)$

we know that the ground crew now has two obligations that we have assumed are not violated. In this case we have two new facts: $desk\_opens$

and *request_man*. By **(MP)** again, applied to *desk_opens* and
> ($desk\_opens \Rightarrow \mathbb{O}\ provide\_man$),

we can derive the obligation of the airline to provide the passenger manifest, $\mathbb{O}\ provide\_man$.

One of the main problems logicians face when formalising normative notions is to avoid so-called *puzzles* and *paradoxes*. Some are logical paradoxes, i.e., the formal system allows to deduce contradictory actions, others are practical paradoxes (including the so-called "practical oddities") where we can get counterintuitive conclusions. For instance, given some facts in SDL, we can deduce the obligation of doing something and at the same time not doing it. This can be shown in the partying example.

Example 34.2: Reasoning on John's Obligations

Let us consider the formalisation of the agreement between Mary and John as presented in Example 34.1. Using the SDL axiomatic system, we can derive the following:

$$(\mathbb{O}\ tell \land \mathbb{O}\ \neg tell)$$

This statement claims that John ought to tell, and at the same time ought not to tell that he goes to the party. It is known as *Chisholm's paradox* [Chi63]. We refrain from giving a formal Hilbert-style derivation and rather describe the reasoning in plain English.

- From 2 and K_O, by MP, we get that $(\mathbb{O}\ party \Rightarrow \mathbb{O}\ tell)$
- From 1 and the above, by MP, we get $\mathbb{O}\ tell$
- From 3 and 4, by MP, we get $\mathbb{O}\ \neg tell$
- So, we can infer $(\mathbb{O}\ tell \land \mathbb{O}\ \neg tell)$

So, we can conclude that John is obliged to tell that he is coming to the party, and at the same time that he must not tell about it.

This example shows that deontic logic as it was defined originally may not be adequate in every context. However, it set the basis for subsequent development of variants of the logic for specifying situations and properties where paradoxes are avoided.

Let us finish our presentation on deontic logic with a discussion on some philosophical issues concerning the logic. It has been observed that norms as prescriptions for conduct are not *true* nor *false*. Now, if norms have no truth-value, how can we reason about them and detect contradictions and define logical consequence? This was a question that bothered many logicians since the very beginning, as for many a logic should be concerned with a formalisation of truth. Von Wright (summarising his own point of view and interpreting early works in the area [Wri99]) has argued that logic has a wider reach than truth, and thus norms and valuations are still subject to a logical view. Another interesting issue is the difference between *prescriptive* and *descriptive* statements: while properties (e.g., as expressed in temporal logics) are descriptive, norms (as expressed in deontic logic) are prescriptive. In this introduction, we did not address the question how to represent what happens when an obligation is not fulfilled or a prohibition is violated. This question is relevant not only on normative (legal) system but also in software engineering: we know that software systems are not only concerned with normal (expected) behaviour, but also with exceptional (alternative) ones, very often representing violations.

There are many variants of deontic logic, trying to address the different issues discussed above. However, as of today there is no logic for normative systems fully addressing all these problems. In Chap. 6 we will see a logic which is especially suited for contract specifications.

2.5.3 Temporal Logic

In *temporal logic* the modal operators $\Diamond$ and $\Box$ are interpreted with respect to the flow of time. That is, $\Diamond\varphi$ means "it will eventually be the case that φ holds", and $\Box\varphi$ stands for "it is always the case that φ holds".[5]

Before giving the formal definitions, we start with a classical example.

Example 35: Dining Philosophers

Dijkstra introduced the famous *dining philosophers problem* [Dij71].

Five philosophers, numbered from 0 to 4, are living in a house where the table is laid for them, each philosopher having his own place at the table.

Their only problem—besides those of philosophy—is that the dish served is a very difficult kind of spaghetti, that has to be eaten with two forks. There are two forks next to each plate, so that presents no difficulty: as a consequence, however, no two neighbours may be eating simultaneously.

There are various ways that a philosopher could behave if hungry. For example, a strategy would be to wait for the left fork to become available, take it, wait for the right fork to become available, take it, eat, and then put both forks back again. It is easy to see that if all

---

[5] Some authors and most tools use **F** and **G** as a notation instead of $\Diamond$ and $\Box$. In some papers, **F** and **G** are used in addition to $\Diamond$ and $\Box$, which then have a different meaning.

philosophers strictly follow such a strategy it might end up in a situation where all five philosophers are holding their respective left fork and are waiting for the right fork to become available.

Another way would be to wait until both forks are available before taking them. However, with this protocol it could be the case that some philosopher never gets to eat, since either the left or right fork are always taken.

The challenge is to find a way for the philosophers to take forks and put them back again such that nobody starves.

In the formalization of such an example, temporal aspects play a crucial role. Typical properties are that each philosopher will *eventually* be able to eat, or that *always* at least two philosophers are not eating. More complex properties include that a philosopher cannot start eating unless his neighbors stop eating and release their forks.

Originally, temporal logic included only the operators $\Diamond$ and $\Box$ from modal logic, with a temporal interpretation. Thus, $\Diamond\varphi$ means that φ is *eventually* true, or that φ holds at some time in the future. Dually, $\Box\varphi$ means that φ is *always* true, or that φ holds in all future time points.

In computer science, time is often regarded as progressing in discrete steps. In particular, in a computation, in each state there is a *next* state which the machine assumes. Thus, in this context it may be necessary to talk about the next time point. In temporal logic, this was included as a separate modal operator $\bigcirc$ (pronounced "next").[6]

In his Ph.D. thesis, Kamp [Kam68] introduced a binary operator $\mathcal{U}$ and showed that it is more expressive than the modal operators introduced above. Intuitively, $(\varphi\,\mathcal{U}\,\psi)$ means that φ will hold *until* the next time when ψ holds. This operator can not be defined with $\Diamond$, $\Box$ and $\bigcirc$. However, it is possible to define the $\Diamond$-operator by $\Diamond\varphi \iff (\top\,\mathcal{U}\,\varphi)$. That is, φ will eventually hold if and only if *true* holds (which is always the case) until some time when φ holds.

Definition 31 (*Syntax of linear temporal logic*) Given a propositional signature $\mathcal{P}$, the syntax of linear temporal logic can be defined as follows.

$$\text{LTL}_{\mathcal{P}} ::= \quad PL_{\mathcal{P}} \mid (\text{LTL}_{\mathcal{P}} \Rightarrow \text{LTL}_{\mathcal{P}}) \mid \bigcirc \text{LTL}_{\mathcal{P}} \mid (\text{LTL}_{\mathcal{P}}\,\mathcal{U}\,\text{LTL}_{\mathcal{P}})$$

As described above, $\Diamond\varphi$ is an abbreviation of $(\top\,\mathcal{U}\,\varphi)$. As in modal logic, $\Box\varphi$ stands for $\neg\Diamond\neg\varphi$ (which in turn stands for $\neg(\top\,\mathcal{U}\,\neg\varphi)$). That is, φ is always true if it is not eventually false. Additionally, sometimes a "release" operator is defined by $(\varphi\,\mathcal{R}\,\psi) \triangleq \neg(\neg\varphi\,\mathcal{U}\,\neg\psi)$. Informally, this states that ψ must hold up to and including the next occurrence of φ, if it exists; otherwise, ψ must hold continuously. Thus, the occurrence of φ releases the obligation that ψ must hold.

---

[6] Some authors write $\mathbf{X}\varphi$ instead of $\bigcirc\varphi$.

Linear temporal logic formulae usually are evaluated on infinite sequences of propositional models. Assume that $\mathcal{M} = (\mathcal{M}_0, \mathcal{M}_1, \ldots)$ is such a sequence, where each $\mathcal{M}_i$ defines a mapping $\mathcal{P} \mapsto \{true, false\}$. Then, $\mathcal{M}^i$ denotes the suffix of the sequence $\mathcal{M}$ starting with the ith element. With this, the semantics of LTL can be defined as follows.

Definition 32 (*Semantics of linear temporal logic*)

- $\mathcal{M} \models p$ if and only if $M_0(p) = true$ for $p \in \mathcal{P}$.
- $\mathcal{M} \not\models \bot$, and $\mathcal{M} \models (\varphi \Rightarrow \psi)$ if and only if $\mathcal{M} \models \varphi$ implies $\mathcal{M} \models \psi$
- $\mathcal{M} \models \bigcirc\varphi$ if and only if $\mathcal{M}^1 \models \varphi$
- $\mathcal{M} \models (\varphi \,\mathcal{U}\, \psi)$ if and only if there exists $i \geq 0$ such that $\mathcal{M}^i \models \psi$, and $\mathcal{M}^j \models \varphi$ for all $0 \leq j < i$.

From this definition it follows that

- $\mathcal{M} \models \Diamond\varphi$ if and only if there exists an $i \geq 0$ such that $\mathcal{M}^i \models \varphi$,
- $\mathcal{M} \models \Box\varphi$ if and only if for all $i \geq 0$ it holds that $\mathcal{M}^i \models \varphi$, and
- $\mathcal{M} \models (\varphi \,\mathcal{R}\, \psi)$ if and only if for all $i \geq 0$ it holds that $\mathcal{M}^i \models \psi$, or for some $i \geq 0$ it holds that $\mathcal{M}^i \not\models \psi$, and $\mathcal{M}^j \models \varphi$ for some $0 \leq j < i$.

In linear temporal logic, time-dependent properties of computational systems can be elegantly formalized.

Example 35.1: Temporal Properties of the Dining Philosophers

To specify properties of the dining philosophers scenario, assume that $\mathcal{P}$ contains for each $i \in \{0 \ldots 4\}$ the proposition symbols $\{phil_i\, eating,\ fork_i\, available,\ phil_i\, hasLeftFork,\ phil_i\, hasRightFork\}$.

(Ph_1) "If philosopher 0 has both left and right fork, in the next moment (s)he will be eating":

$$((phil_0\, hasLeftFork \land phil_0\, hasRightFork) \Rightarrow \bigcirc phil_0\, eating).$$

(Ph_2) "Whenever philosopher 0 has the left fork, in the next state (s)he will eat or drop the fork":

$$\Box(phil_0\, hasLeftFork \Rightarrow \bigcirc(phil_0\, eating \lor fork_0\, available)).$$

(Ph_3) "If philosopher 0 is eating, (s)he will do so until making the forks available":

$$(phil_0\, eating \Rightarrow (phil_0\, eating\, \mathcal{U}\, (fork_0\, available \land fork_1\, available)).$$

(Ph_4) "Philosopher 0 taking the right or philosopher 1 the left fork releases the availability of fork 0":

$$((phil_0\, hasRightFork \lor phil_1\, hasLeftFork)\, \mathcal{R}\, fork_0\, available).$$

(Ph_5) "Always at most one of {philosopher 0, philosopher 1} is eating":

$$\Box\neg(phil_0\, eating \land phil_1\, eating).$$

(Ph_6) "Philosopher 0 will eventually be eating":

$$\Diamond phil_0\, eating.$$

(**Ph_7**)   "Philosopher 0 will always eventually be eating":
$$\Box \Diamond phil_0 \, eating.$$
That is, this philosopher will be eating infinitely often.

As axiomatic calculus for LTL, we can use the following axioms and rules:

(**PL**)   all propositionally valid formulae
(**K**) $(\bigcirc(\varphi \Rightarrow \psi) \Rightarrow (\bigcirc\varphi \Rightarrow \bigcirc\psi))$
(**U**) $(\neg \bigcirc p \iff \bigcirc \neg p)$
(**Rec**) $((\psi \vee \varphi \wedge \bigcirc(\varphi \mathcal{U} \psi)) \Rightarrow (\varphi \mathcal{U} \psi))$
(**N**) $\varphi \vdash \bigcirc \varphi$
(**MP**) $\varphi, (\varphi \Rightarrow \psi) \vdash \psi$
(**Ind**) $((\psi \vee \varphi \wedge \bigcirc(\varphi \mathcal{U} \psi)) \Rightarrow \chi) \vdash ((\varphi \mathcal{U} \psi) \Rightarrow \chi)$

This axiomatic system is sound and complete, cf. [KM08].

Moreover, it is minimal in the sense that every axiom can be shown to be independent from the others. However, it is not very convenient to use in practical derivations. Without proof, we state that the following formulae are derivable.

(**F1**) $((\bigcirc\varphi \Rightarrow \bigcirc\psi) \iff \bigcirc(\varphi \Rightarrow \psi))$
(**F2**) $(\Box(\varphi \Rightarrow \psi) \Rightarrow (\Box\varphi \Rightarrow \Box\psi))$
(**F3**) $((\Diamond\varphi \Rightarrow \Diamond\psi) \Rightarrow \Diamond(\varphi \Rightarrow \psi))$
(**F4**) $((\Box\varphi \wedge \Diamond\psi) \Rightarrow \Diamond(\varphi \wedge \psi))$
(**F5**) $((\Box\Diamond\varphi \wedge \Box\Diamond\psi) \iff \Box\Diamond(\varphi \wedge \Diamond\psi))$
(**n**) $(\Box(\varphi \Rightarrow \bigcirc\varphi) \Rightarrow (\varphi \Rightarrow \Box\varphi))$
(**F6**) $(\Diamond\varphi \iff (\varphi \vee \bigcirc\Diamond\varphi))$
(**F7**) $(\Box\varphi \iff (\varphi \wedge \bigcirc\Box\varphi))$
(**F8**) $(\varphi \mathcal{U} \psi) \iff (\psi \vee (\varphi \wedge \bigcirc(\varphi \mathcal{U} \psi)))$
(**F9**) $(\varphi \mathcal{R} \psi) \iff (\psi \wedge (\varphi \vee \bigcirc(\varphi \mathcal{R} \psi)))$

The last four of these formulae are recursive characterizations of the temporal operators. Since the formulae are valid, they can be used as additional axioms to derive many interesting properties.

Example 35.2: A Temporal Logic Derivation

For example, we show that under the assumption (**Ph_5**) "Always at most one of {philosopher 0, philosopher 1} is eating", it holds that if philosopher 0 is eventually continuously eating, philosopher 1 will starve:

(1) $\Box\neg(phil_0 \, eating \wedge phil_1 \, eating)$ (ass)
(2) $\Box(phil_0 \, eating \Rightarrow \neg phil_1 \, eating)$ $(1, \text{PL})$
(3) $(\Box phil_0 \, eating \Rightarrow \Box\neg phil_1 \, eating)$ $(2, \text{K})$
(4) $(\Diamond\Box phil_0 \, eating \Rightarrow \Diamond\Box\neg phil_1 \, eating)$ $(3, \text{K})$
(5) $(\Diamond\Box\neg phil_1 \, eating \Rightarrow \neg\Diamond\Box phil_1 \, eating)$ (K)
(6) $(\Diamond\Box phil_0 \, eating \Rightarrow \neg\Box\Diamond phil_1 \, eating)$ $(4, 5, \text{PL})$

Subsequently, we show how temporal logic can be used in the analysis of reactive systems. We use LTL to specify and verify properties of systems modeled in PAT, the Process Analysis Toolkit [PAT20]. PAT is a model checker which uses CSP as a modelling language for systems, and LTL as a specification language for properties.

In this chapter, we only give a first example. A more detailed presentation of PAT for the verification of human-computer interfaces can be found in Chap. 7.

Example 35.3: Dining Philosophers in PAT

Using a textual representation of CSP terms which is accepted by PAT, the problem of the dining philosophers can be formulated as follows (the code and description is mostly from the PAT online resources [PAT12]).

```
#define N 5;
Phil(i) = get.i.(i+1)%N -> get.i.i -> eat.i ->
          put.i.(i+1)%N -> put.i.i -> Phil(i);
Fork(i) = get.i.i -> put.i.i -> Fork(i) []
          get.(i-1)%N.i -> put.(i-1)%N.i -> Fork(i);
College() = ||x:{0..N-1} @ (Phil(x) || Fork(x));
#assert College() deadlockfree;
#assert College() |= []<> eat.0;
```

In this code, the global constant N, of value 5, denotes the number of philosophers and forks. There are two sets of objects in the system, i.e., the philosophers and the forks. Each object is modelled as one process; philosophers and forks are numbered from 0 to N-1. A philosopher is described by the process Phil(i), where Phil is the process name and i is a process parameter.

Event eat.i models the event of i-th philosopher starting to eat. This event makes the proposition $phil_i eating$ true. Event get.i.i models the event of ith philosopher picking up the fork on his left hand side. This event makes $phil_i hasLeftFork$ true and $fork_i available$ false. Event put.i.i models the event of putting down the fork from the left hand side. This event makes $phil_i hasLeftFork$ false and $fork_i available$ true. % is the standard modulo-operator; therefore, event get.i.(i+1)%N models the event of ith philosopher picking up the fork on his right hand side. Similarly, event put.i.(i+1)%N models the event of philosopher i putting down the right fork from the right hand side. These events toggle the truth value of the propositions $phil_i hasRightFork$ and $fork_{(i+1)\%N} available$, respectively.

Informally, process Phil(i) describes that the philosopher picks up the fork to the right, then the fork to the left, eats, and then puts down the forks in the same order.

The forks are modeled using the (external) choice operator []. Informally, it states that any fork can be picked up by the philosopher on the left or the one on the right. Notice that the events in processes `Fork(i)` are the same as those in processes `Phil(i)`.

Process `College()` models the whole system. `||` is the parallel composition operator which indicates that all `Phil(x)`- and `Fork(x)`-processes execute in parallel, synchronising on the common events.

The first assertion states that process `College()` is deadlock-free, where `deadlockfree` is a reserved keyword. The second assertion is an LTL formula which states that within process `College()`, always eventually event `eat.0` occurs. This is equivalent to stating that the respective philosopher will not starve to death.

With this input, PAT will quickly find and output a situation where the system deadlocks (and, thus, everybody starves). If we change the order in which forks are picked up, i.e.,

```
Phil(i) = get.i.i -> get.i.(i+1)%N -> \dots
```

then the system is deadlock-free, but anyone philosopher may starve. PAT will find such an execution sequence within a few milliseconds.

Notice that in PAT, events like `eat.0` are used in formulae instead of propositions like $phil_0 \, eating$. We refrain from a formal semantics here and rely on the reader's intuition instead.

2.6 Closing Remarks

In this chapter, we gave an overview of different logics for modelling, specifying, and analysing computational systems. Starting with propositional logic, we introduced its syntax, semantics and methods for derivation and model checking. We then gave an institutional treatment of propositional logic, in order to illustrate the systematic construction underlying any specific logic with this basic case. First- and second-order logic were introduced as natural extensions, emphasizing which properties can and can not be expressed. We described a particular logic as the one underlying the common algebraic specification language CASL. Finally, we gave a brief account of some non-classical logics, viz. (multi-)modal, deontic, and temporal logics.

One may wonder why it is necessary to study such a variety of logics, why not have "one logic for all purposes". There are at least three reasons why it is useful to know different logics:

- Firstly, as we have seen, logics greatly vary in their expressiveness. On the one hand, a more expressive logic may be necessary to formalize a certain property of interest. On the other hand, as a rule of thumb more expressiveness leads to more complex reasoning methods.

- Secondly, logics also vary in the intended application domain. Depending on which aspect of a system is to be modelled, different logics may be adequate. Thus, usability is an important criterion.
- Finally, tool support is essential for practical applications. There are SAT and SMT solvers, automatic or interactive theorem provers, model checkers, model finders, consistency checkers, etc. These tools at least partially automatise the reasoning methods available for a specific logic.

This chapter illustrates that the different logics presented allow to concisely and precisely specify system properties. In general, choosing the "right" specification logic for a problem at hand can be a challenge. With the help of the material presented above, the reader should be well prepared for this task.

2.6.1 Annotated Bibliography

The history of logic is often traced back to ancient Greece, in particular to Aristotle (384-322 BC). That said, the use of logical proof and reasoning was used as early as the 6th century BC both in the West and East. Indeed, there were pre-Socratic philosophers in Greece who were using logical proofs (e.g., Thales and Pythagoras), and logical reasoning was already known in China and India in that period.

An influential work connecting mathematics and logic was Boolean algebra introduced by George Boole (1815–1864) in his first book *The Mathematical Analysis of Logic* (1847), and further developed in his book *An Investigation of the Laws of Thought* (1854).

Following Boole, the development of modern "mathematical" (or "symbolic") logic started in the late 19th century and early 1900s with the *logicism* programme, which considers that some or all of mathematics may be reduced to logic, or at least be modelled in logic. Advocators of this tradition where Frege, Russell and Whitehead, based on pioneering work by Dedekind and Peano. In 1888, Richard Dedekind (1831–1916) proposed an axiomatization of natural-number arithmetic, and Giuseppe Peano (1858–1932) published a simplified version of such axiomatization in his book *The principles of arithmetic presented by a new method* (1889). This axiomatization became known as *Peano axioms*. Peano is also well-known for providing a rigorous and systematic treatment of mathematical induction. Gottlob Frege (1848–1925) developed a formal system for logic in his book *Begriffsschrift* (1879). The book *Foundations of Arithmetic* (1884) is considered a seminal work for the logicist programme. Bertrand Russell (1872–1970) and Alfred Whitehead (1861–1947) wrote *Principia Mathematica* (a three-volume work published between 1910 and 1913), a landmark in classical logic with the aim to set the logical basis for mathematics.

Between 1910 and 1930s *metalogic* was developed, that is the study of the *metatheory* of logic. Pioneers include David Hilbert (1862–1943), Leopold Löwenheim (1878–1957) and Thoralf Albert Skolem (1887–1963), and most notably Kurt Gödel (1906–1978) and Alfred Tarski (1901–1983) who worked on the combination of logic and metalogic. One of the greatest achievements in the history of logic is Gödel's incompleteness theorem (1931).

Claude Shannon (1916–2001) made the connection between Boolean algebra (logic) and electrical circuits of relays and switches, advancing the development of the new discipline concerning hardware circuit design. This was presented in his master thesis *A Symbolic Analysis of Relay and Switching Circuits* (1937).

Two classical books on propositional and first-order logic are the *Introduction to Mathematical Logic* by Church [Chu96] and *First-order Logic* by Smullyan [Smu68]. For a comprehensive presentation of dynamic logic see the book by Harel et al. [HTK00]. Concerning deontic logic see articles by von Wright [Wri51, Wri99].

The use of temporal logic in computer science was introduced by Kröger [Krö76] and Pnueli [Pnu77], who took inspiration from the work done in philosophy by Prior and others. For a more extensive presentation of temporal logic, including a proposal of a deductive system for the logic, see the book series by Manna and Pnueli [MP92], and the book by Kröger and Merz [KM08].

One of the first books on model checking was written by Clarke et al. [CGP01], and a few more books have appeared after that presenting different techniques for software verification (e.g., by Baier and Katoen [BK08] and by Peled et al. [PGS01]).

A good source for surveys, introductory presentations, and more advanced material concerning many different logics, are the volumes and corresponding chapters of various handbooks: Handbook of Logics in Computer Science [AGM95], Handbook of Philosophical Logic [GG04], Handbook of Modal Logic [BvBW07], and Handbook of Model Checking [CHVB18]. Finally, for a historical presentation of logic, see the different volumes (1-11) of the Handbook of the History of Logic [GW09].

2.6.2 Current Research Directions

Current research in the field of logics for software engineering is well-represented by the LICS (Logic in Computer Science) and CSL (Computer Science Logic) conference series. There are also more specific conferences, covering only some aspects of this chapter, such as deduction methods (IJCAR, CAV, LPAR, TABLEAUX, FORTE, RTA, SAT), or non-classical logics (AiML, DL, ICTL, TIME). More general conferences are dealing with formal methods (SEFM, FM, iFM) and theoretical foundations of computer science (FOCS, MFCS, STACS, STOC, ETAPS). The European and North American Summer Schools in Logic, Language and Information (ESSLLI and NASSLLI) provide good opportunities for beginning Ph.D. students to get an insight into special topics.

According to our personal view, the following major trends in this area can be identified:

- Firstly, there is the never ending quest for new methods and tools in the analysis of computational systems: algorithms for model checking, rewriting, satisfiability solving, game-based methods, etc.
- Secondly, researchers strive for more general, abstract results, e.g., by means of category theory, to get a better understanding of general principles.
- Furthermore, there is a lot of research extending the scope of logical analysis to new types of computation: multi-agent systems, machine learning and artificial intelligence, symmetric computation and quantum computing, etc.
- Finally, researchers are designing methods and tools for practical use to solve problems which are of industrial interest, e.g., for security analysis.

Undoubtedly, logic is a necessary foundation for almost all research in Formal Methods, and even computer science in general. Thus, studying logic is certainly beneficial, even if the reader wants to make a contribution in a different area.

References

[AGM95] Samson Abramsky, Dov M. Gabbay, and T. S. E. Maibaum, editors. *Handbook of Logic in Computer Science (Vol. 3): Semantic Structures.* Oxford University Press, Inc., 1995.

[BBH+09] A. Biere, A. Biere, M. Heule, H. van Maaren, and T. Walsh. *Handbook of Satisfiability: Volume 185 Frontiers in Artificial Intelligence and Applications.* IOS Press, NLD, 2009.

[BdRV01] Patrick Blackburn, Maarten de Rijke, and Yde Venema. *Modal Logic*, volume 53 of *Cambridge Tracts in Theoretical Computer Science.* Cambridge University Press, 2001.

[BK08] Christel Baier and Joost-Pieter Katoen. *Principles of model checking.* MIT Press, 2008.

[BM04] Michel Bidoit and Peter D. Mosses. *CASL User Manual – Introduction to Using the Common Algebraic Specification Language.* Springer, 2004.

[Boo47] George Boole. *The Mathematical Analysis of Logic: Being an Essay Towards a Calculus of Deductive Reasoning.* Cambridge Library Collection - Mathematics. Cambridge University Press, 1847.

[BT87] J.A. Bergstra and J.V. Tucker. Algebraic specifications of computable and semicomputable data types. *Theoretical Computer Science*, 50(2):137–181, 1987.

[BvBW07]   Patrick Blackburn, J. F. A. K. van Benthem, and Frank Wolter, editors. *Handbook of Modal Logic*, volume 3 of *Studies in logic and practical reasoning*. North-Holland, 2007.

[CGP01]   Edmund M. Clarke, Orna Grumberg, and Doron A. Peled. *Model checking*. MIT Press, 2001.

[Chi63]   Roderick M. Chisholm. Contrary-to-duty imperatives and deontic logic. *Analysis*, 24:33–36, 1963.

[Chu96]   A. Church. *Introduction to Mathematical Logic*. Annals of Mathematics Studies. Princeton University Press, reprint, revised edition, 1996.

[CHVB18]   Edmund M. Clarke, Thomas A. Henzinger, Helmut Veith, and Roderick Bloem, editors. *Handbook of Model Checking*. Springer, 2018.

[Dij71]   Edsger W. Dijkstra. Hierarchical ordering of sequential processes. *Acta Informatica*, 1:115–138, 1971.

[Fia05]   José Luiz Fiadeiro. *Categories for software engineering*. Springer, 2005.

[For16]   Peter Forrest. The Identity of Indiscernibles. In Edward N. Zalta, editor, *The Stanford Encyclopedia of Philosophy*. Metaphysics Research Lab, Stanford University, winter 2016 edition, 2016.

[GB92]   J.A. Goguen and R.M. Burstall. Institutions: Abstract model theory for specification and programming. *Journal of the ACM*, 39:95–146, 1992.

[GG04]   Dov M. Gabbay and Franz Guenthner, editors. *Handbook of Philosophical Logic*, volume 11. Springer, 2004.

[GW09]   Dov M. Gabbay and John Woods, editors. *Handbook of the History of Logic*, volume 5. Elsevier, 2009.

[HTK00]   David Harel, Jerzy Tiuryn, and Dexter Kozen. *Dynamic Logic*. MIT Press, Cambridge, MA, USA, 2000.

[Kam68]   Johan Anthony Willem Kamp. Tense logic and the theory of linear order, 1968. PhD Thesis.

[KM08]   Fred Kröger and Stephan Merz. *Temporal Logic and State Systems*. Texts in Theoretical Computer Science. An EATCS Series. Springer, 2008.

[Kri59]   Saul Kripke. A completeness theorem in modal logic. *J. Symb. Log.*, 24(1):1–14, 1959.

[Krö76]   Fred Kröger. Logical rules of natural reasoning about programs. In *Third International Colloquium on Automata, Languages and Programming*, pages 87–98. Edinburgh University Press, 1976.

[Mos97]   Peter D. Mosses. CoFI: The common framework initiative for algebraic specification and development. In *TAPSOFT'97*, LNCS 1214, pages 115–137. Springer, 1997.

[Mos04]   Peter D. Mosses. *CASL Reference Manual: The Complete Documentation Of The Common Algebraic Specification Language*. Springer, 2004.

[MP92]   Zohar Manna and Amir Pnueli. *Temporal Logic of Reactive and Concurrent Systems: Specification, The*. Springer-Verlag New York, Inc., 1992.

[MSRR06]   Till Mossakowski, Lutz Schröder, Markus Roggenbach, and Horst Reichel. Algebraic-coalgebraic specification in CoCasl. *J. Log. Algebraic Methods Program.*, 67(1-2):146–197, 2006.

[PAT12]   Process analysis toolkit (PAT) 3.5 user manual, 2012. https://www.comp.nus.edu.sg/~pat/OnlineHelp/index.htm.

[PAT20]   Process analysis toolkit (PAT) 3.5 user manual, 2020. https://pat.comp.nus.edu.sg.

[PGS01]   Doron A. Peled, David Gries, and Fred B. Schneider. *Software Reliability Methods*. Springer-Verlag, 2001.

[Pnu77]   Amir Pnueli. The temporal logic of programs. In *FOCS'77*, pages 46–57. IEEE Computer Society Press, 1977.

[Smu68]   Raymond R. Smullyan. *First-Order Logic*, volume 43 of *Ergebnisse der Mathematik und ihrer Grenzgebiete. 2. Folge*. Springer, 1968.

[VW]   Volkswagen, `https://www.volkswagen.co.uk/en/configurator.html`, Last accessed: August 2021.

[Wri51]   Georg Henrik Von Wright. Deontic logic. *Mind*, 60:1–15, 1951.

[Wri99]   Georg Henrik Von Wright. Deontic logic: A personal view. *Ratio Juris*, 12(1):26–38, 1999.

Chapter 3
The Process Algebra CSP

Markus Roggenbach, Siraj Ahmed Shaikh, and Antonio Cerone

Abstract Concerning distributed systems, process algebra plays a role similar to the one lambda calculus takes for sequential systems: phenomena such as deadlock, fairness, causality can be made precise and studied within process calculi. Furthermore, process algebra is applied in the modelling and verification of industrial strength systems. This chapter introduces syntax, semantics, and methods of the process algebra CSP and studies tool support for CSP including simulation, model checking, theorem proving and code-generation. Examples studied include an automated teller machine, a jet engine controller, a fault tolerant communication system, and a self-stabilising system in the form of a mathematical puzzle. Advanced material on the semantic models of CSP, process algebraic equations, denotational semantics for recursive equations, and refinement based proof methods for deadlock, livelock, and determinism conclude the chapter.

3.1 Introduction

Let's start our discussion of CSP by taking a break. By the way, CSP stands for "Communicating Sequential Processes", and we will see in the subsequent sections why this is an excellent name for this modelling language. Anyway, to properly enjoy your free time, you go over to this nice, cosy bistro for, say, a cup of coffee. This requires some interaction between you and the waiter. The two of you have to arrange for ordering an item ("I want a cup of

Markus Roggenbach
Swansea University, Wales, United Kingdom

Siraj Ahmed Shaikh
Coventry University, Coventry, United Kingdom

Antonio Cerone
Nazarbayev University, Nur-Sultan, Kazakhstan

M. Roggenbach et al., *Formal Methods for Software Engineering*,
Texts in Theoretical Computer Science. An EATCS Series,
https://doi.org/10.1007/978-3-030-38800-3_3

coffee, please."), getting the item, handing over the coffee ("Here you are."), informing you about the price ("That's 2 Euros."), handing over an amount ("Here you are."), and, possibly, giving out change ("Here is your change.").

Both of you run a 'protocol'. Should your protocols fail to fit, the purchase will not take place successfully. For instance, you might first want the item, before to pay—while the waiter insists on payment first: that would lead to a *deadlock*. Or, maybe to get the item, the waiter goes to the kitchen, gets engaged in a never ending chat, and forgets to serve you. That would lead to a *livelock*. Finally, the bistro might have the policy to serve tea instead of coffee in case the kitchen runs out of coffee beans. That would make the protocol *non-deterministic*.

Process algebra allows to model such protocols in a precise way. Furthermore, it provides definitions of mathematical strength for the above mentioned phenomena deadlock, livelock, and determinism. This is one requirement for the formal analysis of safety-critical systems, where deadlock and livelock could have serious consequences. Additionally, process algebra has tool support in the form of simulators, model checkers and theorem provers. Finally, there are implementation strategies for process algebra: given a system, modelled and analyzed within process algebra, this protocol can be transformed into, say, Java code such the Java program exhibits the same properties as the process algebraic model.

Typical applications of process algebra include protocols. These can, e.g., describe the interaction of humans with computers. Here, CSP is utilised, e.g., to detect cognitive errors, see Chap. 7 on "Formal Methods for Human-Computer Interaction". Furthermore, CSP plays an important role in security analysis, see Chap. 8 on "Formal Verification of Security Protocols": using CSP, in 1995, Gavin Loewe exhibited a flaw within the Needham Schroeder security protocol for authentication and showed how to correct it. Further applications of CSP include control systems (see Sect. 3.2.2 on modelling a jet engine controller), distributed systems such as credit card systems [GRS04] or railways [IMNR12], as well as distributed or parallel algorithms [IRG05]. On the more fundamental level, for studying distributed systems, process algebra plays a role like the lambda calculus does for sequential systems: phenomena such as deadlock, fairness, causality can be made precise and studied within process calculi.

In this chapter, we introduce the process algebra CSP, see, e.g., [Hoa85, Ros98, Sch00, Ros10], in an example-driven approach. Section 3.2 shows how to model systems in CSP, discusses how to give semantics to CSP, and how to use CSP's notions of refinement for verification purposes. In Sect. 3.3, we discuss various CSP tools for simulation, model checking, theorem proving and code generation. Finally, Sect. 3.4 provides advanced material on CSP, including a presentation of the three standard models, algebraic laws, giving semantics to recursion by using fixed point theory, and refinement based proof methods for deadlock, livelock, and determinism analysis.

3.2 Learning CSP

In this section we provide a set of case studies that illustrate CSP's range of applications. These case studies introduce core concepts of the language in an example driven way: modelling an automated teller machine introduces CSP's *syntax* in an intuitive way and provides a basic understanding of CSP's constructs—see Sect. 3.2.1. While modelling a jet engine controller, we develop first insights into CSP's *operational and denotational semantics*—see Sect. 3.2.2. Modelling buffers and later a communication system in CSP, we discuss the concepts of *refinement* and *verification*—see Sect. 3.2.3.

3.2.1 ATM Example: CSP *Syntax*

We use the example of an Automated Teller Machine (ATM) in order to introduce the language CSP. Starting with a simple ATM, we add functionality step by step, while staying roughly on the same level of abstraction. We consider the description of the ATM in natural language, which we then model in CSP. This allows us to illustrate the subtleties that are often missed in natural language descriptions during system modelling.

The natural language description is italicised and is followed by a CSP specification. At every stage we also present the grammar used to describe the syntax of the language CSP.

Ordering Events, Recursion, and Process Names

We start with a description of the ATM in natural language.

Example 36: A Simple ATM

Initially the ATM shows a ready screen. A customer then proceeds to

- *insert a bank card, and*
- *enter the pin for the card*

following which the ATM will

- *deliver cash to the customer, and*
- *return the bank card,*

before getting ready for a new session.

The first step in CSP modelling is to extract the events that define system evolution. Example 36 explicitly mentions four events. We name them as follows: *cardI* represents the insertion of the card by the customer and the machine accepting it, *pinE* represents the user entering a pin to authenticate themselves to the machine, *cashO* represents the machine delivering the cash, and *cardO* represents the machine returning the card. A *ready* event is used to represent a fresh session of the ATM when it is ready to serve a new customer. These events constitute the alphabet of communications Σ that the system can perform. For our model of the ATM we have

$$\Sigma = \{cardI, pinE, cashO, cardO, ready\}.$$

The above specification prescribes a certain order on these events. So, for example, delivering cash requires the customer to insert a bank card first. However, there is no order between the delivery of the cash and the return of the card. CSP allows for specifying causal relationships using *action prefix*. The CSP process *Stop* stands for a system that does not perform any events.

We are now ready to specify our first ATM process.

Example 36.1: Modelling Basic Cash Withdrawal

A single session of our simple ATM can be modelled as follows

$$ATM0 = ready \rightarrow cardI \rightarrow pinE \rightarrow cardO \rightarrow cashO \rightarrow Stop$$

ATM0 starts with a *ready* event followed by the events *cardI*, *pinE*, *cardO* and *cashO* in this order, before it stops to engage in any further event. There is no other order in which *ATM0* is willing to engage in the events of the alphabet Σ. This means the specified causal order is a total one, in which any two events are related. This fails to be traceable back to the natural language specification where there is no order prescribed on events *cardO* and *cashO* as discussed above. However, this *over specification* in *ATM0* is justified as most ATMs operate as such.

ATM0 allows for a single session only. We overcome this by using recursion in *ATM1*:

$$ATM1 = ready \rightarrow cardI \rightarrow pinE \rightarrow cardO \rightarrow cashO \rightarrow ATM1$$

ATM1 performs the same activities as *ATM0* however after *cashO* it starts over again.

CSP distinguishes between events and processes. Events are instantaneous and mark system evolution. Processes represent system states.

When modelling in CSP, we are abstracting from time: the delay between two events is unspecified. We only know that events are instantaneous and atomic in the sense that it is not possible to specify that other events may occur 'while event e is happening'.

The process *Stop* does not engage in any event. Given an event a and a process P we write $a \to P$ to form a new process which is willing to engage in event a and then behave as P. We write $N = P$ to define the behaviour of the process name N to be identical with the behaviour of the process P. Should P contain the process name N the equation becomes a recursive one. Recursion is how CSP expresses infinite behaviour.

The CSP syntax for processes seen so far can be summarised in the following grammar, which is defined relatively to an alphabet of events Σ and a set of process names PN.

$$
\begin{aligned}
P ::= \quad &Stop \\
| \; &a \to P \\
| \; &N
\end{aligned}
$$

where $a \in \Sigma$, $N \in PN$. A process equation takes the form

$$N = P$$

where $N \in PN$ and P as described by the above grammar.

Structuring the Alphabet Using Channels

Up to now, the alphabet of events is a plain set without any further structure. In the case of the ATM, one might want to 'tag' the events with the name of the interface at which they take place. Yet another motivation to give structure to the alphabet is when one uses the same datatype in different contexts. For example, on a more concrete level of abstraction, the value of a PIN and the amount of cash might both be modelled as integers. Here, one would like to add a tag in order to indicate whether an integer value is a PIN or the amount of cash to dispensed. CSP offers the construct of a *channel* for this purpose.

We give a bit more detail on how an ATM is structured:

Example 36.2: Distinguishing Communication Channels

An ATM has several interfaces: there is a Display, which can show the ready message; there is a KeyPad, which allows the user to enter the pin; there is a CardSlot, which takes the card in or gives the card back; there is a CashSlot, which delivers the money.

We encode these different devices as channel names and specify:

$$ATM2 = Disply.ready \rightarrow CardSlot.cardI \rightarrow KeyPad.pinE \rightarrow$$
$$CardSlot.cardO \rightarrow CashSlot.cashO \rightarrow ATM2$$

ATM2 displays *ready* message at the *Display*, receives a *cardI* over the CardSlot, etc.

When using channels, one forms composed events consisting of a channel name c and the event e to be communicated. Syntactically, these components are separated by ".."

Every channel c has a set of 'basic' events as its type $T(c)$. In the above example, the channel *CardSlot* has the type $T(CardSlot) = \{cardI, cardO\}$. In CSP it is standard to write

$$events(c) \triangleq \{c.x \mid x \in T(c)\}$$

for a set of events associated with channel c. Given a list of channels $c_1, \ldots, c_n$, $n \geq 1$, their combined events are given by

$$\{\mid c_1, \ldots, c_n \mid\} \triangleq events(c_1) \cup \ldots, \cup events(c_n).$$

Using this notation, we can write the alphabet of *ATM2* in a structured way:

$$\Sigma = \{\mid Display, CardSlot, KeyPad, CashSlot \mid\}$$

Communicating an event e over a channel c adds to the CSP grammar the following primitives

$$P ::= \quad \ldots$$
$$\mid c.e \rightarrow P$$

Process Termination and Sequential Composition

We enrich our ATM example by considering a 'cancel' button:

Example 36.3: Canceling a Session by Interrupt

The ATM behaves as before. Additionally, any time after inserting the bank card and before retrieving the cash, the costumer has a choice to cancel the session. Upon cancellation, the ATM returns the bank card and is ready for a fresh session.

It is straightforward to model the effect of the cancel button:

$$SessionCancel = KeyPad.cancel \rightarrow CardSlot.cardO \rightarrow ATM3$$

In order to integrate *SessionCancel* into the ATM, we make use of the CSP interrupt operator $\triangle$, the CSP process *Skip* that indicates successful termination, and the operator ⑨ for sequential composition of processes.

$$
\begin{aligned}
Session &= (KeyPad.pinE \rightarrow Skip) \,\triangle\, SessionCancel \\
SessionEnd &= CardSlot.cardO \rightarrow CashSlot.cashO \rightarrow ATM3 \\
ATM3 &= Display.ready \rightarrow CardSlot.cardI \rightarrow Session\,\text{⑨} \\
&\quad\ SessionEnd
\end{aligned}
$$

Before and after the event *KeyPad.pinE*, the customer can activate the cancel button—see process *Session*. When *Session* terminates, i.e., there was no interrupt, control is passed over to *SessionEnd*, which returns the card and gives out the cash.

The process $P \triangle Q$ behaves as the process P unless a first event from Q is performed after which Q determines the further behaviour. In CSP, the only way to end the possibility of an interrupt is to *terminate* the process on the left hand side of the interrupt operator, in our case the process *Session*. After using termination for turning off the interrupt, we still want to continue. In CSP, the process *Skip* does nothing but terminate. Sequential composition, written as P⑨Q, behaves like process P, should P terminate, control is passed over to process Q. In case P does not terminate, Q never gets activated.

In terms of methodology, using the interrupt operator has the advantage that the interrupt behaviour is defined only once, in our example in the process *SessionCancel*. Furthermore, there is a clear syntactic distinction between normal behaviour and what shall happen in an exceptional case. In order to define the 'region' in which the interrupt shall be possible we introduced process names to denote system states. Giving names to system states is a typical technique in specifying in process algebra.

One purpose of sequential composition is to resolve bindings, be it the binding of values to variables (see Sect. 3.2.1 below) or the binding of an interrupt to a process. Here the process *Skip* allows one to define the end point of a binding. Furthermore, this process is useful in deadlock analysis when one wants to indicate successful termination of a finite system run—in contrast to getting stuck in a deadlock, which is equivalent to *Stop*.

The CSP operators of this section expand the grammar as follows:

$$
\begin{aligned}
P, Q ::=\ &\ldots \\
&|\ Skip \\
&|\ P \,\text{⑨}\, Q \\
&|\ P \,\triangle\, Q
\end{aligned}
$$

In the following sections we refrain from dealing with the cancel button in order to keep the example focused upon the new operators used.

Offering Choice to the Environment

Interactive systems, such as an ATM, usually offer several services. In the case of an ATM, these services can include mobile phone top-up, printing a mini-statement, on-screen balance, to name just a few. Here we show how to deal with such choices.

Example 36.4: ATM with Customer Choice

Initially the ATM shows a ready screen. A customer then proceeds to insert a bank card and enters the pin for the card following which the ATM will offer the choice between cash withdrawal and checking the balance.

In case of cash withdrawal, the ATM will deliver the cash to the customer, return the bank card, and get ready for a new session.

In case of checking the balance, the ATM will display the account balance, return the bank card, and get ready for a new session.

CSP external choice, written as $P \ \square \ Q$, offers the initial events from both these processes. Should the environment choose an initial event from P, $P \square Q$ behaves like P, should the initial event be from Q, $P \ \square \ Q$ behaves like Q.

Example 36.5: ATM with Customer Choice in CSP

We extend our alphabet by suitable events, e.g., by adding the events *menu* and *accountBalance* to the type of channel *Display* and by adding a channel *Buttons*, and specify:

$$
\begin{aligned}
ATM4 = \ &Display.ready \rightarrow CardSlot.cardI \\
&\rightarrow KeyPad.pinE \rightarrow Display.menu \\
&\rightarrow (\quad (Buttons.checkBalance \rightarrow Display.accountBalance \\
&\qquad \rightarrow CardSlot.cardO \rightarrow ATM4) \\
&\quad \square \\
&\quad (Buttons.withdrawCash \rightarrow CardSlot.cardO \\
&\qquad \rightarrow CashSlot.cashO \rightarrow ATM4) \\
&\quad)
\end{aligned}
$$

After displaying the menu, $ATM4$ offers two different behaviours. If the customer presses the *checkBalance* button on the *Keypad*, the account balance is displayed, the card returned, and the process starts over again. If, however, the customer presses the *withdrawCash*, the process behaves as seen before.

External choice can also be regarded as reading input from the environment. When the input is read over one channel c, this can be expressed in

CSP by $c?x \rightarrow P(x)$: initially, this process offers to engage in any event $c.e \in events(c)$, when the choice is made it binds x to e and behaves like P . P's behaviour can depend upon the value of the variable x. Such a variable x can be used, e.g., in the condition *cond* of the CSP conditional **if** *cond* **then** P **else** Q, where P and Q are processes. Should *cond* evaluate to true, **if** *cond* **then** P **else** Q behaves like P, otherwise like Q. When one wants to express that one is 'sending' the event e over the channel c, one can write $c!e$ instead of $c.e$.

Example 36.6: Alternative Formulation in CSP

Replacing the external choice operator with channel input and conditional, we specify a process which is equivalent to ATM4:

$$ATM5 \quad = Display!ready \rightarrow CardSlot.cardI$$
$$\rightarrow KeyPad.pinE \rightarrow Display!menu$$
$$\rightarrow Buttons?x$$
$$\rightarrow \textbf{if } x = checkBalance \textbf{ then } Balance \textbf{ else } Withdrawal$$

$$Balance \quad = Display!accountBalance \rightarrow CardSlot!cardO \rightarrow ATM5$$

$$Withdrawal = CardSlot!cardO \rightarrow CashSlot!cashO \rightarrow ATM5$$

Note how we again use process names in order to break up the process *ATM5* into sensible units. Rather than having names for all states, we give names only for those which matter.

Yet another possibility to express the customer choice is the CSP prefix choice operator $?x : A \rightarrow P(x)$, which allows the environment to choose any event e from the set A, binds x to this event, and then behaves like $P(e)$. Using this formulation, we can equivalently rewrite ATM5:

Example 36.7: Third Formulation in CSP

$$ATM5' = Display!ready \rightarrow CardSlot.cardI$$
$$\rightarrow KeyPad.pinE \rightarrow Display!menu$$
$$\rightarrow ?x : \{|Buttons|\}$$
$$\rightarrow \textbf{if } x = Buttons.checkBalance \textbf{ then } Balance$$
$$\textbf{else } Withdrawal$$

Note how the type of x changes in $ATM5'$ compared to $ATM5$.

Tool support varies with regards to the prefix choice operator: CSP-Prover supports it, while FDR does not.

Our above claim that *ATM4* and *ATM5* are equivalent can be established, e.g., by using a model checker, see Sect. 3.3 on tool support of CSP. Such a proof, however, relies on a semantically founded understanding what it means for two processes to be equivalent—see our discussion on semantics below, especially on the three standard models of CSP, see Sect. 3.4.1. Though *ATM4* and *ATM5* are equivalent w.r.t. the three standard models, in general these models have different distinguishing powers.

In this section, we extend our CSP grammar by the following primitives

$$P, Q ::= \quad \ldots$$
$$| \ ?x : A \to P$$
$$| \ c?x \to P$$
$$| \ c!e \to P$$
$$| \ P \ \Box \ Q$$
$$| \ \textbf{if} \ cond \ \textbf{then} \ P \ \textbf{else} \ Q$$

Here, A is a set of events, c is a channel, x is a variable, e is an event, and *cond* is a condition in a logic of choice (not determined by CSP).

Internal Choice, System Composition, Abstraction by Hiding

In the previous section we discussed how to let the environment make a choice among offered events and how these choices determine how a process proceeds. In contrast to this is the situation in which a process makes an independent choice and thereby influences its environment.

We discuss this situation in the context of our ATM example:

Example 36.8: ATM with PIN Verification

Additionally to the above described functionality that only concerns the dialog with the customer, an ATM includes a subsystem which determines if the entered PIN is a valid one. Should the entered PIN be wrong, the ATM informs the user about this, and returns to the ready state. In case the PIN is valid, the ATM proceeds as normal.

Our chosen level of abstraction focuses on the order of events rather on details concerning data. Thus, in the context of the ATM, it is not desirable to model an algorithm for PIN verification; we are interested only in the outcomes, namely if the entered PIN is valid or not. Whenever a choice between different options cannot be made on the information available, it becomes a non-deterministic one, i.e., a choice which the process takes internally, without a rationale that could be derived from the process' observable history. In CSP, the process $P \sqcap Q$ models such behaviour: its behaviour is either as prescribed by process P or as prescribed by process Q.

Example 36.9: Abstract PIN Verification in CSP

Again, we assume a suitable alphabet to be available, e.g., that it includes an event *requestPCheck* and that there is a channel *Check* on which we can communicate events *pinOK* and *pinWrong*. In such a setting *PinVerification* can be formulated as follows:

$$PinVerification = requestPCheck \rightarrow comparePinWithCard$$
$$\rightarrow ((Check.pinOK \rightarrow PinVerification)$$
$$\sqcap$$
$$(Check.pinWrong \rightarrow PinVerification))$$

Upon request the process *PinVerification* decides if a PIN is correct and is immediately ready for the next such check. Here, one can think of the event *requestPCheck* as abstracting the 'data transfer' necessary to provide enough information to perform the PIN verification.

The Oxford English dictionary defines a 'system' as "a set of things working together as part of ... an interconnecting network". Process algebra offers *parallel composition* as a powerful means to assemble a system. In our ATM example, one can consider PIN verification and user dialog as two 'things' connected by a 'network' for 'working together'; coordination is needed on the decision if a PIN is valid, however, for PIN verification it is insignificant if the customer—after entering a valid PIN—chooses to withdraw cash or to check the balance. In the CSP general parallel operator $P \,[\![\, A \,]\!]\, Q$ the *synchronisation* set $A \subseteq \Sigma$ contains the events that the processes P and Q have to agree upon: in order to perform an event $e \in A$ the processes P and Q must both be willing to engage in e, however, P and Q can independently engage in any event $e \notin A$.

Other CSP parallel operators are:

- interleaving $|\!|\!|$: the synchronisation set is empty;
- synchronisation $\|$: the synchronisation set is the whole alphabet; and
- alphabetised parallel $P \,[\, A \,\|\, B \,]\, Q$: the process P can only engage in events from the set A, Q can only engage in events from the set B, the processes P and Q synchronise on the events in the set $A \cap B$.

With regards to concurrency, CSP takes the point of view that it is impossible to observe the parallel execution of events, only one event can happen at any time. This approach is called *interleaving semantics*, in contrast to *true concurrency semantics*.

Example 36.10: System Composition in CSP

The process *UserDialog* is similar to *ATM5* from above:

$$
\begin{aligned}
UserDialog = \; & Display.ready \rightarrow CardSlot.cardI \\
& \rightarrow KeyPad.pinE \rightarrow requestPCheck \\
& \rightarrow ((Check.pinOK \rightarrow Services) \\
& \quad \square \\
& \quad (Check.pinWrong \rightarrow Display.messagePinWrong \\
& \quad \rightarrow UserDialog))
\end{aligned}
$$

The difference is that *UserDialog* requests a PIN check after the PIN has been entered on the keypad. It then offers the environment the choice if the PIN was OK or not. Note that here we use the external choice operator. In the positive case *Services* are offered, in the negative case the user is informed that a wrong PIN was entered, and the *UserDialog* goes back to the ready state. *Services* are as before:

$$
\begin{aligned}
Services \quad\quad &= Display.menu \\
& \rightarrow (CashWithdrawal \; \square \; BalanceCheck)
\end{aligned}
$$

$$
\begin{aligned}
BalanceCheck \quad &= Buttons.checkBalance \rightarrow Display.accountBalance \\
& \rightarrow CardSlot.cardO \rightarrow UserDialog
\end{aligned}
$$

$$
\begin{aligned}
CashWithdrawal &= Buttons.withdrawCash \rightarrow CardSlot.cardO \\
& \rightarrow CashSlot.cashO \rightarrow UserDialog
\end{aligned}
$$

Having these two 'things' specified, namely *PinVerification* and *User-Dialog*, we compose *ATM6* to be the 'system', or, in the language of CSP, the process that runs *UserDialog* and *PinVerification* in parallel:

$$
\begin{aligned}
ATM6 = \; & UserDialog \\
& [| \{| requestPCheck, Check |\} |] \\
& PinVerification
\end{aligned}
$$

In order to define the synchronisation set, we make use of the $\{|\_|\}$ operator, which extracts the events from a given list of channels—see Sect. 3.2.1 above. By abuse of notation, in the CSP context it is common (and also supported by tools) to apply this operator also to single events such as *requestPCheck*.

In *ATM6* the communication between the user dialog and the PIN verification is visible to and possibly open to manipulation by the outside world. In order to avoid this, we encapsulate the dialog by hiding all events in the synchronisation set between these processes:

$$
ATM7 = ATM6 \setminus \{| requestPCheck, Check, comparePinWithCard |\}
$$

The process $P \setminus A$ behaves like the process P where the events from the set $A \subseteq \Sigma$ are turned into internal ones, which are invisible to the outside.

In this section, we extend our CSP grammar by the following process constructions:

$$P, Q ::= \ \dots$$
$$| \ P \sqcap Q$$
$$| \ P \, [| \, A \, |] \, Q$$
$$| \ P \, [\, A \parallel B \,] \, Q$$
$$| \ P \, ||| \, Q$$
$$| \ P \parallel Q$$
$$| \ P \setminus A$$

where $A, B \subseteq \Sigma$ are a sets of events.

Parametrised Processes

Most ATMs offer the customer several attempts to enter the correct PIN.

Example 36.11: Three Attempts to Enter the Correct PIN

The ATM behaves as before, but now the customer has three attempts to enter the correct PIN: in case that the PIN is valid, the customer is offered the functionality as described before; in case that the PIN is incorrect and the customer still has an attempt left, the ATM displays that the entered PIN was wrong and gives the customer another opportunity to enter the PIN; in case that the PIN is incorrect and the customer has no attempt left, the ATM displays that the entered PIN was wrong, informs the customer that the card is kept, and returns to the ready screen.

We model this behaviour in CSP by defining the process *PinCheck* which takes the number of attempts that the customer has left as its parameter:

$$UserDialog = Display.ready \rightarrow CardSlot.cardI$$
$$\rightarrow PinCheck(3)$$
$$PinCheck(n) = KeyPad.pinE \rightarrow requestPCheck$$
$$\rightarrow ((Check.pinOK \rightarrow Services)$$
$$\square$$
$$(Check.pinWrong$$
$$\rightarrow \textbf{if} \ (n = 1)$$
$$\textbf{then} \ Display.messagePinWrong$$
$$\rightarrow Display.cardSwallowed$$
$$\rightarrow UserDialog$$
$$\textbf{else} \ Display.messagePinWrong$$
$$\rightarrow PinCheck(n - 1))$$
$$)$$

All other processes are as above, as is the overall system composition.

$ATM8 = UserDialog$
$\qquad [|\,\{|\,requestPCheck,\ Check\,|\}\,|]$
$\qquad PinVerification$

$ATM9 = ATM8 \setminus \{|\,requestPCheck,\ Check,\ comparePinWithCard\,|\}$

Process parameters structure the name space relative to which the language CSP is defined. In the above example n ranges over the integers, i.e., we introduce the (countably many) names

$\qquad\ldots,\ Pincheck(\text{-}1),\ PinCheck(0),\ PinCheck(1),\ \ldots$

For each of these, the equation $PinCheck(n) = \ldots$ defines how the process behaves. In $ATM8$, however, only finitely many of these are reachable, namely $PinCheck(1)$, $PinCheck(2)$, and $PinCheck(3)$. Processes can have an arbitrary but finite number of parameters.

Any datatype can serve as a process parameter. The values of this type can be part of the alphabet of communications, but—as the above example demonstrates—this is not necessarily the case.

Process parameters use expressions of a datatype-specific sublanguage, with type-specific operands and operators, as we know them from primitive data types (int, bool, ...) and abstract data types like lists, stacks, sets, bags. The general convention is that the type-specific sublanguage is not part of the CSP syntax. This is similar to UML/SysML, where expressions for guards, operation bodies, and general actions are not part of the UML/SysML, but 'opaque expressions' that are interpreted in the context of the expression language.

As illustrated in the example, the process parameter x of a process name N on the left hand side on a process equation $N(x) = P(x)$ can be used as a variable in the process $P(x)$ on the right hand side.

It is not necessary to expand our CSP grammar at this point. Introducing process parameters provides more detail how the elements of the set of process names PN look like. As the CSP grammar is defined relatively to PN, it needs no change.

Renaming—More Than Just an Adaption of Names

Some ATMs can deal with different currencies, e.g., those placed at airports.

Example 36.12: Different Currencies via Renaming

The ATM behaves as before, but the user dialog offers the choice to withdraw money in Euro, Sterling, or Dollar.

This feature can be modelled using the CSP renaming operator
$[\![\_]\!]$. Here, we expand the process on the righthandside of the equation for *CashWithdrawal* by three options to rename the event *Buttons.withdrawCash*:

$$CashWithdrawal = (Buttons.withdrawCash \rightarrow CardSlot.cardO$$
$$\rightarrow CashSlot.cashO \rightarrow UserDialog)[\![R]\!]$$

where the relation R is given as

$$R = \{(Buttons.withdrawCash, Buttons.withdrawEuro),$$
$$(Buttons.withdrawCash, Buttons.withdrawSterling)$$
$$(Buttons.withdrawCash, Buttons.withdrawDollar)\}$$

The effect is that the customer can choose among the *Buttons* to withdraw money in *Euro, Sterling,* or *Dollar*. In each case, the card is returned, cash is delivered, and the ATM starts over with the User-Dialog.

This new process *CashWithdrawal* is the only change that we make to *ATM8* and *ATM9* in order to obtain *ATM10* and *ATM11*: it nicely illustrates the power of a one to many renaming.

Looking at the example from a point of modelling, naturally there are other, equivalent ways of how the desired effect can be achieved: as with programming languages, also in specification languages there are many different, equivalent ways to express one behaviour. Concerning the chosen level of abstraction, one might criticise that the resulting ATMs are kind of 'imbalanced': while the user can choose between different currencies, the machine just returns 'cash', not further differentiated into different currencies. The reader might want to try to remedy this and develop the example further.

In specification in general, renaming allows one to adjust the name-space from one development to the name-space used in a different development, i.e., renaming supports the re-use of specifications. In the context of CSP, renaming can also be used to duplicate (see above) or to reduce behaviour.

Given a binary relation R on the alphabet of communications Σ, the process $P[\![R]\!]$ behaves like P, however, whenever P is willing to engage in an event $a \in \Sigma$, $P[\![R]\!]$ is willing to engage in all the events in $\{e \in \Sigma \mid (a,e) \in R\}$. The process $P[\![R]\!]$ is well-formed only for relations R with the property that for all $e \in \Sigma$ there exists $e' \in \Sigma$ with $(e,e') \in R$. In practice, one often defines the relation R by stating only those pairs $(e,e') \in R$ where $e \neq e'$.

In this section, we extend our CSP grammar by one process construction:

$$P ::= \quad \dots$$
$$\mid P[\![R]\!]$$

with $R \subseteq \Sigma \times \Sigma$ such that for all $e \in \Sigma$ there exists $e' \in \Sigma$ with $(e,e') \in R$.

Replicated Processes

Upon starting an ATM first determines how many notes there are in its various cash cartridges.

Example 36.13: ATM with Initialisation

The ATM has different cartridges for £5, £10 and £20 notes. Each cartridge can determine how many notes it holds. When the ATM is started, it first reads from the cartridges how many notes are available. Then it behaves as described in Example 36.12.

For a single denomination d, say for £5 notes, the counter process just communicates the quantity of notes of denomination d.

$$CartridgeCounter(d) = Quantity.d \rightarrow CartridgeCounter(d)$$

The cartridges work in parallel, they are independent of each other, i.e., we can interleave the processes. We instantiate the process *Cartrige-Counter(x)* for all elements of the set of denominations $\{five, ten, twenty\}$ and combine the resulting processes by interleaving:

$$CartridgeCounterS = |||_{x:setofDenominations} CartridgeCounter(x)$$

We do not care about the order in which the ATM reads from the cartridges. However, we want to ensure that it reads from all of them once. The parameter X of the process *Init* is the set of denominations, for which the quantity has still to be read. We instantiate the process $Quantity.x \rightarrow Init(X \backslash \{x\})$ for all elements of the set X and combine the resulting processes by external choice:

$$Init(X) = (\Box_{x:X} \ Quantity.x \rightarrow Init(X \backslash \{x\}))$$
$$\Box$$
$$(X = \{\}) \ \& \ ATM7$$

When the Boolean guard $X = \{\}$ becomes true, the process behaves as *ATM7*.

The final system consists of the cartridges running in parallel with the initialisation process:

$$ATM12 = CartridgeCounterS$$
$$[| \{| Quantity |\} |]$$
$$Init(setofDenominations)$$

From a modelling point of view, again this example can be criticised to be 'imbalanced': first we collect information on how many bank notes the ATM has in different cartridges, later this information is not used at all. The reader might want to try to remedy this and develop the example further.

Replicated process operators combine a family of (parametrised) processes $(P(x))_{x \in I}$ over some index set I into one process, extending the underlying binary operator to an operator that takes $|I|$ many arguments. Replication is possible for a number of binary CSP operators, including external choice, internal choice, interleaving, general parallel, and alphabetised parallel.

The general pattern is that, given a binary operator op for which replication is possible, an expression

$$op_{x:\{a_1, \ldots, a_n\}} P(x)$$

expands to

$$P(a_1) \; op \; \ldots \; op \; P(a_n) \; op \; Q$$

where Q is some suitably chosen 'neutral' process—in the case of the general parallel we have $Q \triangleq Skip$—see Sect. B.2.2 for the details.

Note that the set index set $\{a_1, \ldots, a_n\}$ might vary during process execution. This is the case in the above process $Init(X) : X$ is the set of cartridges that still have to communicate the number of notes available.

In the case of parallel operators, replicated process operators create $|I| + 1$ many parallel processes. In contrast to this, in the case of choice operators, replicated process operators create choice among $|I| + 1$ many processes. After this choice has been resolved, there is only one line of execution.

As syntactic sugar, CSP also includes a Boolean guard process $cond \,\&\, P$ which expands to **if** $cond$ **then** P **else** $Stop$.

In this section, we extend our CSP grammar by the following process constructions:

$$P, Q ::= \;\; \ldots$$
$$| \; cond \,\&\, P$$
$$| \; \square_{x:I} \; P(x)$$
$$| \; \sqcap_{x:I} \; P(x)$$
$$| \; |||_{x:I} \; P(x)$$
$$| \; [\![\, A \,]\!]_{x:I} \; P(x)$$
$$| \; [\![\, A(x) \,]\!]_{x:I} \; P(x)$$

where $cond$ is a condition in a logic of choice (not determined by CSP), I is an index set, $A \subseteq \Sigma$ is a set of events, $(P(x))_{x \in I}$ is a family of processes, $(A(x))_{x \in I}$ is a family of sets with $A(x) \subseteq \Sigma$ for all $x \in I$.

The process $\sqcap_{x:A} P(x)$ is well-formed for $A \neq \emptyset$. The processes $|||_{x:I} P(x)$, $[\![\, A \,]\!]_{x:I} P(x)$, and $[\![\, A(x) \,]\!]_{x:I} P(x)$ are wellformed only for finite index sets I. The process $[\![\, A(x) \,]\!]_{x:I}$ additionally requires that, for all $x \in I$, the alphabet of $P(x)$ is a subset of $A(x)$.

In the case of external and internal choice, replication allows to combine infinite families of processes into one process. In the case of the replicated parallel operators, suitable algebraic laws concerning commutativity and associativity ensure that the order in which the processes are combined does not matter.

3.2.2 Understanding the Semantics—Modelling a Jet Engine Controller

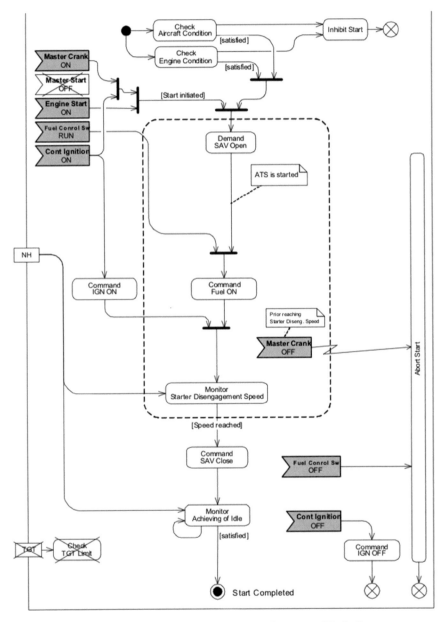

Fig. 3.1 Activity diagram for manual ground start—Courtesy of Rolls-Royce

Closely following an industrial case study on a jet engine controller [HKRS09], we develop CSP's operational and the denotational semantics.

Example 37: The Starting System of a Jet Engine

Starting a physical jet engine essentially involves three steps:
1. A motor—the so-called Starter—which is mechanically coupled to a shaft of the engine, starts to rotate the engine.
2. When the shaft has reached a sufficient rotational speed, fuel is allowed to flow to the combustion chamber and the engine lights up.
3. When the rotational speed of the engine reaches a threshold the engine start is complete.

In modern aeroplanes, these steps are initiated and monitored by an *Electronic Engine Controller* (EEC). Such an EEC encapsulates all signalling aspects of the engine; it controls, protects, and monitors the engine. Its control loop involves: reading data from sensors and other computer systems in the aircraft, receiving commands from the pilot, calculating new positions of the engine actuators, and issuing commands to the engine actuators. In its monitoring function it transmits data about the engine condition and information on any failures diagnosed on the electronics back to the aircraft. Here we focus on the *Starting System*, one of the EEC's many functionalities.

Example 37.1: Manual Ground Start

Figure 3.1 shows the internal logic of a so-called manual ground start in the form of an activity diagram. These activity diagrams are formulated in an informal, graphical specification language. This language was specifically developed by Rolls-Royce in order to describe engine controllers. This language uses symbols with the following meaning:

●	Start point of the activity diagram
◉	End point of the activity diagram
⬭	Box–Used for encoding states as well as activities
↓ ↓ / ↓	Transition–checks for conditions
⊗	Error state
▷	Switch in the cockpit
⊠	Switch in the cockpit, ignored by this activity digram
▭	Displayed signal in the cockpit
┊╌╌╌┊	Control flow in the EEC

This table was established by discussion with the Rolls-Royce engineers.

In system design, one often considers what steps are needed in order to reach a certain goal. In our case, one is interested in the sequence of events that actually starts the engine and in the possibilities to interrupt this procedure:

Example 37.2: Use Cases of Starting the Jet Engine

Typical use-cases of this activity diagram are:

Use-Case 1: "Normal use"   Upon switching the *Engine Start* to On, the EEC will command the Starter Air Valve (*SAV*) to be opened and the starter motor is activated. If the pilot now switches the *Fuel Control Switch* to Run the EEC commands the fuel to flow. If at this point the *Continuous Ignitions* is still On the EEC ignites the motor (not shown in the Figure) and begins to monitor the shaft speed of the engine. Should this speed reach a certain threshold the starting procedure is complete.

Use-Cases: "Interruptions"   While the starting procedure is executed by the EEC, the pilot can abort it by switching the *Master Crank* or the *Fuel Control* to Off. If the pilot switches the *Continuous Ignitions* to Off the starting procedure ends in an error state.

Let us analyse, what the physical entities of our system are:

Example 37.3: Physical Components

The Manual Ground Start diagram in Fig. 3.1 involves the interaction of different, independently acting physical components: the first component consists of the buttons, which the pilot switches in the cockpit. The second component is the aeroplane electronics, which checks for the aircraft condition and for the engine condition. The third component is the decision flow to be implemented in the EEC.

In the following, we model the first and the third component in the language CSP_M [Sca98]. Our model includes an interface for reading reports from the aeroplane electronics. CSP_M is a machine readable version of CSP which also includes language constructs for the description of data. This means especially that the alphabet of communications needs to be constructed as well. The FDR webpages https://cocotec.io/fdr/ provide a syntax reference for CSP_M.

We begin with the first component, namely the buttons that the pilot can use to initiate or abort the starting process. Some of these are *Switch Buttons*. Such buttons have two states: ON and OFF. Pressing a button in state OFF will turn it ON, releasing a button in state ON will turn it OFF. In CSP_M we can model such a button in the following way:

> **Example 37.4: Switch Buttons in CSP**
>
> ```
> channel press, release
> ButtonOFF = press -> ButtonON
> ButtonON = release -> ButtonOFF
> ```

After the CSP$_M$ keyword `channel`, one can declare a list of events. In our example, we declare the events `press` and `release`. Each channel declaration augments the currently defined alphabet by adding the defined constants to it.

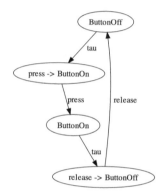

Fig. 3.2 Transition system of `ButtonOFF`

Our CSP specification behaves like the automaton shown in Fig. 3.2. This connection is made more precise by the operational CSP semantics. The CSP operational semantics takes CSP processes as states and defines transition rules between them using firing rules:

$$\frac{}{(a \to P) \xrightarrow{a} P} \qquad \frac{}{N \xrightarrow{\tau} P} \qquad \text{if there is an equation } N = P$$

Provided that the preconditions (the text above the line and besides the rule) of such a rule are fulfilled, there is a labelled transition between the two states shown in the conclusion (the text below the line). There is exactly one state for each process. The *operational semantics*, cf. Sect. 1.1.2, of a given CSP process expression P is the smallest automaton generated from P by the given rules. Often we call this automaton also a transition system. Here, $\tau \notin \Sigma$ is the so-called 'silent' event. τ represents a non observable, i.e., internal, step of the automaton. The transition from the state representing a process name, take for example `ButtonOff`, to the state representing the rhs of its defining equation, in our example `press -> ButtonOn`, is considered to be an non-observable one, i.e., we label it with τ. Section B.3 summarises the CSP firing rules.

Following the edges of this automaton, we obtain the following observations:

1. $\langle\rangle$,
2. $\langle\text{press}\rangle$,
3. $\langle\text{press, release}\rangle$,
4. $\langle\text{press, release, press}\rangle$,
5. ...

Here, we treat the label τ like the empty word with regards to concatenation, i.e., it never appears in the list of observed events. Summarising, the observations are

$$Obs = \{w \mid w \text{ is prefix of a word in } (\text{press release})^*\}$$

where $(\text{press release})^*$ is a regular expression with * as notation of Kleene's star. The *denotational semantics*, cf. Sect. 1.1.2, of a given CSP process expression P describes the set of all possible observations. In our case, we are interested in the sequences or traces that a process can perform. To this end, we define a function $traces_M(P)$ by structural induction on the grammar of process terms. As a parameter, it takes an interpretation M of the process names $N \in PN$:

$$traces_M(a \to P) \triangleq \{\langle\rangle\} \cup \{\langle a\rangle \frown s \mid s \in traces_M(P)\}$$
$$traces_M(N) \quad \triangleq \quad M(N)$$

In the context of CSP, one usually writes the sign $\frown$ for the concatenation of two traces. As seen above, $\langle\rangle$ stands for the empty trace.

For a process equation $N = P$ we require that

$$traces_M(N) = M(N) = traces_M(P).$$

Section 3.4.3 will discuss the semantics of such (potentially) recursive equations. As semantics is defined using an equation, the typical three questions appear:

1. Is there a solution to the equation?
2. Is the solution unique?
3. How can we construct the solution (should it exist)?

In the case that the solution exists, we write the semantics of N of a process name as

$$traces(N)$$

i.e., without referring to the process interpretation M.

Coming back to our case study, we see that the pilot has several switch-buttons in order to control the engine start. In order to model these, we instantiate `ButtonON` and `ButtonOFF` to form the different switch buttons available in the cockpit for the Starting System. This is done by simply using the CSP renaming operator, e.g., for `MasterCrank`:

Example 37.5: Master Crank Button in CSP.

```
channel mc_press, mc_release
MasterCrank =
  ButtonOFF[[press <- mc_press, release <- mc_release]]
```

The operational semantics of renaming is

$$\frac{P \xrightarrow{a} P'}{P[\![R]\!] \xrightarrow{c} P'[\![R]\!]} \; if \, aRc \qquad\qquad \frac{P \xrightarrow{\tau} P'}{P[\![R]\!] \xrightarrow{\tau} P'[\![R]\!]}$$

Its traces are given as

$$traces_M(P[\![R]\!]) \triangleq \{t \mid \exists s \in traces_M(P) . sR^*t\}$$

Here, sR^*t is the lifting of R from events to traces of events, see Sect. B.4.1 for its definition.

Example 37.6: Independent Buttons in CSP

The pilot can arbitrarily press and release the buttons shown in Fig. 3.1. They are independent of each other. Thus, we model them as interleaved processes:

```
Buttons =  MasterCrank ||| MasterStart ||| EngineStart
           ||| FuelControl ||| ContIgnition
```

The textual explanation of the diagram in the Rolls-Royce documentation makes it clear that `EngineStartOn` is a push button with only one state:

```
EngineStart = engineStartOn -> EngineStart
```

Here, we see one typical benefit of formal modelling: the used Formal Methods forces one to clarify details. In our case, modelling in CSP requires us to answer the question: "which events can happen in the system?" In case of a Switch Button the answer is: `press` and `release`, in case of a Press Button the answer is: `press` only. This leads to the insight, that the diagram in Fig. 3.1 depicts the buttons in a wrong way: it shows all of them with the same rendering, although they are different.

We give here the firing rules for the general parallel operator. As seen above, the interleaving operator can be defined in terms of general parallel operator: $P \,|||\, Q \triangleq P \,[\![\emptyset]\!]\, Q$.

Internal events can be performed independently:

$$\frac{P \xrightarrow{\tau} P'}{P\,[\!|\,A\,|\!]\,Q \xrightarrow{\tau} P'\,[\!|\,A\,|\!]\,Q} \qquad \frac{Q \xrightarrow{\tau} Q'}{P\,[\!|\,A\,|\!]\,Q \xrightarrow{\tau} P\,[\!|\,A\,|\!]\,Q'}$$

Events outside the synchronisation set A can be performed independently:

$$\frac{P \xrightarrow{a} P'}{P\,[\!|\,A\,|\!]\,Q \xrightarrow{a} P'\,[\!|\,A\,|\!]\,Q}\,a \in \Sigma\backslash A \qquad \frac{Q \xrightarrow{a} Q'}{P\,[\!|\,A\,|\!]\,Q \xrightarrow{a} P\,[\!|\,A\,|\!]\,Q'}\,a \in \Sigma\backslash A$$

Events in the synchronisation set A need to be synchronised:

$$\frac{P \xrightarrow{a} P' \quad Q \xrightarrow{a} Q'}{P\,[\!|\,A\,|\!]\,Q \xrightarrow{a} P'\,[\!|\,A\,|\!]\,Q'}\,a \in A$$

The denotational semantics is defined as follows:

$$traces_M(P\,[\!|\,A\,|\!]\,Q) \triangleq \bigcup\{s\,[\!|\,A\,|\!]\,t \mid s \in traces_M(P) \wedge t \in traces_M(Q)\}$$

For each pair of traces $s \in traces_M(P)$ and $t \in traces_M(Q)$ one forms a set $s\,[\!|\,A\,|\!]\,t$. Such 'overloading' of operators is typical for denotational semantics in general. The above clause uses the notation $\bigcup\{x_i \mid i \in I\} \triangleq \bigcup_{i \in I} x_i$. The function $\_\_\,[\!|\,A\,|\!]\,\_\_$ on traces has to mirror all possible combinations of when a process can make progress. It is inductively defined by:

$$\langle x\rangle \frown t_1\,[\!|\,A\,|\!]\,\langle x\rangle \frown t_2 \triangleq \{\langle x\rangle \frown u \mid u \in t_1\,[\!|\,A\,|\!]\,t_2\}$$
$$\langle x\rangle \frown t_1\,[\!|\,A\,|\!]\,\langle x'\rangle \frown t_2 \triangleq \emptyset$$
$$\langle x\rangle \frown t_1\,[\!|\,A\,|\!]\,\langle y\rangle \frown t_2 \triangleq \{\langle y\rangle \frown u \mid u \in \langle x\rangle \frown t_1\,[\!|\,A\,|\!]\,t_2\}$$
$$\langle x\rangle \frown t_1\,[\!|\,A\,|\!]\,\langle\rangle \triangleq \emptyset$$
$$\langle y\rangle \frown t_1\,[\!|\,A\,|\!]\,\langle x\rangle \frown t_2 \triangleq \{\langle y\rangle \frown u \mid u \in t_1\,[\!|\,A\,|\!]\,\langle x\rangle \frown t_2\}$$
$$\langle y\rangle \frown t_1\,[\!|\,A\,|\!]\,\langle y'\rangle \frown t_2 \triangleq \{\langle y\rangle \frown u \mid u \in t_1\,[\!|\,A\,|\!]\,\langle y'\rangle \frown t_2\} \cup$$
$$\{\langle y'\rangle \frown u \mid u \in \langle y\rangle \frown t_1\,[\!|\,A\,|\!]\,t_2\}$$
$$\langle y\rangle \frown t_1\,[\!|\,A\,|\!]\,\langle\rangle \triangleq \{\langle y\rangle \frown u \mid u \in t_1\,[\!|\,A\,|\!]\,\langle\rangle\}$$
$$\langle\rangle\,[\!|\,A\,|\!]\,\langle x\rangle \frown t_2 \triangleq \emptyset$$
$$\langle\rangle\,[\!|\,A\,|\!]\,\langle y\rangle \frown t_2 \triangleq \{\langle y\rangle \frown u \mid u \in \langle\rangle\,[\!|\,A\,|\!]\,t_2\}$$
$$\langle\rangle\,[\!|\,A\,|\!]\,\langle\rangle \triangleq \{\langle\rangle\}$$

where $x, x' \in A \cup \{\checkmark\}$, $y, y' \notin A \cup \{\checkmark\}$, $x \neq x'$, and $t_1, t_2 \in \Sigma^* \cup \Sigma^{*\checkmark}$. The event $\checkmark$ will be discussed below. From now on, we refrain from presenting the semantical clauses for the CSP traces semantics and refer the reader to Sect. B.4.

Going on with our case study, we check next if the second component gives an OK for the start:

Example 37.7: Communicating with the Aeroplane Electronics

```
channel aircraftCondition: Bool
channel engineCondition: Bool
channel inhibitStart, startOK

CheckConditions =
          aircraftCondition ? ac
       -> engineCondition ? ec    -> Checking(ac,ec)
    []    engineCondition ? ec
       -> aircraftCondition ? ac -> Checking(ac,ec)

Checking(ac,ec) = if (ac and ec)
                  then startOK -> SKIP
                  else InhibitStart

InhibitStart = inhibitStart -> SKIP
```

We read the aircraft condition and the engine condition in arbitrary order. To this end, we use *communication channels*. In CSP_M, such channels have a type, in our case the type `Bool`.

The operational rules on communication involving channels are:

$$\frac{}{(c?x \rightarrow P(x)) \xrightarrow{c.a} P(x)[a/x]} a \in comms(c) \qquad \frac{}{(c!a \rightarrow P) \xrightarrow{c.a} P}$$

Here, *comms* are the values that one can communicate over the channel c, in our example the Booleans; $[a/x]$ stands for substituting a for x. Often, channels are treated as syntactic sugar in CSP.

Example 37.8: Decomposing the Starting Sequence in CSP

The starting sequence can only proceed when the following events happens: (1) the checks for *Aircraft* and *Engine* condition have been successful, (2) the pilot has issued the necessary starting commands. This is captured in the CSP model in the following way:

```
(CheckConditions [| {|startOK|} |] StartInteractionEEC)
  \ { startOK };
((Region ||| EngineStart)  /\ Interrupts)

Interrupts =     fc_release -> abortStart -> STOP
           []  ci_release -> commandIGNoff -> STOP
```

Here, `CheckConditions` is the above process that checks for the *Aircraft* and *Engine* condition. These checks run independently of the `StartInteractionsEEC`, which model the pilot's input, namely the cranking and the ignition commands. `StartInteractionsEEC` is similar to `CheckConditions`, thus we omit its code.

Only when both checks have successfully been completed (signalled by the event `startOK`), the dialogue with the engine can start. The event `startOK` does not belong to the vocabulary of the EEC as shown in Fig. 3.1. Thus, we better hide it. `Region`, finally, captures the actual starting sequence. We will discuss its details below.

We interpret interrupt events `fc_release`, and `ci_release` to become 'active' only when the starting sequence has begun. We will discuss the semantics of the interrupt operator below.

Naturally, the pilot will be able to press the button `EngineStartOn` at any point of time during the starting sequence. From here onwards, however, it does not have an effect at all: we capture this be adding this button's behaviour as an interleaving process.

Termination is an observable event in CSP. It is written $\checkmark$ (pronounced "tick"). The process *Skip* terminates immediately:

$$\overline{Skip \xrightarrow{\checkmark} \Omega}$$

where Ω is a special process term that is intended to represent any terminated process. Using these notations, we can extend the operational semantics of the renaming operator and the parallel operator.

Renaming has no effect on termination:

$$\frac{P \xrightarrow{\checkmark} P'}{P[\![R]\!] \xrightarrow{\checkmark} \Omega}$$

Concerning the parallel operator, we state first of all that each process can terminate independently:

$$\frac{P \xrightarrow{\checkmark} \Omega}{P [\![A]\!] Q \xrightarrow{\tau} \Omega [\![A]\!] Q} \qquad \frac{Q \xrightarrow{\checkmark} \Omega}{P [\![A]\!] Q \xrightarrow{\tau} P [\![A]\!] \Omega}$$

If both processes have terminated, indicated by the special state Ω, their parallel composition can do so as well:

$$\overline{\Omega [\![A]\!] \Omega \xrightarrow{\checkmark} \Omega}$$

Now let's study sequential composition $P \,\mathring{,}\, Q$. As long as P can perform events different from $\checkmark$, this is what the combined process performs. Only, when P terminates, the second process takes over:

$$\frac{P \xrightarrow{x} P'}{P \,\mathring{,}\, Q \xrightarrow{x} P' \,\mathring{,}\, Q} x \neq \checkmark \qquad \frac{P \xrightarrow{\checkmark} P'}{P \,\mathring{,}\, Q \xrightarrow{\tau} Q}$$

Note, that we again use τ in order to provide a label for the control flow from P to Q.

Hiding is simple when it comes to the operational semantics. Events to be hidden are turned into a τ, all others remain:

$$\frac{P \xrightarrow{x} P'}{P \setminus A \xrightarrow{\tau} P' \setminus A} x \in A \qquad \frac{P \xrightarrow{a} P'}{P \setminus A \xrightarrow{a} P' \setminus A} a \notin (A \cup \{\checkmark\})$$

Hiding has no effect on termination, but we follow again the convention that Ω represents all terminated processes (this is necessary, as the operational rule for the termination of the parallel operator recognizes the termination of the subprocesses by pattern matching with Ω):

$$\frac{P \xrightarrow{\checkmark} P'}{P \setminus X \xrightarrow{\checkmark} \Omega}$$

Example 37.9: The Actual Starting Procedure in CSP

After the pilot initiated the start and the aeroplane electronics agreed that it is OK to start, the actual start procedure can follow:

```
datatype SAVMode = open | close
channel sav:SAVMode

StartInit =  sav.open -> fc_press -> Fuel
          [] fc_press -> sav.open -> Fuel

channel commandFuelON, commandIgnON

Fuel =  commandFuelON -> commandIgnON  -> SKIP
     [] commandIgnON  -> commandFuelON -> SKIP

Region = (StartInit /\
             (mc_release -> abortStart -> STOP));
         (MasterSpeed
          |||
          (ButtonON  [[press < - mc_press,
                       release < - mc_release]]
          )
         )
```

The expected flow of events is that sav.open and fc_press happen in any order. Then the process Fuel shall take over. Finally, the MasterSpeed process is called, which models the final phase of the starting procedure.

While the `StartInit` process is running, it is possible to interrupt it with a `mc_release` command. However, when `StartInit` process has terminated, this interrupt becomes disabled. The pilot, however, still can release the MC, button— however now with no effect anymore: as above, we model this by adding a process in interleaving.

In CSP_M it is possible to create new types using the keyword `datatype`. In our simple case, this type consists of the two constants `open` and `close`. The vertical bar | separates the alternatives of the CSP_M datatype construct.

The external choice operator chooses based on a visible event $b \in \Sigma \cup \{\checkmark\}$:

$$\frac{P \xrightarrow{b} P'}{P \,\Box\, Q \xrightarrow{b} P'} \qquad\qquad \frac{Q \xrightarrow{b} Q'}{P \,\Box\, Q \xrightarrow{b} Q'}$$

An internal event leaves the choice unresolved:

$$\frac{P \xrightarrow{\tau} P'}{P \,\Box\, Q \xrightarrow{\tau} P' \,\Box\, Q} \qquad\qquad \frac{Q \xrightarrow{\tau} Q'}{P \,\Box\, Q \xrightarrow{\tau} P \,\Box\, Q'}$$

At this point we can also illustrate the difference between the external choice and the internal choice operator. The latter resolves the choice by performing an internal event:

$$\frac{}{P \sqcap Q \xrightarrow{\tau} P} \qquad\qquad \frac{}{P \sqcap Q \xrightarrow{\tau} Q}$$

Note that in the traces semantics it is not possible to distinguish between internal and external choice. The semantical clauses are identical:

$$traces_M(P \,\Box\, Q) \triangleq traces_M(P) \cup traces_M(Q)$$
$$traces_M(P \sqcap Q) \triangleq traces_M(P) \cup traces_M(Q)$$

More sophisticated denotational CSP semantics, e.g., the failures/divergences semantics $\mathcal{N}$ and the stable failures semantics $\mathcal{F}$, make more concise observations and are both able to distinguish between the two choice operators.

In an interrupt situation $P \triangle Q$, the control remains with P as long as the events are carried out with P. The moment, an observable event comes from Q, however, the process Q takes over:

$$\frac{P \xrightarrow{x} P'}{P \triangle Q \xrightarrow{x} P' \triangle Q} x \in \Sigma \cup \{\tau\} \qquad\qquad \frac{Q \xrightarrow{y} Q'}{P \triangle Q \xrightarrow{y} Q'} y \in \Sigma$$

This results in a non-deterministic situation, should P offer the first events possible for Q. The moment P or Q terminate, so does $P \triangle Q$:

$$\frac{P \xrightarrow{\checkmark} P'}{P \triangle Q \xrightarrow{\checkmark} \Omega} \qquad\qquad \frac{Q \xrightarrow{\checkmark} Q'}{P \triangle Q \xrightarrow{\checkmark} \Omega}$$

This brings us to the end of our case study: we have seen all ingredients necessary to model the manual ground start functionality of an EEC. Here, we summarise our modelling approach:

Example 37.10: Modelling Rules

1. Commands stated in oval boxes are modelled as events.
2. Transitions are checked in arbitrary order, i.e., it does not matter if the left event or the right event happens first.
3. Components are modelled only once, i.e., there is one push button process, which is then duplicated via renaming.
4. Same behaviour is modelled identically ("Monitor disengagement speed" and "Monitor achieving idle" have the same control flow, are depicted differently in Fig. 3.1, however, are modelled identically in the CSP code).
5. The buttons in the cockpit are independent of each other. Furthermore, the buttons, the EEC, the aeroplane electronics, and the physical engine are independent entities.
6. Interrupts are realised via the CSP interrupt operator.

Such kind of rule set is useful whenever one models a concrete system. It allows one to trace modelling decisions. Furthermore, it can serve as a reference point when discussing the question: "where do the axioms come from?"

Concerning our findings with regards to the jet engine controller, we can state:

Example 37.11: Evaluation

Rolls-Royce uses activity diagrams as shown in Fig. 3.1 merely as memos. The engineers share a common understanding of jet engines, the activity diagrams serve more to trigger knowledge how the control software works. Here, we list some of the shortcomings that we encountered during the modelling:

- Although the *Engine Start* is a momentary button and *Master Crank* is a push button with two states both are shown with the same symbol in the activity diagram. That the *Engine Start* is a momentary button becomes clear from the textual description of the activity diagram. This explains also why there is no interrupt related to this button.
- Although the commands *Command IGN ON* and *Command IGN OFF* appear at first sight to be related, they are not: the command *Command IGN ON* is given by the pilot in the cockpit while the command *Command IGN OFF* is sent by the EEC to the engine. Therefore, we model these commands via two different channels.
- As there is a command *Command FUEL ON* one would expect command *Command FUEL OFF* to appear in the activity diagram, e.g., when aborting the start. However, this is not the case.

Concerning the *suitability* of CSP, on the positive side we can mention that the various CSP operators came very handy in the modelling process. The interleaving operator, the sequential composition and the

---

hiding operator allowed us to capture many system aspects in an elegant way. On the negative side, the global state approach of CSP forced us to explicitly have one process name per transition (arrow in the activity diagrams). This allowed us to take care of or ignore state changes of the buttons while following the control flow of the activity diagram. Overall, however, CSP served well in modelling such a controller.

Modelling a system happens usually with a purpose. In our case the purpose was to create a formal model in order to test the controller. For more details, see [HKRS09]. Purposes for modelling include:

- Better understanding of the system (by, e.g., making it accessible to simulation).
- Clarification of an informal specification.
- Testing.
- Verification.
- Performance analysis.

In this section, we were using predominately the operational semantics to provide an understanding of the various CSP operators. However, "Historically, the operational semantics of CSP was created to give an alternative view to the already existing denotational models rather than providing the intuition in the original design as it has with some other process algebras such as CCS." [Ros98].

3.2.3 Understanding Refinement—Modelling Buffers

Buffers are a commonly used notion in the theory of computer science. Albeit simple, the concept is crucial to the design of data processing and communication. Buffers essentially serve as memory, temporarily holding some data while being transferred from one place to another, usually as part of some distributed computation within a local or networked system. Such a transfer may involve data being moved to or from an I/O device, or transmitting or receiving data on a communication network:

Example 38: Buffers

A buffer is typically characterised to serve data in the order of arrival, useful where some queue of data is to be transferred. It operates on the principles of first in first out (FIFO), assuming no reordering or loss. Data may also be read and written at a different and variable rate. This stands in contrast with related notions of cache and stack: while cache is designed to allow data being written once and read multiple times, stack operates on a last in first out (LIFO) principle.

The rest of this section adopts the following approach: first, the characteristic behaviour of buffers is formally specified. This helps to make explicit the defining properties of a buffer. A most simple buffer is then described to offer such properties by design. This serves as a generic buffer specification for other complicated systems to satisfy. The notion of refinement is introduced as a means to check against process-oriented specifications. Finally a piped communication system is presented as a case for demonstration, where the system is only deemed to be a buffer if it proves to be a refinement of a simple characteristic buffer.

Characteristic Behaviour

For a process to pass as a buffer it must satisfy the defining properties:

Example 38.1: Buffers—Defining Properties

For some buffer B storing elements of a set M, channels *read* and *write* serve the purpose of passing messages to and from it. The three properties understood as such for B are
1. input of messages on *read* channel and output on *write* channel without loss or reordering,
2. available input of any messages on the *read* channel when the buffer is empty, and
3. available output of some message on the *write* channel when the buffer is non-empty.

The above can be stated formally. The first property can be expressed using traces as observations:

Example 38.2: Buffers—First Defining Property

Given a CSP process B, it requires that for any trace the sequence of values to appear on the channel *write* are a prefix and in the same order of the values that appear on the channel *read*.

$$tr \in traces(B) \implies tr \downarrow write \leq tr \downarrow read \qquad (3.1)$$

Given a channel c, the function $\_\_ \downarrow \_\_$ is inductively defined as:

$$\langle \rangle \downarrow c \triangleq \langle \rangle$$
$$(\langle a \rangle \frown tr) \downarrow c \triangleq \begin{cases} m \frown (tr \downarrow c) & ; a = c.m \text{ for some } m \in comms(c) \\ tr \downarrow c & ; else \end{cases}$$

The second property insists that when the queue is empty the process must allow for any message on the input. This property can't be expressed with traces anymore. Traces record which observations can happen, however, not which observations must happen. The process $P = Stop \sqcap (a \rightarrow Stop)$, for example, has the trace $\langle a \rangle$, however, if P internally decides to take the left branch, a can't happen.

Thus, we need a means to observe processes more specifically. To this end, the theory of CSP considers the *refusal sets* of a process. Such a refusal set is a set of events that a process can fail to engage in however long the events are offered.

Consider once more the transition system of the process `ButtonOFF` as shown in Fig. 3.2, cf. page 133. In such transition systems we distinguish between *stable* and *unstable* states. A stable state has no outgoing transition labelled with τ or $\checkmark$. A stable state s refuses any set $X \subseteq \Sigma \cup \{\checkmark\}$ of visible events that are not the label of an outgoing transition. Such a set X is called a *refusal set*. In our example, the process `ButtonOFF` has the alphabet $\Sigma = \{\texttt{press}, \texttt{release}\}$. Here, we obtain:

- `ButtonOff` is unstable.
- `press -> ButtonOn` is stable and has the four refusal sets $\{\}$, $\{\texttt{release}\}$, $\{\checkmark\}$ and $\{\texttt{release}, \checkmark\}$.
- `ButtonOn` is unstable.
- `release -> ButtonOff` is stable and has the four refusal sets $\{\}$, $\{\texttt{press}\}$, $\{\checkmark\}$ and $\{\texttt{press}, \checkmark\}$.

In order to deal with termination, for an unstable state with an outgoing $\checkmark$ transition we record all $X \subseteq \Sigma$ as refusal sets.

The definition of refusal sets has the following consequence: if X is a refusal set of state s and $Y \subseteq X$, then also Y is a refusal set of s. There is not necessarily a largest refusal set w.r.t. set inclusion. The process $(a \rightarrow Stop) \sqcap (b \rightarrow Stop)$ has two maximal initial refusal sets: $\{a, \checkmark\}$ and $\{b, \checkmark\}$.

Now let us study which events the process `ButtonOFF` can refuse after performing a certain trace. Starting at state `ButtonOff`, after observing the empty trace $\langle \rangle$, the process is either in the state `ButtonOff` or—by performing τ—in the state `press -> ButtonOn`. `ButtonOFF` has no refusal sets, however `press -> ButtonOn` has the four refusal sets $\{\}$, $\{\texttt{release}\}$, $\{\checkmark\}$ and $\{\texttt{release}, \checkmark\}$. We combine this information to so-called *failures*:

- $(\langle \rangle, \{\})$,
- $(\langle \rangle, \{\texttt{release}\}$,
- $(\langle \rangle, \{\checkmark\})$, and
- $(\langle \rangle, \{\texttt{release}, \checkmark\}$.

A failure is a pair consisting of a trace and a refusal set. In our example we have: after performing $\langle \texttt{press} \rangle$, we are either in state `ButtonOn` or in state `release -> ButtonOff`. `ButtonOn` has no refusal set, however `release -> ButtonOff` has four refusal sets. Thus, the set of all failures with trace $\langle \texttt{press} \rangle$ is

$$\{ (\langle \texttt{press} \rangle, \{\}), (\langle \texttt{press} \rangle, \{\texttt{press}\},$$
$$(\langle \texttt{press} \rangle, \{\checkmark\}), (\langle \texttt{press} \rangle, \{\texttt{press}, \checkmark)\}\}.$$

The set of all failures of a CSP process P is denoted with

$$failures(P).$$

With the failures of a process at hand, we can express that a certain event, say press, has to be possible after observing a certain trace, say $\langle\rangle$: press is not an element of the refusal sets of the process ButtonOFF after observing the trace $\langle\rangle$, formally:

$$(\langle\rangle, X) \in failures(\text{ButtonOFF}) \implies \text{press} \notin X.$$

Fig. 3.3 Transition system of Div

In the context of failures, there is a process called Div that plays a special role. We add it to our grammar as one further option

$$P ::= \quad \cdots \\ \mid \quad Div$$

though specifiers use it seldomly. Figure 3.3 shows the transition system associated with it. As the only transition it has is labelled with τ, the only observation one can make about Div is the empty trace, i.e., $traces(Div) = \{\langle\rangle\}$. As the only state it has is unstable and has no outgoing $\checkmark$ transition, it has no stable failures, i.e., $failures(Div) = \{\}$.

The failures of a process can either be obtained from the operational semantics—as done above—or computed in a denotational way. We demonstrate here that failures are powerful enough to distinguish between the internal and the external choice operator:

- In the case of internal choice, we simply take the union of the failures of the constituent processes:

$$failures(P \sqcap Q) = failures(P) \cup failures(Q).$$

After performing a τ event in the transition system of $P \sqcap Q$, we reach state P or state Q.
- In the case of external choice, initially all options of P and Q are available—i.e., only the common refusal sets are those of the combined process. When a non-empty trace has been performed, we take the refusals of the individual processes after this trace. Finally, we have to deal with termination: if P or Q can terminate, we can refuse the whole alphabet.

$$\begin{aligned} \mathit{failures}(P \Box Q) = \quad & \{(\langle\rangle, X) \mid (\langle\rangle, X) \in \mathit{failures}(P) \cap \mathit{failures}(Q)\} \\ \cup \ & \{(s, X) \mid (s, X) \in \mathit{failures}(P) \cup \mathit{failures}(Q) \wedge s \neq \langle\rangle\} \\ \cup \ & \{(\langle\rangle, X) \mid X \subseteq \Sigma \wedge \langle\checkmark\rangle \in \mathit{traces}(P) \cup \mathit{traces}(Q)\}. \end{aligned}$$

Section B.4 compiles the clauses for the other operators.

Using failures, we can express the two missing buffer properties:

Example 38.3: Buffers—Second and Third Defining Property

Recall that the communications on the *read* and on the *write* channel are collected in some set M, i.e., $\mathit{comms}(\mathit{read}) = \mathit{comms}(\mathit{write}) = M$. When all messages read by the buffer have appeared as output, i.e., $tr \downarrow \mathit{write} = tr \downarrow \mathit{read}$, and hence the buffer is empty, then the buffer can't refuse any element of $\mathit{read}.M = \{\mathit{read}.m \mid m \in M\}$:

$$(tr, X) \in \mathit{failures}(B) \wedge tr \downarrow \mathit{write} = tr \downarrow \mathit{read} \implies \\ \mathit{read}.M \cap X = \{\} \tag{3.2}$$

If the messages read by the buffer have not yet all appeared as output, $tr \downarrow \mathit{write} < tr \downarrow \mathit{read}$, and hence the buffer is still not empty, then some element of $\mathit{write}.M$ is available for output:

$$(tr, X) \in \mathit{failures}(B) \wedge tr \downarrow \mathit{write} < tr \downarrow \mathit{read} \implies \\ \mathit{write}.M \nsubseteq X \tag{3.3}$$

Here, $s < t$ means: the trace s is a proper prefix of the trace t.

Property (3) is somewhat weaker than (2) as it insists on *some output* to be available from the buffer as opposed to *any input* to it. This serves to distinguish what the environment provides the buffer as an input from what the buffer may offer as output. Characteristically the buffer should output messages in the same sequence as they were input, which is ensured by (1). Combining (1) and (3) it is possible to predict the sequence of message that appear as output.

Essentially, requirement (1) is a *safety property* (slogan: "nothing 'bad' will ever happen"). It ensures that the sequence appearing as input is the same sequence (or a prefix of) that appeared as output. Formulated as a safety property one could state: it will never happen that the input sequence (or a prefix of it) will re-ordered as an output. Requirements (2) and (3) are *liveness properties* (slogan: "eventually, something 'good' will happen"). Formulated as a liveness property, requirement (2) says: when the buffer is empty, eventually it will read an input. Similarly, for (3) we can formulate: when the buffer is non-empty, eventually, some output will happen. See the classical 1985 paper "Defining Liveness" [AS85] for a thorough discussion of the subject. Note the use of traces and failures in formalising these requirements. With traces we state that *undesirable events should not happen*, with failures we formulate that *desirable events could happen*.

Up to now we have formulated in the semantic domains of traces and failures what a buffer should do. Here, we give a CSP process which has the desired properties.

Example 38.4: A Simple *BUFFER* Process

The following process offers properties (1)--(3) by design.

$$BUFFER(\langle\rangle) = read?m \to BUFFER(\langle m \rangle)$$
$$BUFFER(\langle x \rangle \frown q) =$$
$$(read?m \to BUFFER(\langle x \rangle \frown q \frown \langle m \rangle) \sqcap Stop)$$
$$\Box \; write!x \to BUFFER(q)$$

The buffer process $BUFFER(\_)$ has one parameter, which is the sequence of messages currently in the queue. m and x range over the set of allowable messages M; q ranges over all finite sequences over M.

An empty buffer always allows a new message to be read in. A non-empty buffer either accepts a new message on *read* or stops working, however, it is not able to refuse to output on *write*. Requirement (2), and thus Eq. (3.2), does not concern a non-empty buffer. Hence the use of nondeterministic choice between accepting a fresh input or refusing it.

Initially, the queue of read message is empty:

$$BUFFER = BUFFER(\langle\rangle)$$

It is one thing to claim that a process has certain properties. Yet, one better proves that this is the case indeed:

Example 38.5: *BUFFER* has the Prefix Property

Property (3.1) is a consequence of the following invariant: for all traces $tr \in traces(BUFFER)$ and the state q of $BUFFER(\_)$ after performing the trace tr it holds that

$$(tr \downarrow write) \frown q = tr \downarrow read.$$

Proof By induction on the length of tr.

In the case $tr = \langle\rangle$ the property holds obviously.

Now let $tr = tr' \frown a$. By induction hypothesis, we have $(tr' \downarrow write) \frown q' = tr' \downarrow read$, where q' is the state of the buffer after tr'.

If $a = read.m$ for some $m \in M$, we obtain $q = q' \frown \langle m \rangle$ and

$$(tr \downarrow write) \frown q$$
$$= (tr' \downarrow write) \frown q' \frown \langle m \rangle$$
$$= (tr' \downarrow read) \frown \langle m \rangle$$
$$= (tr' \frown \langle read.m \rangle) \downarrow read$$
$$= tr \downarrow read$$

If $a = write.m$ for some $m \in M$, we know that $q' = \langle m \rangle \frown q$. We compute

$$(tr \downarrow write) \frown q$$
$$= (tr' \frown \langle write.m \rangle) \downarrow write) \frown q$$
$$= (tr' \downarrow write) \frown \langle m \rangle \frown q$$
$$= (tr' \downarrow write) \frown q'$$
$$= tr' \downarrow read$$
$$= (tr' \frown \langle write.m \rangle) \downarrow read$$
$$= tr \downarrow read$$

Now we look into the liveness properties:

Example 38.6: *BUFFER* **has the Liveness Properties**

For Property (3.2) and Property (3.3) we compute the failures of the interesting states.

To Property (3.2): let tr be a trace of *BUFFER*. Let $tr \downarrow write = tr \downarrow read$. Then the state q of $BUFFER(\_\_)$ after tr is $\langle \rangle$. The failures of $BUFFER(\langle \rangle)$ are

$$failures(read?m \rightarrow BUFFER(\langle m \rangle)) =$$
$$\{(\langle \rangle, X) \mid \{read.m \mid m \in M\} \cap X = \{\}\} \cup \{(\langle read.m \rangle \frown s, X) \mid \ldots\}$$

I.e., Property (3.2) holds.

To Property (3.3): let tr be a trace of *BUFFER*. Let $tr \downarrow write < tr \downarrow read$. Then the state of $BUFFER(\_\_)$ after tr is of the form $\langle x \rangle \frown q$. We calculate the failures (s, X) with $s = \langle \rangle$ of this process. Let

- $P \triangleq (read?m \rightarrow BUFFER(\langle x \rangle \frown q \frown \langle m \rangle) \sqcap Stop)$ and
- $Q \triangleq write!x \rightarrow BUFFER(q)$.

  Using these abbreviations we calculate:

- Consider P : We observe that $Stop$ has all refusal sets: $failures(Stop) = \{(\langle \rangle, X) \mid X \subseteq A^\checkmark\}$. As the internal choice operator takes the union of the failures, for $\langle \rangle$ we have all refusal sets for P.
- The failures of Q are $\{(\langle \rangle, X) \mid write.x \notin X\} \cup \{(\langle write.x \rangle \frown s, X) \mid \ldots\}$.
- Combining these failure sets we obtain:

$$
\begin{aligned}
&\mathit{failures}(P \,\square\, Q) \\
&= \quad \{(\langle\rangle, X) \mid (\langle\rangle, X) \in \mathit{failures}(P) \cap \mathit{failures}(Q)\} \\
&\quad \cup \{(s, X) \mid (s, X) \in \mathit{failures}(P) \cup \mathit{failures}(Q) \land s \neq \langle\rangle\} \\
&\quad \cup \{(\langle\rangle, X) \mid X \subseteq A \land \langle\checkmark\rangle \in \mathit{traces}(P) \cup \mathit{traces}(Q)\} \\
&= \quad \{(\langle\rangle, X) \mid (\langle\rangle, X) \in \mathit{failures}(Q)\} \\
&\quad \cup \{(s, X) \mid \ldots s \neq \langle\rangle\} \\
&\quad \cup \{\}.
\end{aligned}
$$

Now assume that there exists some failure $(\langle\rangle, X) \in \mathit{failures}(P \,\square\, Q)$ with $\mathit{write}.M \subseteq X$. As failures sets are downward closed, then also $(\langle\rangle, \{\mathit{write}.x\}) \in \mathit{failures}(P \,\square\, Q)$. This, however, is not the case as seen above. Thus, property (3.3) holds.

Refinement

There is an advantage to using a simple process description such as the one for *BUFFER*: if the behaviour demonstrated by the process, in terms of the traces and failures that it exhibits, is acceptable, then the process itself can be used as a specification. Such a specification can then be used to check typically larger or more complicated systems whether they demonstrate the same behaviour. This is the general idea underlying the notion of *refinement*.

For a given specification described by some process P, another process Q is said to meet the specification P if any trace of Q is also a trace of P. One also says: P is refined by Q. Formally, this is defined as

$$P \sqsubseteq_{\mathcal{T}} Q \triangleq \mathit{traces}(Q) \subseteq \mathit{traces}(P)$$

where $P \sqsubseteq_{\mathcal{T}} Q$ is pronounced Q *trace-refines* P. The fewer traces Q has the more 'refined' it is. Q has fewer behaviours that can violate the specification described by P. Following from this $P \sqsubseteq_{\mathcal{T}} \mathit{Stop}$ for any P, i.e., *Stop* is the most trace refined process. Doing 'nothing' is always safe.

The notion of refinement extends to failures: if every trace tr of Q is possible for P and every refusal after this trace is possible for P then Q *failures-refines* P. Formally,

$$P \sqsubseteq_{\mathcal{F}} Q \triangleq \mathit{traces}(Q) \subseteq \mathit{traces}(P) \land \mathit{failures}(Q) \subseteq \mathit{failures}(P)$$

CSP refinement obeys a number of laws including

$$
\begin{aligned}
&P \sqsubseteq P &&(\textit{reflexive}) \\
&P \sqsubseteq Q \land Q \sqsubseteq R \implies P \sqsubseteq R &&(\textit{transitive}) \\
&P \sqsubseteq Q \land Q \sqsubseteq P \implies P = Q &&(\textit{anti-symmetric})
\end{aligned}
$$

As these laws can be proven to hold in all CSP standard models, we omit the index indicating the specific model. Furthermore, refinement is compositional such that it is preserved by all operations of CSP. This means for any CSP context $F(.)$ where a process can be substituted,

$$P \sqsubseteq Q \implies F(P) \sqsubseteq F(Q) \quad (substitution)$$

Let's look at an example refinement. We first specify a concrete system B:

Example 38.7: A Two-element Buffer

$$
\begin{aligned}
B &= read?x \rightarrow B_{one}(x) \\
B_{one}(x) &= read?y \rightarrow B_{two}(x, y) \ \square \ write!x \rightarrow B \\
B_{two}(x, y) &= write!(x) \rightarrow B_{one}(y)
\end{aligned}
$$

While it is intuitively clear that B is a two element buffer, the question is: do we have $BUFFER \sqsubseteq_F B$?

In order to answer this question, we state some algebraic laws of CSP, which all are immediate consequences of the semantic clauses and the definition of refinement. Over the model $\mathcal{F}$, the following algebraic laws hold for refinement:

1. $P \sqcap Q \sqsubseteq_{\mathcal{F}} Q$ (int-choice refinement)
2. $P \ \square \ Stop = P$ (ext-choice unit)
3. $P \sqsubseteq_{\mathcal{F}} Div$ (div most refined process over $\mathcal{F}$)
4. $c?x \rightarrow P(x) = c?y \rightarrow P(y)$ (variable renaming)

For further discussions on the model $\mathcal{F}$ see Sect. 3.4.1. This chapter also includes a discussion of algebraic laws, see Sect. 3.4.2.

Note that laws 2. and 4. both give rise to refinement laws. As refinement is reflexive ($P \sqsubseteq P$), from $P \ \square \ Stop = P$ we obtain $P \ \square \ Stop \sqsubseteq P$, and also from $c?x \rightarrow P(x) = c?y \rightarrow P(y)$ we obtain $c?x \rightarrow P(x) \sqsubseteq c?y \rightarrow P(y)$.

For dealing with recursion, we state the following proof rule of *fixed point induction*:

Let $P = F(P)$ and $Q = G(Q)$ be two systems of process equations, where $P = (P_i)_{i \in I}$ and $Q = (Q_i)_{i \in I}$ are (possibly infinite) vectors of process names over some index set I, and $F = (F_i)_{i \in I}$ and $G = (G_i)_{i \in I}$ are vectors of component functions F_i and G_i, resp., which map a vector of process names to CSP processes over a these names. Let $P \sqsubseteq_X Q$ be defined as $\forall i \in I. P_i \sqsubseteq_X Q_i$ for $X \in \{\mathcal{T}, \mathcal{F}\}$. Then it holds:

- If $F_i(P) \sqsubseteq_{\mathcal{T}} G_i(P)$ for all $i \in I$, then $P \sqsubseteq_{\mathcal{T}} Q$.
- If $F_i(P) \sqsubseteq_{\mathcal{F}} G_i(P)$ for all $i \in I$, then $P \sqsubseteq_{\mathcal{F}} Q$.

The interested reader is referred to Roscoe's book [Ros98], Sect. 9.2, for the proof and further discussion of these rules.

We illustrate this formulation of process equations by the following example:

Example 38.8: Formulation as System of Process Equations

We can formulate both, the $BUFFER$ process and the two element buffer B as systems of process equations.

Both buffers can store elements from some set M. The words over M, i.e., the set M^* provides the index set common to both systems of process equations.

In the case of the general $BUFFER$ process, we take $BUF = (BUF_w)_{w \in M^*}$ as the vector of names. We define the component functions with the help of a vector of variables $X = (X_w)_{w \in M^*}$ as follows.

$$
\begin{aligned}
F_{\langle\rangle}(X) &= read?m \to X_{\langle m \rangle} \\
F_{\langle x \rangle \frown q}(X) &= \quad (read?m \to X_{\langle x \rangle \frown q \frown \langle m \rangle} \sqcap Stop) \\
&\quad \Box \; write!x \to X_q
\end{aligned}
$$

with $x \in M, q \in M^*$.

Similarly, in the case of the two element buffer B, we take $B = (B_w)_{w \in M^*}$ as the vector of names. We define for the two element buffer with the help of a vector of variables $Y = (Y_w)_{w \in M^*}$ the following system of equations:

$$
\begin{aligned}
G_{\langle\rangle}(Y) &= read?x \to Y_{\langle x \rangle} \\
G_{\langle x \rangle}(Y) &= read?y \to Y_{\langle x,y \rangle} \Box \; write!x \to Y_{\langle\rangle} \\
G_{\langle x,y \rangle}(Y) &= write!(x) \to Y_{\langle y \rangle}
\end{aligned}
$$

Now, we are missing out on equations for G_w for $|w| \geq 3$: the two element buffer does not prescribe any behaviour in the case that three or more elements have been stored. These states are no reachable from the initial state of an empty buffer. However, formally we are required to provide equations, when we want to relate the BUF process with our two element buffer. Here we choose:

$$
G_w(Y) = Div
$$

for $w \in M^*, |w| \geq 3$. As Div is the most refined process over $\mathcal{F}$, this definition will ease our refinement proof.

Formulating our processes in the 'right' format for the fixpoint induction rule involved

- finding a common index set,
- transforming process parameters to indices in order to formally obtain a vector of process names,

- explicitly formulating component functions, possibly adding such functions.

Often, these transformations are obvious and there is no need to actually carry them out. But in the beginning it helps to 'stick to the rules'.

```
theorem BUFFER_ref_B: "ProcBUF <=F ProcB"
  apply (unfold ProcBUF_def ProcB_def)
  apply (rule cspF_fp_induct_cpo_ref_right[of _ _ "B_to_BUF"])
  apply (simp_all)
  apply (induct_tac p)
  apply (simp_all)

(* length 0 *)
  apply (cspF_unwind)
  apply (cspF_hsf)+

(* length 1 *)
  apply (cspF_unwind)
  apply (cspF_hsf)+
  apply (rule cspF_Int_choice_left1)
  apply (rule cspF_decompo, auto)

(* length 2 *)
  apply (cspF_unwind)
  apply (cspF_hsf)+
  apply (rule cspF_Int_choice_left2)
  apply (simp)
  done
```

Fig. 3.4 The proof from Example 38.9 as proof script in CSP-Prover

With these formulations, we are now prepared to carry out the proof:

Example 38.9: A Refinement Proof

With the notation established above, we want to show that
$$F_w(X) \sqsubseteq_{\mathcal{F}} G_w(X) \text{ for all } w \in M^*.$$

To this end, we consider the following four cases, which cover all $w \in M^*$:

Case 1, $|w| = 0, i.e., w = \langle \rangle$:

$$\begin{aligned}
F_{\langle \rangle}(X) &= read?m \to X_{\langle m \rangle} \\
&= read?x \to X_{\langle x \rangle} \text{ by (variable renaming)} \\
&= G_{\langle \rangle}(X).
\end{aligned}$$

<u>Case 2, $|w| = 1$</u> : Let $w = \langle x \rangle$ for some $x \in M$.

$$
\begin{aligned}
F_{\langle x \rangle}(X) &= \quad (read?m \rightarrow X_{\langle x \rangle ^\frown \langle m \rangle} \sqcap Stop) \\
&\quad\ \square\ write!x \rightarrow X_{\langle \rangle} \\
&\sqsubseteq_{\mathcal{F}} \quad (read?y \rightarrow X_{\langle x, y \rangle}) \\
&\quad\ \square\ write!x \rightarrow X_{\langle \rangle} \\
&\qquad \text{by (int choice refinement), (substitution),} \\
&\qquad\qquad\qquad\qquad\qquad\qquad \text{(variable renaming)} \\
&= \quad G_{\langle x \rangle}(X)
\end{aligned}
$$

<u>Case 3, $|w| = 2$</u> : Let $w = \langle x, y \rangle$ for some $x, y \in M$.

$$
\begin{aligned}
F_{\langle x, y \rangle}(X) &= \quad (read?m \rightarrow X_{\langle x, y \rangle ^\frown \langle m \rangle} \sqcap Stop) \\
&\quad\ \square\ write!x \rightarrow X_{\langle y \rangle} \\
&\sqsubseteq_F \quad Stop\ \square\ write!x \rightarrow X_{\langle y \rangle} \\
&\qquad \text{by (int choice refinement) and (substitution)} \\
&= \quad write!x \rightarrow X_{\langle y \rangle} \qquad \text{by (ext-choice unit)} \\
&= \quad G_{\langle x, y \rangle}(X)
\end{aligned}
$$

<u>Case 4, $|w| \geq 3$</u> : Let $w = \langle x \rangle ^\frown q$ for some $x \in M$, $q \in M^*$, $|q| \geq 2$.

$$
\begin{aligned}
F_{\langle x \rangle ^\frown q}(X) &= \quad (read?m \rightarrow X_{\langle x \rangle ^\frown q ^\frown \langle m \rangle} \sqcap Stop) \\
&\quad\ \square\ write!x \rightarrow X_q \\
&\sqsubseteq_F \quad Div \text{ by (div most refined process over } \mathcal{F}) \\
&= \quad G_{\langle x \rangle ^\frown q}(X)
\end{aligned}
$$

The above proof also be carried out in CSP-Prover, cf. Fig. 3.4. Rule `cspF_fp_induct_cpo_ref_right` refers to fixec point induction. In the case distinctions, (`cspF_unwind`) represents the unfolding of the equations (i.e., going from the lhs of component function F to the rhs), (`cspF_hsf`)+ is a tactic that automatically applies a number of algebraic laws, however, sometimes one needs to be specific which algebraic law shall be applied and to which argument (`cspF_Int_choice_left1`— chooses the first argument of the internal choice operator on the lhs of an equation).

As refinement holds in all four cases, we know that $BUF \sqsubseteq_F B$. Now let us consider the three defining properties of a buffer in the context of process B:

For the first defining property, we have: let $tr \in traces(B)$. As $BUF \sqsubseteq_F B$ we have $tr \in traces(BUF)$. For the traces of the BUF process it holds that $tr \downarrow write \leq tr \downarrow read$. Thus, the first defining property holds for tr. The proofs for the second and third property are analogous.

Thus, B has all the defining properties of a buffer.

The general idea a refinement $P \sqsubseteq Q$ in CSP is that properties established for P are inherited by Q, as was the case in our buffer example. The general pattern is: safety properties ('nothing bad will ever happen') are inherited in trace refinement $\sqsubseteq_T$, while safety properties and liveness properties ('eventually, something good will happen') are inherited in failures refinement $\sqsubseteq_F$. As model checking for trace refinement has a lower complexity than model checking failures refinement, it is worth to have both refinements.

In a CSP refinement $P \sqsubseteq Q$,

- one can see P as the *specification of a property*, and Q as the system model, i.e., CSP captures properties and system model within one language. We will discuss how to specify livelock, deadlock, and determinism in CSP in Sect. 3.4.4.
- however, one can also think of P as the *first model* that one builds of a system and analyses. Having established that P has the desired properties, e.g., that P is deadlock free, one develops Q and shows via refinement that Q keeps the desired properties. As P is less complex than Q, the hope would be that analysing P is 'easier' than analysing Q.
- finally, one can perceive P as an *abstraction* of a system model Q, where we leave out details from Q. As P is 'less complex', it will be easier to analyse P.

A Communication System

Communication systems often form an important component of larger distributed processing systems where reliable movement of data from one part of the system to the other is critical. From an abstract point of view, the communication system shall behave like a buffer. In the following, we will give a specific communication system and prove that it exhibits the buffer properties described above.

Example 39: A Communication System

A typical communication system consists of three components: a sender, a medium (e.g., a wireless network), and a receiver. The medium transports the messages, usually in an unreliable way.

Here, we give an example of a communication system $COMM\_SYSTEM$ with an imperfect medium, which is designed to communicate bits:

Example 39.1: A Communication System in CSP

$$COMM\_SYSTEM = (SNDR \, [\! | \, ToM \, | \!] \, MEDIUM) \, [\! | \, FromM \, | \!] \, RCVR$$

where the communication sets are given as

- $ToM \triangleq \{toMedium.0, toMedium.1\}$ and
- $FromM \triangleq \{fromMedium.0, fromMedium.1\}$.

  Processes $SNDR$ and $RCVR$ serve to interface with the medium:

  $$SNDR = read?x \rightarrow toMedium!x \rightarrow SNDR$$
  $$RCVR = fromMedium?x \rightarrow write!x \rightarrow RCVR$$

  The $MEDIUM$ is described below

  $$MEDIUM_0 = toMedium?m \rightarrow$$
  $$(\ fromMedium!m \rightarrow MEDIUM_0$$
  $$\sqcap fromMedium!(1 - m) \rightarrow MEDIUM_2)$$
  $$MEDIUM_{n+1} = toMedium?m \rightarrow fromMedium!m \rightarrow MEDIUM_n$$
  $$MEDIUM = MEDIUM_0$$

The medium available is unreliable but somewhat predictable. Designed to transmit a single bit at a time, it may or may not corrupt the transmitted bit. Here, the expression $1 - m$ encodes flipping the bit. If the medium corrupts the bit, it will transmit the next two bits correctly before being able to corrupt another one. In reality, a medium might not be predictable to such an extent. A natural question is this $COMM\_SYSTEM$ behaves like a buffer with the above stated properties (3.1), (3.2), and (3.3). In CSP, this question can be expressed as a refinement assertion. Concretely, we might try to find out if the system behaves like a two place buffer, i.e., do we have $B \sqsubseteq_F COMM\_SYSTEM \setminus$ $(FromM \cup ToM)$? Note that this refinement could fail for two reasons: the first is that the system fails to return the same element, the second is that the system has a different capacity. These reasons are not exclusive.

Here, we use the hiding operator to abstract from the interaction with the $MEDIUM$. Should this refinement hold, we inherit the properties stated for $BUFFER$, i.e., all traces of the abstract $COMM\_SYSTEM$ have the "no loss & no reorder" property; if the $MEDIUM$ is available, the abstract $COMM\_SYSTEM$ offers to $read$ a bit; after the $MEDIUM$ has delivered a message, the abstract $COMM\_SYSTEM$ offers to $write$ a bit.

Using a model checker, e.g., FDR, one obtains:

Example 39.2: The Communication System Is Flawed

We have $B \not\sqsubseteq_F COMM\_SYSTEM \setminus (FromM \cup ToM)$ as the process

$$COMM\_SYSTEM \setminus (FromM \cup ToM)$$

has the trace

$$\langle read.1, write.0 \rangle$$

i.e., the medium corrupts the transmitted message such that 1 is read in but 0 is written out.

    I.e., the first of the two reasons discussed above occurred.

To address this problem, we redesign the *COMM_SYSTEM*. Even if it is taken to be certain that every one of the three bits transmitted through the medium is corrupted, the interfacing processes could be adapted to take the *best-of-three* values. That is, necessarily assuming one out of three bits is corrupted, any two values that match denote the original value transmitted. To this end we modify the two interfacing processes:

Example 39.3: A Corrected Communication System

$$MSNDR = read?x \rightarrow input!x \rightarrow input!x \rightarrow input!x \rightarrow MSNDR$$
$$MRCVR = output?x \rightarrow output?y \rightarrow output?z \rightarrow$$
$$\textbf{if } x == y \textbf{ then } write!x \rightarrow MRCVR$$
$$\textbf{else } write!z \rightarrow MRCVR$$

With these modified processes, we form a new communication system— without changing the *MEDIUM*:

$$MCOMM\_SYSTEM =$$
$$(MSNDR \,[\! |\, ToM \,|\!] \, MEDIUM) \,[\! |\, FromM \,|\!] \, MRCVR$$

Using the model checker FDR, we obtain for this system:
1. $B \sqsubseteq_F MCOMM\_SYSTEM \setminus (FromM \cup ToM)$ and
2. $MCOMM\_SYSTEM \setminus (FromM \cup ToM) \sqsubseteq_F B$,
   i.e., the abstracted modified system behaves like a two place buffer.

3.3 The Children's Puzzle or What CSP Tools Can Do

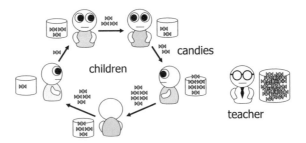

Fig. 3.5 Children's Puzzle: illustration with five children

In this chapter we analyze the Children's Puzzle with various tools for CSP. The puzzle belongs to the lore of mathematical riddles. It appears to be impossible to name its inventor. One reference is [BPFS].

Example 40: The Children's Puzzle

There are $n \geq 2$ children sitting in a circle, each holding an even number of candies.

The following two steps are repeated indefinitely:

Step 1: Every child passes half of their candies to the child on her left.

Step 2: Any child who ends up with an odd number of candies is given another candy by the teacher.

Figure 3.5 depicts an instance of the puzzle after Step 1.

We illustrate this puzzle by exploring it in a concrete setting:

Example 40.1: Example run

Say, there are three children with the names Berta, Emma, and Hugo. Berta has Emma to her left, Emma has Hugo to her left, and Hugo has Berta to his left. Initially, Berta does not hold any candies, Emma has 2 candies, and Hugo has 4 candies. One possible run looks as follows:

In the first round, in Step 1,
• Berta can't pass any candies to Emma,
• Emma passes one candy to Hugo, and
• Hugo passes 2 candies to Berta.
Note that the puzzle does not prescibe any order in which the candies are passed in Step 1.

In the first round, after Step 1,

• Berta holds 2 candies
  (0 candies initially, 0 candies given away, 2 received from Hugo),
• Emma holds 1 candy
  (2 candies initially, 1 given away to Hugo, 0 received from Berta), and
• Hugo holds 3 candies
  (4 candies initially, 2 given away to Berta, 1 received from Emma).

In the first round, in Step 2, the teacher

• does not give any candy to Berta as she holds an even number of candies,
• gives one candy to Emma as she holds an odd number of candies, and

- gives one candy to Hugo as he holds an odd number of candies.

In the first round, after Step 2,

- Berta holds 2 candies
  (2 candies after Step 1, no candies from the teacher),
- Emma holds 2 candy
  (1 candy after Step 1, one candy from the teacher), and
- Hugo holds 4 candies
  (3 candies after Step 1, one candy from the teacher).

As steps are repeated indefinitely, this is only the beginning of one run.

The puzzle does not prescribe any 'synchronisation' between the steps. Our above 'snapshots' "after Step 1" and "after Step 2" are possible in a run, however, do not happen in every run. In particular with many children, a system run can inlude a 'configuration' in which children have completed several rounds (i.e., passed their candies and received a refill from the teacher), while there are children who not yet have passed any candy. It is this lack of synchronisation that makes it a challenge to analyse the puzzle.

Natural questions on the puzzle include:

Example 40.2: Questions to the puzzle

- Will the teacher keep handing out more and more candies?
- Will an unequal distribution of candies eventually become an equal one?

Our puzzle exhibits many characteristics typical of concurrent systems: *Scalability*: The puzzle's size, i.e., the number n of children involved, can vary; *Parameterisation:* The initial distribution of candies can be chosen; *Local activity only:* The children interact with their direct neighbours only, there is no 'broadcast'; *Global result:* The properties of interest concern the system as a whole—one would like to prove them independent of size and parameter.

3.3.1 The Arithmetic Side of the Puzzle

Usually, the Children's Puzzle is analysed as a function, which is iterated on a given initial distribution of candies. We observe:

Lemma 1 *The maximum number of candies held by a single child never increases; the minimum number of candies held by a single child never decreases.*

Proof (The maximum never increases) Let $2M$ be the maximum number of candies a child holds before Step 1 is carried out. Let c be one of the children. Then child c holds $2K \leq 2M$ candies. In Step 1, child c gives away K candies, keeps $K \leq M$ candies, and receives $L \leq M$ candies. Thus, afterwards child c holds $K + L$ candies. Carrying out Step 2, i.e., filling up from the teacher yields:

- If $K = L = M$ or $K = L = M - 1$, then $K + L$ is even. The child is not given another candy. Thus, child c holds $\leq 2M$ candies.
- If $K = M - 1$, $L = M$ or $K = M$, $L = M - 1$, then $K + L$ is odd. The child is given one candy. Thus, child c holds $2M$ candies.
- If $K, L < M - 1$, then child c holds $K + L < 2M - 2$ candies after Step 1. As a child receives at most one candy, child c holds $< 2M$ candies.

(The minimum never decreases.) Analogously.   ∎

Corollary 1 *The teacher will eventually stop handing out candies.*

Lemma 2 *Let $2m$ ($2M$) be the minimum (maximum) number of candies held by a single child. Let $m < M$. The number of children holding $2m$ candies decreases.*

Proof We first argue that a child which holds $2k > 2m$ candies before Step 1 is carried out, never holds the minimum number after Step 1 and Step 2: let c be such a child. In Step 1, child c gives away k candies, keeps $k > m$ candies, and receives $l \geq m$ candies. Thus, afterwards child c holds $k + l > 2m$ candies. Step 2 can only increase this number.

Now we study what happens with a child which holds $2m$ candies before Step 1 is carried out. Let c be a such a child. In Step 1, child c gives away m candies, keeps m candies, and receives $l \geq m$ candies. If $l = m$, child c holds $2m$ candies, which is an even number. Thus, after Step 2 child c still holds $2m$ candies. If $l > m$, child c holds $m + l > 2m$ candies. Step 2 can only increase this number.

As $m < M$ there exists at least one child holding the minimum number of candies with a neighbour to the right, who does not hold the minimum.   ∎

Corollary 2 *Eventually, all children will hold the same number of candies.*

Proof Let m and M be as in Lemma 2. If $m = M$, there is nothing to prove.

If $m < M$, we know that after a finite number of carrying out Step 1 and Step 2, the number of children holding m candies will be zero. Lemma 1 states that the minimum never decreases. Thus, the new minimum must be larger than m. Eventually, this will lead to an even distribution, as according to Lemma 1 the maximum does not increase.   ∎

Reflecting on our analysis, we observe: the proof arguments use arithmetic only; concurrency does not play a role at all. Our analysis is based on two (silent) assumptions: there is a global clock which synchronizes Step 1 and Step 2; the exchange of candies is never blocked. If one wants to make the underlying machine model explicit, one needs to formalise the puzzle.

3.3.2 An Asynchronous Model of the Puzzle in Csp

In order to capture its concurrent aspect properly, we model our puzzle in
Csp. Here is the Csp$_M$ code for three children who can hold up to four
candies:

```
fill(n) = if (n%2==0) then n else n+1

-- Communication in the system:
--        first parameter: position
--        second parameter: number of candies passed

channel c: {0..2}.{0..2}

-- The child processes:
--        p - position,
--        i - number of candies

Child(p,i) =
    (    c.(p+1)%3 ! (i/2)
      -> c.p        ? x     -> Child(p, fill(i/2+x)))
[] (    c.p        ? x
      -> c.(p+1)%3 ! (i/2) -> Child(p, fill(i/2+x)))

-- 3 Children, holding 2*p candies each:

Children = || p:{0..2} @
                 [ {| c.p, c.(p+1)%3 |} ] Child(p,2*p)
```

Fig. 3.6 Csp$_M$ model of the children's puzzle

Example 40.3: The Puzzle in Csp$_M$

Figure 3.6 provides a Csp$_M$ model of the children's puzzle. Each of
the three children is represented by a parametrised process `Child`. The
first parameter p of `Child` is the identity or position of the child in
the system. The second parameter i of `Child` represents the number of
candies that it currently holds.

Passing a number k of candies from one child to another is modelled
as a communication of the value k via a channel c. There are three
channels c.0, c.1, and c.2, each connecting two children. Channel c.0
allows `Child(2,_)` to pass candies to `Child(0,_)`, channel c.1 allows
`Child(0,_)` to pass candies to `Child(1,_)`, and channel c.2 allows
`Child(1,_)` to pass candies to `Child(2,_)`. Here we use _ as a wildcard symbol for the number of candies: the channel connections are
independent of the number of candies the `Child` processes hold. As we
know from our mathematical proof, if initially the maximum number

of candies a child holds is 4, then the number of candies that can be passed is a number in the set $\{0, 1, 2\}$.

We implement the role of the teacher as a function `fill`. Note that in CSP_M, the *mod* operator is written as % sign.

The children can pass candies in an arbitrary order: either, a child first hands out half of her candies and then receives from the child to her left, or the other way round.

The system `Children` then consists of the three processes. These are `Child(0,0)`, `Child(1,2)`, and `Child(2,4)`. They synchronise on the channels 'to their right' to receive candies and 'to their left' to pass on candies. We use the parametrised, general parallel operator in order to form this system.

Note that our model is asynchronous, there is no global clock. The number of times that the child processes have carried out Step 1 and Step 2 can vary.

3.3.3 Analysing the CSP Model with Tool Support

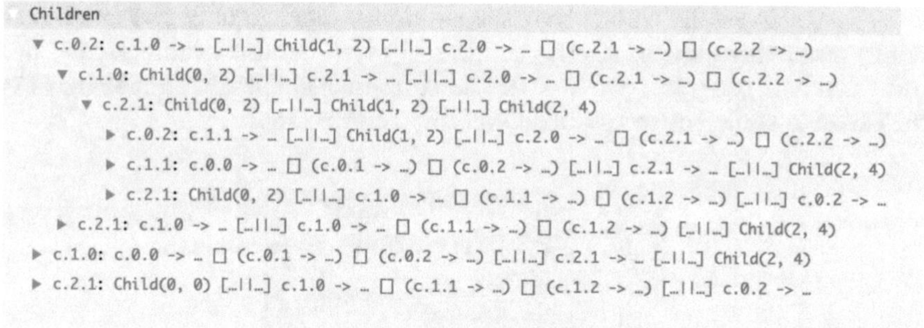

```
Children
▼ c.0.2: c.1.0 -> _ [_||_] Child(1, 2) [_||_] c.2.0 -> _ □ (c.2.1 -> _) □ (c.2.2 -> _)
  ▼ c.1.0: Child(0, 2) [_||_] c.2.1 -> _ [_||_] c.2.0 -> _ □ (c.2.1 -> _) □ (c.2.2 -> _)
    ▼ c.2.1: Child(0, 2) [_||_] Child(1, 2) [_||_] Child(2, 4)
      ▶ c.0.2: c.1.1 -> _ [_||_] Child(1, 2) [_||_] c.2.0 -> _ □ (c.2.1 -> _) □ (c.2.2 -> _)
      ▶ c.1.1: c.0.0 -> _ □ (c.0.1 -> _) □ (c.0.2 -> _) [_||_] c.2.1 -> _ [_||_] Child(2, 4)
      ▶ c.2.1: Child(0, 2) [_||_] c.1.0 -> _ □ (c.1.1 -> _) □ (c.1.2 -> _) [_||_] c.0.2 -> _
    ▶ c.2.1: c.1.0 -> _ [_||_] c.1.0 -> _ □ (c.1.1 -> _) □ (c.1.2 -> _) [_||_] Child(2, 4)
  ▶ c.1.0: c.0.0 -> _ □ (c.0.1 -> _) □ (c.0.2 -> _) [_||_] c.2.1 -> _ [_||_] Child(2, 4)
▶ c.2.1: Child(0, 0) [_||_] c.1.0 -> _ □ (c.1.1 -> _) □ (c.1.2 -> _) [_||_] c.0.2 -> _
```

Fig. 3.7 Simulation using the `:probe` command of FDR

With a *simulator* such as the tool ProBe we can explore a single run of one instance of the puzzle. Figure 3.7 shows a run of the puzzle in ProBe. Here, the puzzle consists of three children who initially are holding zero, two, and four candies, respectively. After choosing the three exchanges `c.0.2`, `c.1.0`, and `c.2.1`, the children hold two, two, and four candies. Selecting furthermore always the first choice offered, eventually, we reach a state in which all three children hold four candies.

Example 40.4: Simulating the Puzzle

The simulation shows: for a specific size, for a specific initial candy distribution, and for a specific execution, eventually, all children hold four candies. From the rules of our puzzle we know that exchanging further candies will not change this state anymore. Thus, we can conclude: for the specific instance of the puzzle and a selected run, we reach a 'stable state'.

We formulate the idea of 'stability' in CSP as follows:

```
StableAfter (n) = if n>0 then  c.0 ? x -> StableAfter (n-1)
                          [] c.1 ? x -> StableAfter (n-1)
                          [] c.2 ? x -> StableAfter (n-1)
                  else Stable
```

```
Stable = c.0!2->Stable [] c.1!2->Stable [] c.2!2->Stable
```

For the first n steps, an arbitrary number of candies can be exchanged. Then, however, the exchange of candies is restricted to two. The refinement statement

```
                assert StableAfter(3) [T= Children
```

over the traces model claims that the children exchange only two candies (if at all) after three steps. As the children's puzzle is also deadlock-free, i.e., always makes progress, proving the trace refinement `StableAfter(n) [T= Children` is sufficient to establish stability after n steps.

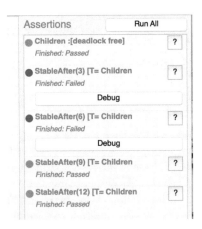

Fig. 3.8 Model checking in FDR

> **Example 40.5: Model Checking the Puzzle**
>
> With a *model checker* such as the tool FDR we can analyse all executions possible in a specific instance of the puzzle. Typical questions are: is our CSP model self stabilised, deadlock free, livelock free, or deterministic? Fig. 3.8 shows the answers that FDR gives to these questions. Our model does not stabilise after three or six steps, however, after nine and twelve steps it is stable. Our model is deadlock free, it is livelock free, and it is deterministic.

```
theorem EventuallyStable_CircChild:
  "[| 1 < length s ; allEven s |]
    ==> EventuallyStable s <=F CircChild s"
apply (rule cspF_tr_left_ref2)
apply (rule EventuallyStable_CircSpec)
apply (simp_all)
apply (rule CircSpec CircChild)
ISO8-----XEmacs: UCD_proc2.thy        (Isar script XS:isar/s Font Scripting )--

proof (prove): step 0

goal (1 subgoal):
  1. [|1 < length s; allEven s|] ==> EventuallyStable s ⊑F CircChild s
```

Fig. 3.9 Interactive proof in CSP-Prover

> **Example 40.6: Analysing the Puzzle with an Interactive Theorem Prover**
>
> With an *interactive theorem prover* such as CSP-Prover [IR05, IR], finally, we can gain a result as strong as Corollary 2, however, for the asynchronous CSP model: the puzzle will stabilise for all numbers $n \geq 2$ of children and for all initial candy distributions, see [IR08]. Figure 3.9 shows part of the proof.

This result comes at high costs though: the proof needs to be carried out interactively, i.e., the user enters proof commands, which the proof engine then executes in order to gain a new proof state. Developing the proof script published by Yoshinao Isobe and Roggenbach [IR08] took about one man month. The final script consists of several thousand lines of proof commands. Running the script takes about half an hour of computation time on a 2.5 GHz Intel Core 2 Duo processor.

We have seen: with simulation we can analyse a single run of a system. With model checking we can verify a single system. With interactive theorem we can verify a parameterised class of systems.

3.3.4 A Transformational Approach to Implementation

It is a long standing and still open research problem of how to link specifications written, e.g., in CSP with implementations written, e.g., in the programming language C++. Here, 'Invent & Verify' and 'Transformation' are the two standard approaches. In the first approach, the implementor 'invents' a program and then establishes either by testing or by mathematical argument that the implementation 'fits' to the specification. Testing approaches to CSP are discussed, e.g., by Ana Cavalcanti and Temesghen Kahsai [CG07, KRS07]. The second approach, which we will follow here, 'translates' a given specification from a specification language into a programming language, in our case from CSP into C++.

In the context of CSP, several techniques have been developed as link to a programming language. They differ in methodology, preservation of semantics, degree of automation, the supported language constructs, target language, to name just a few criteria:

JCSP extends the programming language Java by a library of CSP constructs such as processes, (unidirectional) channels, parallel composition. JCSP allows its user to program Java using CSP constructs to express concurrency. For further information see [Wel, WBM07].

Magee/Kramer provide a set of informal translation rules in their book "Concurrency" [MK06]. They link the process algebra FSP (a CSP dialect designed for teaching purposes) to Java. The provided rules allow one to develop Java programs with 'the same properties' as the FSP specifications.

CSP++ [Gar15, Gar05] is a tool which translates a sublanguage of CSP_M into the programming language C++. Figure 3.10 shows the synthesis process supported by CSP++: in order to capture the concurrent behaviour of the system, the user writes a CSP specification. This specification can be validated by simulation and analysis using standard CSP tools as discussed above. This specification is then translated into CSP++ source code, which is a C++ program using primitives provided by a C++ header files. The CSP++ code can be compiled and linked in order to run on the target computer. As a specialty, CSP++ offers to attach C++ functions to CSP events.

Example 40.7: The Puzzle in CSP++

It is necessary to re-formulate our specification of the Children's Puzzle in the CSP sublanguage supported by CSP++. We have to fix the number of children to three and
- to replace the replicated parallel by the general parallel operator,
- to replace the indexed channels c.p by three individual channels, and

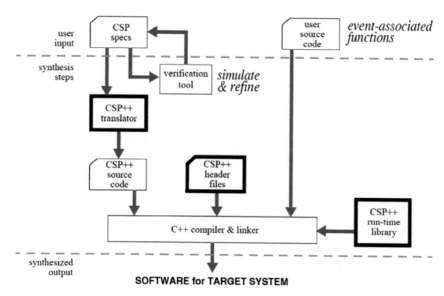

Fig. 3.10 Software synthesis with CSP++ [Gar05]

- to implement the arithmetic given by the `fill` function by a process
  `Compute`

  Furthermore, we add a channel `d` in order observe the communications of the system:

```
channel c0, c1, c2 : {0..4}
channel d : {0..2}.{0..4}

Child(0,x) =
    (c1!x/2 -> d.1.x/2 -> c0?y -> Compute(0,x/2,y))
 [] (c0?y -> c1!x/2 -> d.1.x/2 -> Compute(0,x/2,y))

Child(1,x) =
    (c2!x/2 -> d.2.x/2 -> c1?y -> Compute(1,x/2,y))
 [] (c1?y -> c2!x/2 -> d.2.x/2 -> Compute(1,x/2,y))

Child(2,x) =
    (c0!x/2 -> d.0.x/2 -> c2?y -> Compute(2,x/2,y))
 [] (c2?y -> c0!x/2 -> d.0.x/2 -> Compute(2,x/2,y))

Compute(p,x,y) = if        x+y==0 then Child(p,0)
                 else if x+y==1 then Child(p,2)
                 else if x+y==2 then Child(p,2)
                 else if x+y==3 then Child(p,4)
```

```
                    else                      Child(p,4)

SYS =        (Child(0,0) [|{|c1|}|] Child(1,2))
           [|{|c0,c2|}|]
                Child(2,4)
```

Using FDR, one can establish stability on this model as well.

Translation with CSP++ results in a C++ file of about 200 lines of code. Running the resulting program yields the following output:

$$\langle d.1.0, d.2.1, d.0.2, d.1.1, d.2.1, d.0.2, d.1.2, d.2.1, d.0.2, d.1.2, d.2.2, d.0.2 \rangle$$

As expected, the system stabilizes after nine steps. The channels c0, c1 and c2 are treated by CSP++ as internal events. There is a fixed schedule selecting one possible execution path through the system.

Semantically, there are several fundamental questions concerning such transformational approaches:

1. How does the approach select an subset of CSP processes? As CSP processes can be non-deterministic and computers work in a deterministic fashion, there are clearly CSP processes which can't be implemented—the most general livelock free process $DivF_A\Sigma$, see Sect. 3.4.4, is an example.
2. Is there a well-defined semantical relation (refinement, simulation, ...) between the CSP specification and the implementation?
3. Does the translation preserve essential properties such as deadlock freedom and livelock freedom?

Unfortunately, such questions are scarcely discussed in the literature on transformations.

3.4 Semantics and Analysis

In this section, we are looking deeper into the 'machinery' of CSP.

Traditionally, the semantics of CSP is given in denotational style. In Sect. 3.4.1 we discuss the semantic domain, the notion of refinement, as well as selected semantical clauses for the traces model $\mathcal{T}$, the failures/divergences model $\mathcal{N}$, and the stable failures model F and relate these models with each others.

In Sect. 3.4.2, we give a brief overview on the algebraic laws that hold within CSP. These algebraic laws can be used to give CSP an axiomatic semantics and are often used when analysing processes. The tool CSP-Prover is based on these.

Within CSP denotational semantics, recursive processes obtain their meaning via fixed points. In Sect. 3.4.3, we discuss some basics of this approach.

Finally, in Sect. 3.4.4, we discuss how to analyse CSP processes for livelock, deadlock, and determinism. The 'trick' behind these methods is always the same: one encodes the property under discussion as a CSP process *Prop* and shows that any process that refines *Prop* 'inherits' the desired property. Thus, besides discussing important properties of processes, this section provides 'blueprints' of how to analyse CSP processes.

3.4.1 The Three Standard Models

Traditionally, the semantics of CSP is given in denotational style. To this end, one chooses a set $\mathcal{D}$ of possible observations and defines a semantic function $[\![\_]\!]_{\mathcal{D}}$ from the CSP syntax into $\mathcal{D}$. Each CSP process P is assigned a set of observations $[\![P]\!]_{\mathcal{D}} \in \mathcal{D}$. Often, one speaks of a model $\mathcal{D}$ to denote both, the domain and the semantic function.

Relatively to such a model $\mathcal{D}$ one can answer the questions:

- Are (the denotations) of processes P and Q equal? This equality is written as

$$P =_{\mathcal{D}} Q.$$

- Is process P refined by process Q (with respect to $\mathcal{D}$)? This refinement statement is written as

$$P \sqsubseteq_{\mathcal{D}} Q.$$

Both these questions can be formulated in terms of each other:

Lemma 3 *Let $\mathcal{D}$ be a CSP model in which both two properties*

$$P \sqsubseteq_{\mathcal{D}} Q \iff [\![Q]\!]_{\mathcal{D}} \subseteq [\![P]\!]_{\mathcal{D}} \tag{3.4}$$

$$[\![P \sqcap Q]\!]_{\mathcal{D}} = [\![P]\!]_{\mathcal{D}} \cup [\![Q]\!]_{\mathcal{D}}. \tag{3.5}$$

hold. Then the following equivalences are true:

1. *$P =_{\mathcal{D}} Q$ if and only if $P \sqsubseteq_{\mathcal{D}} Q \wedge Q \sqsubseteq_{\mathcal{D}} P$ and*
2. *$P \sqsubseteq_{\mathcal{D}} Q$ if and only if $P \sqcap Q =_{\mathcal{D}} P$.*

Proof Ad 1: clearly $[\![P]\!]_{\mathcal{D}} = [\![Q]\!]_{\mathcal{D}}$ if and only if $[\![P]\!]_{\mathcal{D}} \subseteq [\![Q]\!]_{\mathcal{D}} \wedge [\![Q]\!]_{\mathcal{D}} \subseteq [\![P]\!]_{\mathcal{D}}$.

Ad 2: "$\Rightarrow$" Let $P \sqsubseteq_{\mathcal{D}} Q$. Then $[\![Q]\!]_{\mathcal{D}} \subseteq [\![P]\!]_{\mathcal{D}}$. Thus $[\![P \sqcap Q]\!]_{\mathcal{D}} = [\![P]\!]_{\mathcal{D}} \cup [\![Q]\!]_{\mathcal{D}} = [\![P]\!]_{\mathcal{D}}$. "$\Leftarrow$" By Contraposition. Let $\neg(P \sqsubseteq_{\mathcal{D}} Q)$, i.e., $\neg([\![Q]\!]_{\mathcal{D}} \subseteq [\![P]\!]_{\mathcal{D}})$. Then there exists an observation $q \in [\![Q]\!]_{\mathcal{D}}$ with $q \notin [\![P]\!]_{\mathcal{D}}$. For q it holds that $q \in [\![P]\!]_{\mathcal{D}} \cup [\![Q]\!]_{\mathcal{D}} = [\![P \sqcap Q]\!]_{\mathcal{D}}$. Thus $P \sqcap Q \neq_{\mathcal{D}} P$. ∎

In all three standard models of CSP, the traces model $\mathcal{T}$, the failures/divergences models $\mathcal{N}$, and the stable failures model $\mathcal{F}$, properties (3.4) and (3.5) both hold.

Concerning binary relations such as refinement, the following definition is standard.

Definition 1 (*Partial order, bottom element*) A *partial order* $(PO, \leq)$ consists of

- a set PO and
- a relation $\leq \ \subseteq PO \times PO$ which is
    - reflexive, i.e., for all $x \in PO : x \leq x$,
    - antisymmetric, i.e., for all $x, y \in PO : x \leq y \wedge y \leq x \Rightarrow x = y$, and
    - transitive, i.e., for all $x, y, z \in PO : x \leq y \wedge y \leq z \Rightarrow x \leq z$.

A *partial order with bottom element* $(PO, \leq, \bot)$ consists of a *partial order* $(PO, \leq)$ and

- an element $\bot \in PO$ with $\bot \leq x$ for all $x \in PO$.

It turns out that refinement in all three standard models is a partial order.

In the following, we provide an overview of the three standard models following the scheme below:

- **Domain:** There is a domain of possible observations, say the domain is the set of all traces. However, not all sets of traces can be the observation of a CSP process. Thus, so-called *healthiness* conditions exclude such impossible observations.
- **Refinement:** The notion of refinement $P \sqsubseteq_{\mathcal{D}} Q$ over $\mathcal{D}$ is defined. Some domains have a most refined process or a least refined process, i.e., a bottom or a top element w.r.t. the refinement order.
- **Semantical clauses:** Here we explain some typical semantical clauses of the model.
- **Relation to other models:** We relate each new model to the models introduced previously.

The Traces Model $\mathcal{T}$

Domain. The traces model observes the finite sequences of events that a process can engage with. The empty observation $\langle \rangle$ is always possible. If there was an observation $o = \langle a_1, a_2, \ldots, a_{k-1}, a_k \rangle$, obviously prior to o is was possible to observe $o' = \langle a_1, a_2, \ldots, a_{k-1} \rangle$. These considerations lead to two healthiness conditions on trace observations: for all observations T it holds that

T1 $T \neq \emptyset$ and T is prefix-closed, i.e., $\forall t \in T.s \leq t \Rightarrow s \in T$.

Here, $s \leq t$ stands for s is a prefix of t.

In contrast to other process algebras, CSP treats termination of processes: the process *Skip* does nothing but to terminate; after termination of P the process $P \, ; Q$ passes control to Q. On the semantic side, termination is represented as a special event $\checkmark$ (pronounce "tick"). Concerning observations, $\checkmark$ can only happen at the end of a trace: after termination, a process does not engage in any further events. Thus, the possible observations over an alphabet Σ are given by

$$\Sigma^{*\checkmark} \triangleq \Sigma^* \cup \{s \,^\frown \langle \checkmark \rangle \mid s \in \Sigma^*\}.$$

Together with the two healthiness conditions, we obtain the CSP traces domain as

$$\mathcal{T} \triangleq \{T \subseteq \Sigma^{*\checkmark} \mid T \text{ fulfills T1}\}.$$

Refinement over $\mathcal{T}$ is defined as

$$P \sqsubseteq_\mathcal{T} Q \iff [\![Q]\!]_\mathcal{T} \subseteq [\![P]\!]_\mathcal{T}$$

i.e., $\mathcal{T}$ has property (3.4) stated in Lemma 3. We observe:

Theorem 1 $\sqsubseteq_\mathcal{T}$ *is a partial order.*

The definition of $\sqsubseteq_\mathcal{T}$ can be read: a process Q refines a process P if Q exhibits at most P's executions. This is why the model $\mathcal{T}$ is concerned with safety: if one has already proven for P that 'nothing bad can happen' (there is no trace representing system failure), then any process Q that refines P will have this property as well. Consequently, the process *Stop*, which has only the empty trace as its denotation, refines all processes. The process

$$RUN\_With\_Skip_\Sigma = (?x : \Sigma \to RUN\_With\_Skip_\Sigma) \sqcap Skip$$

has $\Sigma^{*\checkmark}$ as its denotation, i.e., it is the least refined process over $\mathcal{T}$.

Semantical clauses. In the context of the trace model $\mathcal{T}$, one usually writes $traces(\_\_)$ instead of $[\![\_\_]\!]_\mathcal{T}$. $traces(\_\_)$ is defined inductively over the process syntax. We present here some typical clauses, the full model is given in Sect. B.4.

The process *Stop* admits only the empty trace as observation. Of the terminating process *Skip* one can either observe the empty trace—before termination—or the trace consisting of the termination symbol $\checkmark$.

$$traces(Stop) \triangleq \{\langle \rangle\}$$
$$traces(Skip) \triangleq \{\langle \rangle, \langle \checkmark \rangle\}$$

The multiple prefix operator $?x : X \to P(x)$ offers to the environment all events in X. When the environment has chosen a specific event, say a, the process behaves as $P(a)$:

$$traces(?x : X \to P) \triangleq \{\langle \rangle\} \cup \{\langle a \rangle \,^\frown s \mid s \in traces(P[a/x]), \, a \in X\}$$

Here, CSP uses the substitution $[a/x]$ on the syntactic level in order to resolve the binding of a to the variable x.

CSP internal choice is mapped to set union:

$$traces(P \sqcap Q) \triangleq traces(P) \cup traces(Q)$$

i.e., $\mathcal{T}$ has also property (3.5) of Lemma 3.

The clause for the CSP general parallel operator

$$traces(P \, [\![\, A \,]\!] \, Q) \triangleq \bigcup \{s \, [\![\, A \,]\!] \, t \mid s \in traces(P) \wedge t \in traces(Q)\}$$

is defined in terms of an interleaving function on the traces domain $\mathcal{T}$—see Sect. B.4 for its definition. This style is typical for denotational semantics in general: for every operator in the syntax one defines a function on the semantic domain.

Sequential composition $P \, \mathring{,} \, Q$ has all traces of P which do not terminate. This is ensured by intersection with Σ^*. Furthermore, one obtains all traces which are composed of terminating traces from P—but with $\checkmark$ removed—and traces from Q:

$$traces(P \, \mathring{,} \, Q) \triangleq (traces(P) \cap \Sigma^*)$$
$$\cup \{s \, \hat{} \, t \mid s \, \hat{} \, \langle \checkmark \rangle \in traces(P), t \in traces(Q)\}$$

The conditional operator exhibits yet another CSP specialty. CSP assumes that there is a logic that allows one to formulate conditions φ. This logic, however, is never made explicit. When CSP is integrated with data, as, e.g., in CSP_M [Sca98] or in CSP-CASL [Rog06], this logic and its evaluation rules are made explicit. Here, we formulate only:

$$traces(\textbf{if } \varphi \textbf{ then } P \textbf{ else } Q) \triangleq \begin{cases} traces(P) \text{ if } \varphi \text{ evaluates to } true \\ traces(Q) \text{ if } \varphi \text{ evaluates to } false \end{cases}$$

Like parallel composition, CSP renaming is defined utilising an operation on the the traces domain:

$$traces(P[\![R]\!]) \triangleq \{t \mid \exists s \in traces(P).sR^*t\}$$

Given a relation $R \subseteq \Sigma \times \Sigma$, $R^* \subseteq \Sigma^{*\checkmark} \times \Sigma^{*\checkmark}$ is defined to be the smallest set satisfying:

- $(\langle \rangle, \langle \rangle) \in R^*$
- $(\langle \checkmark \rangle, \langle \checkmark \rangle) \in R^*$
- $(a, b) \in R \wedge (t, t') \in R^* \Rightarrow (a \, \hat{} \, t, b \, \hat{} \, t') \in R^*$

Semantical clauses need to respect the healthiness conditions of the domain, i.e., given that the arguments are in the domain, their result needs to be in the domain as well. Thus, the definition of each semantical clause comes with proof obligations. Let us consider these proof obligations in the simple case of the prefix operator:

$$traces(a \rightarrow P) \triangleq \{\langle\rangle\} \cup \{\langle a\rangle \frown s \mid s \in traces(P)\}$$

Healthiness of this clause means to establish the following theorem:

Theorem 2 $traces(a \rightarrow P) \in \mathcal{T}$ for $traces(P) \in \mathcal{T}$.

Proof First, we show $traces(a \rightarrow P) \subseteq \Sigma^{*\checkmark}$, i.e., the symbol $\checkmark$ appears only at the end of the traces, if at all. Let $t \in traces(a \rightarrow P)$. Then t takes one of the following forms:

- $t = \langle\rangle$. We have $\langle\rangle \in \Sigma^*$ and therefore $\langle\rangle \in \Sigma^{*\checkmark}$.
- $t = \langle a\rangle \frown s$. As $s \in traces(P) \subseteq \Sigma^{*\checkmark}$, we have $s \in \Sigma^*$ or $s \in \Sigma^* \frown \langle\checkmark\rangle$. $s \in \Sigma^*$ implies $\langle a\rangle \frown s \in \Sigma^*$; $s \in \Sigma^* \frown \langle\checkmark\rangle$ implies $\langle a\rangle \frown s \in \Sigma^* \frown \langle\checkmark\rangle$. Thus, in both cases we obtain $t \in \Sigma^{*\checkmark}$.

  It remains to show that both parts of condition T1 hold for $traces(a \rightarrow P)$.

To part 1: By definition $\langle\rangle \in traces(a \rightarrow P)$. Thus, $traces(a \rightarrow P) \neq \emptyset$.
To part 2: Let $t \in traces(a \rightarrow P)$. Then t takes one of the following forms:

- $t = \langle\rangle$. Then the only prefix of t is $\langle\rangle$, which by definition is in $traces(a \rightarrow P)$.
- $t = \langle a\rangle \frown s$. Then a prefix p of t can take one of the following forms:

  - $p = \langle\rangle$. By definition $\langle\rangle$ is in $traces(a \rightarrow P)$.
  - $p = \langle a\rangle \frown s'$ for some prefix s' of s. As $s \in traces(P)$ and $traces(P)$ is prefix-closed, we have that $s' \in traces(P)$. Thus, $\langle a\rangle \frown s' \in traces(a \rightarrow P)$. ∎

As demonstrated by the example of buffers, cf. Sect. 3.2.3, the traces domain does not suffice to express all properties on traces, in which we might be interested in. Thus, there have been a number of refined models that allow for more detailed observation of processes.

The Failures/Divergences Model $\mathcal{N}$

Domain. The failures/divergences model takes a closer look on processes than the traces model $\mathcal{T}$. It specifically records, which divergences a process has. In a transition system, cf. Sect. 3.2.2, a *divergent state* is one, from which it is possible to take an infinite number of τ transitions. If a system has reached such a divergent state, it is possible that the system is internally 'alive' and changes it state forever, however without ever engaging in any event observable from the outside. This is the reason why in such situation one also speaks of a *livelock*. In denotational semantics, a *divergence of a process* is a trace, which in the transition system of this process leads to a divergent state.

The model $\mathcal{N}$ takes the point of view that a process can't be controlled anymore after it has reached a divergent state. On the practical side, this decision makes it easier to work with the model: there is no need to distinguish between processes after they have diverged. The price for this has to be paid on the theoretical side: the semantical clause for the hiding operator is problematic (hiding is only defined for processes with a finitely branching transition system), and the fixed point theory for partial orders, see Sect. 3.4.3, does not work as expected ($\mathcal{N}$ is a complete partial order only for finite alphabets).

In $\mathcal{N}$, the possible observations on a process P over an alphabet Σ are given by a pair $(F, D) = (failures_{\perp}(P), divergences(P))$ with

- $F \subseteq \Sigma^{*\checkmark} \times \mathcal{P}(\Sigma \cup \{\checkmark\})$ and
- $D \subseteq \Sigma^{*\checkmark}$.

As the model $\mathcal{N}$ has only one denotation after a divergence, its process denotations (F, D) have to fulfill the healthiness conditions:

D1 D is extension closed, i.e., $\forall s \in D \cap \Sigma^*, t \in \Sigma^{*\checkmark}.s \frown t \in D$.

D2 F includes all divergence related failures, i.e.,

$$\forall s \in D, X \subseteq (\Sigma \cup \{\checkmark\}).(s, X) \in F.$$

Furthermore, we do not distinguish how processes behave after successful termination:

D3 $s \frown \langle \checkmark \rangle \in D \Rightarrow s \in D$.

Given the failures of a process P, one can extract the traces that P can perform:

$$traces_{\perp}(P) \triangleq \{s \mid \exists X.(s, X) \in failures_{\perp}(P)\}$$

The index $\perp$ in $traces_{\perp}(P)$ and $failures_{\perp}(P)$ indicates, that— thanks to the conditions D1 and D2, resp.—we might record more traces and more refusal sets than actually present in the transition system of the process P.

The following healthiness condition captures the interplay between traces and refusals:

F3 Events, that can't happen, need to be refused, i.e.,

$$\forall (s, X) \in F, \ Y \subseteq \Sigma \cup \{\checkmark\}.$$
$$(\forall a \in Y.s \frown \langle a \rangle \notin traces_{\perp}(F)) \implies (s, X \cup Y) \in F.$$

Here, by abuse notation, we apply $traces_{\perp}(\_\_)$ to a failure set.

The model $\mathcal{N}$ has many more healthiness conditions, see Sect. B.4 for the complete list.

Refinement over $\mathcal{N}$ is defined as

$$P \sqsubseteq_{\mathcal{N}} Q \iff failures_\perp(Q) \subseteq failures_\perp(P) \wedge$$
$$divergences(Q) \subseteq divergences(P)$$

i.e., $\mathcal{N}$ has property (3.4) stated in Lemma 3. We observe:

Theorem 3 $\sqsubseteq_{\mathcal{N}}$ *is a partial order.*

The above refinement definition can be read as:

- Q exhibits at most $P's$ executions (thanks to the definition of $traces_\perp(\_)$),
- Q is more deterministic than P (see the definition of determinism in Sect. 3.4.4), and
- Q has less livelocks than P.

There is one 'uncontrollable process' *Div*. The process *Div* does nothing and diverges immediately. Thus, it has all divergences and all failures. Its denotation is

$$failures_\perp(Div) \triangleq \Sigma^{*\checkmark} \times \mathcal{P}(\Sigma \cup \{\checkmark\})$$
$$divergences(Div) \triangleq \Sigma^{*\checkmark}$$

i.e., it is the least refined process in $\mathcal{N}$. In the traces models, *Div* is equivalent to *Stop*, i.e., $traces(Div) = \{\langle\rangle\}$. There is no most refined process in $\mathcal{N}$.

Semantical clauses In contrast to *Div*, the process *Stop* has refusal sets only for the empty trace and has no divergences:

$$failures_\perp(Stop) \triangleq \{(\langle\rangle, X) \mid X \subseteq \Sigma \cup \{\checkmark\}\}$$
$$divergences(Stop) \triangleq \emptyset$$

Action prefix has the following clauses:

$$failures_\perp(a \rightarrow P) \triangleq \{(\langle\rangle, X) \mid X \subseteq \Sigma \cup \{\checkmark\}, a \notin X\}$$
$$\cup \{(\langle a\rangle \frown s, X) \mid (s, X) \in failures_\perp(P)\}$$
$$divergences(a \rightarrow P) \triangleq \{\langle a\rangle \frown t \mid t \in divergences(P)\}$$

Initially, the process $a \rightarrow P$ can't refuse a, i.e., eventually it has to perform a. After the process has engaged in a, its refusals are those from P. The only divergences of $a \rightarrow P$ are those that arise from P. The set of divergences of $a \rightarrow P$ is empty if $divergences(P) = \emptyset$.

CSP internal choice is mapped to set union in both components:

$$failures_\perp(P \sqcap Q) \triangleq failures_\perp(P) \cup failures_\perp(Q)$$
$$divergences(P \sqcap Q) \triangleq divergences(P) \cup divergences(Q)$$

i.e., $\mathcal{N}$ has also property (3.5) of Lemma 3.

The clauses for renaming are as follows:

$$failures_\perp(P[\![R]\!]) \triangleq \{(s', X) \mid \exists s.s\,R^*\,s' \wedge (s, R^{-1}(X)) \in failures_\perp(P)\}$$
$$\cup \{(s, X) \mid s \in divergences(P[\![R]\!]) \wedge X \subseteq \Sigma \cup \{\checkmark\}\}$$
$$divergences(P[\![R]\!]) \triangleq \{s' \frown t \mid t \in \Sigma^{*\checkmark} \wedge \exists s \in divergences(P) \cap \Sigma^*.s\,R^*\,s'\}$$

Here, $R^{-1}(X) \triangleq \{a \mid \exists a' \in X.a\,R\,a'\}$. We read the failures clause as follows: X is a refusal set of $P[\![R]\!]$ if its 'reverse image' $R^{-1}(X)$ is refused by P. For the sake of healthiness condition D2, we also include all refusals for any divergent trace. Concerning the divergence clauses, we get: if s is a divergence from P that does not end with a $\checkmark$—encoded via $s \in divergences(P) \cap \Sigma^*$— then we rename it via R^* into s' and obtain a divergence for $P[\![R]\!]$. Note that thanks to D3 we have that $s \frown \langle\checkmark\rangle \in D$ implies $s \in D$.

Over $\mathcal{N}$, it is impossible to provide a general semantical clause that computes the divergences of the process $P \setminus X$. The problem is that hiding introduces a divergence in $P \setminus X$ when P can perform an infinite consecutive sequence of events in X. The difficulty is that over $\mathcal{N}$ we consider only finite traces, not infinite ones. However, it is possible to provide a semantical clause for the subclass of finitely branching processes. For further discussion of this matter see, e.g., [Ros98].

Relation to the traces model $\mathcal{T}$. The traces of the model $\mathcal{T}$ are related with the traces of the model $\mathcal{N}$, also the failures as defined in Sect. 3.2.3 are related with the failures of the model $\mathcal{N}$. Given a process P, we have

$$traces_\perp(P) = traces(P) \cup divergences(P) \tag{3.6}$$

and

$$\begin{aligned}failures_\perp(P) = \ &failures(P) \\ &\cup \{(s, X) \mid s \in divergences(P), X \subseteq (\Sigma \cup \{\checkmark\})\}\end{aligned} \tag{3.7}$$

These connections between the models can be derived, e.g., from the operational semantics of CSP. As the operational models can be proven to coincide with the denotational ones, it is justified to use these connections to establish the following theorem:

Theorem 4 *If $divergences(P) = \emptyset$, then $P \sqsubseteq_\mathcal{N} Q \Rightarrow P \sqsubseteq_\mathcal{T} Q$.*

Proof Let $P \sqsubseteq_\mathcal{N} Q$. As $divergences(P) = \emptyset$, we obtain $divergences(Q) = \emptyset$. Therefore, $traces_\perp(P) = traces(P)$ and $traces_\perp(Q) = traces(Q)$. From $failures_\perp(Q) \subseteq failures_\perp(P)$ we derive that $traces_\perp(Q) \subseteq traces_\perp(P)$. Combining these arguments leads to $traces(Q) = traces_\perp(Q) \subseteq traces_\perp(P) = traces(P)$. ∎

Note that is it not possible to relax the precondition of Theorem 4. As Div is the least refined process over $\mathcal{N}$, we have, e.g., $Div \sqsubseteq_\mathcal{N} a \to Stop$. However, $traces(a \to Stop) = \{\langle\rangle, \langle a\rangle\} \not\subseteq \{\langle\rangle\} = traces(Div)$ i.e., $Div \not\sqsubseteq_\mathcal{T} a \to Stop$.

By considering failures, the model $\mathcal{N}$ is capable of expressing liveness properties. In contrast the model $\mathcal{T}$ is limited to safety properties. Consequently, trace refinement does not imply failure divergences refinement:

Example 41: Safety Does Not Imply Liveness

For our very first example on ATMs, we said it was safe to deliver cash *after* the user has been authenticated by card and pin. We formulated this ordering relation as a CSP process:

$$ATM0 = ready \rightarrow cardI \rightarrow pinE \rightarrow cardO \rightarrow cashO \rightarrow Stop$$

Any process, which is a trace refinement of $ATM0$, is safe in the above sense, as it will respect the ordering of the events. On the intuitive level *Stop* is safe, as it never delivers any cash. Formally, it holds that

$$ATM0 \sqsubseteq_{\mathcal{T}} Stop,$$

as *Stop* is the most refined process over $\mathcal{T}$.

However, the process *Stop* would not be an implementation that anyone would be interested in: it never will deliver any cash. Failures divergences refinement takes liveness into account, namely, that eventually something good will happen. Therefore, the model $\mathcal{N}$ allows us to exclude the process *Stop* from the potential implementations. It holds that

$$ATM0 \not\sqsubseteq_{\mathcal{N}} Stop.$$

We see this by considering the failures of both processes:
- $(\langle\rangle, \{ready\}) \in failures_{\perp}(Stop)$ and
- $(\langle\rangle, \{ready\}) \notin failures_{\perp}(ATM0)$.

Thus, $failures_{\perp}(Stop) \not\subseteq failures_{\perp}(ATM0)$.

As a result of Example 41, we formulate:

Corollary 3 *There exist processes P and Q with $P \sqsubseteq_{\mathcal{T}} Q$ but $P \not\sqsubseteq_{\mathcal{N}} Q$.*

The Stable Failures Model $\mathcal{F}$

Domain. The stable failures model is an alternative to the failures/divergences model. The underlying idea is to ignore that a process might diverge. This can become handy, as it is sometimes possible to establish that a process is divergence free as it does not include any operators that could contribute to divergence. For instance, the divergence clause for the action prefix operator is:

$$divergences(a \rightarrow P) \triangleq \{\langle a \rangle ^\frown t \mid t \in divergences(P)\}$$

I.e., if P has no divergences, then $a \to P$ has no divergences. Thus, action prefix does not contribute to divergence. Ignoring divergences not only simplifies the semantical clauses, but also solves the above mentioned problem with hiding. Furthermore, the treatment of recursive equations becomes 'easier'. Finally, one should mention that over $\mathcal{F}$ we do not record more traces or more refusal sets than actually present in the transition system of the process P—as it was possibly the case over $\mathcal{N}$.

In $\mathcal{F}$, the possible observations on a process P over an alphabet Σ are given by a pair $(T, F) = (traces(P), failures(P))$ with

- $T \subseteq \Sigma^{*\checkmark}$ and
- $F \subseteq \Sigma^{*\checkmark} \times \mathcal{P}(\Sigma \cup \{\checkmark\})$.

The traces component T is of a process P in $\mathcal{F}$ is identical with it semantics over $\mathcal{T}$. Thus, we have as healthiness conditions again:

T1 $T \neq \emptyset$ and T is prefix-closed.

There is an interplay between failures sets and traces:

T2 $(s, X) \in F \Rightarrow s \in T$,

i.e., when we record a refusal X after trace s, then s is a trace recorded in the traces component. Note that the converse it not necessarily true: should a process be unstable after a trace s, we won't record any refusal. Thus, in $\mathcal{F}$ one can't derive the traces of a process from its failures component.

The further healthiness condition deals with termination:

T3 $(s ^\frown \checkmark) \in T \Rightarrow (s ^\frown \checkmark, X) \in F,\ X \in \mathcal{P}(\Sigma \cup \{\checkmark\})$.

After termination, a process has all failures.

See Sect. B.4 for further healthiness conditions.

Refinement over $\mathcal{F}$ is defined as

$$P \sqsubseteq_{\mathcal{F}} Q \iff traces(Q) \subseteq traces(P) \wedge failures(Q) \subseteq failures(P)$$

i.e., $\mathcal{F}$ has property (3.4) stated in Lemma 3. We observe:

Theorem 5 $\sqsubseteq_{\mathcal{F}}$ *is a partial order.*

The above refinement definition can be read as:

- Q exhibits at most $P's$ executions and
- Q is more deterministic than P (see the definition of determinism in Sect. 3.4.4).

It might come as a surprise that the process Div is the most refined process over $\mathcal{F}$. Its semantics is:

$$traces(Div) \triangleq \{\langle\rangle\}$$
$$failures(Div) \triangleq \emptyset$$

i.e., the process does nothing and diverges immediately—as we record only stable failures, Div has an empty failures component.

The process

$$RUN_{\Sigma}^{+} = (?x : \Sigma \to RUN_{\Sigma}^{+}) \sqcap Stop \sqcap Skip$$

has $(\Sigma^{*\checkmark}, \Sigma^{*\checkmark} \times \mathcal{P}(\Sigma \cup \{\checkmark\}))$ as its denotation, i.e., it is the least refined process over $\mathcal{F}$.

Semantical clauses. We already have seen the semantical clauses for the traces component of $\mathcal{F}$. The failure clauses for the CSP operators are essentially the same for F and $\mathcal{N}$: the only difference is that we do not need to close under the healthiness conditions D1 and D2.

For the process $Stop$, there is no change as it does not have any divergences:

$$failures(Stop) \triangleq \{(\langle\rangle, X) \mid X \subseteq \Sigma \cup \{\checkmark\}\}.$$

Similarly, no change is required for action prefix:

$$failures(a \to P) \triangleq \quad \{(\langle\rangle, X) \mid X \subseteq \Sigma \cup \{\checkmark\}, a \notin X\}$$
$$\cup \{(a \frown s, X) \mid (s, X) \in failures((\,)P)\}.$$

Again, CSP internal choice is mapped to set union:

$$failures(P \sqcap Q) \triangleq failures(P) \cup failures(Q)$$

i.e., $\mathcal{F}$ has also property (3.5) of Lemma 3.

Concerning renaming, we get a different clause:

$$failures(P\llbracket R \rrbracket) \triangleq \{(s', X) \mid \exists s.sR^*s' \wedge (s, R^{-1}(X) \in failures(P)\}$$

The reason for the difference is that we do not need to 'close' the failure set for the sake of healthiness condition D2.

Hiding has an astonishing simple clause:

$$failures(P \setminus X) \triangleq \{(s \setminus X, Y) \mid (s, X \cup Y) \in failures(P)\}.$$

For traces, hiding on the syntactic level has a counterpart on the semantics:

$$\langle\rangle \setminus X = \langle\rangle$$
$$(\langle x \rangle \frown t) \setminus X = t \setminus X \qquad (if\,x \in X)$$
$$(\langle y \rangle \frown t) \setminus X = \langle y \rangle \frown (t \setminus X) \quad (if\,y \notin X)$$

Relation to model F and model $\mathcal{T}$. We first establish an algebraic law over $\mathcal{F}$:

Theorem 6 *For all* CSP *processes P it holds:* $P \sqcap Div =_{\mathcal{F}} P$.

Proof As $traces(P)$ is non empty and prefix closed, we have $\langle\rangle \in traces(P)$ and thus $traces(P) \cup traces(Div) = traces(P)$. As $failures(Div) \triangleq \emptyset$, we have $failures(P) \cup failures(Div) = failures(P)$.   ∎

This law does not hold over $\mathcal{N}$. Take, for example, $P = Stop$. We have $divergences(Stop) = \emptyset$, however, $divergences(Stop \sqcap Div) = \emptyset \cup \Sigma^{*\checkmark} = \Sigma^{*\checkmark}$.

Corollary 4 *There exist* CSP *processes P with: $P \sqcap Div \neq_{\mathcal{N}} P$.*

Thanks to equations (3.6) and (3.7), we can prove that the models $\mathcal{N}$ and $\mathcal{F}$ agree on divergence-free processes:

Theorem 7 *If $divergences(P) = divergences(Q) = \emptyset$, then $P \sqsubseteq_{\mathcal{N}} Q \Leftrightarrow P \sqsubseteq_{\mathcal{F}} Q$.*

Proof "$\Rightarrow$" Let $P \sqsubseteq_{\mathcal{N}} Q$. Then $failures_{\perp}(Q) \subseteq failures_{\perp}(P)$.

Concerning the traces component in $\mathcal{F}$ we compute: by the definition of $traces_{\perp}(\_\_)$ we have that $failures_{\perp}(Q) \subseteq failures_{\perp}(P)$ implies $traces_{\perp}(Q) \subseteq traces_{\perp}(P)$. Thanks to (3.6) and $divergences(P) = divergences(Q) = \emptyset$ we have $traces_{\perp}(P) = traces(P)$ and $traces_{\perp}(Q) = traces(Q)$. Thus we have: $failures_{\perp}(Q) \subseteq failures_{\perp}(P)$ implies $traces(Q) \subseteq traces(P)$.

Concerning the failures component in $\mathcal{F}$ we compute: thanks to (3.7) and $divergences(P) = divergences(Q) = \emptyset$ we have $failures_{\perp}(P) = failures(P)$ and $failures_{\perp}(Q) = failures(Q)$. Thus $failures_{\perp}(Q) \subseteq failures_{\perp}(P)$ implies $failures(Q) \subseteq failures(P)$.

Together, these two considerations result in $P \sqsubseteq_{\mathcal{F}} Q$.

"$\Leftarrow$" Let $P \sqsubseteq_{\mathcal{F}} Q$. Then $failures(Q) \subseteq failures(P)$.

Concerning the failures component in $\mathcal{N}$ we compute: thanks to (3.7) and $divergences(P) = divergences(Q) = \emptyset$ we have $failures(P) = failures_{\perp}(P)$ and $failures(Q) = failures_{\perp}(Q)$. Thus $failures(Q) \subseteq failures(P)$ implies $failures_{\perp}(Q) \subseteq failures_{\perp}(P)$.

For the divergence component in $\mathcal{N}$ we have by assumption that the following holds: $divergences(P) = divergences(Q) = \emptyset$.

Together, these two considerations result in $P \sqsubseteq_{\mathcal{N}} Q$.   ∎

As the model $\mathcal{F}$ in its traces component uses the same clauses as the model $\mathcal{T}$, we obviously have:

Theorem 8 $P \sqsubseteq_{\mathcal{F}} Q \Rightarrow P \sqsubseteq_{\mathcal{T}} Q$.

Concerning the inverse direction, with the help of Theorem 7 we can utilise Example 41 again: both, *ATM0* and *Stop* are divergent free processes. Thus, *ATM0* $\not\sqsubseteq_{\mathcal{N}}$ *Stop* implies *ATM0* $\not\sqsubseteq_{\mathcal{F}}$ *Stop*. Following these considerations, Example 41 provides a witness for:

Corollary 5 *There exist processes P and Q such that $P \sqsubseteq_{\mathcal{T}} Q$ but $P \not\sqsubseteq_{\mathcal{F}} Q$.*

Reflection

One might wonder if it really is a good thing that there are these three different and many more semantics for CSP. One can put this question into an even wider context, as Rob van Glabbeek did in his classical account [vG01], in which he studies the relationships between no less than 15 different semantics for process calculi. And there are further more. Here, the question arises how many semantics do we need, is the development of further semantic models of process algebra 'scientific nonsense' / occupational therapy for computer scientists?

The general observation concerning such semantics models is that their development is driven by 'necessities': certain shortcomings of an existing model are 'patched' by providing a new model. However, often this new model comes at a 'price'.

We illustrate this on the example of CSP. The traces model is fine when one wants to study safety properties of systems. However, the traces model is too weak to study liveness properties. Thus, when one wants to look, e.g., into divergences, one needs a 'refined' view on processes. The failures/divergences model provides such more fine grained view. However, this model runs into problems with denotational semantics, e.g., it is impossible to provide a clause for the hiding operator. This can be resolved by utilising the failures model, however, this comes at the price that divergences can't be studied in it, though other liveness properties such as deadlocks can.

In the case of CSP, computer science has not been able to provide a single, satisfactory model that could serve as semantical platform for all the analysis methods one might be interested in. Thus, the pragmatic answer to the above question is: one takes the 'simplest' model that allows one to apply the method that delivers the analysis one is interested in. As there are clear relationships between the models (see, e.g., the above theorems concerning refinements, Theorems 4, 7, and 8), it is possible to combine results achieved in different settings.

This situation is similar to the science of physics. When describing the motion of macroscopic objects, from projectiles to parts of machinery, and astronomical objects, classical mechanics is perfectly fine. However, when it comes to aspects at small (atomic and subatomic) scales, quantum mechanics is required. These two different theories are related. Most theories in classical physics can be derived from quantum mechanics as an approximation valid at large (macroscopic) scale.

3.4.2 Algebraic Laws

Process algebras carry their name thanks to the rich set of algebraic laws that allow one to transform processes, to refine processes, and to prove equality between processes.

CSP has laws specific to certain models. Examples include:

- The traces model $\mathcal{T}$ treats internal choice and external choice in the same way: $P \sqcap Q =_\mathcal{T} P \square Q$.
  This law does not hold in the failures/divergences model $\mathcal{N}$ and the stable failures model $\mathcal{F}$.
- In the stable failures model $\mathcal{F}$ one can never infer that a process is divergence free: $P \sqcap div =_\mathcal{F} P$.
  This law does not hold in the failures/divergences model $\mathcal{N}$.

The index on the equal sign indicates in which model the law holds.

There are also laws which express general properties of CSP. Here, one usually omits the index at the equal sign. The following laws hold in all the three standard models of CSP, namely $\mathcal{T}$, $\mathcal{N}$ and $\mathcal{F}$:

- **Congruence laws.** The various process equivalences = are congruences, e.g., if $Q = R$ then $P \sqcap Q = P \sqcap R$.
- **Laws involving single operators.** The choice operators are idempotent, i.e., $P \square P = P$ and $P \sqcap P = P$, symmetric, associative; external choice has *Stop* as its unit, i.e., $P \square Stop = P$.
- **Interplay of different operators, e.g., distributivity laws.** All CSP operators distribute over internal choice, e.g., $(P \sqcap Q) \setminus X = (P \setminus X) \sqcap (Q \setminus X)$.
- **Step laws.** Each CSP operator has a step law. It gives the initial events of the process formed by the CSP operator in terms of the initial events of the constituent processes. As an example, we consider here the (relatively simple) law of the external choice operator:

$$(?\ x : A \to P(x)) \square (?\ x : B \to Q(x)) =$$
$$?\ x : (A \cup B) \to$$
$$(\text{if } (x \in A \cap B) \text{ then } P(x) \sqcap Q(x)$$
$$\text{else if } (x \in A) \text{ then } P(x) \text{ else } Q(x)).$$

Such laws can be proven to be sound relatively to a given denotational CSP model. As an example, we *manually* prove the above step law of the external choice operator in the stable failures model $\mathcal{F}$. Here, we need to show: for all choices of the alphabet Σ, for all choices of $A \subseteq \Sigma$ and $B \subseteq \Sigma$, and for all interpretations of the basic processes $P(x)$ and $Q(x)$ in the semantic domain of $\mathcal{F}$ we have:

$$traces(Ext_{(A,P,B,Q)}) = traces(Step_{(A,P,B,Q)}) \qquad (\#1)$$
$$failures(Ext_{(A,P,B,Q)}) = failures(Step_{(A,P,B,Q)}) \qquad (\#2)$$

where $Ext_{(A,P,B,Q)}$ $(Step_{(A,P,B,Q)})$ denotes the lhs (rhs) of the step law for external choice. We consider the proof of (#2) only. Here, we compute the sets $failures(Ext_{(A,P,B,Q)})$ and $failures(Step_{(A,P,B,Q)})$ by applying the semantic clauses of the model $\mathcal{F}$. After some simplifications we obtain:

$$
\begin{aligned}
failures&(Ext_{(A,P,B,Q)}) = \\
&\{(\langle\rangle, X) \mid A \cap X = \emptyset\} \cup \{(\langle\rangle, X) \mid B \cap X = \emptyset\} \cup \\
&\{(\langle a\rangle \frown t, X) \mid a \in A \wedge (t, X) \in failures(P(a))\} \cup \\
&\{(\langle a\rangle \frown t, X) \mid a \in B \wedge (t, X) \in failures(Q(a))\} \qquad (\#3) \\
failures&(Step_{(A,P,B,Q)}) = \\
&\{(\langle\rangle, X) \mid (A \cup B) \cap X = \emptyset\} \cup \\
&\{(\langle a\rangle \frown t, X) \mid a \in A \cup B \wedge \\
&\quad if \;\; (a \in A \cap B) \;\; then \;\; (t, X) \in failures(P(a)) \cup failures(Q(a)) \\
&\quad else \; if \;\; (a \in A) \;\; then \;\; (t, X) \in failures(P(a)) \\
&\quad else \;\; (t, X) \in failures(Q(a))\} \qquad\qquad\qquad\qquad (\#4)
\end{aligned}
$$

Using standard arguments on sets, one can show that the sets (#3) and (#4) are indeed equal and that therefore the step law holds in $\mathcal{F}$.

In contrast to this approach, one can mechanise such proofs, e.g., with CSP-Prover. Figure 3.11 shows a proof-script in CSP-Prover proving the above step law. The sets of failures of the both processes, i.e., (#3) and (#4), are automatically derived in CSP-Prover, see the lines 9 and 12. This is a powerful technique. Deriving the denotations of processes according to the semantical clauses of a CSP model is a tedious but error prone and complex task—note that the sets (#3) and (#4) are already simplified versions of the sets derived from the semantical clauses. See the paper by Isobe and Roggenbach [IR07] for a more detailed discussion of such proofs. Mistakes found in published algebraic laws for CSP, see e.g., [IR06], demonstrate that presentations of CSP models and axiom schemes will only be 'complete' once they have been accompanied by mechanised theorem proving.

The above proofs concern the soundness of the laws. Yet another question how many laws one actually needs, if the given laws characterise equality in the chosen CSP model, i.e., if they are complete. Early approaches to completeness restrict CSP to finite non-determinism over a finite alphabet [Ros98]. For the stable failures model $\mathcal{F}$, Isobe and Roggenbach give a completeness result for CSP with unbounded non-determinism over an alphabet of arbitrary size using about 80 algebraic laws [IR06].

3.4.3 Foundations: Fixed Points

We now set out to address our questions concerning the semantics of recursive equations stated in Sect. 3.2.2. We begin our excursion with a seemingly simple example:

```
1    lemma cspF_Ext_choice_step:
2      "(? x:A -> P(x)) [+] (? x:B -> Q(x)) =F[M,M]
3      ? x:(A Un B) -> (IF (x : A Int B) THEN P(x) |~| Q(x)
4                      ELSE IF (x : A) THEN P(x) ELSE Q(x))"
5    apply (simp add: cspF_cspT_semantics)
6    apply (simp add: cspT_Ext_choice_step)
7    apply (rule order_antisym, auto)
8    (* ⊆ *)
9      apply (simp add: in_traces in_failures)
10     apply (auto)
11   (* ⊇ *)
12     apply (simp add: in_traces in_failures)
13     apply (elim disjE conjE exE, force+)
14     apply (case_tac "a : B", simp_all add: in_failures)
15     apply (case_tac "a : A", simp_all add: in_failures)
16   done
```

Fig. 3.11 A proof script in CSP-Prover for the step law of the external choice

---

Example 42: A Recursive Equation

A fundamental challenge to semantics in general is: how can we give semantics to a recursive definition such as

$$P = a \rightarrow P$$

Here, the behaviour of the process P on the left hand side is defined using the behaviour of the process P on the right hand side. One reading of such an equation is: we are interested in a denotation for P which remains unchanged when applying the prefix function to it. In other words: we are looking for a fixed point.

Denotational semantics works with at least two standard fixed point theories in order to deal with recursion: (i) partial orders in combination with Tarski's fixed point theorem and (ii) metric spaces in combination with Banach's fixed point theorem. CSP makes use of both approaches. In this section, we study the above equation in the context of the partial order approach.

We observe that the traces domain $\mathcal{T}$ together with set-inclusion $\subseteq$ as ordering relation forms a partial order with bottom element $(\mathcal{T}, \subseteq, \{\langle\rangle\})$: set-inclusion is reflexive, antisymmetric, and transitive; furthermore, $\{\langle\rangle\} \subseteq T$ for all $T \in \mathcal{T}$ as elements of $\mathcal{T}$ are required to be non-empty and prefix-closed.

Definition 2 (*Upper bound, least upper bound*) Let $(PO, \leq)$ be a partial order, let $X \subseteq PO$ be a set of elements.

1. An element $u \in PO$ is called an *upper bound* for X if for all $x \in X . x \leq u$.
2. An element $lub \in PO$ is called a *least upper bound* for X if it is an upper bound for X and for all $ub \in \{u \in PO \,|\, \forall x \in X . x \leq u\}$ holds: $lub \leq ub$.

In the context of the traces domain $\mathcal{T}$, let

- $S \triangleq \{\langle\rangle, \langle a \rangle\}$—i.e., the denotation of the process $a \to Stop$—and
- $T \triangleq \{\langle\rangle, \langle b \rangle\}$—i.e., the denotation of the process $b \to Stop$.

Now consider the set $X \triangleq \{S, T\}$. It has $U \triangleq \{\langle\rangle, \langle a \rangle, \langle b \rangle\}$— i.e., the denotation of the process $a \to Stop \sqcap b \to Stop$ —as an upper bound. U is also the least upper bound of X: let V be an upper bound of X. Then $\langle\rangle \in V$ and $\langle a \rangle \in V$ thanks to $S \subseteq V$, $\langle\rangle \in V$ and $\langle b \rangle \in V$ thanks to $T \subseteq V$. Thus, $U \subseteq V$.

Definition 3 (ω-*Chain*, ω-*complete partial order*)

1. Let $(PO, \leq)$ be a partial order. Let $C = (c_i)_{i \in \mathbb{N}}$ be a sequence in PO, i.e., $c_i \in PO$ for all $i \in \mathbb{N}$. Such a sequence C is called a ω-*chain*, if $c_i \leq c_{i+1}$ for all $i \in \mathbb{N}$.
2. A partial order $(PO, \leq)$ is called ω -*complete* if every ω-chain has a least upper bound, i.e., for all ω-chains $C = (c_i)_{i \in \mathbb{N}}$ holds: $\{c_i \in C \mid i \in \mathbb{N}\}$ has a least upper bound.

Example 42.1: A Ω-Chain over $\mathcal{T}$ with a Least Upper Bound

Consider the sequence S in the traces domain $\mathcal{T}$ with
- $s_0 = \{\langle\rangle\}$—i.e., the denotation of the process $Stop$.
- $s_1 = \{\langle\rangle, \langle a \rangle\}$ —i.e., the denotation of the process $a \to Stop$.
- $s_2 = \{\langle\rangle, \langle a \rangle, \langle a, a \rangle\}$ —i.e., the denotation of the process $a \to a \to Stop$.
- $\ldots$

This sequence is a ω-chain, as $s_i \subseteq s_{i+1}$ for all $i \in \mathbb{N}$. It has the set a^* as its upper bound. a^* is also the least upper bound: let V be an upper bound of $(S_i)_{i \in \mathbb{N}}$. Let $t \in a^*$. Then t is a sequence of a certain number of a's, say of k a's. Thus, $t \in s_k$. As V is an upper bound of $(s_i)_{i \in \mathbb{N}}$, we know that $s_k \subseteq V$. Consequently, $t \in V$.

Theorem 9 $(\mathcal{T}, \subseteq)$ *is a ω-complete partial order.*

Proof We already proved above that $(\mathcal{T}, \subseteq)$ is a partial order. It remains to consider ω-completeness.

Let $C = (c_i)_{i \in \mathbb{N}}$ be a ω-chain. Define $U \triangleq \bigcup_{i \in \mathbb{N}} c_i$. We claim that U is the least upper bound of C. To this end we have to show that (i) $U \in \mathcal{T}$, (ii) U is an upper bound of C, and (iii) that U indeed is the least upper bound.

To (i): U is non-empty as the c_i are non-empty. Now let $t \in U$ be a trace and let $s \leq t$ be a prefix of t. Then there exists some $i \in \mathbb{N}$ such that $t \in c_i$. As c_i is prefix-closed, we have $s \in c_i$ and thus $s \in U$.

To (ii): by construction we have $c_i \subseteq U$ for all $i \in \mathbb{N}$.

To (iii): let V be an upper bound of C. Let $s \in U$ be a trace. Then there exists some $i \in \mathbb{N}$ such that $s \in c_i$. As V is an upper bound of C, we have $c_i \subseteq V$ and, consequently, also $s \in V$. Thus, $U \subseteq V$. ∎

In order to characterise a class of functions that has fixed points, we define:

Definition 4 (*Continuous function*) Let $(PO, \leq)$ be a ω-complete partial order. A function $f : PO \to PO$ is continuous if for all ω-chains $C = (c_i)_{i \in \mathbb{N}}$ in PO the following holds:

1. $lub(f(C))$ exists and
2. $lub(f(C)) = f(lub(C))$.

Continuous functions are monotonic:

Lemma 4 (Monotonicity) *Let $(PO, \leq)$ be a ω-complete partial order. Let $f : PO \to PO$ be a continuous function. Then $a \leq b$ implies $f(a) \leq f(b)$ for all $a, b \in PO$.*

Proof Let $a \leq b$. Define a ω-chain $C = (c_i)_{i \in \mathbb{N}}$ by setting $c_0 \triangleq a$, $c_j \triangleq b$ for all $j \geq 1$. Clearly, $lub(C) = b$. As f is continuous, $lub(f(C))$ exists and we have $lub(f(C)) = f(lub(C)) = f(b)$. As $f(b)$ is an upper bound of $f(C)$, we have $f(a) \leq f(b)$. ∎

Example 42.2: Action Prefix Is Continuous over $\mathcal{T}$

We want to show that the functions

$$f_a : \begin{cases} \mathcal{T} \to \mathcal{T} \\ X \mapsto \{\langle\rangle\} \cup \{\langle a \rangle \frown s \mid s \in X\} \end{cases}$$

—which underly the semantic clause of the action prefix operator— are continuous for $a \in \Sigma$. In Theorem 2 we already proved that the functions f_a indeed map into $\mathcal{T}$.

Let $a \in \Sigma$. Let $C = (c_i)_{i \in \mathbb{N}}$ be a ω-chain over $\mathcal{T}$. We claim that the observation $U \triangleq \bigcup_{i \in \mathbb{N}} f_a(c_i)$ is the least upper bound of $f_a(C) = (f_a(c_i))_{i \in \mathbb{N}}$. :

To 1.   First, we show that $U \in \mathcal{T}$. Concerning the first condition of T1, U is non-empty as $\langle\rangle \in U$. Concerning the second condition of T1, let $t \in U$ and let $s \leq t$. If $s = \langle\rangle$, we are done as by construction $\langle\rangle \in U$. If $s \neq \langle\rangle$ we have $t \neq \langle\rangle$. Thus, there exists some t' such that $t = \langle a \rangle \frown t'$ with $t' \in c_i$ for some $i \in \mathbb{N}$. As $s \leq t$, there also exists an s' with $s = \langle a \rangle \frown s'$ and $s' \leq t'$. As $c_i \in \mathcal{T}$, c_i is prefix-closed, i.e., $s' \in c_i$. Thus, $s = \langle a \rangle \frown s' \in f_a(c_i) \subseteq U$.

Next, we observe that U is an upper bound for $f_a(C)$: By definition of U, we have that $f_a(c_i) \subseteq U$ for all $i \in \mathbb{N}$.

Finally, let V be an upper bound for $f_a(C)$. Let $t \in U$. Then there exists an $i \in \mathbb{N}$ with $t \in f(c_i)$. As $f(c_i) \subseteq V$, we have $t \in V$. Thus, $U \subseteq V$, i.e., U is the least upper bound of $f_a(C)$.

> To 2.   Using the construction of the least upper bound of C from the proof of Theorem 9, we have to show that $f_a(\bigcup_{i\in\mathbb{N}} c_i) = \bigcup_{i\in\mathbb{N}} f_a(c_i)$. This holds, as on both sides f_a is applied to each element of c_i, $i \in \mathbb{N}$.

Continuous functions on pointed cpos have fixed points:

Theorem 10 (Kleene's Fixed Point Theorem) *Let* $(PO, \leq, \perp)$ *be a ω-complete partial order with bottom element. Let* $f : PO \to PO$ *be a continuous function. Then*

$$lub(((f^n(\perp))_{n\in\mathbb{N}}))$$

is the least fixed point of f.

Proof We first prove that $(f^n(\perp))_{n\in\mathbb{N}}$ is a ω-chain. As $\perp \leq x$ for all $x \in PO$, we have especially $\perp \leq f(\perp)$. Applying Theorem 4 yields: $f^n(\perp) \leq f^{n+1}(\perp)$ for all $n \in \mathbb{N}$. As f is continuous, this ω-chain has a least upper bound, i.e., $lub((f^n(\perp))_{n\in\mathbb{N}})$ exists.

Next we consider that $lub((f^n(\perp))_{n\in\mathbb{N}})$ is indeed a fixed point of f:

$$\begin{aligned} & f(lub((f^n(\perp))_{n\in\mathbb{N}})) \\ = \; & lub((f^{n+1}(\perp))_{n\in\mathbb{N}}) \quad \text{thanks to continuity} \\ = \; & lub((f^n(\perp))_{n\in\mathbb{N}}). \quad \text{adding the bottom element} \end{aligned}$$

Finally, we show that $lub((f^n(\perp))_{n\in\mathbb{N}})$ is the smallest fixed point of f. Let x be a fixed point of f, i.e., $f(x) = x$. We know $\perp \leq x$. Applying Theorem 4 to this yields: $f^n(\perp) \leq f^n(x) = x$ for all $n \in \mathbb{N}$, i.e., x is an upper bound of $(f^n(\perp))_{n\in\mathbb{N}}$. Thus, $lub((f^n(\perp))_{n\in\mathbb{N}}) \leq x$.   ∎

Note that the smallest fixed point of a continuous function f is unique. Assume that u and v are both smallest fixed points of f. Then $u \leq v$, as v a the smallest fixed point. But also $v \leq u$, as u is a smallest fixed point. With antisymmetry, we obtain: $u = v$.

Using Theorem 10, we can study how to give semantics to recursive process definitions:

Example 42.3: Semantics of a Recursive Equation

We are now in the position to give a semantics $M(P)$ to the process name P such that the semantics of P makes the equation $P = a \to P$ true on the semantic level. Recall the semantic clauses:

$$\begin{aligned} traces_M(a \to P) & \triangleq \{\langle\rangle\} \cup \{\langle a\rangle \frown s \mid s \in traces_M(P)\} \\ traces_M(n) & \triangleq M(n) \end{aligned}$$

We want to determine a value $M(P)$ with

$$\begin{aligned} traces_M(P) & = traces_M(a \to P) \\ & = f_a(traces_M(P)) \end{aligned}$$

By choosing $M(P)$ to be the smallest fixed point of f_a we can answer our questions stated in Sect. 3.2.2:

1. Is there a solution to the equation?
   Yes. f_a is continuous and thus has a fixed point according to Theorem 10.
2. Is the solution unique?
   Yes. We are choosing the smallest fixed point.
3. How can we construct the solution (should it exist)?
   Theorem 10 provides a construction of the fixed point. The sequence S from above is the ω-chain that provides the least fixed point, which we computed to be a^*.

Note that fixed points are not necessarily unique: the equation $P = P$ (with identity as its underlying function) has all elements of a semantic domain, say of the traces domain $\mathcal{T}$, as its solutions. Only by taking the smallest one we make the solution unique, namely to be $\{\langle\rangle\}$ over $\mathcal{T}$.

The theory developed so far works for functions in one argument (i.e., action prefix) and one variable (i.e., one equation). In order to cater for more complex situations, one needs to consider products of ω-complete partial orders, continuous functions in several arguments, and the composition of continuous functions. The subject of domain theory provides constructions and solutions to these question of denotational semantics, see e.g., [Win93].

Without proof we quote results as presented in Roscoe's book [Ros98]:

Theorem 11 (Ω-Completeness of domains and continuity of the semantic functions)

- *The model $\mathcal{T}$ is a cpo w.r.t. $\subseteq$ as ordering relation and with $[\![Stop]\!]_{\mathcal{T}}$ as bottom element.*
  The model $\mathcal{F}$ is a cpo w.r.t. $\subseteq \times \subseteq$ as ordering relation and with $[\![Div]\!]_{\mathcal{F}}$ as bottom element.
  If one restricts the alphabet Σ to be finite, the model $\mathcal{N}$ is a cpo w.r.t. $\supseteq \times \supseteq$ as ordering relation and $[\![Div]\!]_{\mathcal{N}}$ as bottom element.
- *Over $\mathcal{T}$ and $\mathcal{F}$, the semantic functions underlying action prefix, external choice, internal choice, general parallel, sequential composition, renaming, and hiding are continuous w.r.t. $\subseteq$ and $\subseteq \times \subseteq$ as respective ordering relations.*
  If one restricts the alphabet Σ to be finite, this holds also over $\mathcal{N}$ w.r.t. $\supseteq \times \supseteq$ as ordering relation. Here, the hiding operator over $\mathcal{N}$ is only defined when applied to finitely non-deterministic processes.

Note that Roscoe works with directed sets rather than with ω-chains. However, as ω-chains are special directed sets, Roscoe's results carry over to our slightly simpler setting.

With this result we conclude: choosing the smallest fixed point guaranteed by Kleene's theorem gives semantics to all recursive process equations over the models $\mathcal{T}$ and $\mathcal{F}$, and also—under some restrictions—over the model $\mathcal{N}$.

3.4.4 Checking for General Global Properties

This section provides analysis methods for general properties of parallel systems. We define what it means for a process to be deadlock free, livelock free, or deterministic. For each of these properties we then present a characterisation in terms of CSP refinement. Full proofs are provided.

Livelock

A process is said to diverge or livelock if it reaches a state from which it may forever compute internally through an infinite sequence of invisible events. This is clearly a highly undesirable feature of the process, described by some as "even worse than deadlock" [Hoa85]. Livelock may invalidate certain analysis methodologies, see, e.g., the one presented in this section concerning determinism. Livelock is often caused by a bug in modeling. The possibility of writing down a divergent process arises from the presence of two constructs in CSP: hiding and recursion. Let us, for example, consider the process $P = a \to P$, which performs an unbounded number of a's. If one now conceals the event a in this process, i.e., one forms $P = (a \to P) \setminus a$, it no longer becomes possible to observe any behaviour of this process.

In CSP, the process Div represents this phenomenon: immediately, it can refuse every event, and it diverges after any trace. In the model $\mathcal{N}$, the denotation of Div is given by

$$\begin{aligned} divergences(Div) &:= \Sigma^{*\checkmark} \\ failures_{\perp}(Div) &:= \Sigma^{*\checkmark} \times \mathcal{P}(\Sigma^{\checkmark}) \end{aligned}$$

Div represents an un-controllable process: it has all possible traces and can refuse all events. Conditions **D1** and **D2** of the failures/divergences model ensure that $s \in divergences(P)$ implies $\{t \mid s \frown t \in divergences(P)\} = \Sigma^{*\checkmark}$ and $\{(t, X) \mid (s \frown t, X) \in failures_{\perp}(P)\} = \Sigma^{*\checkmark} \times \mathcal{P}(\Sigma^{\checkmark})$. This justifies the definition:

Definition 5 (*Livelock-freedom/divergence-freedom*) A process P is said to be *livelock-free* (divergence-free) if and only if $divergences(P) = \emptyset$.

Theorem 12 ($\mathcal{N}$-refinement preserves livelock-freedom) *Let P and Q be processes such that P is livelock-free and $P \sqsubseteq_{\mathcal{N}} Q$. Then Q is livelock-free.*

Proof We show: if $P \sqsubseteq_{\mathcal{N}} Q$ and Q has a livelock, then P has a livelock. Let Q have a livelock. Then $divergences(Q) \neq \emptyset$. As $P \sqsubseteq_{\mathcal{N}} Q$, also $divergences(P) \neq \emptyset$. ∎

The set of all livelock-free processes has a maximal element, namely the process $DivF_\Sigma$:

Definition 6 (*Most general livelock-free process*) Relatively to an alphabet Σ we define the process

$$DivF_\Sigma = \sqcap\{a \rightarrow DivF_\Sigma \mid a \in \Sigma\} \sqcap Skip \sqcap Stop$$

If the alphabet Σ is clear from the context we omit the index Σ and just write $DivF$.

Theorem 13 ($DivF$ is livelock-free) *The process $DivF$ is livelock-free.*

Proof As neither *Stop*, nor *Skip* have any divergences, and the prefix operator does not contribute to divergences, we obtain $divergences(DivF) = \emptyset$. ∎

Theorem 14 ($DivF_A$ is maximal among the livelock-free processes) *Let P be livelock free. Then $DivF \sqsubseteq_\mathcal{N} P$.*

Proof Let $DivF \not\sqsubseteq_\mathcal{N} P$. As $failures(Stop) = \{(\langle\rangle, X) \mid X \subseteq \Sigma^\checkmark\}$ and $failures(P \sqcap Q) = failures(P) \cup failures(Q)$ we have that $failures(DivF) = A^{*\checkmark} \times \mathcal{P}(\Sigma^\checkmark)$. Thus, $failures(P) \subseteq failures(DivF)$. Therefore, we must have $divergences(P) \not\subseteq divergences(DivF) = \emptyset$, i.e., $divergences(P) \neq \emptyset$. Thus, P has a livelock. ∎

The theorems of these section provide a sound and complete proof method for livelock analysis:

Corollary 6 (Livelock analysis)
 A process P is-livelock free if and only if $DivF \sqsubseteq_\mathcal{N} P$.

Deadlock

In CSP, deadlock is represented by the process *Stop*. Let Σ be the communication alphabet. Then the process *Stop* has

$$(\{\langle\rangle\}, \{(\langle\rangle, X) \mid X \subseteq \Sigma^\checkmark\}) \in \mathcal{P}(\Sigma^{*\checkmark}) \times \mathcal{P}(\Sigma^{*\checkmark} \times \mathcal{P}(\Sigma^\checkmark))$$

as its denotation in $\mathcal{F}$, i.e., the process *Stop* can perform only the empty trace, and after the empty trace the process *Stop* can refuse to engage in all sets of events.

Stop denotes an immediate deadlock. In general, a process P is considered to be deadlock free, if the process P after performing a trace s never becomes equivalent to the process *Stop*. More formally:

Definition 7 (*Deadlock-freedom*) A process P is said to be *deadlock-free* if and only if

$$\forall s \in \Sigma^*.(s, \Sigma^\checkmark) \notin failures(P).$$

The omission of all proper subsets of $\Sigma^\checkmark$ is justified as in the model $\mathcal{F}$ the set of stable failures is required to be closed under subset-relation: $(s, X) \in failures(P) \wedge Y \subseteq X \Rightarrow (s, Y) \in failures(P)$. Definition 7 can be read as: before termination, the process P can never refuse all events; there is always some event that P can perform.

Theorem 15 ($\mathcal{F}$-refinement preserves deadlock-freedom) *Let P and Q be processes such that P is deadlock-free and $P \sqsubseteq_\mathcal{F} Q$. Then Q is deadlock-free.*

Proof We show: if $P \sqsubseteq_\mathcal{F} Q$ and Q has a deadlock, then P has a deadlock. Let Q have a deadlock, i.e., there exists $s \in \Sigma^*$ with $(s, \Sigma^\checkmark) \in failures(Q)$. As $P \sqsubseteq_\mathcal{F} Q$, we know that $failures(Q) \subseteq failures(P)$. Thus $(s, \Sigma^\checkmark) \in failures(P)$ and P has a deadlock. ∎

The set of all deadlock-free processes has a maximal element, namely the process DF_Σ:

Definition 8 (*Most general deadlock-free process*) Relatively to an alphabet Σ we define the process

$$DF_\Sigma = \sqcap \{a \rightarrow DF_\Sigma \,|\, a \in \Sigma\} \sqcap Skip$$

If the alphabet Σ is clear from the context we omit the index Σ and just write DF.

Theorem 16 (*DF is deadlock-free*) *The process DF is deadlock-free.*

Proof We calculate the semantics of DF in the stable failures model $\mathcal{F}$. The process has all traces, as it can perform all events as well as $Skip$ at any time:

$$traces(DF) = \Sigma^{*\checkmark}.$$

The failures of action prefix are $failures(a \rightarrow P) = \{\langle\rangle, X) \,|\, X \subseteq \Sigma^\checkmark - \{a\}\} \cup \{\langle a \rangle \frown s, X) \,|\, (s, X) \in failures(P)\}$.

The failures of $Skip$ are $failures(Skip) = \{\langle\rangle, X) \,|\, X \subseteq \Sigma\} \cup \{\langle\checkmark\rangle, X) \,|\, X \subseteq \Sigma^\checkmark\}$.

The failures of the internal choice operator are given by the union of the failures of its component processes. Thus,

$$\begin{aligned} failures(DF) = \quad &\{(t, X) \,|\, t \in \Sigma^*, X \subseteq \Sigma \vee \exists a \in \Sigma.\ X \subseteq \Sigma^\checkmark - \{a\}\} \\ &\cup \{(t \frown \langle\checkmark\rangle, X) \,|\, t \in \Sigma^*, X \subseteq \Sigma^\checkmark\}. \end{aligned}$$

$$(3.8)$$

I.e., after a non-terminating trace s, DF never has $\Sigma^\checkmark$ as its refusal set. Thus, DF is deadlock free. ∎

This said, we know that a process P is deadlock free if $DF \sqsubseteq_\mathcal{F} P$. However, is any deadlock free process P a refinement of DF? This is the case as the following theorem shows:

Theorem 17 (DF_A Is Maximal Among Deadlock-Free Processes) *Let P be deadlock free. Then $DF \sqsubseteq_{\mathcal{F}} P$.*

Proof We have to show that $traces(P) \subseteq traces(DF)$ and that $failures(P) \subseteq failures(DF)$. The first inclusion is trivial, as $traces(DF) = \Sigma^{*\checkmark}$, i.e., $traces(DF)$ consists of all possible traces—see the proof of Theorem 16. The second inclusion holds for a similar reason: let $(s, X) \in failures(P)$. If s is a non-terminating trace, i.e., $s \in \Sigma^*$, then X needs to be a proper subset of $\Sigma^{\checkmark}$—otherwise P would have a deadlock. Thus, we know that either $X \subseteq \Sigma^{\checkmark} - \{a\}$ or $X \subseteq \Sigma$. These failures are included in the $failures(DF)$—see the first line of Eq. 3.8. If s is a terminating trace, i.e., $s = s' \frown \langle \checkmark \rangle$ with $s' \in \Sigma^*$, then $(s' \frown \langle \checkmark \rangle, X) \in failures(DF)$ according to the second line of Eq. 3.8. ∎

The theorems of these section provide a sound and complete proof method for deadlock analysis:

Corollary 7 (Deadlock analysis) *A process P is deadlock free if and only if $DF \sqsubseteq_{\mathcal{F}} P$.*

Determinism

Non-determinism is a phenomenon considered by most Formal Methods for reactive systems. In CSP it is represented by the internal choice operator.

This operator can be of good use, e.g., when specifying properties, or when describing systems on high level of abstraction. For instance, both the processes $DivF$ and DF make use of non-determinism in order to specify the properties livelock-freedom and deadlock-freedom, resp. In system specification, we used non-determinism in the process $PinVerification$, cf. Example 36.9. There, the reason for specifying with non-determinism was the level of abstraction: we refrained from modelling the details of the verification process, i.e., we were leaving its details to further development.

The CSP operator of hiding is often seen as an abstraction operator: when the events leading to specific states are 'removed', the choice between these states can become a non-deterministic one. Take for instance the process $P = (a \rightarrow Stop \,\square\, b \rightarrow c \rightarrow Stop)$. It is deterministic thanks to using the external choice operator. However, over $\mathcal{N}$ and $\mathcal{F}$ we have that $P \setminus \{a, b\} =_{\mathcal{F},\mathcal{N}} Stop \sqcap c \rightarrow Stop$, i.e, hiding turns a deterministic process into a non-deterministic one.

The physical computers that we are using are deterministic machines. Thus, when we want to model on implementation level we usually want our processes to be deterministic ones. In the following, we discuss of how to investigate if a process is deterministic.

Definition 9 A process P is (internally) non-deterministic if there exists a trace $s \in \Sigma^*$ and a communication $a \in \Sigma$ such that

1. $s ^\frown \langle a \rangle \in traces(P)$ and
2. $(s, \{a\}) \in failures(P)$.

I.e., the process P is at the same time willing to engage in a and to refuse a. We say that a process is deterministic if it is not non-deterministic.

Following an idea from Ranko Lazić, published by Bill Roscoe in [Ros03], we express a check for determinism via refinement. Given a livelock free process P, we construct two new processes $Spec$ and $G(P)$ such that P is deterministic if and only if $Spec \sqsubseteq_{\mathcal{F}} G(P)$ holds.

Let Σ be an alphabet of communications. We define two copies of Σ, namely $\Sigma.1 = \{a.1 \,|\, a \in \Sigma\}$ and $\Sigma.2 = \{a.1 \,|\, a \in \Sigma\}$. We define a process $Monitor$ over $\Sigma.1 \cup \Sigma.2$ as

$$Monitor =?x.1 : \Sigma.1 \to x.2 \to Monitor$$

The process $Monitor$ first offers all events in $\Sigma.1$. After it received such an event $a.1$, it communicates its counterpart $a.2 \in \Sigma.2$ and behaves again like $Monitor$.

Let P be a process over Σ. From P we construct a copy $P.1$ over $\Sigma.1$. To this end, we rename all events $a \in \Sigma$ to $a.1 \in \Sigma.1$. Analogously we construct a copy $P.2$ over $\Sigma.2$:

$$P.1 = P \llbracket a.1/a \,|\, a \in \Sigma \rrbracket$$
$$P.2 = P \llbracket a.2/a \,|\, a \in \Sigma \rrbracket$$

With the help of these processes we define

$$G(P) = (P.1 \,|||\, P.2) \parallel Monitor$$

$G(P)$ works as follows: process $P.1$ selects a communication $a.1$. The $Monitor$ receives $a.1$ and sends an $a.2$ to $P.2$. If $P.2$ is deterministic, then $P.2$ accepts $a.2$ and $P.1$ can send the next event. However, if $P.2$ is non-deterministic, then $P.2$ can refuse to engage in $a.2$ which leads $G(P)$ to deadlock.

We exploit this possibility of a deadlock to construct the refinement check. To this end we define a process $Spec$ over $\Sigma.1 \cup \Sigma.2$. $Spec$ can immediately deadlock, it can, however, also perform an event from $\Sigma.1$ and then become $Spec'$. $Spec'$ can perform an event from $\Sigma.2$ and become $Spec$, however, it has no deadlock state. Thus, $Spec$ never has a deadlock after a trace of odd length:

$$Spec = (\sqcap \{a.1 \to Spec' \,|\, a.1 \in \Sigma.1\}) \sqcap Stop$$
$$Spec' = \sqcap \{a.2 \to Spec \,|\, a.2 \in \Sigma.2\}$$

Theorem 18 (Determism check) *Let P be a livelock-free process. Let $G(P)$ and $Spec$ be as above. P is deterministic if and only if $Spec \sqsubseteq_{\mathcal{F}} G(P)$.*

Proof We first determine the stable failures semantics of $Spec$.

Its traces are all sequences in which the elements of $\Sigma.1$ and $\Sigma.2$ alternate:

$$traces(Spec) = \{p \leq t \mid t \in \{\langle x.1 \rangle \frown \langle y.2 \rangle \mid x.1 \in \Sigma.1, \, y.2 \in \Sigma.2\}^*\}$$

After a trace of even length, it has all failures due to the process *Stop*. After a trace of odd length, it can refuse termination, all elements of $\Sigma.1$, and all elements of $\Sigma.2$ but one:

$$
\begin{aligned}
failures(Spec) = \\
\{(s, X) \mid s \in traces(Spec), \, length(s) \, even, \, X \subseteq (\Sigma.1 \cup \Sigma.2)^\checkmark\} \\
\cup \{(s, X) \mid s \in traces(Spec), \, length(s) \, odd, \\
\exists a.2 \in \Sigma.2.X \subseteq (\Sigma.1 \cup \Sigma.2 - \{a.2\})^\checkmark\}
\end{aligned}
$$

Now let us consider the stable failures semantics of $G(P)$:

The traces of $G(P)$ are a subset of the traces of *Monitor* as the out-most operator of $G(P)$ is the synchronous parallel operator with the semantics $traces(Q \parallel R) = traces(Q) \cap traces(R)$. The traces of *Monitor* are

$$traces(Monitor) = \{p \leq t \mid t \in \{\langle a.1 \rangle \frown \langle a.2 \rangle \mid a \in \Sigma\}^*\},$$

which is a subset of $traces(Spec)$. Overall, we obtain:

$$traces(G(P)) \subseteq traces(Monitor) \subseteq traces(Spec).$$

Thus, the traces part of the $\mathcal{F}$-refinement always holds.

Now let us consider the failures of $G(P)$. First we compute the failures for synchronous parallel from the clause from general parallel:

$$
\begin{aligned}
failures(Q \parallel R) \\
= \{(u, Y \cup Z) \mid (u, Y) \in failures(Q) \wedge (u, Z) \in failures(R) \\
\wedge u \in traces(Q) \cap traces(R)\}
\end{aligned}
$$

The failures of *Monitor* are

$$
\begin{aligned}
failures(Monitor) \\
= \quad \{(s, X) \mid s \in traces(Monitor), \, lenght(s) \, even, \\
X \subseteq \Sigma.2^\checkmark\} \\
\cup \{(s, Y) \mid s \in traces(Monitor), \, length(s) \, odd, \\
\exists a.1 \in \Sigma.1.s = s' \frown \langle a.1 \rangle \wedge \\
Y \subseteq (\Sigma.1 \cup \Sigma.2 - \{a.2\})^\checkmark\}.
\end{aligned}
$$

Next, we compute the failures for interleaving from the clause of general parallel:

$$
\begin{aligned}
failures(Q \mathbin{|||} R) \\
= \{(u, Y \cup Z) \mid Y - \{\checkmark\} = Z - \{\checkmark\}, \\
\exists s, t.(s, Y) \in failures(Q) \wedge (s, Z) \in failures(R) \\
\wedge u \in s \mathbin{|||} t\}
\end{aligned}
$$

As *Spec* has all failures after an even length trace—see above—we only need to consider odd length traces in our proof.

"$\Rightarrow$" Let *Spec* $\not\sqsubseteq G(P)$. From the above considerations we know that this can only be the case if $failures((P.1 \,|||\, P.2) \,\|\, Monitor) \not\subseteq failures(Spec)$. Even more precisely we know that the reason for the non-inclusion must have the form $(s, (\Sigma.1 \cup \Sigma.2)^{\checkmark})$ where s is a trace of odd length.

According to *Monitor*, the last element of an odd length trace must end with a communication in $\Sigma.1$. Thus, there exist s' and $a.1$ such that $s = s'^{\frown}\langle a.1 \rangle$ and s' is an even length trace. Due to the trace semantics of *Monitor* we know that $s' = \langle a_1.1, a_1.2, \ldots, a_n.1, a_n.2 \rangle$ for some events $a_1, \ldots, a_n \in \Sigma$, $n \geq 0$. Let $t = \langle a_1, \ldots, a_n \rangle$. Then $t.1 \in traces(P.1)$ and $t.2 \in traces(P.2)$. Furthermore, we know that $t.1^{\frown}\langle a.1 \rangle \in traces(P.1)$. Thus, we obtain $t^{\frown}\langle a \rangle \in traces(P)$.

As the failures of *Monitor* after an odd length trace can maximally have $(A.1 \cup A.2 - \{a.2\})^{\checkmark}$ as their refusal set, $P.1 \,|||\, P.2$ must have contributed the set $\{a.2\}$ as a refusal after s. Thus, $P.2$ must have the failure $(t.2, \{a.2\})$. Consequently P must have the failure $(t, \{a\})$.

Thus, P is non-deterministic.

"$\Leftarrow$" Let P be non-deterministic. Then there exist a trace $s = \langle a_1, \ldots, a_n \rangle$ and a communication a such that $s ^{\frown} \langle a \rangle \in traces(P)$ and $(s, \{a\}) \in failures(P)$. As $P.1$ and $P.2$ are obtained by P via bijective renaming, we know that $s.1 ^{\frown} \langle a.1 \rangle \in traces(P.1)$ and $(s.2, \{a.2\}) \in failures(P.2)$. By construction, $\langle a_1.1, a_1.2, \ldots, a_n.1, a_n.2 \rangle^{\frown}\langle a.1 \rangle \in traces(P.1 \,|||\, P.2)$. $P.1$ refuses all the events from $\Sigma.2$. Thus, $(s.1 ^{\frown} \langle a.1 \rangle, \{a.2\}) \in failures(P.1)$. Consequently,

$$(\langle a_1.1, a_1.2, \ldots, a_n.1, a_n.2 \rangle ^{\frown} \langle a.1 \rangle, \{a.2\}) \in failures(P.1 \,|||\, P.2).$$

Obviously, $\langle a_1.1, a_1.2, \ldots, a_n.1, a_n.2 \rangle ^{\frown} \langle a.1 \rangle \in traces(Monitor)$ and it has odd length. According to our above computation of the semantics for *Monitor* holds:

$$(\langle a_1.1, a_1.2, \ldots, a_n.1, a_n.2 \rangle^{\frown}\langle a.1 \rangle, (\Sigma.1 \cup \Sigma.2 - \{a.2\})^{\checkmark}) \in failures(Monitor).$$

This results in

$$(\langle a_1.1, a_1.2, \ldots, a_n.1, a_n.2 \rangle ^{\frown} \langle a.1 \rangle, (\Sigma.1 \cup \Sigma.2)^{\checkmark})$$
$$\in failures((P.1 \,|||\, P.2) \,\|\, Monitor).$$

Consequently, $failures((P.1 \,|||\, P.2) \,\|\, Monitor) \not\subseteq failures(Spec)$. ∎

3.5 Closing Remarks

This chapter introduced the process algebra CSP as means to model and analyse concurrent systems.

Concerning learning CSP, cf. Sect. 3.2, the ATM example introduced the syntax of the language. Based on the jet engine controller example, the semantic concepts of CSP were sketched, namely its operational and denotational semantics. Using the Buffer example, the need for extending observations from traces to failures was demonstrated. Also, together with the Communication System example, the idea of refinement was introduced.

Utilising the Children's Puzzle, cf. Sect. 3.3, the power of different tools realising the methods part of CSP was discussed: with simulation one can obtain results on a single run, with model checking one can analyse all runs of a system, with theorem proving one can analyse a class of systems. Finally, using tools such CSP++ it is possible to automatically translate CSP processes into programs in a programming language.

The section on semantics and analysis, cf. Sect. 3.4, finally provides pointers to the more advanced theory of CSP: its denotational and axiomatic semantics, and how to develop analysis methods.

3.5.1 Annotated Bibliography

Process algebra emerged as branch of computer science in the 1970s. Tony Hoare's text book "Communicating Sequential Processes" [Hoa85] from 1985 made CSP accessible to a wider audience. Alternative approaches to the subject include Robin Milner's Calculus of communicating systems (CCS) [Mil89] and the "Algebra of Communicating Processes" (ACP) [BK84] by Jan Bergstra and Jan Willem Klop. These three 'classical process algebras' are extended by the topic of mobility in Milner's π-calculus [MPW92].

Main text books for CSP include, besides Hoare's seminal contribution, "The theory and practice of concurrency" [Ros98] (there is improved and corrected version from 2005 available electronically) and "Understanding Concurrent Systems" [Ros10] by Bill Roscoe, and "Concurrent and Real-time Systems: the CSP Approach" [Sch00] by Steve Schneider.

The literature on CSP and process algebra is vast. Concerning process algebra semantics in general, van Glabbeek provides a survey on "The Linear Time-Branching Time Spectrum" [vG01]. In [SNW96], Sassone et al. give a classification using category theory. The timelessness of the various CSP semantics is topic of Roscoe's paper "The Three Platonic Models of Divergence-Strict CSP" [Ros08]. In [Ros03], Roscoe studies how to encode properties into CSP refinement checks. Reports on industrial applications of CSP can be found, e.g, in [BKPS97, BPS98, PB99, Win02, IMNR12]. The applications of CSP in human computer interaction is topic of Chap. 7. The CSP approach to security is documented in Chap. 8.

3.5.2 Current Research Directions

Current research on CSP still concerns its foundations, see the above mentioned paper on "platonic CSP models" [Ros08]. There is also ongoing work on making automatic verification more efficient and thus to make the formal verification of concurrent system scalable, see, e.g., [AGR19]. Another stream is, as for nearly all specification formalisms, the question of how CSP specifications relate to implementations. Transformational approaches include JCSP [Wel, WBM07] and CSP++ [Gar15, Gar05]—see Sect. 3.3.4. Another paradigm is to systematically test from CSP specifications, see, e.g., [CG07, KRS07, KRS08]. The third stream of current research concerns CSP extensions. One longstanding question is how to deal with data in the context of CSP. CSP_M [Sca98] includes a functional programming language for data description. CIRCUS, see, e.g., [OCW09], combines the Z formalism with CSP, CSP-OZ [Fis97] provides a combination with Object-Z. CSP-CASL [Rog06] uses the algebraic specification language CASL (see Chap. 4) for data specification. Another longstanding topic is the extension of CSP by time. Here, Schneider's book [Sch00] presents an established and well worked-out setting. Work by, e.g., Ouaknine and Worrel [OW03], however, demonstrates that fundamental questions concerning timed CSP models are still open. Other CSP extensions offer primitives that allow one to model hybrid systems [LLQ10], probabilistic systems [AHTG11], or to combine event based and state based reasoning [ST05].

References

[AGR19]   Pedro Antonino, Thomas Gibson-Robinson, and A. W. Roscoe. Efficient verification of concurrent systems using synchronisation analysis and SAT/SMT solving. *ACM Trans. Softw. Eng. Methodol.*, 28(3):18:1–18:43, 2019.

[AHTG11] Niaz Arijo, Reiko Heckel, Mirco Tribastone, and Stephen Gilmore. Modular performance modelling for mobile applications. In *Second Joint WOSP/SIPEW International Conference on Performance Engineering*, pages 329–334. ACM, 2011.

[AS85]   Bowen Alpern and Fred B. Schneider. Defining liveness. *Inf. Process. Lett.*, 21(4):181–185, 1985.

[BK84]   Jan A. Bergstra and Jan Willem Klop. Process algebra for synchronous communication. *Information and Control*, 60(1-3):109–137, 1984.

[BKPS97] Bettina Buth, Michel Kouvaras, Jan Peleska, and Hui Shi. Deadlock analysis for a fault-tolerant system. In *AMAST*, LNCS 1349. Springer, 1997.

[BPFS]   Tom Bohman, Oleg Pikhurko, Alan Frieze, and Danny Sleator. Puzzle 6: Uni-
            form candy distribution. http://www.cs.cmu.edu/puzzle/puzzle6.html, Last
            accessed: August 2021.
[BPS98]   Bettina Buth, Jan Peleska, and Hui Shi. Combining methods for the livelock
            analysis of a fault-tolerant system. In *AMAST*, LNCS 1548. Springer, 1998.
[CG07]   Ana Cavalcanti and Marie-Claude Gaudel. Testing for refinement in CSP. In
            9th International Conference on Formal Engineering Methods, LNCS 4789.
            Springer, 2007.
[Fis97]   Clemens Fischer. CSP-OZ: a combination of object-Z and CSP. In *Formal meth-
            ods for open object-based distributed systems*. Chapman & Hall, 1997.
[Gar05]   Bill Gardner. Converging CSP specifications and C++ programming via selec-
            tive formalism. *ACM Transactions on Embedded Computing Systems*, 4(2),
            2005.
[Gar15]   Bill Gardner, 2015. http://www.uoguelph.ca/~gardnerw/csp++/, Last accessed
            August 2021.
[GRS04]   Andy Gimblett, Markus Roggenbach, and Bernd-Holger Schlingloff. Towards a
            formal specification of an electronic payment system in CSP-CASL. In *WADT*,
            LNCS 3423. Springer, 2004.
[HKRS09]   Greg Holland, Temesghen Kahsai, Markus Roggenbach, and Bernd-Holger
            Schlingloff. Towards formal testing of jet engine rolls-royce BR725. In *Proc. 18th
            Int. Conf on Concurrency, Specification and Programming, Krakow, Poland*,
            pages 217–229, 2009.
[Hoa85]   Charles Antony Richard Hoare. *Communicating Sequential Processes*. Prentice
            Hall, 1985.
[IMNR12]   Yoshinao Isobe, Faron Moller, Hoang Nga Nguyen, and Markus Roggenbach.
            Safety and line capacity in railways - an approach in timed CSP. In *Integrated
            Formal Methods*, LNCS 7321. Springer, 2012.
[IR]   Yoshinao Isobe and Markus Roggenbach. Webpage on CSP-Prover. http://
            staff.aist.go.jp/y-isobe/CSP-Prover/CSP-Prover.html, Last accessed:
            August 2021.
[IR05]   Yoshinao Isobe and Markus Roggenbach. A generic theorem prover of CSP
            refinement. In *TACAS 2005*, LNCS 3440. Springer, 2005.
[IR06]   Yoshinao Isobe and Markus Roggenbach. A complete axiomatic semantics for
            the CSP stable-failures model. In *CONCUR 2006*, LNCS 4137. Springer, 2006.
[IR07]   Yoshinao Isobe and Markus Roggenbach. Proof principles of CSP - CSP-Prover
            in practice. In *Dynamics in Logistics*. Springer, 2007.
[IR08]   Yoshinao Isobe and Markus Roggenbach. CSP-Prover - a proof tool for the
            verification of scalable concurrent systems. *Japan Society for Software Science
            and Technology, Computer Software*, 25, 2008.
[IRG05]   Yoshinao Isobe, Markus Roggenbach, and Stefan Gruner. Extending CSP-
            Prover by deadlock-analysis: Towards the verification of systolic arrays. In
            FOSE 2005, Japanese Lecture Notes Series 31. Kindai-kagaku-sha, 2005.
[KRS07]   Temesghen Kahsai, Markus Roggenbach, and Bernd-Holger Schlingloff.
            Specification-based testing for refinement. In *SEFM 2007*. IEEE Computer
            Society, 2007.
[KRS08]   Temesghen Kahsai, Markus Roggenbach, and Bernd-Holger Schlingloff.
            Specification-based testing for software product lines. In *SEFM 2008*. IEEE
            Computer Society, 2008.
[LLQ10]   Jiang Liu, Jidong Lv, Zhao Quan, Naijun Zhan, Hengjun Zhao, Chaochen Zhou,
            and Liang Zou. A calculus for Hybrid CSP. In *Programming Languages and
            Systems*, LNCS 6461. Springer, 2010.
[Mil89]   Robin Milner. *Communication and concurrency*. Prentice Hall, 1989.
[MK06]   Jeff Magee and Jeff Kramer. *Concurrency: State Models & Java Programs*.
            Wiley, 2nd edition, 2006.

[MPW92] Robin Milner, Joachim Parrow, and David Walker. A calculus of mobile processes, I. *Inf. Comput.*, 100(1), 1992.

[OCW09] Marcel Oliveira, Ana Cavalcanti, and Jim Woodcock. A UTP semantics for Circus. *Formal Asp. Comput.*, 21(1-2), 2009.

[OW03] J. Ouaknine and J. Worrell. Timed CSP = closed timed ε-automata. *Nordic Journal of Computing*, 10:1–35, 2003.

[PB99] Jan Peleska and Bettina Buth. Formal methods for the international space station ISS. In *Correct System Design*, LNCS 1710. Springer, 1999.

[Rog06] Markus Roggenbach. CSP-CASL: A new integration of process algebra and algebraic specification. *Theoretical Computer Science*, 354(1):42–71, 2006.

[Ros98] A.W. Roscoe. *The theory and practice of concurrency*. Prentice Hall, 1998.

[Ros03] A. W. Roscoe. On the expressive power of CSP refinement. *Formal Aspects of Computing*, 17, 2003.

[Ros08] A. W. Roscoe. The three platonic models of divergence-strict CSP. In *Theoretical Aspects of Computing*, LNCS 5160. Springer, 2008.

[Ros10] A.W. Roscoe. *Understanding Concurrent Systems*. Springer, 2010.

[Sca98] Bryan Scattergood. The Semantics and Implementation of Machine-Readable CSP, 1998. DPhil thesis, University of Oxford.

[Sch00] Steve A. Schneider. *Concurrent and Real-time Systems: the CSP Approach*. Wiley, 2000.

[SNW96] Vladimiro Sassone, Mogens Nielsen, and Glynn Winskel. Models for concurrency: Towards a classification. *Theor. Comput. Sci.*, 170(1-2):297–348, 1996.

[ST05] Steve Schneider and Helen Treharne. CSP theorems for communicating B machines. *Formal Asp. Comput.*, 17(4):390–422, 2005.

[vG01] R.J. van Glabbeek. The linear time-branching time spectrum I – the semantics of concrete, sequential processes. In *Handbook of Process Algebra*. Elsevier, 2001.

[WBM07] Peter Welch, Neil Brown, James Moores, Kevin Chalmers, and Bernhard Sputh. Integrating and extending JCSP. In *CPA 2007*. IOS Press, 2007.

[Wel] Peter Welch. http://www.cs.kent.ac.uk/projects/ofa/jcsp/, Last accessed: August 2021.

[Win93] Glynn Winskel. *The formal semantics of programming languages: an introduction*. MIT Press, 1993.

[Win02] Kirsten Winter. Model checking railway interlocking systems. *Australian Computer Science Communications*, 24(1), 2002.

Part II
Methods

Chapter 4
Algebraic Specification in CASL

Markus Roggenbach and Liam O'Reilly

Abstract Algebraic specification is a specification technique that provides a formal basis for the systematic development of software systems. By developing a Telephone Database in a formal way, syntax, semantics and pragmatics of the language CASL are introduced by example. As techniques to ensure software quality, we discuss how to prove consistency of requirements in CASL and how to test programs against CASL specifications. On the example of so-called Ladder Logic programs, which are in widespread use in control applications, we demonstrate how CASL can be used for program verification. Finally, we provide an overview of CASL's specification structuring mechanisms, which—in principle—can be applied to any specification language. Typical properties of abstract data types are established through automated theorem proving over structured CASL specifications.

4.1 Introduction

Your aunt Erna has one of her generous days, promises to buy you a smart-phone, and asks: "which phone do you want?" You explain to her what features are important to you, but leave it to her to pick the phone—after all, she is being generous enough. So, you might state that you want a specific brand, you have ideas concerning the tech-spec of the cameras, e.g., how many cameras there should be and and how many pixels they should have, weight and size of the phone, as well as the battery life, etc.

Your smart-phone description can be seen as a typical algebraic specification. Namely, rather than describing how the object of your desire shall be

Markus Roggenbach
Swansea University, Wales, United Kingdom

Liam O'Reilly
Swansea University, Wales, United Kingdom

© Springer Nature Switzerland AG 2022, corrected publication 2022
M. Roggenbach et al., *Formal Methods for Software Engineering*,
Texts in Theoretical Computer Science. An EATCS Series,
https://doi.org/10.1007/978-3-030-38800-3_4

built (exact camera part, which specific battery to put in, how the phone is built internally in order to be light enough, etc.), you are speaking just about the properties that the phone is supposed to have. In terms of algebraic specification, you are stating axioms that shall hold. Any phone that has these properties, i.e., for which the stated axioms hold, would satisfy you. This is typical for algebraic specification, which usually is classified as a 'property-oriented' Formal Method, in contrast to 'model-oriented' methods—further discussion on taxonomy can be found in Sect. 1.2.3. What you have done is to narrow down the set of all smart-phones to a pool of acceptable ones. In terms of algebraic specification, the 'meaning', i.e., semantics of your smart-phone description, is the collection of *all* smart-phones which fulfill your wishes.

After studying the pool of acceptable smart-phones, you might discover that it contains some you actually don't like. In that case, you would start to add further axioms to narrow things down. This is not without risk, however, you might end up with an empty pool because your demands can't be fulfilled. In algebraic specification, in this case one speaks of an empty model class.

For simple specifications of this kind, natural language will suffice. However, for building complex technical systems, tool support is needed to analyze various aspects of a specification, e.g., the model class of what one has specified. In this chapter we will study the algebraic specification language CASL, which allows one to write logical axioms and comes with tool support for parsing, static analysis, theorem proving, and testing—to name just a few.

As illustrated in the introduction to this book in Chap. 1, Formal Methods can play many roles in the software life-cycle. For this chapter, we have chosen some of the many ways that algebraic specification can facilitate software engineering, namely how to formulate concise requirements, how to analyze requirements for consistency, how to apply automated random testing for quality assurance in an invent & verify software production process, and how to use it to verify software designs.

This chapter will emphasise specification practice, where a naive understanding of logic will suffice. A discussion of the theoretical background of the logic underlying CASL can be found in Sect. 2.4.3. CASL uses the logic SubPCFOL$^=$, i.e., we have a first-order logic ("FOL") with subsorting ("Sub"), partiality ("P"), sort generations constraints ("C") and equality (indicated by $=$) available. In this chapter we will not consider subsorting. We will look into the sublanguage PCFOL$^=$ of CASL.

This chapter is organised as follows. First, we introduce CASL syntax and semantics; discuss how to use automated theorem proving to establish properties of specifications; and show how to test Java programs against CASL specifications. Then—inspired by an industrial application of Formal Methods —we illustrate how CASL can be used to verify control programs. We conclude this chapter by discussing some of the various specification structuring mechanisms that CASL offers.

4.2 A First Example: Modelling, Validating, Consistency Checking, and Testing a Telephone Database

Our first example demonstrates that algebraic specification can be similar to functional programming: we use terms in order to build up a system history and—by giving axioms—we define a reduction strategy on these terms (c.f., Sect. 4.2.1). Concerning software engineering (c.f., Sect. 1.2.1) we demonstrate that the resulting CASL specification can serve two different purposes: The CASL specification can be used for

- Validation—in Sect. 4.2.2 we validate our specification against use cases; in Sect. 4.2.3 we check that the requirements stated do not contain contradictions.
- Verification—in Sect. 4.2.4 we automatically test if various Java implementations meet our specification.

4.2.1 Modelling

We start with an informal description of a system that we want to design:

Example 43: A Telephone Database

Write a Java program that implements a telephone database with
- *Name* and
- *Number*

as entries. The operations shall include functions such as *update*, *lookUp*, *delete*, and a test *isEmpty*. The *Name* shall act as a key.

CASL **Signatures**

The first step in algebraic modelling is to develop a suitable signature. What are the symbols needed to describe the system and its intended properties? To this end, one can use the following rules of thumb:

- sort symbols provide a classification of the data involved, they provide a 'type system' for the domain under discussion;
- function symbols represent computations, in CASL they are classified into

  - total function symbols—one assumes in the model that the computation always returns a result—and

– partial function symbols;

• predicate symbols represent tests.

Developing a good signature is a challenging step when modelling with algebraic specification.

Example 43.1: Signature

The above narrative explicitly speaks about different kinds of data, namely about "Name" and "Number". Furthermore, we decide to represent the state of the system as "Database". We choose these to be the sorts involved. Besides the computations *update*, *lookUp*, *delete*, we define a constant "initial" to represent the state of the database in the beginning. Finally, we provide a test "isEmpty" that shall hold if there are no entries stored in the database.

This results in the PCFOL$^=$ signature $\Sigma_{\text{DATABASE}} = (S, TF, PF, P)$ with

• $S = \{$ Database, Name, Number $\}$
• $TF = \{$ initial:Database,
　　　　 update: Database $\times$ Name $\times$ Number $\to$ Database,
　　　　 delete: Database $\times$ Name $\to$ Database$\}$
• $PF = \{$ lookUp: Database $\times$ Name $\to$? Number$\}$
• $P = \{$ isEmpty: Database $\}$

In CASL, we can declare this signature as follows:

spec DATABASESIGNATURE =
　　　sorts *Database, Name, Number*
　　　ops *initial : Database*;
　　　　　　lookUp : Database $\times$ Name $\to$? Number;
　　　　　　update : Database $\times$ Name $\times$ Number $\to$ Database;
　　　　　　delete : Database $\times$ Name $\to$ Database
　　　pred *isEmpty : Database*
end

CASL specifications start with the keyword **spec**. They have a name, in the example DATABASE-SIGNATURE, which is separated with an equal sign from the specification body. After the keyword **sorts** (or, equivalently, **sort**) one declares sort symbols. After the keyword **ops** (or, equivalently, **op**), one declares function symbols including their arity. For partial function symbols, the arrow is decorated with a "?" in order to indicate partiality. After the keyword **pred** (or, equivalently, **preds**) one declares predicate symbols and their arity. It is optional to close a CASL specification with the keyword **end**—see Appendix C for the CASL grammar.

The specification DATABASE-SIGNATURE has the signature $\Sigma_{\text{DATABASE}} = (S, TF, PF, P)$ as part of its so-called static semantics. The CASL semantics is defined in a way that a *phrase* (w.r.t. the CASL grammar) is mapped to a *mathematical structure* (e.g., a set or a tuple of values, a function, a relation) as its semantics: the sort declaration

sorts *Database, Name, Number*

has as the mathematical structure a set of sort symbols as its static semantics, the specification DATABASE-SIGNATURE has as the mathematical structure a PCFOL$^=$ signature as part of its static semantics.

In passing we note that CASL signatures do not include variables as defined for first-order signatures in Definition 8. There are various schools of how to define logics. In the chapter on logic we followed one school (variables are included), the design of CASL follows a different school (variables are not included). It is the case that certain theorems on CASL hold only as variables are not part of the signature. Later on, we will the an example of this in Theorem 1, in Sect. 4.3.2.

CASL **Formulae**

The next step in modelling is to give axioms in the form of PCFOL$^=$ formulae, which define the interplay between the operation and predicate symbols.

Example 43.2: Axioms

With the operation symbols *initial* and *update* we construct a table. This table can be inspected with *lookUp*, manipulated with *delete*, and checked with *isEmpty*.

spec DATABASE =
    sorts *Database, Name, Number*
    ops   *initial* : *Database*;
          lookUp : *Database* × *Name* →? *Number*;
          update : *Database* × *Name* × *Number* → *Database*;
          delete : *Database* × *Name* → *Database*
    pred *isEmpty* : *Database*
    $\forall$ *db* : *Database*; *name, name1, name2* : *Name*;
    number : *Number*
    • $\neg$ *def lookUp*(*initial, name*)   %(non def initial)%
    • *name1* = *name2*
      $\Rightarrow$ *lookUp*(*update*(*db, name1, number*), *name2*) = *number*
                          %(name found)%
    • $\neg$ *name1* = *name2*
      $\Rightarrow$ *lookUp*(*update*(*db, name1, number*), *name2*)
        = *lookUp*(*db, name2*)

$$\%(\text{name not found})\%$$

- $delete(initial, name) = initial$ $\%(\text{delete initial})\%$
- $name1 = name2$
  $\Rightarrow delete(update(db, name1, number), name2)$
      $= delete(db, name2)$

$$\%(\text{delete found})\%$$

- $\neg\ name1 = name2$
  $\Rightarrow delete(update(db, name1, number), name2)$
      $= update(delete(db, name2), name1, number)$

$$\%(\text{name not found})\%$$

- $isEmpty(db) \Leftrightarrow db = initial$ $\%(\text{def isEmpty})\%$

end

spec OneSort =
     sort s
end

Taking the operations *initial* and *update* as constructors, our axioms systematically cover how the operations *lookUp* and *delete* and the predicate *isEmpty* interact with them:

- In the state *initial* the database shall have no entries. Consequently, observing this state with the *lookUp* function yields no result— see %(non_def_initial)%. When there is at least one entry in the database, we might obtain a result when observing it: Should the most recent entry match the *lookUp* request, the number of the entry is to be returned—see %(name_found)%; otherwise, we have to inspect the previous entries—see %(name_not_found)%.
- Deleting an entry in the *initial* state shall have no effect—see %(delete_initial)%. When the database has at least one entry, and the most recent entry is for the name which we want to delete, we take this entry away and look for further, earlier entries—see %(delete_found)%. If the most recent entry, however, was not for the name to be deleted, the most recent entry is preserved and we look for earlier entries—see %(name_not_found)%.
- Finally, only the initial database is empty—see %(def_isEmpty)%.

In our setting, axioms in Casl are PCFOL$^=$ formulae. They are stated in the Casl specification text after a bullet point "•". Variables can be declared in one go for several axioms, e.g., $\forall\ db : Database$; *name, name1, name2 : Name; number : Number*. However, it is equally possible to declare the variables for every axiom separately. For reference, axioms can be given a label, e.g., %(name_not_found)%. In the Casl static semantics, the phrases derivable from the non-terminal **FORMULA** of the Casl grammar yield PCFOL$^=$ formulae.

Definition 1 (CASL *static semantics*) Given a CASL specification *sp*, its *static semantics* is

- a PCFOL$^=$ signature $\Sigma(sp)$ and
- a finite set of PCFOL$^=$ formulae $Ax(sp)$.

Model Semantics of CASL Specifications

Given a signature, a *model* is a mathematical structure which interprets all symbols of the signature: sort symbols have sets as their interpretation—often, we speak of them as *carrier sets*—function symbols have functions as their interpretation, and predicate symbols have predicates as their interpretation—see Sect. 2.4.1 in Chap. 2 for further details. There is a *validation relation* $\models$ that expresses if in a model $\mathcal{M}$ a formula φ is valid; if this is the case, we write $\mathcal{M} \models \varphi$—again, see Chap. 2 for further details. Using these notions, we can define the CASL model semantics:

Definition 2 (CASL *model semantics*) Given the static semantics, the *model semantics* of *sp* is defined as

$$Mod(sp) \triangleq \{\Sigma(sp)\text{-model } \mathcal{M} \mid \mathcal{M} \models Ax(sp)$$
$$\text{and all carrier sets in } \mathcal{M} \text{ are non empty}\}$$

Thus, CASL has *loose semantics*, cf. Sect. 2.3.1.

The requirement of non empty carrier sets simplifies reasoning on CASL specifications. The formula T (for "true") holds in all CASL models. However, if we would allow a model $\mathcal{M}$ with $\mathcal{M}_s = \emptyset$, the formula $\exists x : s \bullet T$ would not hold. That is, for a model with an empty carrier set it is possible that adding a quantification can change truth though the quantified variable does not appear in the formula. This choice of CASL is in line with the definition of a first-order model that we gave in Sect. 2.4, Definition 12.

In passing, we mention that—in strict mathematical terms—the above collection $Mod(sp)$ of the models of a specification *sp* forms a *class*, rather than a *set*. Thus, a CASL specification has a *model class* as its model semantics.

For the interested reader we give a brief discussion of the topic. A class is a collection of sets; the precise definition of a class depends on the foundation context, e.g., Zermelo-Fraenkel set theory (where they appear indirectly only) or Von Neumann-Bernays-Gödel set theory (where they are 'first class' citizens).

An initial motivation for distinguishing between the notions of class and set is Russell's paradox: Let S be the collection of all sets. Consider the "collection"

$$R \triangleq \{X \mid X \in S, X \notin X\}. \tag{4.1}$$

Then R is not a set: assume that R was a set; then either $R \in R$ or $R \notin R$; if $R \in R$, then we have by (4.1) that $R \notin R$—contradiction; if $R \notin R$, then we have by (4.1) that $R \in R$—contradiction as well; thus, our assumption must have been wrong, i.e., R is not a set. Consider now the CASL specification

spec ONESORT =
    sort s
end

which declares one sort only and states no axioms. From the collection of all its models we can form the collection of all sets. It holds that

$$S = \{M(s) \mid M \in Mod(\text{ONESORT})\} \cup \{\emptyset\} \qquad (4.2)$$

Now, $Mod(\text{ONESORT})$ can't be a set: assume it was be a set; then S must be a set too, as it is formed by rules from naive set theory (projection to a component, union of two sets); but then R must be a set as well, as it is again formed according to the rules of naive set theory (forming a new set by restriction); as seen above, R is not a set; thus, $Mod(\text{ONESORT})$ is not a set. Thus, in general $Mod(sp)$ is a class rather than a set.

Historical note. Bertrand Russell (1872–1970), the inventor of this paradox, was a philosopher, mathematician and a global intellectual from an English, aristocratic family, who happened to have been born in South Wales and lived the last part of his life in Snowdonia, in North Wales, UK.

4.2.2 Validating

Above, we have developed a specification, i.e., a description of a class of models of the software we intend to write. The question is: is our specification in agreement with our intentions? To answer this question one should *validate* the specification. One possibility to do so is to (i) populate the specification with concrete entities and then to (ii) check if these entities "behave" in the expected way.

Example 43.3: Validating I—Concrete Use Cases

Following the above approach, we provide constants *Hugo* and *Erna* of sort *Name* and constants *N4711* and *N17* of sort *Number*. As CASL has loose semantics, we have to state an axiom %(Hugo different from Erna)% to ensure that the two symbols are interpreted differently. This first step results in the following specification:

spec USECASESETUP = DATABASE
then ops   *Hugo, Erna* : *Name*;
              N4711, N17 : *Number*

- $\neg\ Hugo = Erna$ %(Hugo different from Erna)%
end

In the second step we can state how we expect our telephone database to behave. Note that the stated properties go well beyond the narrative with which we started. In the narrative, we associate a certain meaning with names such as *update, lookUp, etc.*, where one often refrains from stating this explicitly. When writing a formal specification, this meaning needs to be made precise.

spec UseCase = UseCaseSetUp
then %implies
- $\neg\ lookUp(initial, Hugo) = N17$
%(lookUp on initial not equal to 17)%
- $\neg\ def\ lookUp(initial, Hugo)$ %(lookUp on initial undefined)%
- $lookUp(update(initial, Hugo, N4711), Hugo) = N4711$
%(lookUp stores values)%
- $lookUp(update(update(initial, Hugo, N4711), Erna, N17),$
    $Hugo)$
    $= N4711$ %(update does not overwrite)%
- $\neg\ def\ lookUp(update(initial, Erna, N17), Hugo)$
%(lookUp is not defined without update)%
- $\neg\ isEmpty(update(update(initial, Hugo, N4711), Erna, N17))$
%(updating leads to a non empty database)%
- $isEmpty(initial)$ %(the initial database is empty)%
- $isEmpty(delete(update(initial, Hugo, N4711), Hugo))$
%(deleting all entries leads to an empty database)%
end

Using the Heterogeneous Tool Set Hets [MML07] and the automated theorem prover SPASS, we can prove that our specification has all the intended properties—see Fig. 4.1: in the "Goals" column, all of our intended properties are ticked with a "+", indicating that SPASS could derive them from the given axiomatic basis; the subwindow in the right lower corner displays for the highlighted goal deleting_all_entries_leads_to_an_empty_database, which axioms of the specification SPASS was using in order to establish this goal as a theorem.

The tool Hets[1] reports the proof status as follows: "A proved goal is indicated by a "+", a "-" indicates a disproved goal, a space denotes an open goal, and a "x" denotes an inconsistent specification".[2] An open goal can arise from the fact that first-order logic is undecidable; another reason can

---

[1] Available at http://hets.eu.

[2] See the HETS User Guide, available at https://sefm-book.github.io.

Fig. 4.1 Snapshot of Hets validating the database specification with the theorem prover
SPASS

be that neither the goal nor its negation is a consequence of the specification. If a goal is a consequence of the specification, however, appears open in SPASS this is usually for the reason that the search space that is too large: the theorem prover SPASS could not cover in within the allotted time set by the user or with the available memory.

The above specifications show that in CASL one can structure specifications, namely one can *import* named specifications and then extend them after the keyword **then**. In USECASESETUP we first import the specification DATABASE, which we then extend by entities that populate the sorts. For the time being, we will use the CASL structuring constructs import and extensions in a naive way. They can be flattened out as follows: an import can be resolved by substituting the specification text for the name; the keyword **then** can be removed by stitching specification parts together. In Sect. 4.4 we will discuss the CASL structuring language in detail.

More concisely, regarding the static and the model semantics of CASL, this naive view results in:

- the signature of the above specification is the union of all symbols declared in the various parts;
- the axioms of the specification are given by the union of all axioms stated in the various parts;
- the model class of the specification consists of all models for the signature in which the axioms hold.

We can annotate an extension with the keyword **%implies** provided that this extension consists of axioms only. This is the case in the above example of the specification UseCase. The axioms stated after the **then %implies** do not further restrict the model class. They are expected to be a consequence of the specification up to this point. More precisely it shall hold:

$$Mod(sp) = Mod(sp \textbf{ then \%implies } sp')$$

The annotation triggers tools to treat the axioms stated in sp' as intended consequences of sp, i.e., as theorems that shall be proven using the axioms stated in sp.

Concerning the question of validating a specification, the fundamental question is when to stop validating. Are eight axioms enough? Would twenty axioms give a better assurance that the specification expresses what we want? This question is still an open research topic.

We can, for example, consider an abstract property such as the commutativity of the update operation:

Example 43.4: Validating II—an Abstract Property

We would expect that we can swap the order in which we enter two pieces of data into the database, provided the data concerns different names.

spec ABSTRACTPROPERTY =
    DATABASE
then %implies
    $\forall$ db : $Database$; $name1, name2, name3$: $Name$;
    $number1, number2$: $Number$
    • $lookUp(update(update(db, name1, number1),$
                 $name2, number2),$
        $name3)$
      $= lookUp(update(update(db, name2, number2),$
                      $name1, number1),$
          $name3)$
      $if \neg name1 = name2$
                          %(specialised update commutativity)%
end

By loading the above specification into Hets, SPASS easily proves %(specialised update commutativity)% is a consequence of the database axioms, and hence displays a "+" indicating so.

As a reminder, the model class of a specification consists of all models for the signature in which the axioms hold. With this in mind, given a specification of the form

$$sp \text{ then } \%\text{implies } sp'$$

where—for the sake of argument—sp' contains a single formula φ, there are four possibilities of how φ can relate to the axioms of sp:

$Ax(sp) \models \varphi$	$Ax(sp) \models \neg\varphi$	status of φ
does not hold	does not hold	undetermined
does hold	does not hold	theorem
does not hold	does hold	counterexample
does hold	does hold	irrelevant, as $Ax(sp)$ is inconsistent

The first row concerns a situation, where the specification needs to be further developed in order to let φ either hold or not hold (there are models of the specification where φ holds, and there are models where φ does not hold). The second row concerns the situation which we have been using for validation thus far: we confirmed with theorem proving that a property that we were expecting to hold is actually a consequence of the axioms we wrote. The third row concerns a situation in which validation fails. Rather than confirming the expectation, we have to learn that we got our axioms wrong. The fourth row indicates a situation in which the development has gone wrong: there are no models, i.e., what has been specified cannot be implemented. The specification is said to be inconsistent. We study this situation in the next section.

Note that a timeout when attempting to prove a property does not indicate which row of the table one is in.

We conclude the section with an example illustrating the situation when the status of φ is undetermined.

Example 43.5: Validation III—An Undetermined Property

We study a property concerning equality of databases, which at first glance one might expect to hold.

spec UNDETERMINEDPROPERTY = DATABASE
then %implies
  $\forall\ db : Database$; $name1, name2 : Name$;
  $number1, number2 : Number$
  • $update(update(db, name1, number1), name2, number2)$
    $= update(update(db, name2, number2), name1, number1)$
    $if \neg\ name1 = name2$
                              %(general update commutativity)%
  • $\neg\ (update(update(db, name1, number1), name2, number2)$
      $= update(update(db, name2, number2), name1, number1)$
      $if \neg\ name1 = name2)$
                      %(negation of general update commutativity)%
end

SPASS neither proves %(general update commutativity)% nor its negation because both do not logically follow from the specification (timeout in both cases). This shows that the **%implies** is incorrectly placed here. Either DATABASE should be further developed or the formula %(general update commutativity)% should be revisited.

In the following we give an outline of two models of the specification DATABASE that illustrates the situation.

A model where %(general update commutativity)% does not hold is given when the sort *Database* is represented by lists. Here, we find two databases, i.e., lists, which are different though they give back the same values under *lookUp*:

$$\langle (hugo, 4711), (erna, 17) \rangle \neq \langle (erna, 17), (hugo, 4711) \rangle$$

As lists, they are different. However, they map the same numbers to the names under *lookUp*.

A model where %(general update commutativity)% does hold is given when the sort *Database* is represented by maps:

$$\left\{ \begin{array}{l} erna \mapsto 17, \\ hugo \mapsto 4711 \end{array} \right\}$$

As a map has unique keys with updates overwriting, we can give only one representation, i.e., %(general update commutativity)% holds.

Let us reflect upon the property %(general update commutativity)% that we tried to show in the above example. The property actually concerns the database representation in computer memory, namely that 'semantically' identical databases have a unique representation. The DATABASE specification does not prescribe such a property. It is rather on a more abstract level that is concerned with the interplay of the operators and deliberately leaves representation questions open. Usually, decisions on data representation are taken late in a software design process.

4.2.3 Consistency Checking

Another question is if the specification DATABASE is consistent, i.e., if its model class is non empty. As all formulae are true relative to the empty model class, all the above validation effort would be in vain for an inconsistent specification.

Definition 3 (*Consistent specification*) A CASL specification *sp* is called consistent, if and only if $Mod(sp) \neq \emptyset$.

In order to prove consistency, one thus has to provide a model of the specification. For simple specifications, this can be done manually or by making use of a so-called model finder such as Darwin. For complex specifications, Lüth et al. have implemented a CASL consistency checker [LRS04], which is based on a consistency calculus that allows one to reduce the question of consistency for a complex specification to the question of consistency for a simpler specification.

For our example, we first explicitly define an element of the model class:

Example 43.6: Consistency—Manual Proof

We first provide an algebra for the signature of the specification USE-CASESETUP: The state of the database encodes for which name there has been an update. In the carrier sets, we provide two names only and there is just one number. We define the functions using value tables, where the symbol $\perp$ indicates undefinedness.

- $\mathcal{M}(Database) \triangleq \{empty, h\_stored, e\_stored, e\_and\_h\_stored\}$.
- $\mathcal{M}(Name) \triangleq \{h, e\}$.
- $\mathcal{M}(Number) \triangleq \{*\}$.
- $\mathcal{M}(initial) \triangleq empty$.
- $\mathcal{M}(Hugo) \triangleq h$; $\mathcal{M}(Erna) \triangleq e$.
- $\mathcal{M}(N4711) \triangleq \mathcal{M}(N17) \triangleq *$.
- $\mathcal{M}(isEmpty) \triangleq \{empty\}$.
- The functions are defined in through the following tables:

$\mathcal{M}(lookUp)$	empty	h_stored	e_stored	e_and_h_stored
h	$\perp$	$*$	$\perp$	$*$
e	$\perp$	$\perp$	$*$	$*$

$\mathcal{M}(update)$	empty	h_stored	e_stored	e_and_h_stored
h, *	h_stored	h_stored	e_and_h_stored	e_and_h_stored
e, *	e_stored	e_and_h_stored	e_stored	e_and_h_stored

$\mathcal{M}(delete)$	empty	h_stored	e_stored	e_and_h_stored
h	empty	empty	e_stored	e_stored
e	empty	h_stored	empty	h_stored

Then we prove that all axioms of USECASESETUP hold in $\mathcal{M}$: Clearly, %(Hugo different from Erna)% and %(non_def_initial)% hold in $\mathcal{M}$. Let us now consider the axiom %(name_found)%:

$$name1 = name2 \Rightarrow$$
$$lookUp(update(db, name1, number), name2) = number$$

In order to make *name1=name2* true, we choose without loss of generality $\nu(name1) = \nu(name1) = h$ for the variable evaluation ν. Inspecting the table for *update*, we observe that $\mathcal{M}(update)(d, h, *) \in \{h\_stored, eAndh\_stored\}$ for all $d \in \mathcal{M}(Database) = \{empty, h\_stored, e\_stored, eAndh\_stored\}$. Inspecting the table for *lookUp*, we obtain that $\mathcal{M}(lookUp)(d, h) = *$ for $d \in \{h\_stored, eAndh\_stored\}$. $\mathcal{M}(Number) = \{*\}$ consists of a single value only. Thus we obtain $\nu(number) = *$ as the only possible interpretation. Thus, the axiom %(name_found)% holds in $\mathcal{M}$. The other axioms can be proven similarly.

Thus, by definition of the model semantics of CASL we have that $\mathcal{M} \in Mod(\text{UseCaseSetUp})$, i.e., that $Mod(\text{UseCaseSetUp}) \neq \emptyset$, i.e., UseCaseSetUp is consistent.

The above model $\mathcal{M}$ might have come as a surprise. All its carrier sets are finite. The carrier set of *Database* indicates only if the database is *empty*, or that something is stored for h, indicated by value *h_stored*, etc. One would expect a database to store actual data and not simply a flag indicating that data has been stored.

However, up to now there have been no properties stated on what kind of names or numbers the database shall deal with. Concerning the software development process, the above database specification is still on the requirements level, i.e., many design decisions have still to be taken, e.g., the concrete data formats for the sorts *Name* and *Number*. This level of abstraction is achieved by *under-specification*, i.e., the stated axioms allow 'different' implementations, including the intended one, but also not yet excluding "strange" models such as $\mathcal{M}$. As seen above, already such an under-specified, formal requirement specification is useful: it can be validated to exhibit expected properties, as demonstrated for the specification DATABASE.

Proving that all the axioms hold in a model is a tedious, time consuming, and error prone process. Thus, tool support would be worthwhile. In the following we discuss how to use theorem proving for this task. To this end, we (i) encode a (finite) model in CASL (up to isomorphism, see Sect. 2.3.1), and (ii) prove that this model has the desired properties.

Given a $\Sigma = (S, TF, PF, P)$ and a Σ-model $\mathcal{M}$ for a signature with finite carrier sets, we can represent $\mathcal{M}$ in a CASL specification $sp_{\mathcal{M}}$ as follows:

Algorithm 4: Model Encoding

---

input : Σ-model $\mathcal{M}$ with finite carrier sets
output: CASL specification where all models are isomorphic to $\mathcal{M}$

Encode the carrier sets using so-called free types
Declare all operation symbols
Encode all constant and function interpretations of $\mathcal{M}$
    in the form of value tables entries
Declare all predicate symbols
    and encode all predicate interpretations of $\mathcal{M}$ as equivalences

---

The following example illustrates the algorithm:

Example 43.7: Consistency—Automated Proof

Applying the above algorithm to the model $\mathcal{M}$ from the previous example box results in the following CASL specification.

spec MYMODEL =
    %% encoding the carrier sets using "free types":
    free type *Database* ::= *empty* | *h_stored* | *e_stored*
                     | *e_and_h_stored*
    free type *Name* ::= *h* | *e*
    free type *Number* ::= $*$

    %% declaring "initial" and its interpretation:
    op   *initial* : *Database*
    • *initial* = *empty*
    ...
    %% declaring "lookUp" and encoding its value table:
    op   *lookUp* : *Database* × *Name* →? *Number*
    %% first row:
    • ¬ *def lookUp*(*empty*, *h*)
    • *lookUp*(*h_stored*, *h*) = $*$
    • ¬ *def lookUp*(*e_stored*, *h*)
    • *lookUp*(*e_and_h_stored*, *h*) = $*$
    %% second row:
    • ¬ *def lookUp*(*empty*, *e*)
    • ¬ *def lookUp*(*h_stored*, *e*)
    • *lookUp*(*e_stored*, *e*) = $*$
    • *lookUp*(*e_and_h_stored*, *e*) = $*$
    ...
    %% declaring "isEmpty" and its extent:
    pred   *isEmpty* : *Database*
    • ∀ *d* : *Database* • *isEmpty*(*d*) ⇔ *d* = *empty*
end

Now that we have encoded our model, we need to link it to the axioms. To this end we write a CASL **view**:

view CONSISTENCY : USECASESETUP **to** MYMODEL **end**

This view claims that all axioms of USECASESETUP hold in MYMODEL. Using the theorem prover SPASS, one can discharge all arising proof obligations.

    This results in a second proof that USECASESETUP is consistent: As MYMODEL is consistent by construction, and *Mod*(MYMODEL) ⊨

$Ax(\textsc{UseCaseSetUp})$—established by proving the view to be correct with the a theorem prover—we have that $\textsc{UseCaseSetUp}$ is consistent.

The above example makes use of the CASL **free type** construct. In general, this so-called sort generation constraint is an abbreviation for some signature declarations accompanied by higher order formulae. In our context, we use the free type only as a useful abbreviations for elements of first-order logic, e.g.,

free type $Name ::= h \mid e$

is a shorthand for

sort $Name$
ops $h, e : Name$
- $\neg\, h = e$ %(no confusion)%
- $\forall\, x : Name \bullet x = h \vee x = e$ %(no junk)%

i.e., it declares a sort name $Name$, it declares constants h and e of type $Name$, ensures that the declared constants are pairwise different %(no confusion)% and that the carrier set of $Name$ includes only the interpretation of these constants %(no junk)%. For more details, see Definition 18 in Chap. 2.

Another CASL element is the **view** specification. A view has a name, in our case CONSISTENCY, and links two specifications with each other, a source specification, in our case $\textsc{UseCaseSetUp}$, and a target specification, in our case $\textsc{MyModel}$. A view holds if each model of the target specification is a model of the source specification. In the case that source and target specification have different signatures, one considers the reduct of each model of the target specification.

Thus, we have now seen two ways of producing proof obligations in Hets: the CASL **view** and the CASL extension **then** %implies. It is up to the specifier which element to use. A **view** has the advantage that the properties stated in the source specification can be used in the context of different verifications. In contrast, properties stated after a **then** %implies are part of the specification to be verified.

The final new CASL element are comments to the specification text. They start with %% and last to the end of the line.

Reflecting on the above model encoding algorithm, given a model $\mathcal{M}$ over a signature $\Sigma = (S, TF, PF, P)$, by construction $sp_{\mathcal{M}}$ has the signature $\Sigma' = (S, TF \cup \{x : s \mid x \in \mathcal{M}(s), s \in S\}, PF, P)$, and $\mathcal{M}' \in Mod(sp_{\mathcal{M}})$, where $\mathcal{M}'$ is identical to $\mathcal{M}$ on the symbols from Σ and $\mathcal{M}'(x : s) = x$ for $x \in \mathcal{M}(s)$, $s \in S$. The consistency of $sp_{\mathcal{M}}$ can be proven with the consistency calculus for CASL presented in [RS01]. Yet another approach to consistency can be found in [CMM13].

4.2.4 Testing Java Implementations

ConGu [CLV10] is a tool that allows for automated random testing of Java code from algebraic specifications via run-time monitoring. ConGu defines a sublanguage of CASL, however, with a different concrete syntax.

In general, random testing is considered to be weak compared to more involved testing approaches (see, e.g., Chap. 5). However, when guided by axioms, as in ConGu, it turns out to be a powerful verification technique (as understood in the context of Software Engineering, cf. Sect. 1.2.1). A similar approach is also used by the tool JUnit-Quickcheck [HNSA16].

Testing Setup

When performing automated random testing using ConGu, the following four components are needed:

C-1 An algebraic specification written in the bespoke input language of ConGu;

C-2 a system under test (SUT) in the form of an implementation in Java;

C-3 a refinement mapping that defines the syntactic relationship between the specification and the system under test; and

C-4 a so-called random-runner, which is a Java main method. This runner generates and executes random method calls with random data on the SUT.

In the following we will introduce these components for our example.

Algebraic Specification in ConGu (Component C-1)

ConGu requires for partial functions an explicit characterisation of their domain.

Example 43.8: Adapting CASL to ConGu

We definitionally extend our database specification by a predicate *contains* which holds whenever the partial function *lookUp* is defined:

spec DATABASEFORCONGU =
    DATABASE
then %def
    pred *contains* : *Database* × *Name*
    • ∀ *db* : *Database*; *name* : *Name*
        • *contains*(*db*, *name*) ⇔ def *lookUp*(*db*, *name*)
then %implies
        ∀ *db* : *Database*; *name*, *name1*, *name2* : *Name*;
        number : *Number*
    • ¬ *contains*(*initial*, *name*)   %(contains never holds for initial)%

- $contains(update(db, name1, number), name2)$
  if $name1 = name2 \lor contains(db, name2)$

%(contains behaves like lookUp)%

end

The two new axioms %(contains never holds for initial)% and %(contains behaves like lookUp)% are consequences of this definitional extension and can be proven automatically with Hets.

Now that all partial functions have corresponding definedness predicates, we can translate from CASL to ConGu with simple syntactic changes.

System Under Test—Biven as a Class (Component C-2)

The database can be implemented in Java in multiple ways. However, each function and each predicate in the signature of the specification requires a corresponding public Java method. We collect these in the skeleton class representing the implemented datatype (SUT).

For run-time monitoring, ConGu requires that the SUT overrides the method `equals`, since equality used in the axioms is translated into calls to the Java `equals` method. Implementation of the `Clonable` interface and corresponding clone methods are required in the case that the objects of the class under test are mutable. In our case this involves providing two methods:

- `equals`—This is called by ConGu in run-time monitoring in order to check for equality of objects representing the states of the SUT.
- `clone`—This is used to duplicate a mutable object.

Example 43.9: Skeleton Class File

```java
public class Database implements Cloneable {
    public Database() {...}
    public void delete(String name) {...}
    public void update(String name, String number) {...}
    public String lookup(String name) {...}
    public Database clone() {...}
    public boolean equals(Object other) {...}
    public String toString() {...}
    public boolean contains(String name) {...}
    private Set<String> getNames() {...}
}
```

The constructor `Database()` corresponds to the constant `initial`. `getNames` is a utility method which computes the set of names in a given database. It is used by the `equals` method to test equality of databases. The omitted code within the methods is the implementation of the SUT.

Refinement Msapping (Component C-3)

ConGu uses a symbol mapping from the algebraic to the Java world. This is realised via a so-called refinement map which maps symbols of the algebraic specification (sorts, functions and predicates) to Java identifiers (classes and methods, respectively). This mapping also includes parameters. This allows for different orders of parameters in the specification and the SUT.

Example 43.10: ConGu Refinement Mapping

```
refinement
  Name is String
  Number is String
  Database is Database {
    initial: --> Database
      is Database();
    update: Database name:Name number:Number --> Database
      is void update(String name, String number);
    lookup: Database name:Name -->? Number
      is String lookup(String name);
    delete: Database name:Name --> Database
      is void delete(String name);
    contains: Database name:Name
      is boolean contains(String name);
  }
end refinement
```

The SUT can consist out of several classes. Here, these are the standard String class and our Database implementation.

Random-Runner (Component C-4)

A so-called random-runner is required when ConGu is used to perform automated random testing. The random-runner creates a randomly running SUT that can be monitored. To this end, the runner randomly calls the various methods of the SUT with random data.

Database Implementations: Linked List and Binary Search Tree

One possible implementation of our database could use a linked list to store the entries (pairs of name and phone number). With a correct implementation using the linked list approach one might obtain the following error-free output from the random runner:

Example 43.11: Monitoring of Linked List Implementation

```
Database before: Head -> null
Updating: Hugo to the number 959
Database after: Head -> Hugo:959 -> null

Database before: Head -> Hugo:959 -> null
Updating: Hugo to the number 254
Database after: Head -> Hugo:254 -> null

Database before: Head -> Hugo:254 -> null
Updating: Alice to the number 637
Database after: Head -> Alice:637 -> Hugo:254 -> null
```

With a correct implementation using the binary search tree (BST) approach one might obtain the following error-free output from the random runner:

Example 43.12: Monitoring of BST Implementation

```
Database before: Empty tree
Updating: Fred to the number 290
Database after: Fred:290

Database before: Fred:290
Updating: Erwin to the number 680
Database after: Fred:290
                -> Erwin:680
                -> null

Database before: Fred:290
                -> Erwin:680
                -> null
Updating: Hugo to the number 528
Database after: Fred:290
                -> Erwin:680
                -> Hugo:528
```

(ConGu's output has been slightly adapted for space considerations.)

Linked-Lists and Binary Search Trees are just two of the many data structures that can be used to implement a database. The above examples demonstrate that there can be several different correct implementations for one algebraic specification.

Finding Bugs

There might be a bug in the implementation that uses a Linked List:

Example 43.13: A Bug in the Use of the Linked List

The implemented lookup method returns the phone number of the last node in the list that matches the given name. This works fine as long as there is only one entry per name in the list. However, when the update method does not delete old entries then the invariant is violated under which the lookup method works correctly:

```
public void update(String name, String number) {
    // Bug: The programmer forgets to write the following:
    //      this.delete(name);
    Entry entry = new Entry(name, number);
    this.linkedList.insert(entry);
}
```

ConGu can find such a bug with random testing. In our experiments it was detected after randomly executing about 10 tests. Below is a typical error report:

```
Database before: Head -> Erwin:857 -> Hugo:293 -> null
Updating: Erwin to the number 413
Exception in thread "main"
  runtime.exceptions.PostconditionException: Axiom:
    (lookup(update(db, name1, number), name2) = number)
        if
      (name1 = name2);
  from Database

Context variables:
        name1 : Name;* = "Erwin"
        name2 : Name;* = "Erwin"
        number : Number;* = "413"
        db : Database;* = "Head -> Erwin:857 -> Hugo:293
           -> null"
Context term nodes:
        update(db, name1, number) = "Head -> Erwin:413 ->
        Erwin:857 -> Hugo:293 -> null"

        lookup(update(db, name1, number), name2) = "857"
```

When entering the specification into the ConGu system, the specifier might make a typo such as forgetting a negation or adding one:

Example 43.14: A Bug in the Specification

By forgetting the negation in the condition of the "name not found" axiom, the specifier actually produces an inconsistent specification.

```
// Name not found
lookup(update(db, name1, number), name2) =
    lookup(db, name2) if name1 = name2;
```

ConGu can find such a bug with random testing. In our experiments it was detected after randomly executing about 2 tests. Below is a typical error report:

```
Database before: Head -> Hugo:144 -> null
Updating: Hugo to the number 832
Exception in thread "main"
  runtime.exceptions.PostconditionException: Axiom:
    (lookup(update(db, name1, number), name2) =
      lookup(db, name2)) if (name1 = name2);
  from Database

Context variables:
    name1 : Name;* = "Hugo"
    name2 : Name;* = "Hugo"
    number : Number;* = "832"
    db : Database;* = "Head -> Hugo:144 -> null"
Context term nodes:
    update(db, name1, number) = "Head -> Hugo:832 -> null"
    lookup(update(db, name1, number), name2) = "832"
    lookup(db, name2) = "144"
```

In practice, applying random testing with ConGu appears to be an effective means to detect errors, be them in the implementation or the specification. In our experience, the additional cost of applying ConGu in a programming project is small provided there already is an algebraic specification available.

Random testing (via monitoring) with ConGu is sound, i.e., any error reported is an actual instance of non conformance. This holds under the following assumptions:

- the implementations of the equals and clone methods are correct (clone is only required for mutable SUTs); and
- the ConGu tool has been correctly implemented—see the discussion of tool qualification in Sect. 1.2.4.

However, note that random testing does not come with any guarantee of mathematical strength, that a mistake or underlying bug will be discovered, i.e., it is not complete.

4.2.4.1 ConGu Semantically

Both CASL and ConGu take the approach of so-called loose semantics—see the definition under "Model semantics of CASL specifications" in Sect. 4.2.1. This approach is discussed in Sect. 2.3.1 in the Chap. 2, "Logics for Software

Engineering". A Java SUT represents one model in the class of all possible models. It is said to be correct if it falls within the model class of the specification. The ConGu publications provide technically detailed discussions of the link between the world of abstract data types and their implementation in Java that justify the run-time monitoring approach taken. In Chap. 5 on Testing, we will discuss a different approach for test generation from algebraic specifications, where we detail how to relate specifications and programs.

4.2.5 The Story so Far

We have explored basic CASL specifications (as opposed to structured CASL specifications coming up in Sect. 4.4) in syntax and semantics. To this end, we have run through a full exercise from specifying the system informally as a narrative, through building a formal CASL model, validating the model against informal specification (narrative), asserting the model's healthiness by checking its consistency, and finally, to testing if Java implementations conform to our formal model. These elements form a typical "invent and verify" lifecycle when one builds a system from scratch with Formal Methods.

4.3 Verification of Ladder Logic Programs

This section provides a case study within the propositional sublanguage of CASL. We demonstrate a modelling and verification technique originally developed for the verification of so called railway interlocking computers, see, e.g., [Jam10, JR10, JLM+13]. Here we use as an example a pelican crossing in order to demonstrate ladder logic verification. Our example as well as our discussion of Programmable Logic Controllers closely follows [JR10].

Example 44: Pelican Crossing

A Pelican crossing (PEdestrian LIght CONtrolled Crossing), see Fig. 4.2, allows pedestrians to safely cross a flow of traffic. Such crossings have two masts each with the following components:
- a traffic facing set of lights consisting of 3 coloured lights (red, amber, and green) to control traffic.

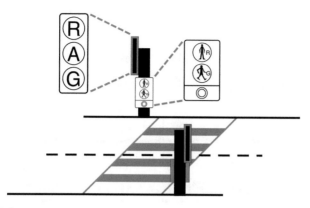

Fig. 4.2 A Pelican crossing. R, A, and G represent red, amber and green lights, respectively

- a pedestrian facing set of lights consisting of 2 coloured lights (normally, a green pedestrian pictogram and a red pedestrian pictogram) to allow pedestrians to cross safely.
- a means for visually impaired pedestrians, e.g., a buzzer, used to indicate it is safe to start crossing.
- a button, used by pedestrians to request to cross the road.

Pelican crossings use the following sequence of lights for traffic:

Active Light	Meaning
Red	Stop
Flashing amber	Continue if crossing is clear
Green	Go
Amber	Prepare to stop

For pedestrians, pelican crossing use the following sequence of lights:

Active Light	Meaning
Green pictogram	Cross the road
Flashing green pictogram	Don't cross as traffic will continue shortly
Red pictogram	Don't cross and wait

The non visual indicator is active when the pedestrian green pictogram is lit (non flashing).

These two sequences are synchronised:

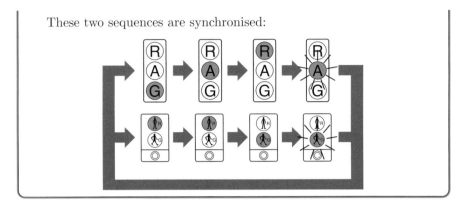

Pelican crossings are safety critical systems: A malfunction, such as showing green on both traffic and pedestrian lights simultaneously, may result in death or serious injury to people. In the following, we demonstrate how to verify in CASL the control program of a pelican crossing for a safety condition.

Example 44.1: Safety Conditions

We consider two safety conditions, one for the pedestrian facing set of lights and one for the traffic facing set of lights: *In each set of lights, there is exactly one light active (i.e., lit or flashing) at any time.*

It is clear that whilst the above are necessary safety conditions they are by no means sufficient. In verification practice, finding the 'right' safety conditions is an art. Risk analysis techniques can support the process of identifying which safety conditions are most important.

4.3.1 Programmable Logic Controllers (PLCs)

A Programmable Logic Controller (PLC), see, e.g., [Bol06], would be a natural choice of how to implement the light sequences of a Pelican Crossing. The operation of a PLC is best described in terms of an imperative program:

Algorithm 5: PLC Operation

| **input** : Sequence of values |
| **output:** Sequence of values |

```
    initialisation
    while (true) do
        read (Input)                                      %% read
(*) State' ← ControlProgram(Input, State)             %% process
        write (Output') & State ← State'                 %% update
```

After initialisation of the system's state, the PLC runs in a non terminating loop. This loop consists of three steps: First, the PLC reads Input, a set of values; based on this Input and the PLC's current State, the PLC computes its next state State' which also includes some Output' values; finally, the PLC writes Output' and updates its state.

Ladder Logic, defined in the IEC standard 61131 [IEC03], is a graphical programming language for PLCs. It gets its name from the ladder like appearance of its programs and is widely used, e.g., in train control systems. From a mathematical point of view, Ladder Logic is a subset of Propositional Logic. In Ladder Logic, Input, State and State' are sets of Boolean variables, where Output is a subset of State (and Output' is a subset of State').

In the context of PLCs programmed in Ladder Logic, there is a variety of common initialisation procedures. One of these is to set all state variables to false and run the program once.

Algorithm 6: PLC Initialisation

---

input : none
output: none

set_to_false (State)
State' ← Program(Input, State)   %% process
State ← State'

Note that, as the input variables are not set in this procedure, usually a Ladder Logic program has several initial states.

4.3.2 Ladder Logic

Avoiding a plethora of syntax definitions and transformations, we refrain from discussing the graphical form of a program in Ladder Logic. We rather define its equivalent in Propositional Logic. Below we give a Ladder Logic control program of a Pelican crossing written in CASL:

Example 44.2: Ladder Logic Control Program

spec TRANSITIONRELATION =
      preds *button, request, old_sh, old_sl, sh, sl, pg, pgf, pr, tg,*
          ta, tr, taf : ()
      preds *button', request', old_sh', old_sl', sh', sl', pg', pgf', pr',*
          tg', ta', tr', taf' : ()
      • $old\_sh' \Leftrightarrow sh$
      • $old\_sl' \Leftrightarrow sl$
      • $sh' \Leftrightarrow (old\_sh' \wedge \neg\, old\_sl') \vee (\neg\, old\_sh' \wedge old\_sl')$
      • $sl' \Leftrightarrow (old\_sh' \wedge \neg\, old\_sl') \vee (\neg\, request \wedge button \wedge \neg\, old\_sl')$

- *request'*
  $\Leftrightarrow (button \wedge \neg old\_sh') \vee (button \wedge \neg old\_sl')$
    $\vee (request \wedge \neg button \wedge \neg old\_sh')$
    $\vee (request \wedge \neg button \wedge \neg old\_sl')$
- $pg' \Leftrightarrow old\_sh' \wedge \neg old\_sl'$
- $pgf' \Leftrightarrow old\_sh' \wedge old\_sl'$
- $pr' \Leftrightarrow \neg old\_sh'$
- $tg' \Leftrightarrow (\neg old\_sh' \wedge \neg old\_sl') \vee (\neg button \wedge \neg request)$
- $ta' \Leftrightarrow \neg old\_sh' \wedge old\_sl'$
- $tr' \Leftrightarrow old\_sh' \wedge \neg old\_sl'$
- $taf' \Leftrightarrow old\_sh' \wedge old\_sl'$

end

The above specification TRANSITIONRELATION declares a number of propositional variables *button, request, …*, each with a primed counterpart *button', request', …*In CASL, propositional variables are 0-ary predicates. This declaration is followed by a sequence of axioms. Each axiom is a faithful representation of one so-called rung of a Ladder Logic program. Details on the encoding of Ladder Logic in Propositional Logic can be found, e.g., in the survey by James et al. [JLM+13].

Explained on an intuitive level, our Ladder Logic program has one input variable, namely *button*. The state of this variable represents if a pedestrian has pressed the button at either pedestrian light during the previous cycle of the PLC. The program also includes a number of pure state variables, namely *request, old_sh, old_sl, sh, sl*. For the output variables, we use the following naming scheme: prefix *p* stands for pedestrian light, prefix *t* for traffic light; the suffixes are given by the following table:

suffix	meaning
g	light shows green
r	light shows red
a	light shows amber
gf	light shows flashing
af	light shows amber flashing

When one of these variables is true, the corresponding light is lit.

As an example of how to read the control program, consider the last axiom or 'rung' of TRANSITIONRELATION, namely

- $taf' \Leftrightarrow old\_sh' \wedge old\_sl'$

This axiom can be read as: if in current cycle of the PLC both state variables *old_sh'* and *old_sl'* are true, then the traffic light will show amber flashing in the next cycle.

> Our program abstracts from the question of how long a traffic light shall stay green or red. In order to deal with time, PLCs offer special boolean variables for setting a timer and for obtaining the information that a fixed time interval has passed.

In the following, we will make use of a number of notations: for a given propositional formula φ, the function *vars* returns the set of propositional variables appearing in φ; we use "prime" to generate a fresh variable. $V' = \{v' \mid v \in V\}$ denotes the set of all fresh variables obtained from a set of variables V.

Definition 4 (*Ladder logic formulae*) A ladder logic formula ψ (relative to a finite set of input variables I and a finite set of state variables C with $I \cap C = \emptyset$) is a propositional formula

$$\psi \equiv ((c_1' \Leftrightarrow \psi_1) \wedge (c_2' \Leftrightarrow \psi_2) \wedge \ldots \wedge (c_n' \Leftrightarrow \psi_n))$$

where $n \geq 0$ and the ψ_i, $1 \leq i \leq n$, are propositional formulae, such that the following conditions hold:

1. For all $1 \leq i \leq n : c_i' \in C'$.
2. For all $1 \leq i, j \leq n$: if $i \neq j$ then $c_i' \neq c_j'$.
3. For all $1 \leq i \leq n : vars(\psi_i) \subseteq I \cup \{c_1', \ldots c_{i-1}'\} \cup \{c_i, \ldots c_n\}$.

Thanks to the three conditions on variables, the equivalence symbol ($\Leftrightarrow$) in a Ladder Logic formula can be read as variable assignment in an imperative program. This is justified, as Ladder Logic programs can be seen as a chain of so-called definitional extensions.

Definition 5 (*Conservative and definitional extension*) In CASL, the annotation

$$sp \textbf{ then } \%\textbf{cons } sp'$$

holds if each model of sp can be expanded to a model of sp **then** sp'. Furthermore, the annotation

$$sp \textbf{ then } \%\textbf{def } sp'$$

holds if each model of sp can be uniquely expanded to a model of sp **then** sp'.

Similar to **%implies**, the annotations **%cons** and **%def** express proof obligations. Hets interfaces with a number of theorem provers to discharge the proof obligations arising from the annotations **%implies**; the CASL Consistency Checker [LRS04] implements a calculus that directly deals with **%cons** and **%def**.

Before we discuss the application of these semantical annotations, we introduce some of CASL's syntactic sugar. For the sake of readability, in CASL it is possible to declare and define predicate symbols in one single step—see the below example. After the keyword **pred** one can write

$$p(v_{1,1}, \ldots, v_{1,m_1} : s_1; \ldots; v_{n,1}, \ldots, v_{n,m_n} : s_n) \Leftrightarrow F \qquad (4.3)$$

($n \geq 0$; $m_i > 0$ and s_i are sort names for $1 \leq i \leq n$; $v_{i,j}$ are variable names for $1 \leq i \leq n, 1 \leq j \leq m_i$; F is a formula in first-order logic) The equivalence is universally quantified over the declared argument variables (which must be distinct, and are the only free variables allowed in F).

Analogously, after the keyword **op** one can write

$$f(v_{1,1}, \ldots, v_{1,m_1} : s_1; \ldots; v_{n,1}, \ldots, v_{n,m_n} : s_n) : s = T$$

The equation is universally quantified over the declared argument variables (which must be distinct, and are the only free variables allowed in T).

Example 44.3: The Ladder Logic Program as a Sequence of Definitional Extensions

Using these new syntactical means, the specification TRANSITIONRELATION can be more concisely written as:

spec TRANSITIONRELATIONALTERNATIVE =
    preds *button, request, old_sh, old_sl, sh, sl, pg, pgf, pr, tg,*
        ta, tr, taf : ()
then %def
    pred *old_sh'()* $\Leftrightarrow$ *sh*;
then %def
    pred *old_sl'()* $\Leftrightarrow$ *sl*;
then %def
    pred *sh'()* $\Leftrightarrow$ *(old_sh'* $\wedge \neg$ *old_sl')* $\vee$ *($\neg$ old_sh'* $\wedge$ *old_sl')*;
...
then %def
    pred *taf'()* $\Leftrightarrow$ *old_sh'* $\wedge$ *old_sl'*;
then %cons
    pred *button'* : ()
end

In the transformation from TRANSITIONRELATION to TRANSITIONRELATIONALTERNATIVE we apply two principles:

- In each definitional extension, we declare and define one new propositional variable. Theorem 1 below proves that the extensions are actually definitional ones.
- In the conservative extension, we declare one new propositional variable and leave its interpretation open. Thus, no models are lost in the extension.

Note that such a transformation is possible for any Ladder Logic program.

The following theorem[3] has been stated as rule (def1) in the consistency calculus by Roggenbach and Schröder [RS01]:

Theorem 1 *If BI is an operation or a predicate definition for a symbol which is new over a specification sp, then the annotation **def** holds in*

$$sp \textbf{ then } \%\textit{def } BI.$$

Proof We prove the theorem only for the case that BI is a predicate definition as given in equivalence 4.3 for a predicate symbol p. The case of an operation definition is analogous.

Let M be a model of sp. Define for each element $e \in M(s_1)^{m_1} \times \ldots \times M(s_n)^{m_n}$ a variable evaluation ν_e with $\nu_e(v_{i,j}) = e_{(\sum_{k=1}^{i-1} m_k)+j}$, i.e., the variable $v_{i,j}$ obtains under ν_e the component with index $(\sum_{k=1}^{i-1} m_k) + j$ from the vector e. Define

- $M'(p)(e) = \nu_e^\sharp(F)$ and
- $M'(x) = M(x)$ for symbols $x \neq p$.

M' clearly is a model of sp **then** $\%$def BI: by construction, equivalence 4.3 holds over M'. Furthermore, there is no other model M'' of the specification of sp **then** $\%$def BI, which is identical with M on the symbols of sp: choosing a value $M''(p)(e)$ different from $M'(p)(e)$ would falsify equivalence 4.3. Thus, the extension is a definitional one. ∎

4.3.3 The Automaton of a Ladder Logic Formula

One can associate an automaton with a ladder logic formula. This automaton has interpretations of the set of propositional variables $I \cup C$ as its states, i.e., the configurations of the PLC. In order to define the automaton's transition relation, we introduce paired valuations. Here, the function *unprime* deletes the prime from a variable.

Definition 6 (*Paired valuations*) Given a finite set of input variables I, a finite set of state variables C, and valuations μ, $\nu : (I \cup C) \to \{0,1\}$ we define the paired valuation $\mu \,\S\, \nu : (I \cup C \cup I' \cup C') \to \{0,1\}$ where

$$\mu \,\S\, \nu(x) = \begin{cases} \mu(x) & \text{if } x \in I \cup C \\ \nu(unprime(x)) & \text{if } x \in I' \cup C'. \end{cases}$$

The models M of our CASL specification are exactly these paired valuations $\mu \,\S\, \nu$.

---

[3] Note that this theorem relies on the fact that CASL signatures do *not* include variables. This is in contrast to signatures as introduced in Chap. 2.

Example 44.4: A Model of our CASL **Specification**

The models of our CASL specification TRANSITIONRELATION are maps

$$M : \{button, request \ldots, taf'\} \rightarrow \{0, 1\}$$

i.e., interpretations of all propositional variables, the primed and the unprimed ones, declared in the specification with the truth values 0—for false—and for 1—true. For example, consider the model M with

$$
\begin{array}{llll}
M(old\_sh) = 0 & M(old\_sl) = 0 \\
M(sh) = 0 & M(sl) = 0 & M(request) = 0 & M(button) = 0 \\
M(tr) = 0 & M(ta) = 0 & M(taf) = 0 & M(tg) = 1 \\
M(pr) = 1 & M(pg) = 0 & M(pgf) = 0
\end{array}
$$

$$
\begin{array}{llll}
M(old\_sh') = 0 & M(old\_sl') = 0 \\
M(sh') = 0 & M(sl') = 1 & M(request) = 1 & M(button') = 1 \\
M(tr') = 0 & M(ta') = 0 & M(taf') = 0 & M(tg') = 1 \\
M(pr') = 1 & M(pg') = 0 & M(pgf') = 0
\end{array}
$$

Model M obviously fulfills the first rung of our specification TRANSITIONRELATION, namely

- $old\_sh' \Leftrightarrow sh$

It is easy but laborious to check that M also fulfills all other axioms. It also is an easy exercise to decompose M into two evaluations μ and ν such that $M = \mu \, \mathring{}\, \nu$.

The automaton of a ladder logic formula is defined as follows:

Definition 7 (*Automaton*) Given a ladder logic formula ψ over $I \cup C$, we define the automaton

$$A(\psi) = (S, S_0, \rightarrow)$$

where

- $S = \{\mu \,|\, \mu : I \cup C \rightarrow \{0, 1\}\}$ is the set of states,
- $\mu \xrightarrow{\nu(I')} \nu$ if $\mu \, \mathring{}\, \nu \models \psi$, defines the transition relation, and
- $S_0 = \{\nu \,|\, \exists \mu : \mu \models \neg C, \mu \, \mathring{}\, \nu \models \psi\}$ gives the set of initial states.
  Here, $\neg C$ expands to $\bigwedge_{c \in C} \neg c$.

The automaton might have more than one start state as the computation of the set of initial states only sets the state variables C, the input variables I can take any value. The automaton $A(\psi)$ is *finite*; it has $2^{|I \cup C|}$ states.

Naturally, a PLC should never stop. In our formalisation in terms of an automaton we can prove this property. Note how the proof makes use of the syntactic structure of a Ladder Logic formula.

Theorem 2 *Let ψ be a ladder logic formula. Let μ be a state in $A(\psi)$. Then there exists a state ν such that $\mu \, \S \, \nu \models \psi$, i.e., it holds that $\mu \xrightarrow{\nu(I')} \nu$.*

Proof (Sketch) By induction on size n of a ladder logic formula. Assume the claim holds for length i. Given an evaluation μ_i for $V_i = I \cup \{c'_1, \ldots, c'_{i-1}\} \cup \{c_i, \ldots, c_n\}$ we set $\mu_{i+1}(x) = \mu(x)$ for $x \in V_i$, $\mu_{i+1}(c'_{i+1}) = 1$ if $\mu_i \models \psi_{i+1}$ and $\mu_{i+1}(c'_{i+1}) = 0$ if $\mu_i \not\models \psi_{i+1}$. ∎

The above proof is just a reformulation of the our observation that Ladder Logic programs can be considered as a sequence of definitional and conservative extensions. Another interpretation of this theorem is: Ladder Logic programs can never state inconsistent requirements on the transition relation. This is in contrast to, e.g., the B method, see, e.g., [Abr10]. B also specifies an automaton. When working with B, however, one needs first to prove that the axioms specifying the transitions do not contradict each other. Coming back to CASL, this theorem means: The specification of a Ladder Logic program is always consistent, i.e., it has a non empty-model class consisting out of all the transitions of the automaton.

Example 44.5: The Automaton for TransitionRelation

Figure 4.3 shows the reachable states of the automaton constructed from TRANSITIONRELATION. For ease ease of reading some variable values have been excluded from the state. The convention is to show the values of $sh, sl, request, button$ as a four-tuple in this order, followed by the names of the those lights which are true in the given state. Initial states are marked by an ingoing arrow starting from a black dot.

4.3.4 Inductive Verification of Ladder Logic Programs

With the associated automaton in mind, we can now formalise safety conditions for our PLC as propositional formulae and make precise, what safety verification shall mean.

Example 44.6: Our Safety Properties

For the Pelican Crossing, we were interested in two safety conditions: *"In each set of lights, there is exactly one light active (i.e., lit or flashing) at any time"*. These can now be expressed as properties in propositional logic that shall hold in each state of the automaton.

For traffic lights, exclusively one out of tg, ta, tr, taf shall be true:
- $(tg \wedge \neg\, ta \wedge \neg\, tr \wedge \neg\, taf) \vee (\neg\, tg \wedge ta \wedge \neg\, tr \wedge \neg\, taf)$
  $\vee (\neg\, tg \wedge \neg\, ta \wedge tr \wedge \neg\, taf) \vee (\neg\, tg \wedge \neg\, ta \wedge \neg\, tr \wedge taf)$

Respectively, for pedestrian lights, exclusively one out of pg, pgf, pr shall be true:

- $(pg \wedge \neg\, pgf \wedge \neg\, pr) \vee (\neg\, pg \wedge pgf \wedge \neg\, pr)$
  $\vee (\neg\, pg \wedge \neg\, pgf \wedge pr)$

We can manually check that this holds for the automaton depicted in Fig. 4.3.

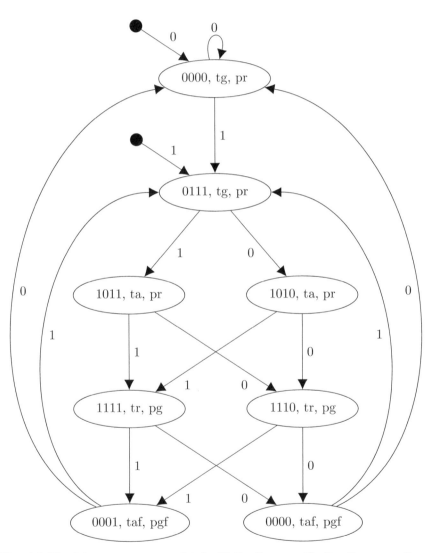

Fig. 4.3 The finite automaton associated with the CASL specification TRANSITIONRELA-TION

More formally, we define the verification problem for a ladder logic formula ψ for a verification condition φ:

Definition 8 (*The verification problem for ladder logic programs*)

$$A(\psi) \models \varphi$$

if and only if for all reachable states μ of $A(\psi)$ we have $\mu \models \varphi$.

There are a number of different approaches of how to solve this verification problem. These include inductive verification, bounded model checking, and temporal induction. We will focus here on inductive verification only, but want to point out that the other two can equally be encoded in CASL.

Inductive verification checks if an over approximation of the reachable state space is safe. The idea is verify two conditions:

1. Show that all initial states are safe.
2. Show that the following holds for all states s: If s is safe, then all successors of s are safe as well.

The over approximation happens in the second step: here one considers all sfae states rather than the *reachable* ones only. This idea makes inductive verification a very efficient approach involving at most two calls to a theorem prover. However, the price of this is that one might have to deal with false positives—i.e., the verification might fail due to an unreachable safe state which has an unsafe successor.

Let Ψ represent the Ladder Logic program, including the variable declarations, let φ be the safety property of concern formulated over unprimed variables, and let φ' be the safety property formulated over primed variables. Then we can encode inductive verification in CASL as follows:

spec INITIALSTATESARESAFE = Ψ **then** $\bullet$ $\bigwedge_{c \in C} \neg c$
then %implies
$\qquad \bullet \ \varphi'$
end

spec TRANSISTIONSARESAFE = Ψ
then %implies
$\qquad \bullet \ \varphi \Rightarrow \varphi'$
end

In the following, we illustrate inductive verification for our example:

Example 44.7: Initial States are Safe

In order to verify that initial states are safe, we negate all state variables—as prescribed in Definition 7—and check if our safety properties hold for all primed variables:

spec INITIALSTATESARESAFE =
    TRANSITIONRELATION
then • $\neg\, request \wedge \neg\, sh \wedge \neg\, sl \wedge \neg\, old\_sh \wedge \neg\, old\_sl \wedge \neg\, pg$
        $\wedge \neg\, pgf \wedge \neg\, pr \wedge \neg\, tg \wedge \neg\, ta \wedge \neg\, tr \wedge \neg\, taf$
then **%implies**
    • $(tg' \wedge \neg\, ta' \wedge \neg\, tr' \wedge \neg\, taf') \vee (\neg\, tg' \wedge ta' \wedge \neg\, tr' \wedge \neg\, taf')$
        $\vee\, (\neg\, tg' \wedge \neg\, ta' \wedge tr' \wedge \neg\, taf')$
        $\vee\, (\neg\, tg' \wedge \neg\, ta' \wedge \neg\, tr' \wedge taf')$
                                                    %(initial states are safe for vehicles)%
    • $(pg' \wedge \neg\, pgf' \wedge \neg\, pr') \vee (\neg\, pg' \wedge pgf' \wedge \neg\, pr')$
        $\vee\, (\neg\, pg' \wedge \neg\, pgf' \wedge pr')$
                                                    %(initial states are safe for pedestrians)%
end

Using Hets, we can see that both safety properties hold.

Some safety conditions hold for the over approximation of the state space:

Example 44.8: The Automaton is Safe w.r.t. Pedestrian Lights

We encode the inductive exploration of the state space as an implication: should the state given through the unprimed variables be safe, then the state given through the primed variables is safe as well.

spec TRANSITIONSANDPEDESTRIANSAFETY =
    TRANSITIONRELATION
then **%implies**
    • $(pg \wedge \neg\, pgf \wedge \neg\, pr) \vee (\neg\, pg \wedge pgf \wedge \neg\, pr)$
        $\vee\, (\neg\, pg \wedge \neg\, pgf \wedge pr)$
        $\Rightarrow (pg' \wedge \neg\, pgf' \wedge \neg\, pr') \vee (\neg\, pg' \wedge pgf' \wedge \neg\, pr')$
            $\vee\, (\neg\, pg' \wedge \neg\, pgf' \wedge pr')$
                                                    %(safety is preserved for pedestrians)%
end

Using Hets, we can see that this safety property holds.

However, for other safety conditions, due to the over approximation of the state space, inductive certification might fail:

Example 44.9: The Induction Step Fails for Vehicles

Using the same encoding method of the induction step for the safety
condition on traffic lights results in:

spec TRANSITIONSANDTRAFFICSAFETY =
    TRANSITIONRELATION
then %implies
- $(tg \wedge \neg\, ta \wedge \neg\, tr \wedge \neg\, taf) \vee (\neg\, tg \wedge ta \wedge \neg\, tr \wedge \neg\, taf)$
  $\vee\, (\neg\, tg \wedge \neg\, ta \wedge tr \wedge \neg\, taf) \vee (\neg\, tg \wedge \neg\, ta \wedge \neg\, tr \wedge taf)$
  $\Rightarrow (tg' \wedge \neg\, ta' \wedge \neg\, tr' \wedge \neg\, taf')$
  $\quad\vee (\neg\, tg' \wedge ta' \wedge \neg\, tr' \wedge \neg\, taf')$
  $\quad\vee (\neg\, tg' \wedge \neg\, ta' \wedge tr' \wedge \neg\, taf')$
  $\quad\vee (\neg\, tg' \wedge \neg\, ta' \wedge \neg\, tr' \wedge taf')$

                              %(safety is preserved for vehicles)%
end

Using Hets, we can see that this safety property does not hold.

Conceptually, there can be many reasons for inductive verification to fail:

1. We have a wrong model: i.e., the translation of the Ladder Logic Program
   into Propositional Logic went wrong.
2. We made a modelling error when expressing the safety condition in Propositional Logic.
3. The failure is due to a false positive.
4. There actually is a mistake in the Ladder Logic program.
5. The programmer of the Ladder Logic program deliberately violated the
   safety condition—e.g., for some optimisation purpose, justified by domain
   knowledge.

It is up to the verifier to carefully work out, what the reason actually is.
That is, scientifically Ladder Logic verification through inductive verification
is solved: we know how to do it and automated tools often scale to real world
problems. However, to be practically applicable, also the error analysis needs
to be automated. The first two items above need to be addressed by thorough engineering. The third item possibly requires verification approaches to
complement inductive verification. One can always make a first attempt with
inductive verification (as it is computationally 'cheap'). Should inductive verification fail, one runs a computationally more expensive approach in order
to further investigate. Such approaches include, e.g., inductive verification
(see below) and the IC3 algorithm [Bra12]. The fourth concerns the quality
control process that we actually want to establish. The fifth is an example
of industrial practices and theoretical considerations in opposition when it
comes to the question of safety.

Inductive Verification with Invariants

As demonstrated in Example 44.9 above, inductive verification might unexpectedly fail (we could verify by manually inspecting the automaton that the safety condition concerning vehicles holds for our ladder logic program). However, this does not necessarily imply that the safety condition does not hold on the reachable states, i.e., the ones we are interested in. In verification practice, one often works with invariants to mitigate such situations. Invariants are formulae that hold in all reachable states (and possibly others). In order to establish that a formula is an invariant, we can, e.g., use inductive verification.

With an invariant, verification becomes:

1. Show that all initial states are safe.
2. Show that the following holds for all states s: If s is safe and the invariant holds in s, then all successors of s are safe as well.

Let I be an invariant, then we encode verification with an invariant as follows:

spec INITIALSTATESARESAFE $= \Psi$ **then** • $\bigwedge_{c \in C} \neg c$
then %implies
     • φ'
end

spec TRANSITIONSARESAFEWITHINVARIANT $= \Psi$
then %implies
     • $\varphi \wedge I \Rightarrow \varphi'$
end

In general, invariants are hard to find. They might arise from the application domain (e.g., a Ladder Logic program reads from a sensor a value in $\{0, 1, \ldots, 5\}$ and uses three boolean variables $v_2 v_1 v_0$ to represent the binary value of the read-in integer—i.e., the values 110 and 111 will never appear in the reachable states) or be a property of the program itself: there is a rich literature on automatic invariant detection of programs.

Example 44.10: Safety Holds Also for Traffic Lights

By analysing our Ladder Logic program with the help of Karnaugh maps, we visually spotted a candidate for an invariant:

$$\neg \left(\neg \; button \wedge \neg \; request \wedge (sh \vee sl) \right)$$

With the help of this formula, we want to establish that safety is preserved for vehicles.

With inductive verification, one can prove that the above formula is an invariant. Concerning Step 1 of inductive verification, we have seen in Example 44.7 that the initial states are safe for vehicles. Thus, it remains to show that safety for vehicles is preserved under the invariant.

spec TRANSITIONSARESAFEUNDERINVARIANT =
     TRANSITIONRELATION
then %implies
- $\neg\,(\neg\ button \wedge \neg\ request \wedge (sh \vee sl))$
  $\wedge\,((tg \wedge \neg\ ta \wedge \neg\ tr \wedge \neg\ taf) \vee (\neg\ tg \wedge ta \wedge \neg\ tr \wedge \neg\ taf)$
    $\vee\,(\neg\ tg \wedge \neg\ ta \wedge tr \wedge \neg\ taf) \vee (\neg\ tg \wedge \neg\ ta \wedge \neg\ tr \wedge taf))$
  $\Rightarrow (tg' \wedge \neg\ ta' \wedge \neg\ tr' \wedge \neg\ taf')$
    $\vee\,(\neg\ tg' \wedge ta' \wedge \neg\ tr' \wedge \neg\ taf')$
    $\vee\,(\neg\ tg' \wedge \neg\ ta' \wedge tr' \wedge \neg\ taf')$
    $\vee\,(\neg\ tg' \wedge \neg\ ta' \wedge \neg\ tr' \wedge taf')$
            %(safety is preserved for vehicles under an invariant)%
end

Using Hets, we can show that this property holds, i.e., safety holds also for traffic lights.

Note that in the above example we carefully established first that the invariant holds in all reachable states. Had we not done this, we might have added a formula that restricts the reachable states. In that case, we could have produced a false negative, i.e., we could have said that the system is safe as we found it to be safe for a subset of the reachable states.

4.4 Structuring Specifications

Programming in the small and programming in the large has been an established topic since the mid 1970s. Paraphrasing DeRemer and Kron [DK76], programming languages require primitives such as assignment, conditional, loop for writing functions and procedures in a module (in the small). But they also need a "module interconnection language" for knitting those modules together (in the large). Within CASL we already have used the **then** construct in order to form larger specifications from smaller ones.

When choosing a programming language for a specific project, not only the programming paradigm is important but also which standard libraries and third party modules are available. The same holds for specification. The specifier does not want to re-invent the wheel. Thus, a specification language ought to support libraries and provide constructs to utilise library contents.

CASL has taken the idea of structured specifications further than any other specification language, and is thus, exemplary. Beyond that, the CASL structuring operators have been developed in such a way that any specification language can re-use them—provided its semantics has been written in a certain style. Namely, in CASL, all structuring operators have 'institution-independent' semantics, c.f. Sect. 2.3. Their definition recurs only on the building blocks that are present in all institutions and refrain from utilising notions specific to a particular institution. Thus, given any concrete institution, these operators are available.

Structuring specifications is considered 'good practice' for many reasons. These include:

- **Separation of concerns:** Different parts or aspects of the system appear in different specifications. This allows specifier and reader to focus on the system part or aspect of concern. It reduces the complexity (measured, e.g., in the numbers of axioms) of a single specification and thus might increase the likelihood to get the specification right.
- **Re-use of specification-text:** Named specification texts can be used in different contexts. This increases efficiency when writing and when maintaining specifications: axioms need to be written only once, and management of change is eased (if an axiom needs updating, there is only one place that needs changing).
- **Theorem proving:** There are proof calculi for structured specifications that distribute proof obligations along the specification structure. When analysing systems, this can help in identifying which part of the overall specification went wrong in the case that a proof fails.

Up to now, we have studied what one might want to call algebraic specification 'in the small', i.e., we have modeled and analyzed systems that we expressed as a single CASL specification. Even in this, for practicality we allowed ourself to 'import' named specifications and to 'extend' these. Now, we shall explore specification 'in the large', i.e., we will study how simple specifications can be combined to form more complex ones.

CASL structuring includes a variety of constructions of which we will study:

- Named specification—to create a named entity for use in different contexts;
- Extension—keyword **then**—to enhance a given specification;
- Union—keyword **and**—to share properties between two specifications;
- Renaming—keyword **with**—to adapt symbol names;
- Hiding—keyword **hide**—to get rid of unwanted symbols;
- Libraries—keyword **library**—to allow collections of named specifications; and
- Parametrisation—to enable the re-use of a construction principle of data types in different situations.

Most of these specification operations do not change expressivity of the language, i.e., there exist 'unstructured', 'flat' specifications with the same model class. However, it turns out that hiding increases the expressivity.

In the following, we will exemplify how one can make good use of these structuring operations. The library of CASL Basic Datatypes, Part V in

[Mos04],[4] provides a rich collection of further examples on the use of these specification building operators, the CoFi Note M-6 "Methodological guidelines"[5] discusses how to make 'good' use of these operators.

4.4.1 Extension

Extensions—keyword **then**—enhance a given specification by adding new symbols and or new axioms to it. We write

$$Sp \text{ then } Sp'$$

for specification Sp' extending specification Sp. It is a CASL convention to bracket sequences of extensions to the left, i.e., Sp **then** Sp' **then** $Sp'' = \{Sp$ **then** Sp' $\}$ **then** Sp''. Here, $\{$ and $\}$ are the CASL symbols to group specifications. Often, Sp' is not a specification on its own but rather a specification fragment. Take for instance the following example:

Example 45: Extending Partial to Total Orders

A partial order is a binary relation, which is reflexive and transitive. In the context of a partial order, one often speaks about the infimum (supremum) of two elements: that largest (smallest) element that is smaller (larger) than these two elements. A typical example of a partial order are sets ordered by subset-inclusion. The infimum of two sets is given by their intersection, the supremum is given by their union. Note that for two elements there might not be an infimum (supremum), i.e., infimum (supremum) is a partial operation.

spec PARTIALORDER =
    sort $Elem$
    pred $\_\_\leq\_\_ : Elem \times Elem$
    $\forall \, x, y, z : Elem$
    • $x \leq x$
    • $x \leq y \wedge y \leq z \Rightarrow x \leq z$
    ops $inf, sup : Elem \times Elem \rightarrow? Elem$
    $\forall \, x, y, z : Elem$
    • $inf(x, y) = z$
        $\Leftrightarrow z \leq x \wedge z \leq y \wedge \forall \, t : Elem \bullet t \leq x \wedge t \leq y \Rightarrow t \leq z$
    • $sup(x, y) = z$
        $\Leftrightarrow x \leq z \wedge y \leq z \wedge \forall \, t : Elem \bullet x \leq t \wedge y \leq t \Rightarrow z \leq t$
end

---

[4] Available at `https://github.com/spechub/Hets-lib`.

[5] Available at `https://sefm-book.github.io`.

Two disjoint sets are not ordered by set inclusion. That is, they are incomparable. This is different for the natural numbers. There we have the situation that any two natural numbers are in an ordering relation. If this property holds of a partial order, one also speaks of a total order. In total orders, it makes sense to speak about the min (max) of two elements. It turns out that—given the extra axiom for total orders—the definitions of inf (sup) and min (max) coincide:

spec TOTALORDER =
    PARTIALORDER
then $\forall \, x, \, y : Elem \bullet x \leq y \vee y \leq x$
    ops $min, \, max : Elem \times Elem \rightarrow Elem$
    $\forall \, x, \, y : Elem$
    • $min(x, \, y) = x$ *when* $x \leq y$ *else* y
    • $max(x, \, y) = x$ *when* $y \leq x$ *else* y
then %implies
    $\forall \, x, \, y : Elem$
    • $min(x, \, y) = inf(x, \, y)$ %(min=inf)%
    • $max(x, \, y) = sup(x, \, y)$ %(max=sup)%
end

Using Hets, we can prove that these two equations hold.

In the above example, the first extension within the specification TOTALORDER extends the specification PARTIALORDER by three axioms and two new symbols. These three axioms make use of the $\leq$ symbol although it is not declared within the first extension—the symbol is imported from PARTIALORDER. In this sense, the first extension is not a specification of its own, but a fragment that only makes sense together with another specification. The same holds for the second extension.

4.4.2 Union

Unions—keyword **and**—share properties between two specifications. We write

$$Sp_1 \text{ and } Sp_2$$

for taking the union of specification Sp_1 and specification Sp_2. The **and** operator is associative.

Example 46: Building Up an Asymmetric Relation with Union

Relations are everywhere. For example, the integers are ordered by the $<$ relation.

spec RELATION =
      sort   *Elem*
      pred __$\sim$__ : *Elem* $\times$ *Elem*
end

No integer is smaller than itself. Such relations are called irreflexive.

spec IRREFLEXIVERELATION = RELATION
then $\forall\, x : Elem \bullet \neg\, x \sim x$
end

For integers it is also the case that its ordering relation $<$ is transitive:

spec TRANSITIVERELATION = RELATION
then $\forall\, x,\, y,\, z : Elem \bullet x \sim y \wedge y \sim z \Rightarrow x \sim z$
end

For integers it also holds that if a number is smaller than the other, the reverse it never the case. Such relations are called 'asymmetric':

spec ASYMMETRICRELATION = RELATION
then $\forall\, x,\, y : Elem \bullet \neg\, x \sim y \; if \; y \sim x$
end

It turns out that our observation on the integers is a general one: whenever a relation is both irreflexive and transitive, it is asymmetric.

spec STRICTORDER =
      IRREFLEXIVERELATION **and**   TRANSITIVERELATION
then %implies
      ASYMMETRICRELATION
end

Using Hets, we we can prove that all axioms stated in the specification ASYMMETRICRELATION hold.

In the above example, we observe that, e.g., the sort symbol *Elem* is declared twice when taking the union, once in the specification IRREFLEXIVERELATION and once in the specification TRANSITIVERELATION. The resulting signature of the union will have the symbol *Elem* only once. The reason for this is that in CASL the signature of the union is obtained by the ordinary union of the signatures (not their disjoint union). Thus all occurrences of a symbol in the specifications are interpreted uniformly (rather than being regarded as homonyms for potentially different entities). This is known in CASL as the 'same name, same thing' principle.

Also note that the union operator has a higher precedence than the extension operator, i.e., Sp_1 **and** Sp_2 **then** $Sp' = \{Sp_1$ **and** $Sp_2\}$ **then** Sp'.

When forming the union of two specifications, naturally the consistency of the resulting specification is a concern. In the above example, we know that there was no problem as the integers with $<$ are a model of both, IRREFLEXIVERELATION and TRANSITIVERELATION. In specification practice, unions are often a source of inconsistency due to 'interaction' between axioms coming from different specifications.

4.4.3 Renaming

Renamings—keyword **with**—change the symbols of a specification. Given a specification Sp, we write

$$Sp \textbf{ with } SY_1 \mapsto SY_1', \ldots, SY_n \mapsto SY_n'$$

$n \geq 1$, for obtaining a new specification, which is like Sp but with symbol SY_i consistently changed into symbol SY_i', for $i = 1 \ldots n$. The $\mapsto$ symbol is expressed in the concrete syntax as |->.

Example 47: Lists Satisfy the Monoid Axioms

Monoids are algebraic structures that are often found in data types. Usually, monoids are written multiplicatively with the * operation:

spec MONOID =
    sort S
    ops $1 : S$;
            $\_\_*\_\_ : S \times S \to S$
    • $\forall\, x : S$ • $1 * x = x$ %(1 is left unit)%
    • $\forall\, x : S$ • $x * 1 = x$ %(1 is right unit)%
    • $\forall\, x, y, z : S$ • $x * (y * z) = (x * y) * z$ %(monoid associativity)%
end

Lists are one of the data types with monoid structure: the empty list $[]$ is the unit element of the append operation $\_\_{+}{+}\_\_$, the append operation is associative:

spec LIST =
    sort $Elem$
    free type $List ::= [] \mid \_\_{::}\_\_(Elem;\ List)$
then %def
    op $\_\_{+}{+}\_\_ : List \times List \to List$
    $\forall\, L, M : List;\ e : Elem$

- $[] ++ L = L$ %%(++ empty list)%%
- $(e :: L) ++ M = e :: (L ++ M)$ %%(++ non−empty list)%%

end

We first generate lists as a free type. Then, we definitionally extend this specification: for each alternative of the free type, we give one axiom, namely one that says how the append operation works with the empty list $[]$, and one axioms that says how the append operation works with non empty lists, i.e., lists which have the form $e :: L$.

Using *renaming*, we can now express our expectation that lists are indeed monoids. First, we check with Hets the result of the renaming:

MONOID **with** $S \mapsto List,\ 1 \mapsto [],\ \_*\_ \mapsto \_++\_$

Inspecting the theory of the resulting specification with Hets, we obtain:

```
sorts List
op [] : List
op __++__ : List * List -> List
forall x : List . [] ++ x = x
forall x : List . x ++ [] = x
forall x, y, z : List . x ++ (y ++ z) = (x ++ y) ++ z
```

Having verified that the renaming produces what was expected (the monoid axioms are now written using the signature elements from lists), we can state that Lists are Monoids:

spec LISTSAREMONOIDS = LIST
then %implies
      MONOID **with** $S \mapsto List,\ 1 \mapsto [],\ \_*\_ \mapsto \_++\_$
end

However, to our disappointment, the theorem prover SPASS only shows a timeout and appears not to be able to prove this property.

The reason why SPASS can't prove the property is that SPASS is a first-order theorem prover while the sort generation constraint from the free type is a higher order formula. When CASL specifications are translated by Hets into the language of SPASS, the higher order axioms of the free types are omitted. Without the higher order axiom it is not possible to prove that Lists are a monoid: the induction principles required are not available.

However, it is safe to add the required induction principles (which are formulae in first-order logic) to the LIST specification: we know that the induction principles are a consequence of the sort generation constraint coming with the free type.

Example 47.1: Induction Proof with SPASS

Our specification comes in several parts:
1. we import the LIST specification;
2. we add an induction axiom so that SPASS can prove that [] is a right unit to the append operation ++;
3. with the induction axiom around, we expect that SPASS will be able to prove both, induction case and induction step for associativity;
4. then we add an induction axiom so that SPASS can prove that the append operation ++ is associative;
5. with these theorems and induction axioms available, SPASS should manage to prove that lists form a monoid.

spec LISTWITHINDUCTIONPRINCIPLES = LIST
then • ([] ++ [] = [])
      ∧ ∀ M : $List$; e : $Elem$
        • M ++ [] = M ⇒ (e :: M) ++ [] = e :: M)
    ⇒ ∀ M : $List$ • M ++ [] = M
                        %(induction axiom for right unit)%
then %implies
    ∀ L, M, N : $List$; e : $Elem$
    • [] ++ (M ++ N) = ([] ++ M) ++ N
                     %(induction base for assoc)%
    • ∀ K : $List$; e : $Elem$
      • ∀ M, N : $List$
        • K ++ (M ++ N) = (K ++ M) ++ N
         ⇒ (e :: K) ++ (M ++ N) = ((e :: K) ++ M) ++ N
                    %(induction step for assoc)%
then • ((∀ M, N : $List$ • [] ++ (M ++ N) = ([] ++ M) ++ N)
      ∧ ∀ K : $List$; e : $Elem$
        • ∀ M, N : $List$
         • K ++ (M ++ N) = (K ++ M) ++ N
         ⇒ (e :: K) ++ (M ++ N) = ((e :: K) ++ M) ++ N)
    ⇒ ∀ L, M, N : $List$ • L ++ (M ++ N) = (L ++ M) ++ N
                     %(induction axiom for assoc)%
then %implies
    MONOID **with** S ↦ $List$, 1 ↦ [], $\_\_*\_\_$ ↦ $\_\_++\_\_$
end

The first thing to note is that the two inductions schemes that we add in part 2 and in part 4 do not change the model class: they are consequences of the free type that we used in order to specify the sort *List*. However, SPASS is not capable of proving them, as SPASS is a first-order prover. With these additional axioms in place, SPASS proves the desired result that lists are monoids.

This specification demonstrates two 'tricks' in first-order theorem proving.

The *first trick* is to add suitable induction axioms. This is applied when considering the right unit property. For a manual proof by induction we would consider the two cases:

1. Base Case: $[] ++ [] = []$.
2. Induction Step: provided it holds that $M ++ [] = M$ for all lists M, we can show that $(e :: M) ++ [] = (e :: M)$ for all $e \in Elem$, i.e., for lists that are longer by one element.

Axiom %(induction axiom for right unit)% states that proving base case and step case suffices to establish the required law.

We apply the first trick also for associativity. For a manual proof by induction we would consider the two cases:

1. Base Case: $[] ++ (M ++ N) = ([] ++ M) ++ N$ for all lists M, N.
2. Induction Step: provided it holds that $K ++ (M ++ N) = (K ++ M) ++ N$ for all lists K, M, N, we can show that $(e :: K) ++ (M ++ N) = ((e :: K) ++ M) ++ N$ for all $e \in Elem$.

Axiom %(induction axiom for assoc)% states that proving base case and step case suffices to establish the required law.

The *second trick* is to ask SPASS to prove suitable intermediate lemmas. Axioms %(induction base for assoc)% and %(induction step for assoc)% are such lemmas. The idea is to reduce the search space for SPASS and to provide the prover with suitable intermediate results that help to establish the desired theorem. While it might be natural for a human to decompose the premise of %(induction axiom for assoc)% and to prove both cases separately, SPASS fails to spot this. However, SPASS is capable to first prove both these axioms, and then to prove from them that the append operation is associative.

Reflecting on the second trick, one could think of kind of a 'distance' between the given axiomatic basis and the proof goal. If the distance is small enough, SPASS can automatically find the proof. If the distance is too big, one needs to split it into several parts. If the splitting is done in a suitable way, SPASS will be capable to prove the overall goal in a step wise manner.

The above example shows that 'automated theorem proving' might not be as automated as one would like it to be. In theorem proving practice, it often is the case that one needs to find suitable lemmas and with these pave the way for the theorem prover. Nonetheless, the example also demonstrates that the theorem prover is of great help: none of the base or step cases needed to be manually proven—it was rather enough to state them. Thus, loads of tedious steps were taken care of by the theorem prover. It was enough to 'sketch' the proof.

Coming back to the CASL structuring operations, it should be noted that renamings are not required to be injective, i.e., it is possible to collapse symbols, e.g., two different sorts can be united. This can be useful: e.g., at an early development stage, one establishes several categories of identifiers, say, for post codes and the name of people; at a later design stage, one decides that both of these identifiers should simply be strings and not to be treated in a different way. Naturally, such non injective renamings are a potential source of inconsistency.

4.4.4 Libraries

Libraries—keyword **library**—collect named specifications. Specifications within one library can refer to each other. In a library, it is also possible to import named specifications from another library.

Given a string LN, and named CASL specification definitions $LI_1, \ldots, LI_n$, $n \geq 0$, then

$$\textbf{library } LN \; LI_1 \ldots LI_n$$

is a library formed out of these specifications.

In order to import named specifications $IN_1, \ldots, IN_n$, $n \geq 1$, from another library LN, we write

$$\textbf{from } LN \textbf{ get } IN_1, \ldots, IN_n \textbf{ end}$$

where the keyword **end** is optional.

There is also the possibility to change the names of the imported specifications via a renaming. Such imports can appear anywhere in a library, i.e., a library, with regards to the concepts that we consider here, is an arbitrary sequence of named specifications and import statements. However, note that CASL libraries work with linear visibility, e.g., given a sequence $IN_1, \ldots, IN_n$, of named specifications, the name of specification IN_i can only be used in specifications IN_j with $j > i$.

CASL comes with a comprehensive library of basic data types. These include standard types such as numbers, characters, lists, bags, graphs, ..., for more details see [Mos04].[6]

Example 47.2: Lists with a Length Function

We have compiled all specifications of this section into one library with the name *ExamplesForStructuringOperations*.

A typical operation on lists is to compute how many items they contains. This requires the sort of natural numbers to be available. Here,

---

[6] Available at `https://github.com/spechub/Hets-lib`.

we can utilise the natural numbers as defined in the CASL library of Basic Datatypes, c.f. [Mos04].

The natural number specification from the CASL Basic Datatypes is monomorphic, i.e., it has only one model up to isomorphism, see Sect. 2.3.1. It comes with operations and predicates as expected, i.e., there are operations for basic arithmetic, *min* for the minimum, *max* for the maximum, and the ordering relations $<$, $\leq$, $>$, and $\geq$ are defined, besides many others.

library *ExamplesForStructuringOperations*

...

from *Basic/Numbers* **get** NAT

spec LISTWITHLENGTH = LIST **then** NAT
then %def
      op   *length* : *List* $\rightarrow$ *Nat*
      $\forall\, L : List;\ e : Elem$
      • $length([]) = 0$ %(length empty list)%
      • $length(e :: L) = length(L) + 1$ %(length non−empty lists)%
end

4.4.5 Parameterisation and Instantiation

Specifications can take other specifications as their formal parameter. A generic specification is written:

$$\textbf{spec } SN[SP_1]\dots[SP_n] = SP \textbf{ end}$$

where SN is a specification name, SP_i, $i = 1\dots n$, $n \geq 0$ and SP are CASL specifications. The semantics of this generic specification is given by translation to already discussed structuring constructs, namely it is the semantics of

$$\{SP_1 \textbf{ and } \dots \textbf{ and } SP_n\} \textbf{ then } SP.$$

Example 47.3: Sorted Lists

Sorted lists provide a typical example of why one would like to write a parameterised specification: neither the list construction, nor the property 'sorted' depends on what kind of data the lists are formed of. The only requirement is that the data which we organise as lists is totally ordered: syntactically the axioms concerning the property 'sorted' depend on an ordering relation $\leq$; semantically, lists will be totally ordered if the data is.

spec SORTEDLIST[TOTALORDER] = LIST
then %def
      pred *sorted* : *List*
      $\forall\ e, f : Elem;\ l : List$
      • $sorted([\,])$
      • $sorted(e :: [\,])$
      • $sorted(e :: (f :: l)) \Leftrightarrow e \leq f \wedge sorted(f :: l)$
end

The specification SORTEDLIST takes the specification TOTALORDER from Example 45 as a parameter and builds upon the specification LIST from Example 47. Note that both specifications, TOTALORDER and LIST, are declaring the sort *Elem*—this poses no problem: CASL applies the principle "same name same thing", i.e., several declarations of the same symbol name are legitimate.

For simplicity, we consider specification instantiation only for specifications with one formal parameter, i.e., specifications of the form

$$\textbf{spec } SN[FP] = SP \textbf{ end}$$

In order to instantiate this parameterised specification with a CASL specification AP as the actual parameter, we write

$$SN[AP \textbf{ fit } SY_1 \mapsto SY_1', \ldots, SY_n \mapsto SY_n' \,]$$

$n \geq 0$, where the symbol SY_i from the signature of the formal parameter specification shall be 'identified' with the symbol SY_i' from the signature of the actual parameter specification, $0 \leq i \leq n$. Again, the semantics of instantiation can be expressed by translation to already discussed structuring constructs, namely it is the semantics of

$$\{\{FP \textbf{ then } SP\} \textbf{ with } SY_1 \mapsto SY_1', \ldots, SY_n \mapsto SY_n'\} \textbf{ and } AP.$$

Instantiating a formal parameter FP with an actual parameter AP comes with a proof obligation, namely one needs to show that the model class of the actual parameter is included in the model class of the formal parameter (after suitable renaming), i.e., one needs to discharge the proof obligation

$$AP \textbf{ then } \%\textbf{implies } \{FP \ \textbf{ with } SY_1 \mapsto SY_1', \ldots, SY_n \mapsto SY_n'\}.$$

See the CASL reference manual [Mos04] for a thorough discussion of instantiation, including the general case of instantiating specifications with several parameters.

Example 47.4: Instantiation of Sorted Lists

We can instantiate our parameterised specification SORTEDLIST, e.g., with the natural numbers as specified in the CASL Basic Datatypes:

spec SORTEDNATLIST =
    SORTEDLIST
      [NAT **fit sort** *Elem* $\mapsto$ *Nat*, **ops** *inf* $\mapsto$ *min*, *sup* $\mapsto$ *max*]
end

We need to map the sort symbol ELEM from TOTALORDER to the sort symbol *Nat* from NAT, further the specification TOTALORDER includes the operation symbols *inf* and *sup* which—as proven in Example 45—can be identified with *min* and *max*, resp.

    We refrain from discharging the proof obligations arising from this instantiation: using SPASS they could be proven inductively, following the ideas discussed in Example 47.1.

One subtle point with parameterisation and instantiation concerns the requirement that the actual parameter and the body of the parameterised specification must not share symbols. We briefly illustrate this by example and hint at the CASL construct that deals with this situation.

Example 47.5: Sorted Lists with Length

In Example 47.2, we gave gave a specification of lists equipped with a *length* operation. Naturally, we could write a parameterised specification of sorted lists, which also includes this operation:

spec SORTEDLISTWITHLENGTHNONAT[TOTALORDER] =
    LISTWITHLENGTH **and** SORTEDLIST[TOTALORDER]
end

Note, that again the "same name same thing" principle applies to this specification: there are numerous elements declared and stated several times in SORTEDLISTWITHLENGT.

    When we now try to instantiate SORTEDLISTWITHLENGTNONAT with the the specification NAT, we obtain the error message:

> Symbols shared between actual parameter and body must be in formal parameter.

The specification NAT is the actual parameter. Also, it is part of LISTWITHLENGTH, i.e., part of the body of SORTEDLISTWITHLENGTH. Thus, all symbols of NAT are shared.

In order to allow for such instantiations in CASL, one has to provide a specification that includes the shared symbols as a special 'parameter' after the keyword **given**:

spec SORTEDLISTWITHLENGTH[TOTALORDER] **given** NAT =
  LISTWITHLENGTH **and** SORTEDLIST[TOTALORDER]
end

Here, we take the specification NAT itself as the specification with the shared symbols. With this construction we can write the desired instantiation:

spec SORTEDNATLISTWITHLENGTH =
  SORTEDLISTWITHLENGTH
  [NAT **fit sort** *Elem* $\mapsto$ *Nat*, **ops** *inf* $\mapsto$ *min*, *sup* $\mapsto$ *max*]
end

4.4.6 Hiding

The previously shown operators for structuring specifications share the property that they do not change the expressivity of the specification language used, i.e., for all specifications formed with these structuring operators, there exists a specification without structuring operators which has the same model class. This is not the case for the hiding operator. The hiding operator actually increases the expressivity of CASL, i.e., it allows to specify data types that can't be specified in CASL without hiding.

Hiding—keyword **hide**—removes symbols of a specification. Given a specification Sp, we write

$$Sp \text{ hide } SY_1 \ldots, SY_n$$

$n \geq 1$, for obtaining a new specification, which is like Sp but with the symbols SY_1 to SY_n removed. Note that sometimes more symbols than listed will be removed in order to obtain a signature after hiding. Consider, for instance, the following example:

spec HUGO =
  sorts *s*, *t*
  op   *o* : *s* → *t*
end

spec ERNA = HUGO **hide** *t* **end**

Here, specification ERNA is defined by hiding sort t in specification HUGO. The signature of ERNA consists only out of the sort symbol s, the operation symbols o is hidden as well, although it is not part of the symbol list. The reason for this is that applying o results in an element of sort t—which has been removed from the signature.

In passing we mention that there are several interpretations possible to what is means to 'remove' a symbol from the theory. The distributed ontology, modelling, and specification language DOL has four constructs for this [MCNK15].

It has been a long standing question in algebraic specification, what data types, i.e., classes of algebras closed under isomorphism—see Sect. 2.3.1—can be expresses using algebraic specification. Bergstra and Tucker give systematic answers to this in their paper "Algebraic specification of computable and semi-computable data types" [BT87]. They show as Theorem 4.1 the following result:

Example 48: Algebra Without Specification in Equational Logic

In CASL, one can specify natural numbers with a square function:

spec NEWNAT =
> **free type** $Nat ::= 0 \mid suc(Nat)$
> **op** $\quad 1 : Nat = suc(0)$
> **ops** $\quad \_\_+\_\_, \_\_*\_\_ : Nat \times Nat \to Nat$
> **op** $\quad square : Nat \to Nat$
> $\forall\ n,\ m : Nat$
> - $0 + n = n$
> - $suc(m) + n = suc(m + n)$
> - $0 * n = 0$
> - $suc(m) * n = n + (m * n)$
> - $square(n) = n * n$

end

Using the hiding operator, we can remove the $+$ and the $*$ operation from the signature

spec MYDATATYPE =
> NEWNAT **hide** $\_\_+\_\_, \_\_*\_\_$

end

Utilising Hets we can obtain the Theory of MYDATATYPE and see that indeed MYDATATYPE has the desired signature:

```
sort Nat
op 0 : Nat
op 1 : Nat
op square : Nat -> Nat
op suc : Nat -> Nat
```

Bergstra and Tucker prove now that there is no specification in equational logic (with initial semantics) that has the same model class as MyDataType [BT87].

4.5 Closing Remarks

In this chapter, we discussed the Formal Method algebraic specification using the example of the language CASL and its tooling environment Hets. We presented CASL *syntax* through various examples and discussed how to model systems and how to validate specifications. CASL *semantics* is given by the model class of a specification. Such a model class can be empty. In this context, we introduced the notion of consistency. Furthermore, specifications can be extended. Extensions can have different properties which CASL captures through the annotations **implies**, **def**, and **cons**. These properties are expressed over the model classes of structured CASL specifications. Algebraic specification *methods* discussed include

- model encoding in order to prove consistency,
- automated theorem proving, and
- random testing.

Concerning the relation between specification and programming, we demonstrated that they overlap: the CASL specification of a Ladder Logic program and the program itself are semantically 'the same'. The section on structured specification demonstrated how many operations one can think of in order to 'knit' small specifications together to larger ones.

The methods of this chapter are universal. With the basics of algebraic specification—signatures as interfaces, axioms as 'rules' for invoking operations, algebras as implementations—we have an intellectual means of a mathematical nature to design, model, and analyse data wherever we find it, now and in the future. New data arises daily as software engineers develop and maintain applications, as software is invading and automating more and more aspects of life, for example in industrial production, surveillance, medical and environmental practice.

4.5.1 Annotated Bibliography

Classical texts on algebraic specification include the books by Ehrig and Mahr [EM85, EM90] as well as the edited volume by Astesiano, Kreowski, and Krieg-Brückner [AKKB99]. The book by Loeckx, Ehrich and Wolf [LEW97] provides an accessible discussion of concepts central to algebraic specification, with a focus on questions of theoretical nature. Sannella and Tarlecki have written probably the most comprehensive compendium on the subject [ST12].

The theory of institutions, which provides the accepted method of semantics definition in algebraic specification, is discussed in the book by Diaconescu [Dia08] as well as in the book by Sannella and Tarlecki [ST12].

The CASL Reference Manual [Mos04] is the authoritative handbook on CASL; the CASL User Manual [BM04] provides an example based overview on the language. Mossakowski, Meader, and Lüttich provide a scientific discussion of their Hets tool in their publication [MML07]—though Hets dates back to the early 2000s, it is still maintained and subject to further developments. The user interface of Hets represents structured specifications as development graphs [MAH06].

Other algebraic specification languages include ASF-SDF [BHK89], Cafe-OBJ [DF98], and Maude [CDE+07, Ölv17].

4.5.2 Current Research Directions

Some researchers, among them even a few who shaped the field, perceive algebraic specification as a closed chapter in the history of science, coming as a nicely wrapped-up parcel where all conceivable structural results have been achieved and light has been shone into all corners. From our perspective, this is a misconception and large parts of the landscape remain unexplored if only seen from the right angle. In the following, we will point the reader to some of these uncharted territories.

As for nearly all specification languages, the link between algebraic specification and modelling as well as programming could be stronger:

- In a formally based software development, a still open issue is how to semantically combine an algebraic specification language and a behavioural modelling language [RBKR20], or an algebraic specification language and a programming language. The most obvious approach here would be to 'transport' the modelling or programming language into the realm of algebraic specification. To this end, one would capture the semantics of the modelling or programming language as a so-called institution and provide suitable mappings to express the semantic relations between modelling, programming and specification.

- Algebraic specification is strong when it comes to modularisation and architectural specification. However, notions from software engineering such as components, objects and generics still need further consideration for proper integration of algebraic specification, modelling and programming.
- With regards to testing, we considered random testing in ConGu. Others have suggested methods of how to generate test cases based on the axioms stated in an algebraic specification. In Sect. 5.4 "Using Algebraic Specifications for Testing", we describe such an approach based on the work of Gaudel [BGM91]. However, it is still a challenge to generate concrete values for a signature with axioms for applying data driven testing in practice.

The above mentioned 'institutionalisation' of non algebraic specification languages—i.e., capturing the semantics of specification languages in the mathematical framework of a so-called institution—promises a rich perspective in software development [KMR15]. Institutions are the accepted method of semantics definition in algebraic specification. Institutionalising comes with the objectives (1) to allow for semantic preserving translation between different specification formalisms and (2) to enrich specification formalisms with the rich structuring mechanisms available, e.g., in CASL. There are two different driving forces for the first objective: often, one needs different specification languages to capture the distinct aspects of a system—having translations available allows for heterogeneous specification; such translations allow for the comparison of formalisms, e.g., in terms of expressivity. Concerning the second objective, we observe that, while almost all programming languages have structure, too many (non algebraic) specification languages don't and the specifier has to write large, monolithic texts. This makes specification practice hard and limits adoption of Formal Methods.

Concerning methodology, like nearly all Formal Methods, algebraic specification suffers from being applied in a too 'heavy' way. There is a silent expectation that one has to specify the model class to the last detail rather than just to focus on a few but critical properties. This has led to a lack of light-weight integration into tools and methodologies. Examples which would lend themselves to such 'light-weight' algebraic specification include domain engineering [Bjø17] and generic programming [SM09].

References

[Abr10]   Jean-Raymond Abrial. *Modeling in Event-B*. Cambridge University Press, 2010.

[AKKB99] Egidio Astesiano, Hans-Joerg Kreowski, and Bernd Krieg-Brueckner, editors. *Algebraic Foundations of Systems Specification*. Springer, 1999.

[BGM91]   Gilles Bernot, Marie Claude Gaudel, and Bruno Marre. Software testing based
          on formal specifications: A theory and a tool. *Softw. Eng. J.*, 6(6):387–405,
          November 1991.
[BHK89]   J.A. Bergstra, J. Heering, and P. Klint, editors. *Algebraic Specification.* ACM
          Press, 1989.
[Bjø17]   Dines Bjørner. Manifest domains: analysis and description. *Formal Aspects of
          Computing*, 2017.
[BM04]   Michel Bidoit and Peter D. Mosses. *CASL User Manual – Introduction to
          Using the Common Algebraic Specification Language.* Springer, 2004.
[Bol06]   William Bolton. *Programmable Logic Controllers.* Newnes, 2006.
[Bra12]   Aaron R. Bradley. Understanding ic3. In *Theory and Applications of Satisfi-
          ability Testing – SAT 2012*, pages 1–14. Springer, 2012.
[BT87]   J.A. Bergstra and J.V. Tucker. Algebraic specifications of computable and
          semicomputable data types. *Theoretical Computer Science*, 50(2):137 – 181,
          1987.
[CDE+07] Manuel Clavel, Francisco Durán, Steven Eker, Patrick Lincoln, Narciso Martí-
          Oliet, José Meseguer, and Carolyn Talcott. *All About Maude - a High-
          performance Logical Framework: How to Specify, Program and Verify Systems
          in Rewriting Logic.* Springer, 2007.
[CLV10]   Pedro Crispim, Antónia Lopes, and Vasco Thudichum Vasconcelos. Runtime
          verification for generic classes with ConGu 2. In *Formal Methods: Foundations
          and Applications*, LNCS 6527, pages 33–48. Springer, 2010.
[CMM13]   Mihai Codescu, Till Mossakowski, and Christian Maeder. Checking conserva-
          tivity with Hets. In *Conference on Algebra and Coalgebra in Computer Sci-
          ence*, LNCS 8089. Springer, 2013.
[DF98]   Răzvan Diaconescu and Kokichi Futatsugi. *CafeOBJ Report: The Language,
          Proof Techniques, and Methodologies for Object-Oriented Algebraic Specifica-
          tion.* World Scientific, 1998
[Dia08]   Razvan Diaconescu. *Institution-independent Model Theory.* Birkhäuser Basel,
          2008.
[DK76]   Franklin L. DeRemer and Hans H. Kron. Programming-in-the-large versus
          programming-in-the-small. In *Programmiersprachen*, pages 80–89. Springer,
          1976.
[EM85]   Hartmut Ehrig and Bernd Mahr. *Fundamentals of Algebraic Specification 1.*
          Springer, 1985.
[EM90]   Hartmut Ehrig and Bernd Mahr. *Fundamentals of Algebraic Specification 2.*
          Springer, 1990.
[HNSA16] John Hughes, Ulf Norell, Nicholas Smallbone, and Thomas Arts. Find more
          bugs with quickcheck! In *Proceedings of the 11th International Workshop on
          Automation of Software Test*, pages 71–77. ACM, 2016.
[IEC03]   IEC. Programmable Controllers - Part 3: Programming languages. Standard
          61131-3, 2003.
[Jam10]   Phillip James. SAT-based model checking and its applications to train control
          software. Master's thesis, Swansea University, 2010.
[JLM+13] Phillip James, Andy Lawrence, Faron Moller, Markus Roggenbach, Monika
          Seisenberger, Anton Setzer, Karim Kanso, and Simon Chadwick. Verification
          of solid state interlocking programs. In *SEFM 2013 Collocated Workshops*,
          LNCS 8368, pages 253–268. Springer, 2013.
[JR10]   Phillip James and Markus Roggenbach. Automatically verifying railway inter-
          lockings using SAT-based model checking. In *Proceedings of AVoCS'10*. Elec-
          tronic Communications of the EASST, 2010.
[KMR15]   Alexander Knapp, Till Mossakowski, and Markus Roggenbach. Towards an
          institutional framework for heterogeneous formal development in UML – A
          position paper. In *Software, Services, and Systems*, LNCS 8950, pages 215–
          230. Springer, 2015.

[LEW97]   Jacques Loeckx, Hans-Dieter Ehrich, and Markus Wolf. *Specification of Abstract Data Types*. Wiley, 1997.

[LRS04]   Christoph Lüth, Markus Roggenbach, and Lutz Schröder. CCC – the Casl Consistency Checker. In *WADT*, LNCS 3423, pages 94–105. Springer, 2004.

[MAH06]   Till Mossakowski, Serge Autexier, and Dieter Hutter. Development graphs - proof management for structured specifications. *J. Log. Algebr. Program.*, 67(1-2):114–145, 2006.

[MCNK15]   Till Mossakowski, Mihai Codescu, Fabian Neuhaus, and Oliver Kutz. The distributed ontology, modeling and specification language – DOL. In *The Road to Universal Logic*, volume 2, pages 489–520. Birkhäuser, 2015.

[MML07]   Till Mossakowski, Christian Maeder, and Klaus Lüttich. The heterogeneous tool set, Hets. In *TACAS 2007*, LNCS 4424, pages 519–522. Springer, 2007.

[Mos04]   Peter D. Mosses. *CASL Reference Manual: The Complete Documentation Of The Common Algebraic Specification Language*. Springer, 2004.

[Ölv17]   Peter Csaba Ölveczky. *Designing Reliable Distributed Systems*. Springer, 2017.

[RBKR20]   Tobias Rosenberger, Saddek Bensalem, Alexander Knapp, and Markus Roggenbach. Institution-based encoding and verification of simple UML state machines in CASL/SPASS. In *WADT 2020*, LNCS 12669, pages 120–141. Springer, 2020.

[RS01]   Markus Roggenbach and Lutz Schröder. Towards trustworthy specifications I: Consistency checks. In *WADT 2001*, LNCS 2267. Springer, 2001.

[SM09]   Alexander Stepanov and Paul McJones. *Elements of Programming*. Addison-Wesley, 2009.

[ST12]   Donald Sannella and Andrzej Tarlecki. *Foundations of Algebraic Specification and Formal Software Development*. Springer, 2012.

Chapter 5
Specification-Based Testing

Bernd-Holger Schlingloff and Markus Roggenbach

Abstract In this chapter, we apply Formal Methods to software and systems testing. After some introductory remarks on the importance of software testing in general, and formal rigour in particular, we give a typical example of a computational system as it occurs as part of a bigger system. We show how to formally specify and model such a system, how to define test cases for it, and how to monitor testing results with temporal logic. In order to do so, we use simplified state machines from the unified modelling language UML2. With the example, we describe the underlying methodology of test generation and discuss automated test generation methods and test coverage criteria. We present Tretmans' classical conformance testing theory, and Gaudel's theory of test generation from algebraic specifications. Finally, we discuss available tools, and point to research topics in the area of specification-based testing.

5.1 The Role of Testing in Software Design

Imagine that you have a friend who is an engineer and a hobby pilot. She invites you to a trip on her brand-new self designed airplane. When you look puzzled, she tells you not to worry—although never tested, the whole plane had been thoroughly simulated during development. With some reluctance, you agree to join her on the maiden flight, so the two of you take off. In the air, the left wing makes some funny noises. When you analyze the source, you notice that the wing flaps are frequently deployed and undeployed. You land and check the flap motors and cables, but cannot find any problem. So, the problem must be in the control software. You wonder how to locate the problem. Together with your friend, you set up a hardware-in-the-loop test

Bernd-Holger Schlingloff
Humboldt University and Fraunhofer FOKUS, Berlin, Germany

Markus Roggenbach
Swansea University, Wales, United Kingdom

© Springer Nature Switzerland AG 2022, corrected publication 2022
M. Roggenbach et al., *Formal Methods for Software Engineering*,
Texts in Theoretical Computer Science. An EATCS Series,
https://doi.org/10.1007/978-3-030-38800-3_5

environment, where the aircraft sensors (wind speed, angle of attack, and lift coefficient) and actuators (flap motors) are connected to the software in a simulation. Yet, all of your random simulation runs do not exhibit the motor behaviour which you experienced while flying the plane. Thus, you begin to think about systematic ways to construct test cases.

This chapter is concerned with the application of Formal Methods to software and systems testing. For a long time, testing had been considered to be an inherently informal activity: after a program was written by a team of proficient programmers, the testers would have to sit down and do some experiments with it in order to detect potential errors. Program design and implementation was considered to be a superior activity which required high skills and formal rigour, whereas testing was seen as an inferior activity which was left to those who were not capable of writing good program code. With ever increasing complexity, competition, and quality demands, this situation has changed. Nowadays, often programming and bug fixing is considered to be something which can be left to third parties, but testing of complex interactions and making sure that the delivered software is free of faults is of utmost importance. With this change of attitude, the need for formal rigour in software testing began to rise. The term "software crisis" had been coined in the late 1960s, when the cost of producing software exceeded that of buying hardware. Today, we are facing a similar "software quality crisis": For many systems, the cost of verification and validation exceeds the cost of actual programming by a large amount. Thus, improved and more efficient methods are needed for guaranteeing that a program meets its specification.

One of the most common methods for quality assurance of computational systems is testing. Often, it is stated that testing is just a substitute for formal verification with the claim that "testing can be used to show the presence of bugs, but never to show their absence" [Dij70]. However, this viewpoint is not quite correct. Firstly, as we will show in Sect. 5.4 later on, in combination with theorem proving, testing can very well be used to show the absence of errors in a particular piece of code.

Secondly, each verification activity only considers a certain aspect of a computational system. So far, no computational system exists for which all constituent layers (application program, middleware, operating system, hardware, and development tools) have been formally specified, let alone verified. Thus, the verification of a particular component only shows the absence of errors *in that component*, not in the complete system. As an example, even if we have verified a particular sorting algorithm, we cannot be sure whether a payroll program using that algorithm will issue the correct paychecks. Testing considers a system within its environment; i.e., testing the sorting algorithm within the payroll program can reveal all sorts of problems caused by the interaction of components.

Thirdly, contrasting verification and testing per se is not adequate, because the two techniques try to answer different questions. In formal verification, an algorithm or program is shown to be correct with respect to a formal specification; i.e., one mathematical object is compared to another one. In testing, an actual system (i.e., a material object in the physical world) is examined as to whether it conforms to the user's expectations. Thus, in verification we try to answer the question whether two mathematical objects are equivalent, whereas in testing we are concerned with the question whether the behaviour of a physical system matches our ideas about it.

It follows that testing and verification are not mutually replaceable: formal verification is necessary in order to increase the confidence in a particular software module or algorithm, whereas testing is necessary in order to predict the behaviour of the actual code on a computer or a computational device (i.e., a physical system). Thus, testing and verification are complementary techniques in quality assurance.

We favour the following definition:

Definition 1 Testing is the process of systematically experimenting with a material object (in the physical world) in order to establish its quality.

According to this definition, testing is a *dynamic* activity, where a *subject* (the tester) interacts with an *object* (the system under test, SUT). To achieve some results, the SUT must be executed. This puts testing in contrast to other activities in software quality assurance such as static analysis, abstract interpretation, formal verification, or model checking. In these, the program code is analysed as a mathematical object.

In this book, the system under test always is some information-processing device or some executable binary code on a computer. However, much of the theoretical background also applies to other sorts of testing, e.g., the testing of cable cars or musical instruments.

In general, we distinguish between an *experiment*, which is a singular activity not necessarily related to quality, and a *test*, which is a systematic set of experiments to find out the quality of a system. According to standard definitions, *quality* is "the degree of accordance to the intention or specification". That is, there is no absolute notion of quality (and, probably, no "best" quality), but quality is always relative to somebody's conception of an ideal appearance or workmanship. Correspondingly, there are many possible quality measures:

- functionality, usefulness, usability,
- efficiency (with respect to time, space or money),
- reliability, availability, maintainability, safety, and security,
- robustness/stability,
- portability, modularity, extensibility,
- ...

For each of these quality measures and any given system, a value can be determined via testing. Hence there are many variants of testing: functional testing, performance testing, robustness testing, etc. The most important quality criterion of software, however, is *correctness*, i.e., the absence of failure. Thus, in this chapter we only consider testing a computational system for correctness.

Fig. 5.1 Errors, faults and failures

To do so, we need to discuss the meaning of the words 'correctness' and 'failure'. A *failure* is a deficit with respect to the intended functionality of a system. That is, a failure is a deviation of the actual behaviour of the system from the specified or required one. A failure of the system may cause an *incident* or *accident*, i.e., a negative effect onto the environment of the system, especially onto people. It is caused by a *fault* or *defect*, which is a wrong state of the system, due to a flaw in the design or manufacturing process. Each fault can be traced back to some human *error* or *mistake*, that is, a misconception about the system to be built or operated. As a mnemonic, 'an error can lead to a fault, and a fault can lead to a failure" (see Fig. 5.1).[1]

In order to find errors via testing, it is important that the 'right' conception is made explicit. That is, it is impossible to test the correctness of a program without being given a *specification* which describes the intended 'correct' behaviour. Often, specifications are given only implicitly, or imprecisely. Examples of badly formulated requirements are "the system shall never crash", "it must always react to user input", "there should not be any error messages", and similar. Here, it is unclear to which time period 'never' and 'always' refer to (not in a hundred years?); thus, such requirements can not be tested. Other bad (untestable) formulations include "the system should be as fast as possible", "the system must achieve a feasible cost/benefit ratio", or "the security of the system must be properly maintained". Here, we do not know what is 'possible' and 'feasible', or which security threats need to be considered.

Hence in order to set up a proper test for a system, it is necessary to provide a specification which is precise, unambiguous, and has a clear semantics. In the previous chapters, formal languages were introduced which allow to formulate systems properties in such a way. In this chapter, we will show how these formalisms can be used for testing.

In the above definition, testing was defined to be a systematic experiment, that is, there must be some systematics which the experiments follow. There are two main paradigms for such a systematics: The structure of the test suite can be derived from the structure of the SUT, or from the structure of the specification. The former paradigm in known as *code-based testing*, the latter is called *specification-based testing*. Code-based testing is also known as *white-box testing*, since the code must be revealed to the test developers.

---

[1] Here, we deviate from some parts of the literature where the words 'error' and 'fault' are interchanged; for us, an error (occurring in the human mind) is more fundamental than a fault (occurring in an artefact).

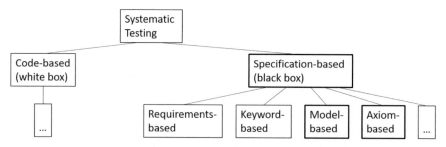

Fig. 5.2 Classification of testing methods

Accordingly, specification-based testing is sometimes called *black-box testing*, since the SUT is treated as a 'black box', whose interior is hidden from the tester's view.

In industrial practice, specification documents are mostly written in natural language and structured into *requirements*. Each requirement consists of one or a few sentences and describes a single functionality of the SUT. Within specification-based testing, most practitioners use this structuring to derive test cases: For each requirement, there is at least one test case to check whether it is correctly implemented. More formal approaches identify certain *keywords* in the specification, describing actors, actions, methods, interfaces, data items etc. Test cases are structured according to the available groups of keywords.

Requirements can be denoted in the form of formal models, in some suitable modelling language such as CSP or UML. In *model-based testing*, such models are used to automatically derive test cases. Requirements can also be formulated as axioms, e.g., in algebraic specifications, as described in Chap. 4. Test generation methods using such specifications sometimes are called *axiom-based testing*. Figure 5.2 gives a rough classification of testing method, according to the source which is used for the generation of test cases.

5.2 State-Based Testing

As we have seen in the previous chapters, there are two dimensions in the formal specification of systems, which could be called space and time. Specification formalisms such as algebraic specification or first-order logic are well-suited to describe the structural aspects of a system. In contrast, process algebras and temporal logics focus on the dynamic aspects, on the change of the system's state in time.

This section deals with the generation of tests for reactive systems from state-based models and specification formalisms. In order to understand how test cases can be generated from such models, we first present a method of modelling reactive systems with UML state machines. Then, we will discuss different test generation algorithms and test coverage criteria.

5.2.1 Modelling Stateful Systems

For a first example, we consider a simple switch as it is contained in many video camera models of a major brand. The switch is used to turn the device on and off, as well as to choose certain settings.

Example 49: VCR Switch

The VCR power switch under consideration is laid out as a non-locking slide switch. From its normal position, it can be pushed in two directions (up and down); it always returns to its normal (middle) position. The manual explains: "You need to slide the POWER switch repeatedly to select the desired power mode to record or play. ...To turn the power on, slide the POWER switch down. To enter the recording or playing mode, slide the switch repeatedly until the respective lamp for the desired power mode lights up. To turn off the power, slide the POWER switch up."

According to this description, the device has several power modes which can be selected with the switch. However, the description leaves open some questions, e.g., which power modes exist, which is the initial mode after turning the power on, or on the exact sequence of modes which is assumed when sliding the switch repeatedly down. Some experiments reveal that the device will always start in "record" mode, and that repeatedly pushing the switch down cycles through the three power modes "memory", "play", and back to "record". Intuitively, record mode is for video recording, memory mode for taking pictures and play mode for viewing recorded material.

This example is typical for a number of similar systems. Their main characteristic is that they are *stateful reactive systems*. That is, the system can be in any one of a number of states. It continuously reacts to stimuli from the environment: In any state, given a certain input, it produces a designated output and takes a transition into a new state. Many different formalisms have been suggested for the modelling of stateful reactive systems, including finite automata, process algebras (see Chap. 3 on CSP), Petri nets, and others. Testing with CSP is discussed, e.g., by Cavalcanti and Hierons [CH13]. For this example, we will use *UML2 state machines*.

The Unified Modeling Language UML is a standardised, general-purpose language for modelling all sorts of computational systems. It comprises a set of graphic notation elements which can be combined according to the syntax rules given in the UML meta model.

- A *UML state machine* consists of a nonempty set of regions.
  - Each *region* contains a number of vertices and transitions.
    - A *vertex* can be a state or a pseudostate.
      - A *state* can be a *simple state*, a *composite state* or a *submachine state*. A submachine state contains again a UML state machine, and a composite state may contain one or more regions. This allows hierarchical structuring of state machines.
      - A *pseudostate* can be used to indicate a special point in a computation. We will use only *initial pseudostates* which are entered upon start of a component; besides that, there are fork, join, junction, choice, entry point, exit point, and terminate pseudostates.
    - A *transition* is a connection from a source vertex to a target vertex. It can contain a number of triggers, a guard, and an effect.
      - A *trigger* references an event, for example, the reception of a message or the execution of an operation.
      - A *guard* can be any UML constraint, i.e., a boolean condition on certain variables (for instance, class attributes).
      - An *effect* can be any UML behaviour. For example, an effect can be the assignment of a value to an attribute, the triggering of an event, or the execution of yet another state machine.
    In the graphical representation of state machines, a transition with trigger t, guard g and effect e is labelled by t [g] /e. All three elements (trigger, guard and effect) are optional, each one can be omitted. An empty trigger means that the transition is to be taken immediately when the state is entered, and an empty guard is equivalent to the guard true.

For specification-based testing, it is important to describe the behaviour of the system under test in a formal way, such that there are no more ambiguities or vaguenesses in the description. In order to come up with such a formal description, we recommend a procedure consisting of three steps:

Step 1. Definition of interfaces;
Step 2. Definition of operating modes; and
Step 3. Definition of transitions.

In **Step 1**, external interfaces of the system under consideration are identified. There are two categories of external interfaces: those where the environment sends a signal or trigger to the system under test, and those where the system sends a response to such a stimulus. (In the context of automated testing it is often confusing to talk about 'input' and 'output', because an input to the system under test is an output by the environment and vice versa. We thus prefer to call the interface categories "ENV2SUT" and "SUT2ENV", respectively.) As a convention, we identify elements of ENV2SUT with a question mark (?), and system actions in SUT2ENV with an exclamation mark (!).

Example 49.1: Modelling the VCR Switch: Interfaces

In the case of the VCR power switch, the SUT is the (software of the) VCR; we consider the user interface consisting of the physical slide switch and the LEDs as belonging to the environment of the SUT.

The user has two possible actions to perform: pushing the switch up or down. We call these events up and dn, respectively. Thus ENV2SUT= {up?, dn?}.

The system can react by transitioning into the respective camera mode and lighting up appropriate lights. In our case, there are three LEDs, each of which can be either on or off. This can be expressed in different ways: There could be signals "LEDi is turned on" and "LEDi is turned off", or "on/off for LEDi is toggled" ($i \in \{1, 2, 3\}$). Alternatively, we could imagine that the LED control is a vector of three boolean variables, which is set by a single command. For sake of demonstration, we choose the last alternative. That is, SUT2ENV= {000!, 100!, 010!, 001!}. Here, 000! means that all LEDs are off, 100! indicates that exactly the first LED is on, etc.

Step 2 is to fix the components and major operating modes of the system. A component may be a physically or logically coherent part of the SUT. Each mode is characterized by the main functions performed during operation in this mode, and by the range of possibilities offered to the user.

Example 49.2: Modelling the VCR Switch: Modes

The video camera contains, amongst other components, the switch and the LEDs as described above. It can be on or off, and when it is on, it can be in mode rec, mem or play. We take these as the modes of the VCR switch. Each of the modes offers different functions to the user: In record mode the video head is spinning and it is possible to record a video signal. In memory mode it is possible to take a photo. In play mode the LCD screen is activated and it is possible to view the recordings.

In **Step 3** we have to group the modes, identify which transitions occur, and to construct a state machine diagram for the system.

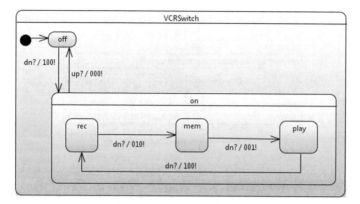

Fig. 5.3 UML state machine for the camera switch

Example 49.3: Modelling the VCR Switch: Transitions

According to the informal specification, event dn leads from off to on, and up leads from on to off. Furthermore, repeated occurrences of dn cycle through rec, mem and play. In Fig. 5.3 we decided to place the three operating modes rec, mem and play as states in a region within state on. That way, the event up leads to off from any of these states.

Note that the semantics of UML determines that 'unexpected' events are skipped. That is, if the machine is, e.g., in state off and an up-event is received, it just stays in this state and the event is discarded.

The model in Fig. 5.3 can be seen as a formalisation of the informal requirements given in Example 49. It describes the intended behaviour of the switch, giving a precise meaning to phrases like 'repeatedly' or 'desired mode'. Usually, a UML state machine is an abstraction of an actual target system (in our case, a video camera recorder). Such an abstraction can be used in two ways:

- for constructing the target system by a stepwise refinement process, and
- as a source for the generation of test cases for the target system.

The first of these uses is known as 'model-based design' (MBD), whereas the second one has been called 'model-based testing' (MBT).[2] In this chapter, MBT is discussed.

---

[2] Note that the use of the word 'model' significantly differs here from its use in logic. In MBD/MBT, a model is defined to be a purposeful abstraction of some target system, whereas in logic a model is a semantical structure for the evaluation of formulae.

SUT Characteristics	Test Case
functional	tuple (input, output)
reactive	sequence
nondeterministic	decision tree
parallel	partial order
interactive	test script or program
real-time	timed event structure
hybrid	set of real-valued functions

Fig. 5.4 Different SUT aspects and corresponding test cases

5.2.2 Test Generation for State-Based Systems

There are various notions of what a test case is. In the most general sense, a *test case* is the description of a (single) experiment with the SUT. A *test suite* is a set of test cases; it describes a cohesive set of experiments. Depending on the aspect of a system under test that is to be considered, test cases can have several forms—see Fig. 5.4, which is taken from a survey by one of the authors [WS11].

For the time being, we restrict our attention to the testing of deterministic reactive systems. Thus, in this section, we consider test cases which are sequences. (Later on, for conformance testing of nondeterministic systems, we will generate test cases which are trees. In Sect. 5.4, we will generate input values as test cases for functional programs.) In the most general setting, a test case for a reactive system is a sequence of events. Here, an event can be any action from the tester, which serves as a stimulus for the SUT. Additionally, an event could also be an observable reaction of the SUT: an expected response from the SUT, an observable behaviour, a visible transition, state or configuration, etc.

For our purposes, a *test case* is a finite path of input and output actions in the state machine. In other words, a state machine is transformed into a directed graph. In this graph, we consider finite paths from the initial state. The labels on the transitions of such a path form a test case. The test generator constructs a test suite for a predefined set of *test goals*. A test goal could be, e.g., to reach a certain state or transition in the machine. A goal is *covered* by a test suite if there is a test case in the suite such that the goal is contained in the test case. Different test generators support different coverage criteria; a detailed discussion can be found in Sect. 5.2.4 below. For example, for the criterion "all-states" a test suite with only one test case is sufficient:

$\{$ (dn?, 100!, dn?, 010!, dn?, 001!) $\}$.

The criterion "all-transitions" is still satisfied by a one-element test suite:

$\{$ (dn?, 100!, dn?, 010!, dn?, 001!, dn?, 100!, up?, 000!) $\}$.

If we require "decision coverage", a test generator will yield a test suite where all branches in the model are contained:

{ (dn?, 100!, up?, 000!),
 (dn?, 100!, dn?, 010!, up?, 000!),
 (dn?, 100!, dn?, 010!, dn?, 001!, up?, 000!),
 (dn?, 100!, dn?, 010!, dn?, 001!, dn?, 100!) }.

These test suites can be executed on any implementation of the VCR switch. Each test case serves as an experiment which the test system performs with the system under test, providing it with stimuli and observing the responses. For example, in our VCR switch a test execution would consist of subsequently pushing the switch in the specified directions and noting down the resulting LED patterns. The execution of a test case *passes*, if the SUT shows the specified behaviour; otherwise it *fails*.

With such a setting, there are some issues to be considered. Firstly, not all test generators provide test cases in the form of input-output sequences. Some tools might not allow to view outputs, but to observe the *configuration* in which the SUT currently is. (A *configuration* differs from a state in the fact that qualified names of substates are used.) That is, the "all-states" test case from above is given as

{ (off, dn?, on.rec, dn?, on.mem, dn?, on.play) }.

In order to execute such a test case, we have to conceive means in the SUT to observe in which state or configuration it is. In the example, this can be inferred from the state of the LEDs; in general we might have to provide additional test interfaces to the SUT.

Secondly, and more important, we have not taken into account *nondeterminism*. A state machine is called *deterministic*, if for every configuration and every input there is at most one enabled transition; i.e., the successor configuration is uniquely determined. If the specification is nondeterministic, then the SUT may have several alternatives how to react to a given stimulus in a given state. Therefore, it may be necessary to consider several alternative test cases at the same time while executing tests. This may cause computational overhead; it can be hard to determine according to which of the alternatives the SUT actually reacts. To avoid this overhead at test execution time, often it is required that test specifications must be deterministic.

Thirdly, in *specification-based* testing we assume that the internal implementation of the SUT is hidden from the tester. Test cases are derived from the specification, without resorting to the program code of the SUT. Sometimes, this is referred to as 'black-box testing' as opposed to 'white-box testing' or 'code-based testing'. Therefore, testing can never be used to give any guarantees about a system's behaviour. A malicious implementer might always design the SUT such that all specified test cases pass, yet the system shows some unwanted behaviour in untested parts.[3] As a less drastic example, if the SUT is implemented using a nondeterministic choice between two alternative paths, it may choose one of those arbitrarily often. Then, the

---

[3] This fact has been used in the past by some car manufacturers to illegally program different behaviour for the carburettor when the car is under test or on the road.

tester can never be sure whether the other path has been exercised. However, this situation changes if certain assumptions about the SUT can be made. We will discuss this so-called *conformance testing* in Sect. 5.3 below.

5.2.3 Monitoring of Execution Traces

So far, we have derived different sets of test cases from a state machine model of the SUT. The next step is to defined how the execution of a test case is to be evaluated. In Sect. 5.3, we will assume that a test case yields the "pass" verdict if and only if the observations described in the test case can be made during the execution. Here, we will take the more general point of view of *specification-based testing*. In this view, desired properties of the system under test are formulated in a logical specification language. Whether the execution of a test case at an SUT passes or fails is determined by this specification and is independent of the way of how the test case has been generated.

An *execution trace*, or simply *trace*, is the sequence of events obtained by executing a test case. In general, the *test oracle* is the part of the test system which determines whether a trace satisfies the given properties or not. In other words, the test oracle issues a test verdict for a given trace.

For example, we might want to check that "whenever the switch is pushed up, the power will be turned off", "The operating mode 'play' is reached by repeatedly sliding the POWER switch down at most three times", or "whenever the VCR switch is pushed down, one of the lamps is lit until it is slid up again".

Such properties can be conveniently denoted in linear temporal logic (LTL, cf. Chap. 2). In order to formalize the properties in LTL, we have to fix the proposition alphabet $\mathcal{P}$. Here, again, we have to consider which elements of the SUT can be observed by the tester. One approach is to use the interfaces "ENV2SUT" and "SUT2ENV" as the set of basic propositions. Thus, e.g., 000! indicates that the VCR turns off, and 001! that it changes into play mode. With these, the above properties can be written as follows.

$$\Box(\text{up?} \Rightarrow \bigcirc 000!)$$

$$\Box(\text{dn?} \Rightarrow \bigcirc(001! \vee \bigcirc(\text{dn?} \Rightarrow \bigcirc(001! \vee \bigcirc(\text{dn?} \Rightarrow \bigcirc 001!)))))$$

$$\Box(\text{dn?} \Rightarrow \bigcirc(\neg 000! \, \mathcal{W} \, \text{up?}))$$

In the last of these formulae, we are using an *unless*-operator rather than an *until*, since we cannot guarantee that the generated test sequences always end in the off-state.

The test oracle takes these formulae and checks whether they are satisfied by the execution sequences which are obtained by running the tests on the SUT. Subsequently, we give an algorithm for checking a test execution trace

with an LTL formula. The algorithm is a straightforward translation of the
semantics of LTL, where the temporal formulae are 'unwound' according to
the recursive characterizations

$$\Diamond\varphi \Leftrightarrow (\varphi \vee \bigcirc\Diamond\varphi), \text{ and}$$
$$(\varphi\,\mathcal{U}\,\psi) \Leftrightarrow (\psi \vee (\varphi \wedge \bigcirc(\varphi\,\mathcal{U}\,\psi))).$$

Algorithm 7: Monitoring test executions

---

function monitor(σ, φ) // Precondition: $length(\sigma) \geq 1$
  if $\varphi = \bot$ **then return** $\bot$
  else if $\varphi = p$ **then return** $(\sigma_0 = p)$ // σ_0 is the first element of σ
  else if $\varphi = (\varphi_1 \Rightarrow \varphi_2)$ **then**
    if monitor $(\sigma, \varphi_1) = \bot$ **then return** $\top$
    else return monitor (σ, φ_2)

  else if $\varphi = \bigcirc\varphi_1$ **then**
    if $length(\sigma) = 1$ **then return** $\bot$
    else return monitor $(\sigma^{(1)}, \varphi_1)$ // $\sigma^{(1)}$ is σ without the first element

  else if $\varphi = \Box\varphi_1$ **then**
    if monitor $(\sigma, \varphi_1) = \bot$ **then return** $\bot$
    else return monitor $(\sigma^{(1)}, \Box\varphi_1)$

  else if $\varphi = \Diamond\varphi_1$ **then**
    if monitor $(\sigma, \varphi_1) = \top$ **then return** $\top$
    else if $length(\sigma) = 1$ **then return** $\bot$
    else return monitor $(\sigma^{(1)}, \Diamond\varphi_1)$

  else if $\varphi = (\varphi_1\,\mathcal{U}\,\varphi_2)$ **then**
    if monitor $(\sigma, \varphi_2) = \top$ **then return** $\top$
    else if monitor $(\sigma, \varphi_1) = \bot$ **then return** $\bot$
    else if $length(\sigma) = 1$ **then return** $\bot$
    else return monitor $(\sigma^{(1)}, (\varphi_1\,\mathcal{U}\,\varphi_2))$

In this basic form, with nested temporal formulae (e.g., $(\varphi_1\,\mathcal{U}\,(\varphi_2\,\mathcal{U}\,\varphi_3))))$)
the algorithm may traverse the given execution sequence several times. It
can be improved somewhat by separating the present- and future-part of φ
in each step and traversing σ only once. More precise, every LTL formula
φ can be written as a boolean combination of formulae φ_i and $\bigcirc\psi_j$, where
the φ_i do not contain temporal operators and thus can be evaluated in σ_0. If
this yields no result, then ψ_j can be evaluated recursively on $\sigma^{(1)}$. Havelund
and Rosu [HR01] describe an efficient implementation of this idea with the
rewriting tool MAUDE.

The result of Algorithm 7 is not necessarily the same as the result of model
checking the formula with the state machine. Model checking determines
whether the formula is satisfied for all possible paths of the state machine. In
testing, we are evaluating the oracle formulae with actual runs of the SUT,
not with some abstract model.

5.2.4 Test Generation Methods and Coverage Criteria

There are various ways how to generate test cases from UML state machines or other formal models. Let us assume that the purpose of test generation is to find a test suite where each state of the UML state machine appears at least once. That is, we are trying to satisfy the "all-states" coverage criterion. Thus, for each state we have to find a sequence of events which will trigger transitions leading into that state. There are several ways of doing so: We can employ a forward-directed search, starting with the initial configuration and employing a depth-first or breadth-first search for the goal. Alternatively, we can use a backward search, starting with the goal and stepwise going backward in order to find the initial configuration. Dijkstra's single-source shortest path algorithm assigns for each node in a graph its distance from a particular node via a greedy search. Such an assignment can be used in the backward search: Instead of choosing any predecessor, we select one with a minimal distance to the initial state. Of course, if a state is already covered, it is not necessary to cover it twice. Therefore, we have to maintain a list of covered test goals.

The pseudocode of this algorithm is given in Algorithm 8 on the next page. In this code, `length` is a mapping (e.g., an array) from states to $\mathbb{N}_0 \cup \infty$, giving for each state s the length of the shortest path from the initial state s_0 to s. `prev` and `trans` are mappings from states to states and transitions, respectively, giving for each state the previous state from which, and the transition by which it is reached on the shortest path. In the first phase, the algorithm calculates the values for `length`, `prev` and `trans` via greedy breadth-first search. Then, in the second phase, the algorithm outputs the test cases, starting with the path to the most distant state which has not been covered.

Obviously, since in general a model may contain cycles, it may have infinitely many runs. If there are finitely many states, then there is a finite test suite which covers all of these states. However, potentially infinitely many test cases could be derived from a cyclic model. In order to be able to execute the generated test suite within a finite amount of time, a finite subset of all possible test cases must be selected. Model coverage criteria can help to estimate to which extent the generated test suite reaches a certain testing goal. Typical model coverage criteria for UML state machines or Finite State Machine testing models are

- all-states: for each state of the state machine, there is a test case which contains this state,
- all-transitions: the same for each transition of the state machine,
- all-events: the same for each event in the alphabet, which is used in at least one transition,

---

Algorithm 8: Test generation by Dijkstra's shortest path calculation

---

input : state machine with states S, initial state s_0
output: set of test sequences TS
data : prev[s]: previous node in path from s_0 to s,
          trans[s]: transition from prev[s] to s,
          length[s]: length of shortest path from s_0 to s,
          U is the set of states which have not been treated yet.

U ← S   `// all states initially in U ;`
foreach $s \in$ U **do**
   | prev[s] ← *undef*   `// previous state unknown;`
   | trans[s] ← *undef*;
   | length[s] ← ∞   `// unknown length from s0 to s;`

length[s_0] ← 0   `// length from s0 to s0 is 0;`

while(U ≠ ∅) select $s \in$ U such that length[s] is minimal
(i.e., $\forall s' \in$ U (length[s'] ≥ length[s])) // (initially, $s = s_0$);
remove s from U   `// s has been treated;`
foreach *transition t from s to s'* **do**
   | **if** $s' \in$ U **then**
      | // s' is reached from s via t with length[s] + 1;
      | **if** length[s] + 1 < length[s'] **then**
         | // A shorter path to s' has been found;
         | prev[s'] ← s ;
         | trans[s'] ← t ;
         | length[s'] ← length[s] + 1

TS ← ∅; U ← S;   `// Re-initializing U ;`
while(U ≠ {s_0}) select $s \in$ U such that length[s] is maximal
`// for minimizing the number of test cases;`
create new empty test sequence σ;
while (// (or, equivalently, prev[s] ≠ *undef*))$s \neq s_0$ remove s from U;
insert trans[s] at the front of σ TS ← TS ∪ σ;
return TS;

- depth-n: for each run $(s_0, a_0, s_1, a_1, \ldots, a_{n-1}, s_n)$ of length at most n from the initial state there is a test case containing this run as a subsequence, and
- all-n-transitions: for each run of length n from any state $s \in S$, there is a test case which contains this run as a subsequence (all-2-transitions is also known as all-transition-pairs; all-1-transitions is the same as all-transitions, and all-0-transitions is the same as all-states).

Algorithm 8 constructs a test suite satisfying the all-states criterion. It can be easily modified to the all-transitions and all-events criterion. The all-events criterion can be regarded as a minimum in black-box testing processes. It requires that every input is activated at least once, and every possible output is observed at least once. If there are input actions which have never been tried, we cannot say that the system has been thoroughly tested. If there are specified output reactions which could never be produced during testing, chances are high that something is wrong with the implementation. The all-states and all-transitions criteria are related; clearly, if the state machine is connected, than any test suite satisfying the all-transitions criterion also satisfies the all-states criterion. In technical terms, all-transitions subsumes all-states. Likewise, all-transitions subsumes all-events. In a state machine with n states and m events, the all-transitions criterion can require up to $m *$ $(n-1)$ test cases. For practical purposes, besides the all-transitions criterion often the depth-n criterion is used, were n is set to the diameter of the model, i.e., the length of the longest run in which no state appears twice. Alternatively, n can be set to the (estimated) diameter of the SUT. The criterion all-n-transitions is quite extensive; for $n \geq 3$ this criterion often results in a huge test suite. Clearly, all-n-transitions subsumes depth-n, and all-$(n+1)$-transitions subsumes all-n-transitions.

Automated test generation algorithms strive to produce test suites satisfying a certain coverage criterion. Therefore, the choice of the coverage criterion has significant impact on the particular algorithm and the resulting test suite. However, none of the above criteria uniquely defines a test suite. For the criterion depth-n there is a unique minimal test suite, namely the set of all runs of length n, plus the set of all maximal runs (which end in a state from which there is no transition) of length smaller than n. This set can be easily constructed via depth-first search.

For the other coverage criteria mentioned above, the existence of a minimal test suite can not be guaranteed. For the actual execution of a test suite, its size is an important figure. The size of a test suite can be measured in several ways:

- the number of all events, i.e., the sum of the length's of all test cases,
- the cardinality, i.e., the number of test cases in the test suite,
- the number of input events, or
- a combination of these measures.

At first glance the complexity of the execution of a test suite is determined by the number of all events which occur in it. However, often it is a very costly operation to reset the SUT after one test in order to run the next test; hence it is advisable to minimize the number of test cases in the test suite. Likewise, for manual test execution, the performance of an input action can be much more expensive than the observation of output reactions; hence the number of (manual) inputs must be minimized.

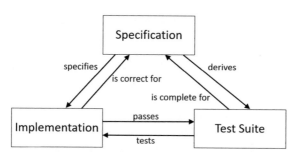

Fig. 5.5 The validation triangle

5.3 Conformance Testing

In Sect. 5.2.1 we remarked that for an SUT which is completely unknown we can never hope to prove that a test suite exercises all parts of the implementation. A main question raised by this observation is the following.

When could we say that a test suite is 'complete'?

To answer this question, we assume for the moment the viewpoint that the intended behaviour of an SUT is entirely defined by the specification. That is, an implementation is *correct* for a specification if and only if it exhibits all of the specified behaviour, and nothing else.

Having fixed a suitable definition of correctness, we can formulate two fundamental properties a test suite should possess:

- each correct implementation should pass the test suite (*Soundness*), and
- each implementation which is not correct with respect to the specification should fail (*Exhaustiveness*).

A test suite which is sound and exhaustive is called *complete* for a specification.[4] In other words, given an implementation im, specification sp and a test suite ts which is complete for sp, im is correct for sp if and only if im passes ts. We depict this so-called "validation triangle" in Fig. 5.5.

Each test suite ts induces an equivalence relation on the set of all implementations: im_1 is *ts-equivalent* to im_2 if im_1 passes ts if and only if im_2 passes T. Given a specification sp, im_1 is *testing-equivalent* to im_2 if for any complete test suite ts for sp, im_1 is ts-equivalent to im_2. Testing equivalence gives a behavioral way to characterize the class of implementations which are correct with respect to a given specification.

Yet, in order to make the definitions precise, we need to clarify what the observable, specified behaviour of an implementation is. When should an SUT "pass" or "fail" a test case? In the previous subsection, we defined model

---

[4] Note that this convention differs slightly from the use of the terms in logic, where a calculus is called *sound* if all provable statements are true, and *complete* if all true statements are provable.

coverage criteria for state machine testing models. That is, the specification consists of a state machine, from which a test suite is derived. We assumed that the observable behaviour is defined by the states and transition of the specification. That is, we assumed that we can observe in which state of the testing model the SUT currently is. A complete test suite would consist, e.g., of all sequences of states and transitions in the specification. A test case passes, if the respective sequence is observable while running the SUT. This assumption leads to a very strict correctness notion: Basically, here an SUT is correct for a specification if and only if its behaviour can be described by a state machine which is isomorphic to the specification, i.e., has the same states and transitions. If the SUT itself is given as a state machine, then it must be isomorphic to the specification. (Here, we disregard the case of two 'identical' transitions connecting the same states with the same trigger, guard and effect.)

However, the assumption of observability of all internal states of an SUT may be too strong. We might, e.g., be only able to observe the transition labels (trigger, guard and effect). In this case, a complete test suite checks if implementation and specification are *trace-equivalent*, i.e., have the same executions. A question which still has to be discussed in this setting is whether we can observe *silent* transitions, which have no trigger and effect.

Or, we might be able to observe whether a trace is *completed* in the specification and/or the implementation. Yet another option is whether we can observe that specification or implementation *refuse* to do certain actions at some stage. This leads to yet another equivalence relation induced by the notion of a complete test suite. There have been many different testing equivalences defined and analyzed in the literature, e.g., [dNH83, Abr87, CH93].

The most general view on testing is that the SUT is a black box to which a tester can send arbitrary inputs, and observe only the outputs produced as a reaction. Thus, for UML, certain triggers are declared to be inputs from the tester to the SUT, and certain actions are declared to be outputs to the tester. On one hand, the tester can send an input to the SUT at any time, but does not know how this input is processed, or whether it is processed at all. On the other hand, the tester can observe the outputs of the SUT when they are produced, but does not know whether the SUT will emit an output or not. Alternatively, we might say that by a suitable timeout mechanism the tester is also able to observe whether the SUT sends an output or not, i.e., whether the SUT is *quiescent*.

In his dissertation and subsequent work [Tre93, Tre96], Tretmans formalized this approach in terms of *Input-Output Transition Systems* (IOTS).

Definition 2 (*Input-Output Transition System, IOTS*) An IOTS **S** is a structure $\mathbf{S} = (S, I, O, \Delta, s_0)$, where

- S is a countable, nonempty set of *states*,
- $L = I \cup O$ is a countable set of *labels* or *observable actions*, where $i \in I$ is an *input* and $o \in O$ is an *output* to the transition system ($I \cap O = \emptyset$),

- $\Delta \subseteq S \times (L \cup \{\tau\}) \times S$ is the transition relation, $\tau \notin L$ is the *silent event*, and
- s_0 is the initial state.

In this definition, the silent event τ stands for some non-observable, internal action of the system. A transition $(s, \mu, s') \in \Delta$ is denoted as $s \xrightarrow{\mu} s'$. A *computation* of the IOTS **S** is a finite sequence of transitions $s_0 \xrightarrow{\mu_1} s_1 \xrightarrow{\mu_2} \cdots \xrightarrow{\mu_{n-1}} s_{n-1} \xrightarrow{\mu_n} s_n$. As a shorthand, we write $s_0 \xrightarrow{\mu_1 \cdots \mu_n} s_n$. The relation $s \xrightarrow{\mu^*} s'$ is naturally extended to arbitrary states $s, s' \in S$. (For the definition of $*$ see Example 2 in Chap. 1). We require that any IOTS contains only *reachable* states, i.e., for all $s \in S$ there exists a computation $s_0 \xrightarrow{\mu^*} s$ ending in s. For technical reasons, we also require that there are no infinite silent computations in a transition system, i.e., no sequences $s_0 \xrightarrow{\tau} s_1 \xrightarrow{\tau} s_2 \xrightarrow{\tau} \cdots$. In particular, this means that there are no τ-loops, i.e., no state s such that $s \xrightarrow{\tau^*} s$. A *trace* is the sequence of non-τ labels of a computation; it captures the observable aspects only. Formally, $trace(s_0 \xrightarrow{a_1 \cdot \mu_2 \cdots \mu_n} s_n) = a_1 \cdot trace(s_1 \xrightarrow{\mu_2 \cdots \mu_n} s_n)$, and $trace(s_0 \xrightarrow{\tau \cdot \mu_2 \cdots \mu_n} s_n) = trace(s_1 \xrightarrow{\mu_2 \cdots \mu_n} s_n)$. The set of traces from state $s \in S$ is denoted by $traces(s)$. The traces of the IOTS (S, I, O, Δ, s_0) are those from the initial state: $traces((S, I, O, \Delta, s_0)) = traces(s_0)$. Two IOTS (S, I, O, Δ, s_0) and $(S', I, O, \Delta', s_0')$ with the same set of labels are called *trace-equivalent*, if $traces(s_0) = traces(s_0')$. Following Tretmans [Tre96], we use the subsequent notation.

$$s \xRightarrow{a} s' \text{ iff } \exists s_1, s_2 (s \xrightarrow{\tau^*} s_1 \wedge s_1 \xrightarrow{a} s_2 \wedge s_2 \xrightarrow{\tau^*} s')$$
$$s \xRightarrow{a_1 \cdots a_n} s' \text{ iff } \exists s_0, \ldots, s_n (s = s_0 \wedge s_0 \xRightarrow{a_1} s_1 \xRightarrow{a_2} \cdots \xRightarrow{a_n} s_n \wedge s_n = s')$$

(Here, as always, 'iff' is short for 'if and only if'.) Thus, $traces(s) = \{\mu^* \in L^* \mid \exists s'(s \xRightarrow{\mu^*} s')\}$. Talking about IOTS, we are not only interested in what the system can do, but also in what it can not do. An IOTS is called *(strongly) input-enabled*, if in any state any input signal can be sent to it. Formally, this holds if for all $s \in S$ and $i \in I$ there is an $s' \in S$ such that $s \xrightarrow{i} s'$. Thus, in an input-enabled IOTS, all inputs are possible in any state. Even if it is input-enabled, a system may *refuse* to give certain outputs after a computation. This is denoted as follows.

$(s \text{ after } \sigma \text{ refuses } A)$ iff $\exists s'(s \xRightarrow{\sigma} s' \wedge \neg \exists \mu \in A, s'' \in S.(s' \xRightarrow{\mu} s''))$.

In testing, we can assume that it is possible to observe that a system refuses to give certain outputs. Two IOTS s and s' are called *testing equivalent*, if for all $\sigma \in L^*$ and $A \subseteq O$ it holds that $(s \text{ after } \sigma \text{ refuses } A)$ iff $(s' \text{ after } \sigma \text{ refuses } A)$. Testing equivalence is a stronger notion than trace equivalence: If two systems are testing equivalent, then they have the same traces. There are efficient graph-search algorithms to determine whether two given IOTS are testing equivalent [Tre08].

Often, we are not interested in the equivalence of two specifications. We are given a specification sp, which is explicitly stated as an IOTS, and an implementation im, which is a black box. Then, im correctly implements sp, if any observable behaviour of im is allowed by sp. A first approach on formalizing this notion is the so-called *trace preorder*:

$$(im \leq_{tr} sp) \text{ iff } (traces(im) \subseteq traces(sp)).$$

However, this definition does not take into respect refusals. An implementation which refuses to do anything would be correct for every specification. Assuming that we can observe the absence of outputs after a sequence of interactions, we may revise this notion. A trace is *quiescent* if it may lead to a state from which the system cannot proceed without inputs from its environment, i.e., a state where it refuses all outputs. Formally, the set of quiescent traces from state s is defined by $Qtraces(s) = \{\sigma \in L^* \mid (s \textbf{ after } \sigma \textbf{ refuses } O)\}$. The *input-output-testing preorder* is defined as follows:

$$(im \leq_{iot} sp) \text{ iff } (im \leq_{tr} sp) \text{ and } (Qtraces(im) \subseteq Qtraces(sp)).$$

We introduce the special label $\delta \notin L$ to denote quiescence. Given any state $s \in S$, we define the *possible outputs* at this state by $out(s) = \{\delta\}$, if there is no $s' \in S$ and $o \in O$ such that $s \xrightarrow{o} s'$, else $out(s) = \{o \in O \mid \exists s' \in S.(s \xrightarrow{o} s')\}$.

Furthermore, the possible outputs after performing trace σ are given by $out(s \textbf{ after } \sigma) = \bigcup\{out(s') \mid s \xrightarrow{\sigma} s'\}$. It is not hard to prove that for input-enabled im and sp,

$$(im \leq_{iot} sp) \text{ iff } \forall \sigma \in L^* (out(im \textbf{ after } \sigma) \subseteq out(sp \textbf{ after } \sigma)).$$

Tretmans defines several other, similar testing preorders. The first one is called *I/O-conformance* and defined by

$$(im \textbf{ ioconf } sp) \text{ iff } \forall \sigma \in traces(sp) (out(im \textbf{ after } \sigma) \subseteq out(sp \textbf{ after } \sigma)).$$

The intuition is that an implementation can conform to a specification, even if the specification is not input-enabled. If, for example, $\sigma \notin traces(sp)$, where $\sigma = \sigma'i$ and $i \in I$, then $out(sp \textbf{ after } \sigma) = \emptyset$, whereas $out(im \textbf{ after } \sigma)$ could be nonempty. Thus, $(im \leq_{iot} sp)$ would not hold, whereas $(im \textbf{ ioconf } sp)$ could still be the case. Hence $(im \textbf{ ioconf } sp)$ does not imply $(im \leq_{iot} sp)$. On the other hand, $(im \leq_{iot} sp)$ clearly implies that $(im \textbf{ ioconf } sp)$.

I/O-conformance assumes that the tester can observe quiescence at the end of a test. By a suitable timeout mechanism, it may also be possible to observe quiescence during the execution of a test case. In order to formalize an appropriate testing preorder, Tretmans introduces *suspension transitions*. Given an IOTS $\mathbf{S} = (S, I, O, \Delta, s_0)$, we define the *suspension transition system* $\mathbf{S}^\delta = (S, I, O^\delta, \Delta^\delta, s_0)$, where $\delta \notin L$ is the new symbol for quiescence, $O^\delta = O \cup \{\delta\}$, and $\Delta^\delta = \Delta \cup \{(s, \delta, s) \mid \neg \exists o \in O, s' \in S.(s \xrightarrow{o} s')\}$. That is, if state s is quiescent in $\mathbf{S}$, then $out(s) = \{\delta\}$ in $\mathbf{S}^\delta$. With this definition, the *suspension traces* of $\mathbf{S}$ are just the traces of $\mathbf{S}^\delta$. The testing preorder $\textbf{ioco}$ is defined via the suspension transition system:

$$(im \textbf{ ioco } sp) \text{ iff } \forall \sigma \in traces(sp^\delta) (out(im^\delta \textbf{ after } \sigma) \subseteq out(sp^\delta \textbf{ after } \sigma)).$$

Intuitively, $(im \textbf{ ioco } sp)$, if and only if

- if *im* may produce output *o* after trace σ, then *sp* can produce *o* after σ, and
- if *im* cannot produce any output after trace σ, then *sp* cannot produce any output after σ.

Since $\mathbf{S}^\delta$ is an 'enhanced' version of $\mathbf{S}$, the **ioco** relation is finer than **ioconf**: (*im* **ioco** *sp*) implies (*im* **ioconf** *sp*), but not vice versa.

Example 49.4: VCR Switch as IOTS

In order to demonstrate some of these concepts, we give an abstract model of a switch related to our VCR switch Example 49.1 (see also Fig. 5.3). The abstract model has only two modes: off and on. The events dn? and up? turn the device on and off, and trigger the LED settings 100! and 000!, respectively. In the on mode, further dn? events change the settings of the LEDs (amongst other things). Since in contrast to UML state charts, an IOTS has no complex transition labels, we have to insert additional states. This leads to the following specification *sp_abstract*.

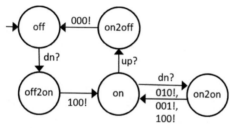

This specification is not input-enabled; however, we could easily make it input-enabled by declaring that unexpected inputs are simply neglected. That is, for every state $s \in S$ and $i \in I$ such that there is no transition $s \xrightarrow{i} s'$, we add a transition $s \xrightarrow{i} s$.

In the next step, we can refine the on mode to include minor modes. This results in the specification *sp_refined*.

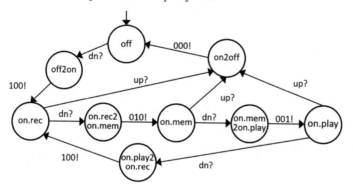

Now, the following can be checked:

- $traces(sp\_refined) \subseteq traces(sp\_abstract)$.
- $(sp\_refined \leq sp\_abstract)$ for $\leq\, \in \{\leq_{ior}, \leq_{iot}, \mathbf{ioconf}, \mathbf{ioco}\}$.
- None of the inverse relations holds.

Here, we consider $sp\_refined$ as an implementation for the specification $sp\_abstract$, and vice versa. This is justified, as we are seeing an implementation as a black box—and, naturally, an IOTS can serve as a black box.

For a further negative example, consider that we add a transition on $\xrightarrow{\tau}$ on2off to $sp\_abstract$. This could model, e.g., a battery failure, when the machine automatically shuts down and switches off all LEDs. Now, even though trace-preorder still holds, the refined specification no longer conforms to this modified abstract specification.

The **ioco** relation can be used to define test sets for a given specification and black-box implementation. As mentioned above on Sect. 5.2.2, we assume that the specification is *deterministic* in the sense that for any $s \in S$ and $a \in L$ there is at most one s' such that $s \xRightarrow{a} s'$. (Note that both $sp\_abstract$ and $sp\_refined$ in the above example are deterministic.)

A *test case* for the IOTS $\mathbf{S} = (S, I, O, \Delta, s_0)$ is a finite tree, where the leafs are labelled by $\{pass, fail\}$, and the non-leaf nodes are labelled by states from S. Formally, a test case for $\mathbf{S}$ is a finite transition system $\mathbf{T} = (S^T, O \cup \{\delta\}, I, \Delta^T, s_0^T)$ such that

1. for every $s \in S^T$ there is a unique finite path from s_0^T to s (tree property),
2. there is a mapping $\iota: S^T \to (S \cup \{pass, fail\})$ with $\iota(s_0^T) = s_0$,
3. if $\iota(s^T) = s$ and $out(s) = \{o_1, \ldots, o_n\}$, where $s \xRightarrow{o_i} s_i$, then s^T has at least n children $s_1^T, \ldots, s_n^T$ such that $s^T \xrightarrow{o_i} s_i^T$ and $\iota(s_i^T) = s_i$ or $\iota(s_i^T) = pass$.
4. if $\iota(s^T) = s$ and $o \in (O \setminus out(s))$, then there is a child $s' \in S^T$ such that $s^T \xrightarrow{o} s'$ and $\iota(s') = fail$.
5. if $\iota(s^T) = s$, then for at most one $i \in I$ such that $s \xRightarrow{i} s_i$, there is a child $s_i^T \in S^T$ such that $s^T \xrightarrow{i} s_i^T$ and $\iota(s_i^T) = s_i$.
6. if $\iota(s^T) = s$ and $out(s) \neq \{\delta\}$, then s^T has a child s^T such that $s^T \xrightarrow{\delta} s_i^T$ and $\iota(s_i^T) = fail$.

A test case thus is a partial 'unfolding' of the specification, and is used to test the implementation. In the test case, the role of input and output is reversed with respect to the specification: each output of the specification (and, thus, the system under test) is an input for the test case, and vice versa. Clause (3) of the above asserts that if the specification/implementation allows to produce an output μ, then the test case can consume this μ as an input. Clause (4) guarantees that an output which is not allowed by the specification, if produced by the implementation, leads to a failed test case.

Clause (5) allows to send an input to the system under test from the test case, which is foreseen by the specification. With clause (6), it is guaranteed that the implementation may not deadlock in a situation where the specification allows an output. This reflects the assumption that a non-reaction of the system under test can be observed via a timeout-mechanism.

Example 49.5: A Test Case for the VCR Switch as IOTS

Consider the example IOTS from above. A test case for this specification could be, e.g., the following.

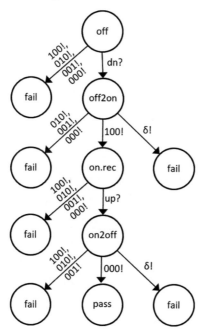

Informally, this test case sends a dn event to the SUT, checks whether it produces the LED pattern 100, and then sends an up and checks for 000. If the SUT shows a wrong LED pattern or does not react at all, the test fails

---

For executing such a tree test case with a black-box implementation, the test harness can

- send an input to the implementation (according to Clause (5), there is at most one possibility to do so),

- wait for an output or timeout from the implementation and proceed accordingly, or
- pass or fail the test execution if a leaf is reached.

Algorithm 9: Conformance test generation

---

input : IOTS S, initial state s_0
output: Test case (tree) T
data : U the set of tree nodes which have not been treated yet.

Start with a one-node tree T consisting of a root s_0^T labelled with s_0;
$U \leftarrow \{s_0^T\}$;
while$(U \neq \emptyset)$ select $s^T \in U$ and let s be the label of s^T;
remove s^T from U;
foreach $o \in O$ **do**

  create a child c^T of s^T such that $s^T \xrightarrow{o} c^T$;
  if $s \xRightarrow{o} s'$ **then**
    begin
      label c^T with s' and put it in U;
    or
      label c^T with **pass**
  else label c^T with **fail**;

if $\exists o \in O, s' \in S.(s \xRightarrow{o} s')$ **then**
  create a child c^T of s^T such that $s^T \xRightarrow{o} c^T$, label c^T with **fail**

if $\exists i \in I, s' \in S.(s \xRightarrow{i} s')$ **then**
  begin
    select some $i \in I$ such that $(s \xRightarrow{i} s')$;
    create a child c^T of s^T such that $s^T \xrightarrow{i} c^T$;
    label c^T with s' and put it in U;
  or
    do nothing

return T;

Clauses (3)–(6) from above can be seen as a (nondeterministic) procedure for generating tests from a specification. A pseudocode formulation of this procedure is given in Algorithm 9. In the algorithm, each of the possible resolutions of the **or** statement gives rise to a different test case. It can be shown that the set of all test cases which can be produced according to this procedure is complete (i.e., sound and exhaustive, cf. Fig. 5.5). That is, if *ts(sp)* is the set of test cases which is derived from specification *sp* with Algorithm 9, then any implementation *im* passes *ts(sp)* if and only if (*im* **ioco** *sp*).

Even if the number of states in a specification *sp* is finite, there may be infinitely many test cases for *sp*, if *sp* contains at least one cycle. That is, in general the size of the complete test suite *ts(sp)* is infinite. This holds since we may cover each state on the cycle an arbitrary number of times. The question arises, which number of test cases is 'sufficient' to establish that an implementation *im* conforms to *sp*? We have remarked above, that without any knowledge of the internal structure of the SUT, we can never be sure that we have covered all relevant parts. However, there are reasonable assumptions which can be made on this structure. One such assumption is that the SUT can faithfully be represented by a finite IOTS. That is, for any unknown system under test, there exists a (still unknown) finite IOTS which exhibits the same behaviour. The length of the maximal path in which no state is repeated is called the *diameter* of the IOTS. Thus, the basic testing hypothesis is that each SUT can be represented by an IOTS which has a finite diameter.

Another assumption which in some cases is reasonable is that the SUT is deterministic, i.e., repeatedly given the same input sequence from the initial state will drive the SUT through the same sequence of internal states and transitions. An observation to be made about this is that if *sp* is finite and acyclic, then only finitely many test cases can be derived from it. Based on this observation, for deterministic implementations, test case selection can be based on unwinding of the specification. In particular, Simao and Petrenko [dSSP14] assume the following:

1. the implementation has at most as many states as the specification,
2. if in any state there is a conflict between an input and an output, then the input is selected (so-called *input-eagerness*).

It is shown that under these assumptions, a finite and complete test suite for **ioco** can be derived by unwinding the specification.

5.4 Using Algebraic Specifications for Testing

A different approach to specification-based testing is via abstract data types. Whereas the above automata-based approach seems to be more oriented towards reactive systems, the algebraic approach is more oriented towards functional computations. The CASL language, see Chap. 4, allows to specify a program by a first-order signature, i.e., with functions and relations, plus additional first-order axioms. In Sect. 4.2.4 we described how to test Java implementations against algebraic specifications. In this section, we elaborate on this approach.

Example 50: Days-in-Month Function

As an example, consider the problem of determining the number of days in a month. That is, the function *"dim"* ("days in month") takes as input a month, i.e., a natural number between 1 and 12, and a year in the Gregorian calendar (i.e., a natural number greater than 1582). It calculates the number of days in that particular month. This result can be used, e.g., to check whether a given birth date is valid or not.

For the definition of the *dim*-function, we need an auxiliary predicate stating whether a given year is a leap year. A year is a leap year if it is evenly divisible by 4; however, if the year can be evenly divided by 100, it is not a leap year, unless the year is also evenly divisible by 400.

To specify this example in CASL, we rely on the sorts NAT and LIST from the standard library. *Month* and *Year* can be defined as subsorts of NAT. *isLeapYear* then is a predicate and *dim* an operation in this specification.

from *Basic/Numbers* **get** NAT
from *Basic/StructuredDatatypes* **get** LIST

spec DAYSINMONTH = NAT **and** LIST[NAT **fit sort** *Elem* $\mapsto$ *Nat*]
then sort   *Month* = $\{n : Nat \bullet 1 \leq n \land n \leq 12\}$
     sort   *Year* = $\{n : Nat \bullet 1583 \leq n\}$
     pred *isLeapYear* : *Year*
     $\forall y : Year \bullet isLeapYear(y)$
          $\Leftrightarrow (y \ mod \ 4 = 0 \land y \ mod \ 100 > 0) \lor y \ mod \ 400 = 0$
     op   *dim* : *Month* $\times$ *Year* $\to$ *Nat*
     $\forall m : Month; y : Year$
         • $m \ \epsilon \ [\ 1, 3, 5, 7, 8, 10, 12\ ] \Rightarrow dim(m, y) = 31$
         • $m \ \epsilon \ [\ 4, 6, 9, 11\ ] \Rightarrow dim(m, y) = 30$
         • $isLeapYear(y) \Rightarrow dim(2 \ as \ Month, y) = 29$
         • $\neg \ isLeapYear(y) \Rightarrow dim(2 \ as \ Month, y) = 28$
end

There are many different ways to implement such a *dim*-function (and many ways to implement it incorrectly). For example, you might want to try the term (cf. [McE14])

$$28 + (m + \lfloor m/8 \rfloor) \ \% \ 2 + 2 \ \% \ m + 2 * \lfloor 1/m \rfloor + (m == 2) * isLeapYear(y)$$

Assume that we are given a program realizing the *dim*-function. However, the program is given as a binary file only, that is, we do not have access to the source code. Thus, we can not verify whether it conforms to the above CASL specification DAYSINMONTH. How could we test this implementation? Which test cases would be appropriate? Which tests would be sufficient? When does a test case pass or fail?

For programs realizing a function, a test case is considered to be a tuple of input values for the function parameters. The expected output is determined by the test oracle. Which tuples are to be considered for testing the *dim*-function? Typical examples would be, e.g., the tuples of input values $(1, 1970)$, $(12, 2000)$, and $(6, 999999)$. Clearly, since the year is given as an integer, there are infinitely many possibilities. One could argue whether a date with a year smaller than 1583 constitutes a valid test case. The specification does not allow such inputs, *dim* is a total function on its input domains. Yet, the infinite complexity still remains. If we restrict attention to a finite interval (say, $[1583..MAX\_VALUE]$ with $MAX\_VALUE = 2^{31} - 1$), there are $12 * (2^{31} - 1584) \approx 2.5 * 10^{10}$ test cases. However, depending on the programming environment, even this large set might be too restricted: In languages like C and Java, we can also input numbers greater than MAX_VALUE, and the result is an integer according to the internal binary encoding. Moreover, some environments allow to pass to a function arguments of wrong type, or more arguments than specified. In such settings, e.g., $dim(-1, "abc")$ and $dim(12, 2000, 17)$ would be valid test cases.

Similar to state-based coverage criteria, data-based test coverage criteria have been defined. A typical method is to split a linearly ordered domain into ranges, and select values from the boundaries of these partitions. For example, it would be possible to split the integer range into $(MIN\_VALUE, 0)$, $[0, 1582)$, and $[1583, MAX\_VALUE)$ to test the parameter "Year". Another method would be to partition a data domain into equivalence classes, and to make sure that at least one representative is selected from each equivalence class. In our example, we could split the domain *Month* into $\{1, 3, 5, 7, 8, 10, 12\}$, $\{4, 6, 9, 11\}$ and $\{2\}$, and choose one month of each partition. Yet another method uses classification trees, where data values are categorized according to different attributes, and representatives from each classification are chosen [GG93]. Here, test cases would be, e.g., "a February in a leap year", "an invalid month in a negative year", etc.

Though testing using such coverage criteria is often applied in practice, this form of testing should not be considered formal: it has little mathematical foundations. In contrast to this, in a seminal paper Gaudel describes a theoretical foundation for test case generation from algebraic specifications [Gau95]. In this approach, test cases are ground instances of the specification axioms.

In logic, a *ground term* is a term without variables, i.e., a term which is formed by using only function and constant symbols (compare Definition 17).

Example 51: Ground Terms in a Specification of N

In the context of our above CASL specification DAYSINMONTH, we import a specification of natural numbers NAT. The specification includes the CASL code:

free type $Nat ::= 0 \mid suc(Nat)$

This line of code declares the constant symbol 0 and a unary function symbol suc. With these symbols, we can form, e.g., the ground terms 0, $suc(0)$ and $suc(suc(0))$. Using additional constants and functions together with parsing annotations, CASL allows to represent such ground terms in the more usual, decimal representation as 0, 1, and 2—see above.

Given a universally quantified logical formula $\varphi = \forall x_1, \ldots, x_n\ \psi$, a *ground instance* of φ is obtained by consistently substituting each occurrence of a variable x_i in ψ by a ground term t_i.

For example, consider the formula

$$\forall\ m : Month;\ y : Year \bullet m \in [\ 4,\ 6,\ 9,\ 11\ ] \Rightarrow dim(m,\ y) = 30$$

Ground instances of this formula are, e.g.,

- $0 \in [\ 4,\ 6,\ 9,\ 11\ ] \Rightarrow dim(0,\ 0) = 30$
- $1 \in [\ 4,\ 6,\ 9,\ 11\ ] \Rightarrow dim(1,\ 2000) = 30$
- $4 \in [\ 4,\ 6,\ 9,\ 11\ ] \Rightarrow dim(4,\ 2001) = 30$

The first of these instances has a mistake in the types of the arguments m and y. In the CASL semantics, 0 *as* $Month$ and 0 *as* $Year$ are undefined, since $Month$ and $Year$ are subtypes of Nat which do not include 0. The result of calling a function on an undefined value is undefined. Therefore, $dim(0, 0)$ is undefined. In the second instance, the antecedent evaluates to $false$. Thus, the formula is $true$ regardless of the result of $dim(1, 2000)$, and there is nothing to test. However, the third instance leads to a 'meaningful' test: We can call the function dim in the implementation with the arguments 4 and 2001, and check whether the result is 30.

In general, the goal of specification-based testing is to determine whether a given specification sp is correctly implemented by a (black-box) system under test im. Subsequently, we discuss what this means if sp is an algebraic specification. Usually, an algebraic specification language like CASL consists of formulae in some extended first-order logic. Recall that a first-order signature $\Sigma = (\mathcal{F}, \mathcal{R}, \mathcal{V})$ consists of a set of function symbols $\mathcal{F}$, relation symbols $\mathcal{R}$, and variable symbols $\mathcal{V}$, see Definition 8. (Predicates are unary relations, and constants are 0-ary functions.) Consider an algebraic specification sp in the signature Σ, and assume that $\mathcal{F}^{sp} \subseteq \mathcal{F}$ and $\mathcal{R}^{sp} \subseteq \mathcal{R}$ are the functions and relations *specified* by sp. A *specification formula* is a universally quantified formula $\varphi = \forall x_1, \ldots, x_n\ \psi(x_1, \ldots, x_n)$ in the specification language. In testing from algebraic specifications, a *test case* for $\mathcal{F}^{sp} \subseteq \mathcal{F}$ and $\mathcal{R}^{sp} \subseteq \mathcal{R}$

is defined to be any ground instance $\psi(t_1, \ldots, t_n)$ of such a specification formula.

For such a test case to be executable, the system under test *im* must provide an implementation f_im for each function $f \in \mathcal{F}^{sp}$, and p_im for each relation in $p \in \mathcal{R}^{sp}$. Here, 'provide' means that the function or relation can be called from the tester with appropriate arguments, and returns a result. In the case of a relation, the result is a boolean value. If the call of the function or relation causes the implementation to go into an infinite loop or throw an error, the result is undefined. (Detecting an infinite loop can be approximated by a suitable time-out mechanism, see also Sect. 5.3 above). In our example, the specification defines the predicate *isLeapYear* and function *dim*. The implementation must provide isLeapYear and dim.

There are also other symbols occurring in the specification formulae, e.g., $f \in (\mathcal{F} \setminus \mathcal{F}^{sp})$, $p \in (\mathcal{R} \setminus \mathcal{R}^{sp})$, boolean connectors, equality, and maybe other primitives of the specification language. These must be available in the testing framework (e.g., JUnit), independent of the system under test. In our example, these other functions include the constants 0, 1, 2001, the *mod* function, testing whether an element is in a list, etc.

Even if the testing framework provides all these functionalities, it may not be able to evaluate all kinds of formulae. If the test case, which is a ground instance of a specification formula, contains free variables, then it is unclear which values to assign to them. Moreover, if the test case contains an existentially quantified subformula, it may be hard to decide whether it holds or not. For example, consider the formula
$$\forall x \exists y (y > x \wedge prime(y) \wedge prime(y+2)).$$
Ground instances of this formula are, for every $n \in \mathbf{N}$
$$\exists y (y > n \wedge prime(y) \wedge prime(y+2)).$$
In order to evaluate such a formula, the tester would have to construct a prime twin bigger than n, which is trivial for $n = 10$, somewhat hard for $n = 10^6$, and a currently open challenge for $n = 10^{10^6}$. Therefore, usually it is required that the test case is a quantifier- and variable-free formula of the specification logic. Such test cases can be evaluated by Algorithm 10.

Here are a few remarks on this algorithm.

- Formally, constants are 0-ary functions; thus, no special clause needs to be given in the algorithm. Similar, as in classical propositional logic, it is sufficient to give evaluation rules for $\perp$ and implication. All other Boolean operators can be defined from these two, see Sect. 2.2.
- The algorithm is closely related to the semantics of first-order terms and predicates (Definitions 13 and 14). Thus, many authors consider the implementation to be a first-order model of a specification. We will come back to this point below.
- We evaluated equality of terms as "equal for the test system". That is, if the test system can not observe any difference in their values, the terms are considered to be equal. In some languages like Java or C, there is a

---

Algorithm 10: Evaluation of a test case in an SUT

Given symbols $f \in \mathcal{F}^{sp}$ and $p \in \mathcal{R}^{sp}$, let $\mathbf{f}_{im}$ and $\mathbf{p}_{im}$ be the implementation of f and p in the system under test;
and for $f \in (\mathcal{F} - \mathcal{F}^{sp})$ and $p \in (\mathcal{R} - \mathcal{R}^{sp})$, let f_{tf} and p_{tf} be the function f and predicate p in the testing framework, respectively

function $\texttt{eval\_term}_{im}(t)$
input : Term t, where $t = f(t_1, \ldots, t_n)$
result : Value of t, or *undef* if no value can be obtained

let $et_1 = \texttt{eval\_term}_{im}(t_1)$, ..., $et_n = \texttt{eval\_term}_{im}(t_n)$;
if one of et_1, ..., et_n is *undef* **then return** *undef*
else if $f \in (\mathcal{F} \setminus \mathcal{F}^{sp})$ (i.e., f is 'built-in') **then**
$\quad$ **return** $f_{tf}(et_1, \ldots, et_n)$

else if $f \in \mathcal{F}^{sp}$ (i.e., f is implemented in the SUT) **then**
$\quad$ call $\mathbf{f}_{im}$ in the SUT with parameters $(et_1, \ldots, et_n)$;
$\quad$ **if** call has a timeout or exception **then return** *undef*
$\quad$ **else return** the result obtained by this call

function $\texttt{eval}_{im}(\psi)$
input : Quantifier- and variable-free formula ψ
result : Evaluation of ψ (*true* or *false*)

if $\psi = \bot$ **then return** *false*
else if $\psi = (\psi_1 \implies \psi_2)$ **then**
$\quad$ **if** $\texttt{eval}_{im}(\psi_1) = \textit{false}$ **then return** *true* **else return** $\texttt{eval}_{im}(\psi_2)$

else if $\psi = p(t_1, \ldots, t_n)$ **then**
$\quad$ **let** $et_1 = \texttt{eval\_term}_{im}(t_1)$, ..., $et_n = \texttt{eval\_term}_{im}(t_n)$;
$\quad$ **if** one of et_1, ..., et_n is *undef* **then return** *false*
$\quad$ **else if** $p \in (\mathcal{R} \setminus \mathcal{R}^{sp})$ *(i.e., p is 'built-in')* **then**
$\quad\quad$ **return** $p_{tf}(et_1, \ldots, et_n)$

$\quad$ **else if** $p \in \mathcal{R}^{sp}$ *(i.e., p is implemented)* **then**
$\quad\quad$ call $\mathbf{p}_{im}$ in the SUT with parameters $(et_1, \ldots, et_n)$
$\quad\quad$ **if** call has a timeout or exception **then return** *false*
$\quad\quad$ **else return** the result obtained by this call

else if $\psi = (t_1 = t_2)$ **then**
$\quad$ **let** $et_1 = \texttt{eval\_term}_{im}(t_1)$ and $et_2 = \texttt{eval\_term}_{im}(t_2)$;
$\quad$ **if** both et_1 and et_2 are defined and equal **then return** *true*
$\quad$ **else if** both et_1 and et_2 are *undef* **then return** *true*
$\quad$ **else return** *false*

---

distinction between a reference to an object (a pointer) and the object itself (e.g., a list). That is, two different references might denote the same object, and the same reference might denote two different objects. Here, we defer the decision of what is considered equal to the test system.

- With respect to undefined values, the evaluation algorithm is based on the semantics of CASL. There are other choices. In particular, in CASL all functions are strict: they return undefined if any argument is undefined. Most programming languages include non-strict operations, where the result is determined by evaluating arguments only when needed. That is, in these languages the function

$$foo(x, y) = \textbf{if} \ (x == 0) \ \textbf{then} \ y \ \textbf{else} \ x$$

is *not* the same as the CASL function *bar* specified by

- $(x = 0) \Rightarrow bar(x, y) = y$
- $\neg(x = 0) \Rightarrow bar(x, y) = x$

since for y undefined, $foo(1, y)$ yields 1 whereas $bar(1, y)$ yields undefined.

- The type system of most specification logics (including CASL) is rather rigid, whereas the type system of the programming environment may be more liberal. For example, in Java, one could implement the *dim* function by a method `int dim(int m, int y)` with integer arguments. This would allow the calls `dim(0,0)` and `dim(-1, 1582)` with parameters outside the range of *Month* and *Year*. As another example, in C, one could write a function `int fun(int n, ...)` which takes a variable number of arguments. Legal function calls would then be, e.g., `fun(1, 2)` and `fun(3, 1, 2, 3)`. In testing from algebraic specifications, such calls, however, will never be issued by the testing framework, as they are no well-formed terms of the specification logic. In other words, robustness testing is not in the scope of this approach.

The *test verdict* is the result of the execution of a test case with a particular system under test. Our algorithm for evaluating a specification formula in the implementation allows to define the test verdict in a natural way. A test case, i.e., a ground instance $\psi(t)$ of a specification formula $\forall x \ \psi(x)$, yields the verdict `pass` for an implementation *im*, if the evaluation $eval_{im}(\psi(t))$ results in `true`. Otherwise it yields `fail`. Therefore, a method for testing from algebraic specifications is to obtain test cases as ground instances from algebraic specifications and to evaluate these with the implementation.

However, this method only makes sense if the implementation follows a certain basic principle. Consider the following (faulty) Java implementation of the function *isLeapYear*:

```
static boolean x = true;
static boolean isLeapYear (int y){
    return ((y%400==0)?true:(y%4==0)?(y%100!=0):(x=!x));}
```

For this implementation, the statement

```
System.out.print(isLeapYear(2001) == isLeapYear(2001));
```

prints `false` to standard output. This is because the implementation of *isLeapYear* has a *side-effect* in `x`: multiple calls of *isLeapYear* with identical arguments might yield different results. However, in the specification the ground equation $isLeapYear(2001) = isLeapYear(2001)$ is always true. Therefore, the system under test must not only provide implementations of all functions and relations in $\mathcal{F}^{ps}$ and $\mathcal{R}^{sp}$, respectively, but these implementations must behave like mathematical functions and relations. This is reflected in the following assumption.

Fundamental testing assumption   When testing a program against an
algebraic specification, the result of evaluating a function or predicate
depends only on the actual parameters provided. That is, a function/pred-
icate always yields the same result when called with the same values, independent of the context.

(Note that in programming language theory, the property that a function has
no side-effects is sometimes called *referential transparency*.)

An algebraic specification *sp* in the signature $\Sigma = (\mathcal{F}, \mathcal{R})$ (with $\mathcal{V} = \emptyset$),
specifies functions $\mathcal{F}^{sp} \subseteq \mathcal{F}$ and predicates (relations) $\mathcal{R}^{sp} \subseteq \mathcal{R}$. The system
under test provides implementations for all function symbols in $\mathcal{F}^{sp}$ and for
all relation symbols in $\mathcal{R}^{sp}$. For test execution, the test system must provide
functions and relations for $(\mathcal{F} \setminus \mathcal{F}^{sp})$ and $(\mathcal{R} \setminus \mathcal{R}^{sp})$, respectively. Thus, the
fundamental testing assumption must hold for both, the test system and the
system under test.

The fundamental testing assumption can only hold if implementation and
test system are deterministic. That is, this testing approach is not suited,
e.g., for the test of parallel programs, where the result of a function call may
depend on the specific interleaving of threads in the scheduler. It is possible
to extend the theory such that this case can be handled as well.

Furthermore, in testing, we assume that the test system itself is correct.
The question how to test a test system is out of scope for the present exposition. However, we need to define what it means for an implementation *im*
to be correct with respect to an algebraic specification *sp*. Recall the notion
of a first-order model $\mathcal{M}$ for a signature $\Sigma = (\mathcal{F}, \mathcal{R}, \mathcal{V})$ from Sect. 2.4.1. This
is a structure $\mathcal{M} = (U, \mathcal{I}, \mathbf{v})$ consisting of a nonempty set U, the universe of
discourse, an interpretation $\mathcal{I}$ for function and relation symbols, and a vari-
able valuation $\mathbf{v}$. We define the first-order model $\mathcal{M}_{im} = (U, \mathcal{I}, \mathbf{v})$ associated
with the implementation *im* as follows:

- The universe U consists of all results which could possibly be obtained by
  the test system (either by using functions of the test system itself or by
  calling functions from the implementation):
  $$U = \{x \mid \exists \text{ ground term } t.(\texttt{eval\_term}_{im}(t) = x)\}.$$
  From this, it follows that U usually contains the undefined value *undef*.
- The interpretation $\mathcal{I}(f)$ of a function symbol is the function implemented
  in the implementation *im* or in the test system:
  $$\mathcal{I}(f)(t_1, \ldots, t_n) = \texttt{eval\_term}_{im}(f(t_1, \ldots, t_n)).$$
  Similarly, the interpretation $\mathcal{I}(p)$ of a relation symbol is the relation imple-
  mented in *im* or in the test system:
  $$\mathcal{I}(p)(t_1, \ldots, t_n) \Leftrightarrow \texttt{eval}_{im}(p(t_1, \ldots, t_n)).$$
- Since an algebraic specification has no free variables, we can select any
  $u \in U$ and let $\mathbf{v}(x) = u$ for all $x \in \mathcal{V}$.

Lemma 1 *(a) For $\mathcal{M}_{im} = (U, \mathcal{I}, \mathbf{v})$ as defined above and any ground term t
it holds that $\mathcal{I}(t) = \texttt{eval\_term}_{im}(t)$. (b) For any ground formula ψ it holds
that $\mathcal{M}_{im} \models \psi$ if and only if $\texttt{eval}_{im}(\psi) = true$.*

This lemma follows from the construction of $\mathcal{M}_{im}$ and the definition of the evaluation algorithm $\texttt{eval\_term}_{im}$ and $\texttt{eval}_{im}$. It relies on the fact that with respect to undefined values, the algorithm reflects the semantics of terms and predicates in CASL. Formally, Lemma 1 can be shown by inductions on t and ψ. Intuitively, it states that im passes exactly those tests which hold in $\mathcal{M}_{im}$.

We can now formulate the notion of correctness of an implementation w.r.t. an algebraic specification. Let $Ax(sp)$ denote the set of all axioms (specification formulae) in sp, as in Definition 1 from Chap. 4.

Definition 3 Implementation im is defined to be correct w.r.t. the algebraic specification sp, if the corresponding model $\mathcal{M}_{im}$ is a model of the specification, i.e., if $\mathcal{M}_{im} \models Ax(sp)$.

Here are some remarks on this definition.

- Since the model associated with an implementation depends on the possible observations which the test system can make, correctness also depends on the test system.
- Our notion of correctness is consistent with the usual notion of refinement between specifications: Specification sp' is a *refinement* of specification sp, if all models of sp' are also models of sp, i.e., if $\{\mathcal{M} \mid \mathcal{M} \models Ax(sp')\} \subseteq \{\mathcal{M} \mid \mathcal{M} \models Ax(sp)\}$. With the implementation im, there is associated the model $\mathcal{M}_{im}$. Now im is correct w.r.t. sp according to our definition, if and only if $\mathcal{M}_{im} \in \{\mathcal{M} \mid \mathcal{M} \models Ax(sp)\}$, that is, if and only if im is a refinement of sp.
- Our above example specification DAYSINMONTH has (up to isomorphism) just one model $\mathcal{M}_{\text{DAYSINMONTH}}$. Therefore, any implementation im of DAYSIN-MONTH is correct, if $\mathcal{M}_{im}$ is isomorphic to $\mathcal{M}_{\text{DAYSINMONTH}}$. However, if we would omit any of the four defining clauses of the dim function, there would be many non-isomorphic models. Hence there would also be many different implementations of this function which are correct. Any implementation which is correct for the original DAYSINMONTH is also correct for such a "relaxed" specification.
- In our definition of $\mathcal{M}_{im}$, the universe consists of all values which can be obtained by the evaluation of a ground term. As a border case, in CASL a specification may have no ground terms at all.

Example 52: Semigroup in CASL

Consider the specification of a semigroup in CASL.

spec SEMIGROUP =
    sort S
    op $\_\_ + \_\_ : S \times S \to S$
    $\forall\, x,\, y,\, z : S \bullet x + (y + z) = (x + y) + z$
end

> Since this specification has no constant symbols, its set of ground terms is empty.

In this case, $\mathcal{M}_{SemiGroup}$ is not well-defined, since the universe U would be the empty set. In specification logics, and in particular in CASL, this is not allowed—for a discussion see Sect. 4.2. Therefore, our definition of correctness of an implementation w.r.t. spec SemiGroup is meaningless. We could as well say that *any* implementation is correct w.r.t. this specification, since there are no ground formulae to test it. Nevertheless, it can be useful to write such a specification. Many data structures in computer science are associative, and proofs about the correctness of a program or program transformations might rely on this fact. In Sect. 4.2.4, we show how to automatically generate random test cases from a specification, which are not based on ground terms. This random testing approach can be used to test implementations of, e.g., SemiGroup. Here, we are aiming at the development of complete test suites.

Recall from Sect. 5.3 above that a test suite *ts* is called *complete*, if it is sound and exhaustive. It is *sound* if each correct implementation passes, and *exhaustive* if each incorrect implementation fails. Fundamental questions in testing are whether complete test suites exist and how they can be constructed. Whether a *finite* complete test suite exists is an additional question.

In test generation from algebraic specifications, a test case is a ground instance of a universally quantified specification formula. Let the 'full' test suite $ts_{full}(sp)$ of specification sp be the set of all these ground instances. Assuming that the set of ground terms is nonempty, we may ask: Does $ts_{full}(sp)$ constitute a complete test suite?

Of course, the answer to this question depends on the type of formulae which are allowed in the specification. It turns out that if we restrict our attention to universally quantified specification formulae $\forall x_1, \ldots, x_n \, \psi$, where ψ is quantifier-free, we can prove completeness of $ts_{full}(sp)$:

Theorem 1 *im is correct w.r.t. sp, if and only if im passes $ts_{full}(sp)$.*

Intuitively, this theorem holds since in term-generated models, a universally quantified formula $\forall x_1, \ldots, x_n \, \psi(x_1, \ldots, x_n)$ is equivalent to the (possibly infinite) set of formulae $\{\psi(t_1, \ldots, t_n) \mid t_1, \ldots, t_n \text{ are ground terms}\}$. However, since the proof is not self-evident, we give a sketch.

Proof Soundness ("only if") is more or less obvious, since it follows from the instantiation principle of first-order logic (see Sect. 2.4.1): $\models (\forall x \, \psi(x) \Rightarrow \psi(t))$. According to Definition 3 above, *im* is correct w.r.t. *sp*, if and only if $\mathcal{M}_{im} \models Ax(sp)$. Assume $\mathcal{M}_{im} \models Ax(sp)$, and let $T \in ts_{full}(sp)$ be a test case. We have to show that *im* passes T. We know that there exists a formula $\varphi = \forall x_1 \ldots x_n \, \psi(x_1, \ldots, x_n) \in Ax(sp)$, such that T is a ground instance $\psi(t_1, \ldots, t_n)$ of φ. Since $\mathcal{M}_{im} \models Ax(sp)$, in particular it holds that

$\mathcal{M}_{im} \models \varphi$. It follows that $\mathcal{M}_{im} \models \psi(t_1, \ldots, t_n)$. According to the definition of $eval_{im}$, this means that $eval_{im}(\psi(t_1, \ldots, t_n)) = true$, i.e., im passes T.

For exhaustiveness, we need the fact that $\mathcal{M}_{im}$ is term-generated. Assume that im passes $ts_{full}(sp)$, i.e., for every $T \in ts_{full}(sp)$ it holds that $eval_{im}(T) = true$. From Lemma 1 it follows that $\mathcal{M}_{im} \models T$, where $\mathcal{M}_{im} = (U, \mathcal{I}, \mathbf{v})$ is constructed as described above. We have to show that $\mathcal{M}_{im} \models Ax(sp)$. Assume for contradiction that $\mathcal{M}_{im} \not\models Ax(sp)$, e.g., $\mathcal{M}_{im} \not\models \forall x_1, \ldots x_n \, \psi(x_1, \ldots, x_n)$. Then there exist $u_1 \ldots u_n \in U$ and $\mathcal{M}'_{im} = (U, \mathcal{I}, \mathbf{v}')$, where $\mathbf{v}' = \mathbf{v}$ except that $\mathbf{v}'(x_1) = u_1, \ldots, \mathbf{v}'(x_n) = u_n$, and that $\mathcal{M}'_{im} \not\models \psi(x_1, \ldots, x_n)$. Since $\mathcal{M}_{im}$ is term-generated, for every $u \in U$ there exists a ground term t with $\mathbf{v}(t) = u$. Since ground terms contain no variables, for any ground term t we have $\mathbf{v}(t) = \mathbf{v}'(t)$. Therefore, for every $u \in U$ there exists a ground term t with $\mathbf{v}'(t) = u$. Thus, we have $\mathcal{M}'_{im} \not\models \psi(t_1, \ldots, t_n)$. Since $\psi(t_1, \ldots, t_n)$ does not contain any of $x_1, \ldots, x_n$ and $\mathbf{v}'$ differs from $\mathbf{v}$ at most in the assignment of $x_1, \ldots, x_n$, it follows that $\mathcal{M}_{im} \not\models \psi(t_1, \ldots, t_n)$. This is a contradiction, since $\psi(t_1, \ldots, t_n) \in ts_{full}(sp)$ and $\mathcal{M}_{im} \models ts_{full}(sp)$. ∎

If the specification sp contains at least one constant, and a function which can be repeatedly applied to this constant, then the set of ground terms is infinite. In such a case, the full test suite $ts_{full}(sp)$ can have infinitely many test cases. Consequently, it cannot be effectively executed.

Thus, the question arises of how $ts_{full}(sp)$ can be reduced to a finite and small size, without sacrificing completeness. To this end, Gaudel et al. [BGM91] introduce a mechanism: by making a reasonable hypothesis on the SUT, the test designer can exclude certain test cases. Starting with a complete test suite and assuming a number of test hypotheses, it is possible to obtain a finite, practically applicable test suite. As the original test suite was complete, and the hypotheses can be justified, the resulting finite test suite is complete as well.

In other words, if an SUT passes the resulting reduced test suite, its correctness w.r.t. the specification is proven. This contradicts Dijkstra's slogan (quoted in the introduction to this chapter) that "program testing can be used to show the presence of bugs, but never to show their absence". With the approach described here, testing can be used to show the absence of bugs in a given program.

In general, a *test hypothesis* is a logical formula which limits the range of a quantifier. Consider the specification formula $\varphi = \forall x \, \psi(x)$, where x is from a certain domain Δ (e.g., the natural numbers). A *regularity hypothesis* is a statement of the form "if ψ holds for all x from a certain subdomain $\delta \subseteq \Delta$, then ψ holds for all $x \in \Delta$". In first-order logic, this can be written as
$$(\forall x(\delta(x) \Rightarrow \psi(x)) \Rightarrow \forall x \, \psi(x)).$$
As a side-remark, note that with the choice of $\delta(x) = \forall y(y < x \Rightarrow \psi(y))$, this formula yields the principle of transfinite induction (cf. Sect. 2.4.2):
$$(\forall x(\forall y(y < x \Rightarrow \psi(y)) \Rightarrow \psi(x)) \Rightarrow \forall x \, \psi(x)).$$
However, usually δ will not hearken back to ψ. A regularity hypothesis can

also be formulated as follows: "if ψ fails for some $x \in \Delta$, then it must also fail for some $x \in \delta$". If this hypothesis holds, then it is sufficient to test $\psi(x)$ for all $x \in \delta$ to find out whether or not $\psi(x)$ holds for all $x \in \Delta$.

To make the hypothesis effective, δ should be 'much smaller' than Δ. However, the choice of δ is not arbitrary, and the validity of the test hypothesis must be shown by some other means than testing.

Example 50.1: Regularity Hypothesis for the *dim*-Function

For example, consider the specification formula
$$\varphi = \forall y (isLeapYear(y) \Rightarrow dim(2, y) = 29).$$
A typical test case (ground instance) would be, e.g.,
$$(isLeapYear(0) \Rightarrow dim(2, 0) = 29).$$
However, testing the function *dim* with input *Year* smaller than 1583 may be irrelevant. The specification restricts valid input values to years after the Gregorian calendar reform. Therefore, we make the assumption:
$$isLeapYear(y) \text{ evaluates to } \textit{false} \text{ for } y < 1583.$$
First, let us convince ourselves that this is indeed a regularity hypothesis: Let $\psi(x) = (isLeapYear(x) \Rightarrow x \geq 1583)$, and $\delta(x) = (x \geq 1583)$. Then $(\forall x(\delta(x) \Rightarrow \psi(x)) \Rightarrow \forall x\ \psi(x))$ becomes $(\forall x((x \geq 1583) \Rightarrow (isLeapYear(x) \Rightarrow x \geq 1583)) \Rightarrow \forall x(isLeapYear(x) \Rightarrow x \geq 1583))$, which is logically equivalent to $\forall x(x < 1583 \Rightarrow \neg isLeapYear(x))$.

There is no guarantee that this regularity hypothesis is correct: It could very well be that a faulty implementation sets, e.g., $isLeapYear(0)$ to *true*. However, if the test hypothesis can be justified (e.g., by looking at the code of *isLeapYear(y)*, then the full test suite can be reduced to values $y \geq 1583$: With the hypothesis, e.g., the test case $(isLeapYear(0) \Rightarrow dim(2, 0) = 29)$ is equivalent to *true* and there is no need to execute it.

An obvious question now is whether we can find another regularity hypothesis, which limits the range of the parameter y towards large numbers. Perhaps surprisingly, the answer is negative. In fact, analysing the specification, we see that *no* implementation on a physical machine can ever be correct for it. This holds for any specification involving the **free type** *Nat*. The specification *requires* that there are infinitely many objects of this type, whereas any physical implementation obviously can provide only finitely many such objects, cf. Definition 18 in Chap. 2.

Thus, we have to change the specification to
$$\textbf{sort } Year = \{n : Nat \mid 1583 \leq n \wedge n \leq MAX\_VALUE\}$$
With this modification of the specification we can make a similar regularity hypothesis as above, and thus obtain a finite and complete test suite.

Yet, this test suite is still not practical; it is way too large to be executed in practice. Thus, we continue developing further hypotheses. Let again be

$\varphi = \forall x \; \psi(x)$, where x is from a certain domain Δ. A *uniformity hypothesis* for φ on a subdomain $\delta \subseteq \Delta$ states that the system under test behaves 'uniformly' with respect to φ for all $x \in \delta$. The claim is that "if $\psi(x)$ holds for any value $x \in \delta$, then $\psi(x)$ holds for all $x \in \delta$". In first-order logic, this is
$$(\exists x(\delta(x) \wedge \psi(x)) \Rightarrow \forall x \; (\delta(x) \Rightarrow \psi(x))), \text{ or, equivalently,}$$
$$\forall x(\delta(x) \Rightarrow (\psi(x) \Rightarrow \forall x \; (\delta(x) \Rightarrow \psi(x)))).$$
If the uniformity hypothesis holds, then it suffices to test any $x \in \delta$ in order to establish ψ for all $x \in \delta$: We can choose an arbitrary representative $x \in \delta$ and test whether $\psi(x)$ passes. If so, then according to the uniformity hypothesis, the test $\psi(x)$ passes for all $x \in \delta$. Thus, there is no need to execute the respective test cases. If not, then we have found a value where $\psi(x)$ fails, thus the test suite fails.

As above, a uniformity hypothesis must be justified by some other means than testing.

Example 50.2: Uniformity Hypothesis for the *dim*-Function

For our example specification, a typical uniformity hypothesis is "*isLeapYear(y)* behaves uniformly for all $y > 1582$, which are divisible by 400".

Here, 'uniformly' means that the same statements in the program graph are executed in the same order, but with different values. A justification for this hypothesis could be given by program analysis. Formally, for this uniformity hypothesis, let $\delta(y) = (y > 1582 \wedge y \bmod 400 = 0)$, and $\psi(y) = (isLeapYear(y) \iff (y \bmod 4 = 0 \wedge y \bmod 100 > 0) \vee (y \bmod 400 = 0))$. Then the hypothesis becomes
$$\exists y > 1582.(y \bmod 400 = 0 \wedge isLeapYear(y)) \Rightarrow$$
$$\forall y > 1582.(y \bmod 400 = 0 \Rightarrow isLeapYear(y)).$$
Thus, it suffices to test, e.g., $isLeapYear(2000)$. If it passes, then $isLeapYear(y)$ must also pass for $y = 1600, 2000, 2400$, etc.

In order to further reduce the size of the test suite, other uniformity hypotheses can be made. An example would be "*isLeapYear(y)* behaves uniformly for all $y > 1582$, which are not divisible by 4". But, we need to make sure that the justification is valid. As a counter-example, consider "*isLeapYear(y)* behaves uniformly for all $y > 1582$, where y is divisible by 100". For a correct implementation, this hypothesis can not be justified, as e.g., $y = 2000$ and $y = 2100$ need to be treated differently.

5.5 Tool Support for Testing

For formal methods in software engineering, tool support is an essential ingredient. Software tools can help in those activities which are too tedious or demanding for humans, or which have to be done over and over again. In professional software testing, several activities have to be performed:

- **Test planning:** It must be determined what the scope of the testing process is, which goals are targeted by the testing activities, and when and where the testing should take place.
- **Test design:** Abstract test cases for the identified testing goals must be described. These abstract test cases can be denoted in informal, semi-formal, formal, or even executable notation.
- **Test development:** Test data must be assembled and/or selected from an existing pool. Furthermore, abstract test cases and concrete test data must be combined into executable test scripts.
- **Test execution:** The test cases must be executed, i.e., the SUT must be connected to the test system, it must be started, stimuli must be sent to the SUT, and responses of the SUT must be observed (and maybe logged).
- **Test evaluation:** For each test, it must be determined whether it has passed or failed. If a test case has failed, appropriate development activities must be triggered.
- **Test assessment:** It must be decided when testing is to be stopped, and whether the determined goals have been reached.
- **Test documentation:** Test results must be documented and archived. Often, specific formats for test documentation are prescribed here.
- **Test lifecycle management:** It must be planned how testing activities integrate into the development process (code-first or test-first approach, agile or phase-based testing, incremental or continuous testing processes, etc.)
- **Test management:** Roles and responsibilities of people involved in the testing must be assigned and administered.
- **Test tool selection:** It must be determined which software tools to use for which testing tasks.
- **Test tool administration:** The selected tools must be procured, installed, and maintained.

When it comes to test automation, most software engineers focus on automated test execution. However, all of the above activities can be supported by appropriate tools, and can be automated to a certain extent. For example, test planning can be supported by project management tools; test development can be supported by automatic test data generators and automated scripting engines; and test evaluation can be automated by an executable test oracle, e.g., based on Algorithm 7 or Algorithm 10.

Popular testing tools can be classified into the following categories:

- Test design tools,
- capture-replay tools, GUI testing tools,
- test drivers, tools for automated test execution,
- test coverage analysis tools,

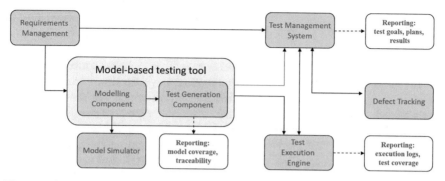

Fig. 5.6 A typical industrial tool landscape for model-based testing

- test suite management tools, application life-cycle management tools,
- defect tracking tools,
- unit testing tools,
- integration testing tools,
- load, stress and performance testing tools, and
- bundled tool suites.

For each of these categories, dozens of tools are available, both commercially and from academic providers. Specification-based testing tools are mostly in the category "test design tools". In fact, test generation from formal specifications can be seen as a kind of automated test design. However, the LTL monitoring approach described in Sect. 5.2.3 is an automated test evaluation.

Model-based testing tools usually consist of two subcomponents:

(a) A graphical editor or modelling environment which allows to draw models and check their syntax, and
(b) a test generator transforming the model into abstract test cases according to certain quality criteria.

A typical test tool landscape including a model-based testing tool is depicted in Fig. 5.6. Core components are shaded in grey, whereas reporting features are depicted in white.

Here, a requirements management tool is used to elicit, document, analyze and control the intended capabilities of a system. These requirements are used as a basis for modelling the system; elements in the model are linked to requirements and vice versa. This allows to trace the use of requirements in the system's development. The model can be executed in a suitable simulator, which allows to animate it and validate whether it conforms to the expectations. The test generation component of the model-based testing tool transforms the model into a test suite. It may contain a reporting component, which measures model or requirements coverage and helps to trace requirements in the test suite. Results of test case generation are handed over to a test management system and a test execution engine. The test execution engine is responsible for the connection of the testing system with the system under test. It replaces abstract events by concrete signals, messages, parameters or procedure calls, invokes the SUT with these concrete stimuli, and records its responses. The test management system keeps track of which tests have been executed with which results, and which requirements have been tested. If a test case fails, then it files a ticket with an appropriate defect tracking system. Optional features of the test management system are automated reporting of test goals, test plans, and test results; the test execution engine may produce log-files of test execution and SUT test coverage.

5.6 Closing Remarks

In this chapter, we looked at the fundamentals of specification-based testing. With the help of a small example, we studied modelling techniques such as labelled transition systems, and UML state machines. We gave specifications in linear temporal logic, and discussed algorithms for test generation and monitoring. Then, we studied conformance testing in more detail. We defined several conformance relations between specification and implementation, and showed how to generate complete test suites from the specification. Subsequently, we considered test generation from algebraic specifications. We showed that ground terms can be used as test cases for functional programs, where the set of all ground equations forms a complete test suite. Then, we showed how to reduce the size of this test suite to finite by appropriate hypotheses about the system under test. Finally, we discussed engineering tools to augment the activities in the testing process.

5.6.1 Annotated Bibliography

In this section, we discuss the background and historical development of some of the concepts in this chapter. The origins of testing are as old as programming itself: From the very beginning, programs had not only to be designed, but also to be tested. However, there was no real distinction between programming, debugging, monitoring, verification, validation and testing of software. Testing was integrated in the overall topic of 'quality control' [Jur51]. Folklore [MG] claims that the first dedicated software testing team was formed in 1958 for the Project Mercury, the first human space flight program of the United States. The 1961 book "Computer Programming Fundamentals" [LW61] reports on the respective experiences in a separate chapter on software testing.

An early technical paper on the systematics is the 1967 IBM white paper "Evaluation of the Functional Testing of Control Programs" [Elm67], which calls for a more scientific approach to software testing. One of the first books dedicated especially to the topic of software testing are the proceedings of the 1972 Chapel Hill Symposium on "Program Test Methods" [Het73].

In 1979, Glenford Myers described software testing as an 'art' [Mye79]. This still very readable book remarks that testing is different from debugging, since it is addressing a different question (not where, but whether a program contains errors). It is also different from verification, since the goal is to detect errors, not to show their absence: "A successful test case is one that detects an as-yet undiscovered error". The main motivation of a tester therefore is to find errors in a program, not to show or bring about its correctness. This observation lead to a revival of research in the foundations of software testing. Today, there is a wealth of textbooks available for software testing. Here, we only mention the undergraduate text by Jorgensen [Jor95], the comprehensive volume by Mathur [Mat08], the classic textbook by Beizer [Bei90], and the introduction by Ammann and Offutt [AO08]. For the interested reader, it is easy to find other literature which is maybe more adept to the personal taste. There are also various curricula in testing, elaborated by the International Software Testing Qualifications Board (ISTQB). These can be used to prepare for a "certified tester" qualification [GVEB08, KL16].

The description of program control structures by state-transition systems dates back to the 1950s [RS59]. The idea to use such formalisms also for test generation appeared already in the 1970s. The main applications in those years were in the field of testing the logic of integrated circuits and telecommunication protocols. The main modelling paradigms were finite state machines, Petri nets and specialized formalisms like the ITU-T Specification and Description Language (SDL). In telecommunications, it is essential that each communicating device conforms to the specified protocol. Thus, the telecommunication standardisation industry was amongst the first to define reference test suites and models for testing.

In the late 1990s it was understood that this approach can be effectively used for practically all types of software, including operating systems, compilers, data base management systems and others [Jor95]. Thus, the term 'model-based testing' was coined [AD97]. This new approach set the task to seek more adequate modelling paradigms for the new classes of target systems. Thus, formalisms like the unified modelling language (UML) and, later on, the UML testing profile (UTP) were defined [OMG13]. In the mid-2000s, model-based testing became a hype both in academia and industry. In 2004, the first workshop dedicated to model-based testing was held in Barcelona [PPS15]. A graduate-level textbook on model-based testing was composed from lectures at a Dagstuhl seminar [BJK+05]. Commercial tools like Conformiq [Hui07], UniTesk [BKKP02], and RT-tester [Pel13] were developed. Consequently, one of the first textbooks on practical model-based testing was based on a tools approach [UL07]. A more scientific collection focussing also on the theoretical background deals with applications im embedded systems [ZSM11].

Conformance testing can be seen as a special branch of specification-based testing. Brinksma and Tretmans [BT00] give an annotated bibliography on the history of conformance testing with transition systems up to the year 2000. A good survey on IOCO testing can be found in [Tre08]. Proof-supported testing from algebraic specifications was proposed by Marie-Claude Gaudel in 1994 [Gau95]. Subsequently, a large number of articles appeared following up on this approach. Since 2007, the TAP conference is devoted to the convergence of proofs and tests [GM07].

5.6.2 Current Research Directions

Within the last decade, specification-based testing, and its industrial variant model-based testing, has become a well-established formal method which is used in practical applications on a regular basis. However, the development of the underlying theory is not yet finished. In particular, topics of ongoing interest include

- the search for specification languages for test generation, which are more abstract, yet easy to understand;
- the adaptation of algorithms and tools to deal with specification formalisms which are widely used in industry;
- more efficient methods for test generation in terms of fault detection capabilities and test execution times;
- the increase of automation in model-based testing, e.g., to automatically generate SUT adapters;
- the integration of specification-based testing with various other formal methods such as static analysis and program verification; and
- the integration of test generation tools into continuous development environments and application lifecycle management tool chains.

Even though the methods described in this chapter are regularly used for industrial systems, there are several areas in which the theory still needs to be developed. One such area is the testing of distributed and collaborative systems. Consider, e.g., a group of (human or machine) agents working together to achieve a common goal. Here, each agent can be modelled and tested according to the methods described in this chapter. The environment of one agent are all other agents, plus the context in which the whole group is working. For testing the interaction between two agents, protocol testing methods can be applied. However, elaborated methods for testing the emerging behaviour of the whole group are lacking.

The problem is made even more complex if the agents are adaptive and change their behaviour over time. This may be the case, e.g., because they employ learning algorithms. In such a case, often it is already a problem how to describe the intended functionality. For example, consider a deep neural network for character recognition. In order to formally specify the system goals, we need to describe the shape of letters in a general way. However, for most characters (e.g., letter 'a') there are border cases which may or may not be recognized as an 'a'. Yet, the test has to assign a verdict whether the letter recognition works as intended. A first approach to test such systems would be to determine a 'corridor of admissible behaviour' and to test whether the system stays within this corridor. However, a formal theory for the test of evolving systems has yet to be developed.

Another 'hot' topic is security testing. Verification of security protocols is discussed in Chap. 8. As systems are becoming more and more complex, there are more and more possible attacks. Testing a system for vulnerabilities is an open-ended game between intruders and defenders: Whenever one hole is fixed, others open. In particular with large-scale distributed systems used by many people, robustness against malicious attacks, confidentiality, authentication and accountability are predominant issues. Currently, most people are relying on processes and policies to assure the security of systems. Security qualification is mostly done with manual or informal methods (audits, reviews, assessments, vulnerability scans, penetration tests etc.). There are some approaches for testing certain security-related features, e.g., in smartcards. An interesting idea in this context is to use techniques from machine learning and artificial intelligence also for the construction of test suites. However, missing is a general framework for the selection and execution of security tests.

References

[Abr87]   Samson Abramsky. Observation equivalence as a testing equivalence. *Theor. Comput. Sci.*, 53:225–241, 1987.
[AD97]   Larry Apfelbaum and John Doyle. Model based testing. In *Software Quality Week Conference*, May, 1997.
[AO08]   Paul Ammann and Jeff Offutt. *Introduction to Software Testing.* Cambridge University Press, 2008.
[Bei90]   Boris Beizer. *Software Testing Techniques.* Van Nostrand Reinhold, 1990.
[BGM91]   Gilles Bernot, Marie Claude Gaudel, and Bruno Marre. Software testing based on formal specifications: A theory and a tool. *Softw. Eng. J.*, 6(6):387–405, November 1991.
[BJK+05]   Manfred Broy, Bengt Jonsson, Joost-Pieter Katoen, Martin Leucker, and Alexander Pretschner. *Model-Based Testing of Reactive Systems: Advanced Lectures.* Springer, 2005.
[BKKP02]   Igor B. Bourdonov, Alexander S. Kossatchev, Victor V. Kuliamin, and Alexander K. Petrenko. UniTesK test suite architecture. In *FME 2002*, LNCS 2391, pages 77–88. Springer, 2002.
[BT00]   Ed Brinksma and Jan Tretmans. Testing transition systems: An annotated bibliography. In *MOVEP*, LNCS 2067, pages 187–195. Springer, 2000.

[CH93]   Rance Cleaveland and Matthew Hennessy. Testing equivalence as a bisimula-
            tion equivalence. *Formal Aspects of Computing*, 5(1):1–20, Jan 1993.
[CH13]   Ana Cavalcanti and Robert M. Hierons. Testing with inputs and outputs in
            CSP. In *FASE 2013*, LNCS 7793, pages 359–374. Springer, 2013.
[Dij70]   Edsger W. Dijkstra. Structured programming. In *Software Engineering Tech-
            niques*, 1970.
[dNH83]   R. de Nicola and M. C. B. Hennessy. Testing equivalences for processes. In
            Automata, Languages and Programming, pages 548–560. Springer, 1983.
[dSSP14]   Adenilso da Silva Simão and Alexandre Petrenko. Generating complete and
            finite test suite for ioco: Is it possible? In *MBT*, volume 141 of *EPTCS*, pages
            56–70, 2014.
[Elm67]   William R. Elmendorf. Evaluation of the functional testing of control programs,
            1967.
[Gau95]   Marie-Claude Gaudel. Testing can be formal, too. In *TAPSOFT '95*, pages
            82–96, Berlin, Heidelberg, 1995. Springer.
[GG93]   Matthias Grochtmann and Klaus Grimm. Classification trees for partition test-
            ing. *Softw. Test., Verif. Reliab.*, 3(2):63–82, 1993.
[GM07]   Yuri Gurevich and Bertrand Meyer, editors. *Tests and Proofs*, LNCS 4454.
            Springer, 2007.
[GVEB08]   Dorothy Graham, Erik Van Veenendaal, Isabel Evans, and Rex Black. *Foun-
            dations of Software Testing: ISTQB Certification*. Thomson, 2008.
[Het73]   William C. Hetzel, editor. *Program test methods*. Prentice-Hall, 1973.
[HR01]   Klaus Havelund and Grigore Rosu. Monitoring programs using rewriting. In
            ASE 2001, pages 135–143. IEEE, 2001.
[Hui07]   Antti Huima. Implementing Conformiq Qtronic. In *Testing of Software and
            Communicating Systems*, pages 1–12. Springer, 2007.
[Jor95]   Paul C. Jorgensen. *Software Testing: A Craftsman's Approach*. CRC Press, 1st
            edition, 1995.
[Jur51]   J.M. Juran. *Quality-control handbook*. McGraw-Hill, 1951.
[KL16]   Anne Kramer and Bruno Legeard. *Model-Based Testing Essentials - Guide to
            the ISTQB Certified Model-Based Tester: Foundation Level*. Wiley Publishing,
            1st edition, 2016.
[LW61]   H.D. Leeds and G.M. Weinberg. *Computer programming fundamentals*.
            McGraw-Hill, 1961.
[Mat08]   Aditya P. Mathur. *Foundations of Software Testing*. Addison-Wesley, 1st edi-
            tion, 2008.
[McE14]   Curtis McEnroe. A formula for the number of days in each month, Friday, 5
            December, 2014. https://cmcenroe.me/2014/12/05/days-in-month-formula.
            html.
[MG]   Joris Meerts and Dorothy Graham. The history of software testing. http://
            www.testingreferences.com/testinghistory.php.
[Mye79]   Glenford J. Myers. *Art of Software Testing*. Wiley, 1979.
[OMG13]   OMG, Object Management Group. UTP, UML Testing Profile, v. 1.2, 2013.
[Pel13]   Jan Peleska. Industrial-strength model-based testing - state of the art and
            current challenges. In *MBT*, volume 111 of *EPTCS*, pages 3–28, 2013.
[PPS15]   Nikolay V. Pakulin, Alexander K. Petrenko, and Bernd-Holger Schlingloff, edi-
            tors. *MBT 2015*, EPTCS 180, 2015.
[RS59]   Michael O. Rabin and Dana Scott. Finite automata and their decision prob-
            lems. *IBM J. Res. Dev.*, 3(2):114–125, April 1959.
[Tre93]   Jan Tretmans. A formal approach to conformance testing. In *Protocol Test
            Systems*, volume C-19 of *IFIP Transactions*, pages 257–276. North-Holland,
            1993.
[Tre96]   Jan Tretmans. Test generation with inputs, outputs, and quiescence. In
            TACAS, LNCS 1055, pages 127–146. Springer, 1996.

[Tre08]   Jan Tretmans. Model based testing with labelled transition systems. In *Formal Methods and Testing*, pages 1–38. Springer, 2008.

[UL07]   Mark Utting and Bruno Legeard. *Practical Model-Based Testing: A Tools Approach*. Morgan Kaufmann, 2007.

[WS11]   Stephan Weißleder and Bernd-Holger Schlingloff. Automatic model-based test generation from uml state machines. In *Model-Based Testing for Embedded Systems*. CRC Press, 2011.

[ZSM11]   Justyna Zander, Ina Schieferdecker, and Pieter Mosterman, editors. *Model-Based Testing for Embedded Systems*. CRC Press, 2011.

Part III
Application Domains

Chapter 6
Specification and Verification of Normative Documents

Gerardo Schneider

Abstract This chapter is concerned with the formal specification and verification of *normative documents*, that is, documents containing what is mandatory, permitted and prohibited. In computer science and software engineering, these documents are usually referred as *contracts*. As the application domain is quite vast, we give a high level description of a general approach to the field, and we provide few motivating examples from different domains, after clarifying on the notion of 'contract'. We proceed then with the presentation of the formal language $\mathcal{CL}$ and we show which kind of analysis may be done on $\mathcal{CL}$ contracts, focusing mainly on the analysis of normative conflicts.

6.1 Contracts: Help or Trouble?

You got your new smart phone and decided to download as many free applications as possible. Since you do not have time nor patience, you simply click on the 'agree' button of all the agreements without reading them. At the end of the month when you got the bill for the use of Internet traffic, you get a shock on the high bill. After a careful enquire you get to know that one particular application was not free after all. You cancel it, but there is a cancelation fee to be paid. Nothing can be done as you have 'consciously and willingly' given your consent.

You travel to Brazil and you brought your smart phone with you. You have agreed to a special offer with your provider on the price and roaming policies when abroad, which stipulated among other things that you could enjoy free Internet access up to 100 Mb after 20:00. You enjoy your life drinking caipirinha, going to the beach, dancing samba, and uploading all your nice photos to Facebook and writing comments about your day, doing it at night benefiting from the offer of your phone provider. However, when you

Gerardo Schneider
University of Gothenburg, Sweden

© Springer Nature Switzerland AG 2022, corrected publication 2022
M. Roggenbach et al., *Formal Methods for Software Engineering*,
Texts in Theoretical Computer Science. An EATCS Series,
https://doi.org/10.1007/978-3-030-38800-3_6

come home and the first after-holidays bill comes, you cannot believe the 4 digits bill you get from your phone provider, mainly due to roaming abroad. When asking for details you were told that the contract stipulated, among other things: (i) Brazil did not qualify for the special offer; (ii) some days you spent a bit more than 100 Mb, and the cost per additional Mb was 10 EUR. After being told that, you spent hours going into the contract and you did indeed find something about the above. It took you even more hours and the help of a lawyer to check out all the conditions.

The above scenarios (both adaptations of real life cases) show the importance, for the bad more than good, of 'contracts'.

Could the bad outcome be foreseen and prevented? Would it have been possible to get a warning about the agreement not being valid in Brazil automatically, or when the amount downloaded comes close to 100 Mb?

Now imagine that you are a lawyer and you need to modify existing contracts, adding or deleting clauses for different clients. Though your work is typically reviewed by you and colleagues to be sure they are correct and without conflictive clauses, it happens from time to time that mistakes are overlooked. After few years without any problem arising from a given contract, you are suddenly in court trying to defend the impossible given that your legal document is flawed. In case of contradictory clauses, which clause should be taken into account? Is the contract suddenly declared void? Who is liable? Would it be possible to have an intelligent editor to help lawyers draft contracts allowing for the possibility to detect inconsistencies, conflictive clauses, etc?

Though we will not be able to address and answer in detail all the above questions, in what follows we present work addressing issues related to the specification and analysis of contracts, or more generally of *normative documents*, that is, documents containing what is mandatory, permitted and prohibited.

In next section we give a high level description of a general approach to the field, and we provide few motivating examples from different domains, after clarifying on the notion of 'contract'. We proceed then with the presentation of the formal language $\mathcal{CL}$ (Sect. 6.4), and we show in Sect. 6.5 which kind of analysis may be done on $\mathcal{CL}$ contracts, focusing mainly on the analysis of normative conflicts. We finish with a reflection on the content of the chapter, suggestion for further reading, an extensive annotated bibliography, and a brief description of current research trends.

6.2 What Are Contracts?

In this section we start with an informal discussion on what we mean by contract in the context of computer science and software engineering. We proceed by presenting three motivating examples concerning the specification and verification of contracts, and we finish with the presentation of an ideal framework concerning contract specification and analysis.

6.2.1 On the Notion of Contract

According to the Webster's online dictionary[1] a *contract* is a "binding agreement between two or more persons that is enforceable by law". In the context of computer science and software engineering, however, the term contract has mostly been used not according to the common meaning of the word. We thus find the use of the term, among other things, to: (i) denote pre-/post-conditions and invariants (e.g., *programming-by-contracts* and JML); (ii) describe how services interact by defining service level agreements (e.g., WS-Agreement); (iii) denote behavioural interfaces component-based systems, to characterise a component based on the way it communicates with its environment.; (iv) regulate parties' ideal mode of interaction in protocols; (v) model and regulate social behaviour in multi-agent systems.

In this chapter we are **not** concerned with *software contracts* in the above sense, as we do not want to reason about programs, services, components, etc., but rather write, analyse and monitor the contracts themselves (which describe and prescribe behaviour *about* software systems).

We aim at having a formalisation at a certain level of abstraction of documents containing prescriptive information in the form of rules or norms. So, in the rest of this chapter we will use the term 'contract' in a very large sense, meaning any document containing norms specifying what are the obligations, prohibitions, and permissions of the different parties involved, as well as what is to be done in case of specific violations. Our definition of contract includes, among others, legal (contractual) documents, requirements in software engineering, workflow and work procedure descriptions, and regulatory documents.

In what follows we will provide a couple of motivating examples to show the generality of this definition.

6.2.2 Motivating Examples

Workflow Specifications

A workflow specification (description of a work) may be considered as a normative text describing the tasks to be performed for a specific job, and

---

[1] http://www.websters-online-dictionary.org.

prescribing procedures of normal and abnormal situations. We show in Example 53 part of the description of an airline ground-crew working procedure.

Example 53: Airline Ground Crew Work Description

The following example is from the Fenech, Pace and Schneider [FPS09a].

1. The ground crew is obliged to open the check-in desk and request the passenger manifest from the airline two hours before the flight leaves.
2. The airline is obliged to provide the passenger manifest to the ground crew when opening the desk.
3. After the check-in desk is opened the check-in crew is obliged to initiate the check-in process with any customer present by checking that the passport details match what is written on the ticket and that the luggage is within the weight limits. Then they are obliged to issue the boarding pass.
4. If the luggage weights more than the limit, the crew is obliged to collect payment for the extra weight and issue the boarding pass.
5. The ground crew is prohibited from issuing any boarding passes without inspecting that the details are correct beforehand.
6. The ground crew is prohibited from issuing any boarding passes before opening the check-in desk.
7. The ground crew is obliged to close the check-in desk 20 min before the flight is due to leave and not before.
8. After closing check-in, the crew must send the luggage information to the airline.
9. Once the check-in desk is closed, the ground crew is prohibited from issuing any boarding pass or from reopening the check-in desk.
10. If any of the above obligations and prohibitions are violated a fine is to be paid.

Among other things, we might be interested to analyse such workflows in order to detect whether the tasks are doable, there are no contradictions, and eventually simulate abnormal situations. Additionally it could be interesting to determine responsibilities for different tasks.

Legal Documents

Legal documents, including different kind of contractual documents, are complex. They must be as precise as possible while at the same time being open to different interpretations in case of specific instantiations. They should also be as complete as possible while keeping in mind that too many exceptions for specific cases might compromise readability. Briefly, it can take an enormous amount of time to write, read, understand and analyse such documents.

The contract shown in Example 54 stipulates the obligations and rights of an internet provider and a client of the service.

Though a complete analysis of legal documents seems to be too ambitious, the same kind of analysis may be practical for other normative documents like workflow specifications or simpler regulations. Moreover, whenever contracts are concerned with *computer-mediated transactions* (i.e., where transactions are performed by software systems), it could be of great help to have a runtime monitor (extracted from the legal description) to ensure that the contract is being satisfied, or otherwise to detect violations, and also to determine liabilities and causalities.

Example 54: Contract Between Internet Provider and Client

The following clauses are from Pace, Prisacariu and Schneider [PPS07].

This deed of **Agreement** is made between:
1. [**name**], from now on referred to as **Provider** and
2. [**name**], from now on referred to as the **Client**.

INTRODUCTION

3. The **Provider** is obliged to provide the **Internet Services** as stipulated in this **Agreement**.
5. DEFINITIONS

    5.1.10. **Internet traffic** may be measured by both **Client** and **Provider** by means of Equipment and may take the two values **high** and **normal**.

OPERATIVE PART

7. CLIENT'S RESPONSIBILITIES AND DUTIES

    7.1. The **Client** shall not:
    7.1.1. supply false information to the Client Relations Department of the **Provider**.
    7.2. Whenever the Internet Traffic is **high** then the **Client** must pay [*price*] immediately, or the **Client** must notify the **Provider** by sending an e-mail specifying that he will pay later.
    7.3. If the **Client** delays the payment as stipulated in 7.2, after notification he must immediately lower the Internet traffic to the **normal** level, and pay later twice (2 ∗ [*price*]).

    [...]

8. CLIENT'S RIGHTS

    8.1. The **Client** may choose to pay either: (i) each month;   (ii) each three (3) months;   (iii) each six (6) months.

9. PROVIDER'S SERVICE

[...]
10. PROVIDER'S DUTIES

    10.1 The **Provider** takes the obligation to return the personal data of the client to the original status upon termination of the present **Agreement**, and afterwards to delete and not use for any purpose any whole or part of it.
    10.2 The **Provider** guarantees that the Client Relations Department, as part of his administrative organisation, will be responsive to requests from the **Client** or any other Department of the **Provider**, or the **Provider** itself within a period less than two (2) hours during *working hours* or the day after.

11. PROVIDER'S RIGHTS

    11.1. The **Provider** takes the right to alter, delete, or use the *personal data* of the **Client** only for statistics, monitoring and internal usage in the confidence of the **Provider**.
    11.2. **Provider** may, at its sole discretion, without notice or giving any reason or incurring any liability for doing so:
    11.2.2. suspend Internet Services immediately if **Client** is in breach of Clause 7.1;

13. TERMINATION

    13.1. Without limiting the generality of any other *Clause* in this *Agreement* the **Client** may terminate this *Agreement* immediately without any notice and being vindicated of any of the Clause of the present Agreement if:
    13.1.1 the **Provider** does not provide the Internet Service for seven (7) days consecutively.
    13.2. The **Provider** is forbidden to terminate the present Agreement without previous written notification by normal post and by e-mail.
    13.3. The **Provider** may terminate the present Agreement if:
    13.3.1 any payment due from **Client** to **Provider** pursuant to this **Agreement** remains unpaid for a period of fourteen (14) days;

16. GOVERNING LAW

[...]

Terms of Service

Probably everybody has had the frustrating experience of having to accept an agreement, called *terms of service* or ToS, before being allowed to download an application on an electronic media (e.g., smart phone). In such cases, the standard behaviour usually is not to read the agreement, but simply click on the 'agree' button, knowing that otherwise it will not be possible to use the application.

Given the time constraints we are under today, it could be highly desirable to have the possibility to simply press a button 'Quick Analysis' before engaging in accepting any agreement, so that a quick analysis of the document is performed. This analysis might for instance highlight our obligations, and describe briefly under which conditions they are enacted. Also, it could allow the user to make a quick query on specific questions, for instance "What is the worst case scenario if I do breach the agreement".

6.3 A Framework for Specification and Analysis of Contracts

We present in this section a conceptual framework to handle electronic (and legal) contracts (see Fig. 6.1). A contract written in natural language is successfully refined towards a formal language which might be analysed statically and at runtime (flow from right to left in the figure). A scenario highlighting a typical use of the framework is described in what follows.[2]

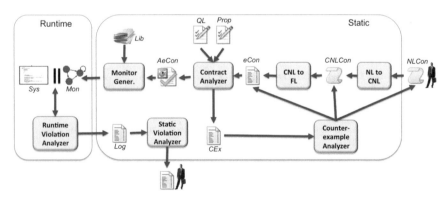

Fig. 6.1 High-level description of a contract analysis framework

---

[2] In subsequent sections we will develop the concepts mentioned in the framework and present them in more details.

1. A contract is written in natural language (NL), as for instance English, by an end-user (e.g., a lawyer) and fed into the framework (*NLCon*).
2. The contract is automatically translated into a structured restricted version of English (*Controlled Natural Language* – CNL), by the module *NL to CNL*, obtaining the contract *CNLCon*.
3. The *CNLCon* contract is then transformed into a formal version by the *CNL to FL* module, into what is called the electronic contract (or *eCon* for short).
4. *eCon* is then analysed by using the *Contract Analyser* module. The analyser will check the contract against predefined properties (e.g., absence of normative conflicts: it does not contain contradictory obligations, prohibitions nor permissions), also allowing the verification of the contract against user-defined properties (*Prop*), as for instance that it is never the case that if the provider is obliged to pay clients based on a number of clicks on advertisements, then the provider always pays the same amount. The user might perform queries in the query language *QL* (e.g., what are all the client's obligations and under which conditions are they enacted).
5. In case of errors (contradictions, properties not satisfied, etc.), the *Contract Analyser* gives a counter-example *CEx*, which is analysed by the *Counter-example Analyser*. This module translates the counter-example into a format understandable by the end-user, showing where the problem *might* be. With the help of the module the user will try to identify the problem in *NLCon*, modifying the original contract and restarting the process again if necessary.
6. If the *Contract Analyser* does not find any problem (and after the user is satisfied with the performed queries), the contract is then approved (*AeCon*), and a runtime monitor *Mon* is automatically obtained from it. This is performed by the *Monitor Generator* module, which might use some predefined libraries *Lib* to compute algorithmic procedures implicitly stated in the contract (e.g., average values, percentages per given time-periods), not possible to be extracted automatically.
7. The Contract Analyser could also be used at runtime whenever the user is confronted for the first time with a contract or agreements to accept. Users will have the possibility to launch queries (as done statically) to quickly highlight what are the important clauses concerning themselves, and decide whether to agree or not. *[Not explicitly represented in the picture.]*
8. The runtime monitor *Mon* is run in parallel with the underlying software system *Sys*.
9. The *Runtime Violation Analyser* detects when the contract is violated and uses the history of the transaction to analyse it at runtime (eventually canceling the transaction, giving a warning, or performing any previously defined operation).
10. The result provided by the *Runtime Violation Analyser*, the file *Log*, will be passed to the *Static Violation Analyser* so the end-user can analyse it off-line to further determine what where the causes and who was responsible

for the contract violation. A decision then must be taken according to the obtained information. The *Log* file will contain all the history of the transaction (not only the violation) making it possible for the end-user to, for instance, determine liabilities and analyse customers behaviour.

Note that the *Runtime* part of the framework is only valid for the case of dealing with computer mediated transactions. The *Static* part is valid for any kind of (legal) contract.

A full realisation of the framework presented above is a challenging task, most of it is still ongoing research. In the rest of this chapter we will present a small part of such long-term vision. We start by presenting the formal language $\mathcal{CL}$, as a candidate for the *eCon* in the framework.

6.4 The $\mathcal{CL}$ Language

We present here the formal language $\mathcal{CL}$, designed with the aim to formalise, at a certain level of abstraction, normative texts containing clauses determining the obligations, permissions and prohibitions of the involved parties. $\mathcal{CL}$ is strongly influenced by dynamic, temporal, and deontic logic (cf. Sect. 2.5). It is an *action-based* language meaning that the modalities are applied to actions and not to *state-of-affairs*. In particular, it is possible to express complex actions in the language by using operators for choice, sequence, conjunction (concurrency) and the Kleene star (see Example 2). Besides, the language allows to specify penalties (*reparations*) associated to the violation of obligations and prohibitions.

6.4.1 Syntax

We start with the syntax of $\mathcal{CL}$ and provide a brief intuitive explanation of its notations and terminology. A discussion on $\mathcal{CL}$ semantics will be given later in this section. A contract in $\mathcal{CL}$ may be obtained by using the syntax shown in Fig. 6.2.

A $\mathcal{CL}$ contract is written as a conjunction of clauses representing (conditional) normative expressions, as specified by the initial non-terminal C in the definition. The first line of the definition shows that a contract can be an *obligation* (C_O), a *permission* (C_P), a *prohibition* (C_F), a conjunction of two clauses, or a clause preceded by the dynamic logic square brackets. $\top$ and $\bot$ are the trivially satisfied and violating contracts, respectively. $\mathbb{O}$, $\mathbb{P}$ and $\mathbb{F}$ are deontic modalities (see Sect. 2.5.2); the obligation to perform a complex action α is written as $\mathbb{O}_C(\alpha)$, showing the primary obligation to perform α, and the reparation contract C if α is not performed. This represents

$$C := C_O \mid C_P \mid C_F \mid C \wedge C \mid [\beta]C \mid \top \mid \bot$$
$$C_O := \mathbb{O}_C(\alpha) \mid C_O \oplus C_O$$
$$C_P := \mathbb{P}(\alpha) \mid C_P \oplus C_P$$
$$C_F := \mathbb{F}_C(\alpha)$$
$$\alpha := 0 \mid 1 \mid a \mid \alpha \& \alpha \mid \alpha.\alpha \mid \alpha + \alpha$$
$$\beta := 0 \mid 1 \mid a \mid \beta \& \beta \mid \beta.\beta \mid \beta + \beta \mid \beta^*$$

Fig. 6.2 $\mathcal{CL}$ syntax

what is usually called a *Contrary-to-Duty* (*CTD*) as it specifies what is to be done if the primary obligation is not fulfilled. The prohibition to perform a complex action α is represented by the expression $\mathbb{F}_C(\alpha)$, that specifies also the contract to be enacted in case the prohibition is violated (the contract C); this is called *Contrary-to-Prohibition* (*CTP*). Both CTDs and CTPs are useful to represent normal (expected) behaviour as well as the alternative (exceptional) behaviour. $\mathbb{P}(\alpha)$ represents the permission of performing a given complex action α. As expected there is no associated reparation, as a permission cannot be violated.

We have so far mentioned 'complex actions' without defining them. At the bottom of Fig. 6.2 it is possible to see the BNF for actions, where α and β represent regular expressions with and without the Kleene star, respectively. The Kleene star ($*$), used to model repetition of actions, is not allowed inside the deontic modalities, though they can be used in dynamic logic-style conditions. Indeed, action β may be used inside the dynamic logic modality (the bracket $[\cdot]$) representing a condition in which the contract C must be executed if β is performed.

The binary constructors ($\&$, ., and $+$) represent *concurrency* (also called *synchrony*), *sequence* and *choice* over basic actions, respectively. Compound actions are formed from basic actions by using these operators. Conjunction of clauses can be expressed using the $\wedge$ operator; the exclusive choice operator ($\oplus$) can only be used in a restricted manner. 0 and 1 are two special actions that represent the *impossible action* and the *skip action* (matching any action), respectively.

The concurrency operator $\&$ should only be applied to actions that can happen simultaneously. To respect this constraint, the software engineer should make explicit which actions are *mutually exclusive*. In $\mathcal{CL}$ this is done by defining the following relation between actions: $a\#b$ if and only it is not the case that $a\&b$. Examples of such actions would be *"the ground crew opens the check-in desk"*, and *"the ground crew closes the check-in desk"*.

An important note on the above is that in $\mathcal{CL}$ it is not possible in general to have a modular (compositional) way of writing specifications. Very often the natural language description has implicit dependencies (in the form of using words as 'it', 'that', 'the above', etc.), or explicit dependencies making reference to specific clauses by name or number. In $\mathcal{CL}$ the user must find and write down such dependencies explicitly.

We show in Example 55 the formalisation of natural language sentences in the formal language $\mathcal{CL}$.

Example 55: Airport Ground Crew

Let us consider the following clause of the ground crew example:

The ground crew is obliged to open the check-in desk and request the passenger manifest two hours before the flight leaves.

Taking a to represent *"two hours before the flight leaves"*, b to be *"the ground crew opens the check-in desk"*, and c to be *"the ground crew requests the passenger manifest"*, then this could be written in $\mathcal{CL}$ as

$$[a]\mathbb{O}(b \& c).$$

Let us now consider the following additional clause imposing a penalty in case the above clause is not respected:

If the ground crew does not do as specified in the above clause, then a penalty should be paid

Assuming that p represents the phrase *"paying a fine"*, we would write all the above, in combination with the previous formalisation in $\mathcal{CL}$, as

$$[a]\mathbb{O}_{\mathbb{O}(p)}(b \& c).$$

6.4.2 Semantics

As we have discussed in the introduction to Formal Methods (cf. Sect. 1.1.2) there are three main ways of giving semantics to a formal language: operational, denotational and axiomatic. The choice on which style to use depends on different factors, ranging from a personal taste to more complex technical reasons.

$\mathcal{CL}$ has been given three different semantics: (i) an encoding into the μ-calculus; (ii) a trace semantics; (iii) a Kripke semantics. The three may be seen as variants of denotational semantics, though the trace semantics might be seen as a low-level operational semantics.

In what follows we will only present in more detail the trace semantics, as this is the one used for the conflict analysis further developed in next section. The other two semantics will be briefly sketched at the end of this section.

6.4.2.1 $\mathcal{CL}$ Trace Semantics

A trace semantics usually gives us information about the traces (or sequences of actions) that are valid, or accepted, by any formula (expression) of a logic (formal language).

Originally, the trace semantics for $\mathcal{CL}$ was developed with the sole purpose of being able to explain monitoring aspects. For that, it was defined over infinite sequences of actions, and did not contain any information on whether the action was mandatory, permitted, or prohibited, or none of the above. However, it was realised later that in order to use the trace semantics for a certain kind of analysis (namely conflict analysis), it should be modified so that:

1. it contains deontic information (including permissions); and
2. it 'accepts' finite prefixes.

We present in what follows a trace semantics for $\mathcal{CL}$ which takes into account the above two requirements.

For a contract with action alphabet Σ, we will introduce the *deontic alphabet* Σ_d consisting of O_a, P_a and F_a for each action $a \in \Sigma$, in order to represent which normative behaviour is enacted at a particular moment. Given a set of concurrent actions α, we will write O_α to represent $\{O_a \mid a \in \alpha\}$.

Given a $\mathcal{CL}$ contract C with action alphabet Σ, the semantics will be expressed in the form $\sigma, \sigma_d \vDash C$, where σ is a finite trace of sets of concurrent actions in Σ and σ_d is a finite trace consisting on sets of sets (needed to distinguish choices from conjunction) of deontic information in Σ_d. The statement $\sigma, \sigma_d \vDash C$ is said to be well-formed if $length(\sigma) = length(\sigma_d)$; in what follows we will consider only well-formed semantic statements.

Intuitively, a well-formed statement $\sigma, \sigma_d \vDash C$ will correspond to the statement that action sequence σ is possible under (will not break) contract C, with σ_d being the deontic statements enforced from the contract.

Example 56 shows some traces for two different contracts.

Example 56: $\mathcal{CL}$ Trace Semantics

Let us consider the contract

$$C = [a]\mathbb{O}(b) \wedge [b]\mathbb{F}(b).$$

According to the contract, we have that $\sigma_d = \langle \{\emptyset\}, \{\{O_b\}\} \rangle$, and given the trace $\sigma = \langle \{a\}, \{b\} \rangle$, it is the case that $\sigma, \sigma_d \vDash C$.

The contract
$$C' = \mathbb{F}(c) \wedge [1](\mathbb{O}(a) \wedge \mathbb{F}(b))$$

stipulates that it is forbidden to perform action c and that after the execution of any action, there is an obligation to perform an

> a (while prohibiting the execution of b). We can thus write $\sigma_d = \langle\{\{F_c\}\}, \{\{O_a\}, \{F_b\}\}\rangle$. The contract allows the execution of actions a and b concurrently, and then a concurrently with c ($\sigma = \langle\{a, b\}, \{a, c\}\rangle$), and we have that $\sigma, \sigma_d \vDash C'$.

Note that the presentation of the traces σ and σ_d suggests that both traces are paired position-wise as they have the same length. Though this is true to some extent, it might help to 'read' the traces in the following way: first consider the first element of the σ_d trace, then the first of the σ trace, and successively. That is, in the first contract given in Example 56, we should read the whole judgment as explained in what follows. At the beginning there is no normative constraint (first element of the trace σ_d being $\{\emptyset\}$). Then after a is executed (first element of the σ trace being $\{a\}$) we have that the contract stipulates an obligation of executing action b (second element of the trace σ_d being $\{\{O_b\}\}$). We then have that b is executed (second element of the σ trace being $\{b\}$), which is in accordance with the corresponding obligation at that moment in σ_d.

In Example 56.1 below we show a trace for a contract containing the obligation of a choice. The choice is expressed at the trace level by having a set containing a set with the normative concepts affected by the choice (in this case, 2 obligations).

Example 56.1: $\mathcal{CL}$ Trace Semantics (continued)

Let us consider the contract

$$C'' = [a]\mathbb{O}(b + c) \wedge [b]\mathbb{F}(b).$$

If we consider first that an a is performed, then only the obligation to do b or c remains (the prohibition is thus discarded since no b did happen). From this point on, any trace continuation performing b or c would be acceptable. Thus, for

$$\sigma = \langle\{a\}, \{b\}\rangle \quad \text{and} \quad \sigma_d = \langle\{\emptyset\}, \{\{O_b, O_c\}\}\rangle$$

we have that $\sigma, \sigma_d \vDash C''$.

Given two traces σ_1 and σ_2, we will use $\sigma_1; \sigma_2$ to denote their concatenation, and $\sigma_1 \cup \sigma_2$ (provided the length of σ_1 is equal to that of σ_2) to denote the point-wise union of the traces: $\langle\sigma_1(0)\cup\sigma_2(0), \sigma_1(1)\cup\sigma_2(1), \ldots\sigma_1(n)\cup\sigma_2(n)\rangle$. $\mathcal{CL}$ trace semantics is shown in Fig. 6.3. We give an intuitive explanation of the semantics in what follows (note that we do not present the trivial cases of actions 0 and 1).

$$\sigma, \sigma_d \vDash C \quad \text{if} \quad length(\sigma) = length(\sigma_d) = 0 \tag{6.1}$$

$$\sigma, \sigma_d \vDash \top \quad \text{if} \quad \sigma_d(0) = \emptyset \text{ and } \sigma(1..), \sigma_d(1..) \vDash \top \tag{6.2}$$

$$\sigma, \sigma_d \vDash C_1 \wedge C_2 \quad \text{if} \quad \sigma, \sigma_d' \vDash C_1 \text{ and } \sigma, \sigma_d'' \vDash C_2 \text{ and } \sigma_d = \sigma_d' \cup \sigma_d'' \tag{6.3}$$

$$\sigma, \sigma_d \vDash C_1 \oplus C_2 \quad \text{if} \quad (\sigma, \sigma_d \vDash C_1 \text{ and } \sigma, \sigma_d \nvDash C_2) \text{ or } (\sigma, \sigma_d \vDash C_2 \text{ and } \sigma, \sigma_d \nvDash C_1) \tag{6.4}$$

$$\sigma, \sigma_d \vDash [\epsilon]C \quad \text{if} \quad \sigma, \sigma_d \vDash C \tag{6.5}$$

$$\sigma, \sigma_d \vDash [\alpha_\&]C \quad \text{if} \quad (\alpha_\& \nsubseteq \sigma(0) \Rightarrow \sigma, \sigma_d \vDash \top) \text{ and} \tag{6.6}$$

$$(\alpha_\& \subseteq \sigma(0) \Rightarrow (\sigma_d(0) = \emptyset \text{ and } \sigma(1..), \sigma_d(1..) \vDash C)) \tag{6.7}$$

$$\sigma, \sigma_d \vDash [\overline{\alpha_\&}]C \quad \text{if} \quad (\alpha_\& \subseteq \sigma(0) \Rightarrow \sigma, \sigma_d \vDash \top) \text{ and} \tag{6.8}$$

$$(\alpha_\& \nsubseteq \sigma(0) \Rightarrow (\sigma_d(0) = \emptyset \text{ and } \sigma(1..), \sigma_d(1..) \vDash C)) \tag{6.9}$$

$$\sigma, \sigma_d \vDash [\beta; \beta']C \quad \text{if} \quad \sigma, \sigma_d \vDash [\beta][\beta']C \tag{6.10}$$

$$\sigma, \sigma_d \vDash [\beta + \beta']C \quad \text{if} \quad \sigma, \sigma_d \vDash [\beta]C \wedge [\beta']C \tag{6.11}$$

$$\sigma, \sigma_d \vDash [\beta^*]C \quad \text{if} \quad \sigma, \sigma_d \vDash C \wedge [\beta][\beta^*]C \tag{6.12}$$

$$\sigma, \sigma_d \vDash \mathbb{O}_C(\alpha_\&) \quad \text{if} \quad \sigma_d(0) = O_{\alpha_\&} \text{ and} \tag{6.13}$$

$$(\alpha_\& \subseteq \sigma(0) \Rightarrow \sigma(1..), \sigma_d(1..) \vDash \top) \text{ and} \tag{6.14}$$

$$(\alpha_\& \nsubseteq \sigma(0) \Rightarrow \sigma(1..), \sigma_d(1..) \vDash C) \tag{6.15}$$

$$\sigma, \sigma_d \vDash \mathbb{O}_C(\alpha; \alpha') \quad \text{if} \quad \sigma, \sigma_d \vDash \mathbb{O}_C(\alpha) \wedge [\alpha]\mathbb{O}_C(\alpha') \tag{6.16}$$

$$\sigma, \sigma_d \vDash \mathbb{O}_C(\alpha + \alpha') \quad \text{if} \quad \sigma, \sigma_d \vDash \mathbb{O}_\top(\alpha) \wedge \mathbb{O}_\top(\alpha') \wedge [\overline{\alpha + \alpha'}]C \tag{6.17}$$

$$\sigma, \sigma_d \vDash \mathbb{F}_C(\alpha_\&) \quad \text{if} \quad \sigma_d(0) = F_{\alpha_\&} \text{ and} \tag{6.18}$$

$$(\alpha_\& \subseteq \sigma(0) \Rightarrow \sigma(1..), \sigma_d(1..) \vDash C) \text{ and} \tag{6.19}$$

$$(\alpha_\& \nsubseteq \sigma(0) \Rightarrow \sigma(1..), \sigma_d(1..) \vDash \top) \tag{6.20}$$

$$\sigma, \sigma_d \vDash \mathbb{F}_C(\alpha; \alpha') \quad \text{if} \quad \sigma, \sigma_d \vDash \mathbb{F}_\bot(\alpha) \text{ or } \sigma, \sigma_d \vDash [\alpha]\mathbb{F}_C(\alpha') \tag{6.21}$$

$$\sigma, \sigma_d \vDash \mathbb{F}_C(\alpha + \alpha') \quad \text{if} \quad \sigma, \sigma_d \vDash \mathbb{F}_C(\alpha) \wedge \mathbb{F}_C(\alpha') \tag{6.22}$$

$$\sigma, \sigma_d \vDash \mathbb{P}(\alpha_\&) \quad \text{if} \quad \sigma_d(0) = P_{\alpha_\&} \text{ and } \sigma(1..), \sigma_d(1..) \vDash \top \tag{6.23}$$

$$\sigma, \sigma_d \vDash \mathbb{P}(\alpha; \alpha') \quad \text{if} \quad \sigma, \sigma_d \vDash \mathbb{P}(\alpha) \wedge [\alpha]\mathbb{P}(\alpha') \tag{6.24}$$

$$\sigma, \sigma_d \vDash \mathbb{P}(\alpha + \alpha') \quad \text{if} \quad \sigma, \sigma_d \vDash \mathbb{P}(\alpha) \wedge \mathbb{P}(\alpha') \tag{6.25}$$

Fig. 6.3 The deontic trace semantics of $\mathcal{CL}$

Basic conditions: Figure 6.3 shows at line (6.1) that empty traces satisfy any contract.

Done, Break: The simplest definitions are those of the trivially satisfiable contract $\top$, and the unsatisfiable contract $\bot$. In the case of $\bot$, only an empty sequence will not have yet broken the contract, while in the case of $\top$, any sequence of actions satisfies the contract (whenever no obligation, prohibition, or permission is present on the trace). See Fig. 6.3 line (6.2).

Conjunctions: For the conjunction of two contracts, the action trace must satisfy both contracts and the deontic traces are combined point-wise, as shown in Fig. 6.3 line (6.3).

Exclusive disjunction: Figure 6.3 line (6.4) displays the case for the disjunction, which is similar to conjunction. Note that the rule is valid only for C_1 and C_2 being both of the form C_O or C_P. In the rest of this chapter we will write $C_1 \oplus C_2$ with the understanding that the above restriction applies.

Conditions: Conditions are handled structurally. It has been shown that there exists a normal form such that concurrent actions can be pushed to the bottom level [KPS08]. See Fig. 6.3 lines (6.5)–(6.12).

Obligations: Obligations, like conditions, are defined structurally on action expressions. The base case of the action simply consisting of a conjunction of actions can be dealt with by ensuring that if the actions are present in the action trace, then the contract is satisfied, otherwise the reparation is enacted. The case for the sequential composition of two action sequences is handled simply by rewriting into a pair of obligations. The case of choice (+) is the most complex one, in which we have to consider the possibility of having either obligation satisfied or neither satisfied, hence triggering the reparation. (Recall that the star operator cannot appear within obligations.) See Fig. 6.3 lines (6.13)–(6.17).

Prohibitions: The semantics for prohibitions is similar to obligations, with the difference that prohibition of a choice is more straightforward to express. See Fig. 6.3 lines (6.18)–(6.22).

Permissions: See Fig. 6.3, (6.23)–(6.25) for the semantics of permissions.

In Sect. 6.5.1 we will see how the trace semantics presented here is suitable for conflict detection.

6.4.2.2 Other Semantics for $\mathcal{CL}$

Why are there so many semantics for $\mathcal{CL}$? There are practical reasons to have different semantics. As we have already discussed, the trace semantics is useful for monitoring purposes, and for certain kind of analysis where a full (Kripke) semantics is not needed, as for the detection of normative conflicts.

The encoding into the μ-calculus (or in general into another logic) is useful to study expressiveness of the language, and to get a way to semantically prove things about the language by using the proof engine (or model-based approach) of the target logic (μ-calculus in this case). Given the non-standard combination of deontic, dynamic and temporal operators with regular expressions over actions, the semantics is in general non-compositional and thus rewriting rules are given so $\mathcal{CL}$ expressions are preprocessed before giving the encoding.

The Kripke semantics is *the* semantics of the language as it unambiguously gives the meaning of each $\mathcal{CL}$ expression not by means of interpreting it into another logic or formal language, but rather by using the standard world semantics used in modal logics. Many of the 'properties' of the language are somehow forced in the semantic definition, making it impossible to derive many of the well known paradoxes inherited from the normative deontic notions. This semantics is thus very complicated and not easy to understand

or manipulate in manual proofs. However, it could be used as the basis of an
ad hoc model checker for $\mathcal{CL}$, provided a decision procedure exists.

6.5 Verification of $\mathcal{CL}$ Contracts

We can perform different kinds of analysis on contracts. For $\mathcal{CL}$ contracts in
particular, there is work showing the feasibility of performing conflict analysis, model checking, and runtime verification. In the rest of this section we
will focus on the first, only giving a short overview on the other two.

6.5.1 Conflict Analysis of $\mathcal{CL}$ Contracts

We show in what follows an informal presentation on the usefulness of the
trace semantics presented above to detect conflicts.

Let $C = [a]\mathbb{O}(b) \wedge [b]\mathbb{F}(b)$ be a contract defined over the action alphabet $\{a, b\}$. Our aim is to check whether contract C has normative conflicts.
Let us assume for the moment that the trace semantics is based on infinite sequences and does not have any information about the deontic modalities associated with actions. So, the traces 'accepted' by the contract C are
$\{\langle a, b, any \rangle \mid any = (a+b)^\omega\} \cup \{\langle b, a, any \rangle \mid any = (a+b)^\omega\}$. According to the
semantics, no trace starting with action $\{a, b\}$ (i.e., with a and b occurring
concurrently) will be accepted by the contract, since this would imply a contract violation due to the enacted conflicting obligation and prohibition. Even
more, due to the lack of deontic information in the trace, it is not possible
to identify the possible occurrence of a normative conflict. When performing
a conflict analysis we would like to have a *witness* of such a conflict, and
in particular a systematic way to obtain an automaton that recognises such
prefixes containing conflicts. So, it is clear that such 'trace semantics' needs
to be modified, namely with:

1. the addition of deontic information (which obligations, permissions and
   prohibitions are satisfied at any moment);
2. the addition of a trace semantics for permission (this is not necessary in
   principle for a trace semantics only for monitoring purposes); and
3. the addition of the possibility to 'accept' certain finite prefixes (in order
   to get the witness for conflicts).

This is exactly what we have done in the previous section. Having the basis
for our conflict analysis, we proceed now with a more detailed explanation
on how the verification is done.

6.5.1.1 Formal Definition of Conflict-Free Contracts

Conflicts in normative texts comes in four different flavours. The first two are when at a given moment the same action is under the scope of two opposite deontic modalities. This happens when one is being obliged and forbidden to perform the same action (e.g., $\mathbb{O}(a) \wedge \mathbb{F}(a)$), and when one is being permitted and forbidden to perform the same action (e.g., $\mathbb{P}(a) \wedge \mathbb{F}(a)$). In the former we would end up in a situation where whatever is performed will violate the contract. The latter case does not necessarily correspond to a real conflict but rather a *potential* one: if the permission is exercised then a conflict will occur.

The remaining two kinds of conflicts occur when certain modalities are applied to two mutually exclusive actions generating real or potential normative conflicts. This happens when obligations are applied to mutually exclusive actions (e.g., $\mathbb{O}(a) \wedge \mathbb{O}(b)$ with $a\#b$), and similarly with permissions and obligations (e.g., $\mathbb{P}(a) \wedge \mathbb{O}(b)$ with $a\#b$). Note that we have not included as potentially conflictive the case of permissions of mutually exclusive actions. Whether this is indeed a case to be considered or not is a philosophical question. In any case this case could be added to our analysis without inconvenience.

Example 57 shows the importance of the (temporal) relation between sentences, and the vulnerability of abstraction in modelling and specification.

Example 57: Normative Conflicts in Airport Ground Crew

Let us consider the two following sentences taken from Example 53:

- *The ground crew is obliged to open the check-in desk and request the passenger manifest from the airline two hours before the flight leaves.*
- *The ground crew is obliged to close the check-in desk 20 min before the flight is due to leave and not before.*

Though these two sentences would be considered to be conflict-free by any human, a machine trying to automatically parse these sentences would have to be 'aware' that the first sentence refers to actions happening *before* the action specified in the second sentence.

If the sentences are represented in a formal language where time is abstracted away, then this time dependency is lost and the sentences would be considered to be in conflict (as the ground crew would be *obliged to open* the check-in desk and *obliged to close* it).

Let us come back to our main aim, that is to detect normative conflicts on $\mathcal{CL}$ contracts. Example 58 shows the connection between traces and conflicts.

Example 58: Traces and Conflicts in $\mathcal{CL}$

Let us consider again the $\mathcal{CL}$ contract

$$[a]\mathbb{O}(b) \wedge [b]\mathbb{F}(b)$$

with allowed actions a and b. It is clear that both traces $\sigma_1 = \langle \{a\}, \{b\} \rangle$ and $\sigma_2 = \langle \{b\}, \{a\} \rangle$ satisfy the contract. However, any trace starting with concurrent actions $\{a, b\}$ (e.g., $\langle \{a, b\}, \{b\} \rangle$) will not be accepted by the contract since any action following it will violate either the obligation to perform b or the prohibition from performing b. In this case, since unspecified, the reparation is the $\perp$ clause which cannot be satisfied regardless of what action is performed.

We have so far reasoned about conflicts in $\mathcal{CL}$ in a completely informal manner. In order to have a formal treatment of conflicts we should first define the notion of *conflict-free contract*. We do so at the semantic level, as shown in the following definition.

Definition 1 For a given trace σ_d of a contract C, let $D, D' \subseteq \sigma_d(i)$ (with $i \geq 0$). We say that D is *in conflict with* D' if and only if there exists at least one element $e \in D$ such that:

$$e = O_a \wedge (F_a \in D' \vee (P_b \in D' \wedge a\#b) \vee (O_b \in D' \wedge a\#b)),$$
$$\text{or} \quad e = P_a \wedge (F_a \in D' \vee (P_b \in D' \wedge a\#b) \vee (O_b \in D' \wedge a\#b)),$$
$$\text{or} \quad e = F_a \wedge (P_a \in D' \vee O_a \in D').$$

A contract C is said to be *conflict-free* if for all traces σ and σ_d such that $\sigma, \sigma_d \vDash C$, then for any $D, D' \subseteq \sigma_d(i)$ $(0 \leq i \leq len(\sigma_d))$, D and D' are not in conflict.

Definition 1 formalises the four cases of conflicts we have previously discussed, presented in a more concise manner (each line of the definition presents the case for each one of the three deontic modalities).

In the following example we show how to see a normative conflict in a $\mathcal{CL}$ expression, by inspecting a trace.

Example 59: Normative Conflicts in $\mathcal{CL}$

Let us consider the contract

$$C = [a]\mathbb{O}(b + c) \wedge [b]\mathbb{F}(b).$$

We can show that C is not conflict-free (i.e., there is at least one conflict) since $\langle \{a, b\}, \{b\} \rangle, \langle \{\emptyset\}, \{\{O_b, O_c\}, \{F_b\}\} \rangle \vDash C$, and there are $D, D' \subseteq \sigma_d(1)$ such that D and D' are in conflict. To see this, let us take $D =$

$\{O_b, O_c\}$ and $e = O_b$. We have then that for $D' = \{F_b\}$, $F_b \in D'$ (satisfying the first line of Definition 1).

In this subsection we have characterised the notion of conflict in contracts by analysing the set of traces accepted by the contract. In order to get a decision procedure (an algorithm) to detect conflicts in $\mathcal{CL}$ contracts, we will take an automata-based approach. In the following subsection we will show how this automata is generated from a $\mathcal{CL}$ formula.

6.5.1.2 Conflict-Aware Automata for $\mathcal{CL}$ Contracts

We now show how to generate a finite-state automaton from a $\mathcal{CL}$ contract C. The automaton is defined so that the language accepted by the automaton corresponds to the traces given by the semantics of the contract. We also define the notion of conflict in the generated automaton.

Definition 2 Given a contract C over an action alphabet Σ and corresponding deontic alphabet Σ_d, we can construct an automaton $A(C) = \langle S, A_{\&}, s_0, T, V, l, \delta \rangle$ where:

- S is the set of states,
- $A_{\&}$ is the set of concurrent actions from Σ,
- s_0 is the initial state,
- $T \subseteq S \times A_{\&} \times S$ is the set of labelled transitions,
- V is a special violation state,
- l is a function labelling states with the $\mathcal{CL}$ clause that holds in that state ($l : S \to \mathcal{CL}$), and
- $\delta : S \to 2^{\Sigma_d}$ is a function labelling states with the set of deontic notions that hold in that state.

We say that a *run* (sequence of states) is accepted by the automaton if none of the states of the run is V. Similarly, we say that the automaton *accepts a word* w, consisting of a sequence of actions, if none of the actions of w is the label of a transition which target state containing V, in which case we write $\texttt{Accept}(A(C), w)$. Note that the automaton is deterministic.

We are not only interested in providing a definition of automata accepting exactly those traces that are 'accepted' by $\mathcal{CL}$, but also in giving an algorithmic procedure to construct such automata, described in what follows.

The construction of our automata uses a *residual contract function* f that, given a $\mathcal{CL}$ formula C and an action α, will return the clause that needs to hold in the following step (after 'consuming' the action). The definition of f is shown in Fig. 6.4.

In this definition, the binary operator α/β (*'tail'*) is used which gives the tail of α if its head matches β. It is inductively defined as follows:

$$\alpha'_\& / \alpha_\& = \epsilon \text{ if } \alpha'_\& \subseteq \alpha_\&, \text{ otherwise } 0$$
$$(0; \alpha)/\alpha_\& = 0$$
$$(1; \alpha)/\alpha_\& = \alpha$$
$$(\alpha; \alpha')/\alpha_\& = (\alpha/\alpha_\&); \alpha'$$
$$(\alpha + \alpha')/\alpha_\& = \alpha/\alpha_\& + \alpha'/\alpha_\&$$

Here are some examples for the tail operator.

- $(a; b)/a = b$
- $((a; b) + (a; c))/a = b + c$.

$$f : \mathcal{CL} \times A_\& \to \mathcal{CL}$$
$$f(\top, \varphi) = \top$$
$$f(\bot, \varphi) = \bot$$
$$f(C_1 \wedge C_2, \varphi) = f(C_1, \varphi) \wedge f(C_2, \varphi)$$

$$f(C_1 \oplus C_2, \varphi) = \begin{cases} \top & \text{if } (f(C_1, \varphi) = \top \wedge f(C_2, \varphi) = \bot) \vee \\ & (f(C_1, \varphi) = \bot \wedge f(C_2, \varphi) = \top) \\ \bot & \text{if } (f(C_1, \varphi) = f(C_2, \varphi) = \top) \vee \\ & (f(C_1, \varphi) = f(C_2, \varphi) = \bot) \\ f(C_1, \varphi) \oplus f(C_2, \varphi) & \text{otherwise} \end{cases}$$

$$f([\alpha_\&]C, \varphi) = \begin{cases} C & \text{if } \alpha_\& \subseteq \varphi \\ \top & \text{otherwise} \end{cases}$$

$$f([\overline{\alpha_\&}]C, \varphi) = \begin{cases} C & \text{if } \alpha_\& \not\subseteq \varphi \\ \top & \text{otherwise} \end{cases}$$

$$f([\overline{\alpha; \alpha'}]C, \varphi) = \begin{cases} C & \text{if } (\alpha; \alpha')/\varphi = 0 \\ [\overline{(\alpha; \alpha')/\varphi}]C & \text{otherwise} \end{cases}$$

$$f([\overline{\alpha + \alpha'}]C, \varphi) = f([\overline{\alpha}]C, \varphi) \wedge f([\overline{\alpha'}]C, \varphi)$$
$$f([\beta; \beta']C, \varphi) = f([\beta][\beta']C, \varphi)$$
$$f([\beta + \beta']C, \varphi) = f([\beta]C \wedge [\beta']C, \varphi)$$
$$f([\beta^*]C, \varphi) = f(C \wedge [\beta][\beta^*]C, \varphi)$$

$$f(\mathbb{O}_C(\alpha_\&), \varphi) = \begin{cases} \top & \text{if } \alpha_\& \subseteq \varphi \\ C & \text{otherwise} \end{cases}$$

$$f(\mathbb{O}_C(\alpha; \alpha'), \varphi) = f(\mathbb{O}_C(\alpha) \wedge [\alpha]\mathbb{O}_C(\alpha'), \varphi)$$

$$f(\mathbb{O}_C(\alpha + \alpha'), \varphi) = \begin{cases} \top & \text{if } f(\mathbb{O}_\bot(\alpha), \varphi) = \top \text{ or } f(\mathbb{O}_\bot(\alpha'), \varphi) = \top \\ C & \text{if } f(\mathbb{O}_\bot(\alpha), \varphi) = \bot \text{ and } f(\mathbb{O}_\bot(\alpha'), \varphi) = \bot \\ \mathbb{O}_C(\alpha + \alpha'/\varphi) & \text{otherwise} \end{cases}$$

$$f(\mathbb{F}_C(\alpha_\&), \varphi) = \begin{cases} C & \text{if } \alpha_\& \subseteq \varphi \\ \top & \text{otherwise} \end{cases}$$

$$f(\mathbb{F}_C(\alpha; \alpha'), \varphi) = f([\alpha]\mathbb{F}_C(\alpha'), \varphi)$$
$$f(\mathbb{F}_C(\alpha + \alpha'), \varphi) = f(\mathbb{F}_C(\alpha) \wedge \mathbb{F}_C(\alpha'))$$
$$f(\mathbb{P}(\alpha_\&), \varphi) = \top$$
$$f(\mathbb{P}(\alpha \cdot \alpha'), \varphi) = f(\mathbb{P}(\alpha) \wedge [\alpha]\mathbb{P}(\alpha'), \varphi)$$
$$f(\mathbb{P}(\alpha + \alpha'), \varphi) = f(\mathbb{P}(\alpha) \wedge \mathbb{P}(\alpha'), \varphi)$$

Fig. 6.4 The residual contract function f

The algorithm for computing the residual function is not very complex, but it has a lot of cases and small subtleties making it difficult to understand. So, instead of explaining it in detail we rather give an illustrative example on a simple $\mathcal{CL}$ formula on how it works (see Example 60).

Example 60: Using the Residual Contract Function

Let us consider the $\mathcal{CL}$ contract $C \triangleq [a]\mathbb{O}(b) \wedge [b]\mathbb{F}(b)$. The application of f to contract C and action a, $f(([a]\mathbb{O}(b) \wedge [b]\mathbb{F}(b)), a)$, will give the conjunction of the application of f to both conjunct subformulae:

$$f(([a]\mathbb{O}(b)), a) \wedge f([b]\mathbb{F}(b)), a)$$

By applying the definition of f again to the first conjunct, we get:

$$f(([a]\mathbb{O}(b)), a) = \mathbb{O}(b)$$

as it matches the first occurrence of $f([\alpha_\&]C, \varphi)$. Continuing now with the second conjunct (same part of the definition applies), we get:

$$f([b]\mathbb{F}(b)), a) = \top$$

We thus have that:

$$f(([a]\mathbb{O}(b) \wedge [b]\mathbb{F}(b)), a) = \mathbb{O}(b)$$

which is in accordance with the intuition that if action a happens, then the obligation to do b is enacted and the rest of the contract is 'forgotten'.

The residual function f is just auxiliary; the automaton is built using the *construction function* f_c shown in Fig. 6.5 This functions takes as argument the initial state of the automaton s_0 (where $l(s_0) = C$). Besides the residual function f, f_c uses function f_d (shown in Fig. 6.6) that adds all the relevant deontic information to each state (we take $\alpha_\&$ to be equal to $a_1 \& \ldots \& a_n$).

$$
\begin{aligned}
f_c(s) = \quad & \text{if } l(s) = 1 \text{ then} \\
& \quad T := T \cup (s, 1, s) \\
& \text{if } l(s) = 0 \text{ then} \\
& \quad V := s \\
& \quad T := T \cup (V, 1, V) \\
& \text{otherwise} \quad \forall a \in A_\& \\
& \quad \text{if } \exists\, s' \in S \text{ s.t. } l(s') = f(l(s), a) \\
& \quad \text{then } T := T \cup (s, a, s') \\
& \quad \text{otherwise} \\
& \quad\quad \text{new } s' \\
& \quad\quad l(s') := f(l(s), a) \\
& \quad\quad S := S \cup s' \\
& \quad\quad T := T \cup (s, a, s') \\
& \quad\quad d(s') := f_d(l(s')) \\
& \quad\quad f_c(s')
\end{aligned}
$$

Fig. 6.5 The construction function f_c

$$
\begin{aligned}
f_d(C_1 \wedge C_2) &= f_d(C_1) \cup f_d(C_2) \\
f_d(\mathbb{O}(\alpha_\&)) &= \{\{O_{a_1}\}, \ldots, \{O_{a_n}\}\} \\
f_d(\mathbb{F}(\alpha_\&)) &= \{\{F_{a_1}\}, \ldots, \{F_{a_n}\}\} \\
f_d(\mathbb{P}(\alpha_\&)) &= \{\{P_{a_1}\}, \ldots, \{P_{a_n}\}\} \\
f_d(\mathbb{O}(\alpha + \alpha')) &= \{x \cup y \mid x \in f_d(\mathbb{O}(\alpha)) \\
&\qquad \text{and } y \in f_d(\mathbb{O}(\alpha'))\} \\
f_d(\text{ otherwise }) &= \emptyset
\end{aligned}
$$

Fig. 6.6 The deontic labelling function f_d

Note that we have omitted the case for $\oplus$ in the deontic labelling function description. In practice, two different automata are created for each one of the choices, and the analysis proceeds as usual. Also note that there is no explicit labelling function for $F(\alpha + \alpha')$ and $P(\alpha + \alpha')$, since these cases are reduced to conjunction.

We will not give a detailed reading of the main and auxiliary algorithms, but rather provide examples to get a feeling on how they work.

Example 61 provides a discussion on the automaton construction for contract $\mathbf{C} \triangleq [\mathbf{a}]\mathbb{O}(\mathbf{b}) \wedge [\mathbf{b}]\mathbb{F}(\mathbf{b})$.

Example 61: Building the Automaton for a Contract

Let us consider again the contract $[a]\mathbb{O}(b) \wedge [b]\mathbb{F}(b)$. The automaton is constructed by applying f_c to the state s_0, where $l(s_0) = [a]\mathbb{O}(b) \wedge [b]\mathbb{F}(b)$. Every possible transition is created (in this case, transitions labelled with a, b and $a\&b$ from this state to a new state labelled with the result of applying function f to the original formula and the label of the transition as parameters. Thus, the state that is reached with the transition labelled with action a is $f([a]\mathbb{O}(b) \wedge [b]\mathbb{F}(b), a) = \mathbb{O}(b)$. If there is another state with the same label, the transition will connect to the existing state and the new one will be discarded (this ensures termination). If there is no such a state, f_c is then recursively called on this new state. Eventually we either reach a satisfying state, a violating state, or a state already labeled with the formula.

The automaton corresponding to the contract of Example 61 is shown in Fig. 6.7. Note that what is written in each state is the *subformula* remaining to be satisfied. Formally speaking, each state will be 'marked' with the deontic information as defined by the function f_d. So, $\mathbb{O}(a)$ is a syntactic expression in $\mathcal{CL}$, while O_a is the corresponding 'marking' at the state saying there is an obligation of doing a.

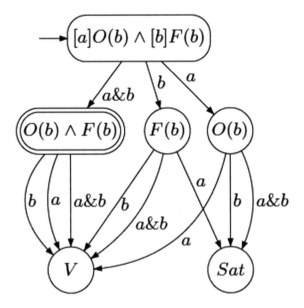

Fig. 6.7 Automaton for $[a]\mathbb{O}(b) \wedge [b]\mathbb{F}(b)$

Since our objective is to find conflicts by analysing the constructed automaton, we need to define what a conflict is at the automaton level. The definition is straightforward and it is very similar to the definition given for $\mathcal{CL}$ traces.

Definition 3 Given a state s of an automaton $A(C)$, let $D, D' \subseteq f_d(s)$. We say that D is *in conflict with* D' if and only if there exists at least one element e of D such that:

$$e = O_a \wedge (F_a \in D' \vee (P_b \in D' \wedge a\#b) \vee (O_b \in D' \wedge a\#b)),$$
$$\text{or } e = P_a \wedge (F_a \in D' \vee (P_b \in D' \wedge a\#b) \vee (O_b \in D' \wedge a\#b)),$$
$$\text{or } e = F_a \wedge (P_a \in D' \vee O_a \in D').$$

An automaton $A(C)$ is said to be *conflict-free* if for every state $s \in S$ and for any $D, D' \subseteq f_d(s)$, D and D' are not in conflict.

Example 62 shows how this definition of conflict is applied to a given automaton.

Example 62: Conflicts on Automata

The automaton shown in Fig. 6.7 is not conflict-free, since there exists a state which is not conflict-free. Consider that s is the double-lined state labelled with $\mathbb{O}(b) \wedge \mathbb{F}(b)$, then $f_d(s) = \{\{O_b\}, \{F_b\}\}$. Using Definition 3, let $e = O_b$. For this state to be conflict-free, any subset $D \in f_s(s)$ should not contain F_b, which is not the case.

6.5.1.3 Conflict Detection Algorithm

The main algorithm takes a contract written in $\mathcal{CL}$ and decides whether the given contract may reach a state of conflict. Once the automaton was generated from the contract as explained above, the conflict detection algorithm simply consists of a standard forward or backward reachability analysis looking for states containing conflicts. This analysis is based on a fixed-point computation as usually done for model checking.[3]

Example 63: Detecting Conflicts on Automata

Performing reachability analysis on the $\mathcal{CL}$ contract we have seen in a previous example, whose automaton is shown in Fig. 6.7, would identify that the state labelled $\mathbb{O}(b) \wedge \mathbb{F}(b)$, reachable from the initial state upon receiving action $a\&b$, is conflictive as it contains the deontic information $\{\{O_b\}, \{F_b\}\}$.

Example 63 shows how the algorithm determines the sequence of actions leading to the contradictory state for the contract discussed in Example 62.

6.5.1.4 Correctness of the Algorithm

We will not prove the correctness and completeness of the algorithm in detail, but we will show the essential steps to do so. We first need to prove the following auxiliary results: (i) the traces accepted by the automaton coincide with those 'accepted' by the contract in $\mathcal{CL}$ (according to the trace semantics); (ii) a contract C in $\mathcal{CL}$ is conflict-free if and only if the generated automaton $A(C)$ is conflict-free. The first part is stated as follows.

Lemma 1 *Given a $\mathcal{CL}$ contract C, the automaton $A(C)$ accepts all and only those traces σ that satisfy the contract:*

$$\sigma, \sigma_d \models C \text{ if and only if } \texttt{Accept}(A(C), \sigma).$$

The proof is based on a long and tedious induction on the structure of the formula, proving that f_c (and the auxiliary functions f and f_d) are complete and correct.

Note that our algorithm checks that no state contains a conflict rather than checking all possible satisfying runs. In order to prove that this is correct we need to prove that we generate only and all the reachable states.

---

[3] In this book, we do not explain how model checking algorithms for temporal logics work; we refer the reader to any standard book on the topic, e.g, [CGP99].

Proposition 1 *The function f_c generates all and only reachable states.*

Based on the above proposition and the definition of conflict at the trace and the automaton level, we can prove that the automata construction function preserves conflict-freedom, and that no spurious conflicts are generated.

Lemma 2 *A contract C written in $\mathcal{CL}$ is conflict-free if and only if the automaton $A(C)$ is conflict-free.*

With the above auxiliary results, and the correctness and completeness proofs of standard forward reachability analysis, we can finally prove our main result.

Theorem 1 *The $\mathcal{CL}$ conflict detection algorithm is correct and complete.*

Termination is trivially guaranteed since the generated automaton is finite and the reachability analysis is based on a standard computation.

6.5.1.5 The Conflict Analyser CLAN

The techniques and algorithms for conflict detection on $\mathcal{CL}$ contracts have been implemented in the prototype tool CLAN [FPS09c]. The user inputs a $\mathcal{CL}$ contract and the set of mutually exclusive actions, and CLAN generates the automaton, performs the analysis described in the previous subsection, and in case of conflicts provides counter-examples.

As a final remark on CLAN, it is worth noting that based on the existing algorithm, CLAN could easily be extended to perform additional analyses, detecting: (i) superfluous clauses; (ii) states labelled with a deontic notion multiple times; (iii) what it is enforced after a sequence of actions; (iv) what actions would lead to a specific obligation; (v) overlapping clauses; and (vi) clauses repeating similar deontic properties.

6.5.2 The AnaCon Framework

AnaCon [ACS12] is a framework where normative texts are written in *Controlled Natural Language* (CNL) and automatically translated into the formal language $\mathcal{CL}$ using the *Grammatical Framework* (GF). In AnaCon, $\mathcal{CL}$ expressions are analysed for normative conflicts by the CLAN tool, which gives counter-examples in cases where conflicts are found.

AnaCon takes as input a text file containing the description of a contract in two parts:

1. the contract itself written in CNL; and
2. a list of mutually exclusive actions.

```
[clauses]
if {the flight} leaves {in two hours} then both
  - {the ground crew} must open {the check-in desk}
  - {the ground crew} must request {the passenger manifest}
[/clauses]
[contradictions]
    {the ground crew} open {the check-in desk} #
    {the ground crew} request {the passenger manifest} ;
[/contradictions]
```

Fig. 6.8 Sample contract file in AnaCon format

Figure 6.8 shows a sample of the input file to the framework, containing part of the description of what an airline ground crew should do before flights leave (more on CNL later in this section).

The system is summarised in Fig. 6.9, where arrows represent the flow of information between processing stages. (For space considerations the picture is shown in 2 columns with the understanding that the flow continues from the bottom of the left figure to the up right part of the right figure.)

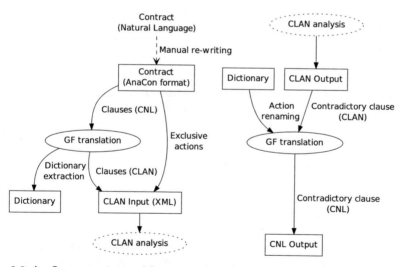

Fig. 6.9 AnaCon processing workflow

AnaCon essentially consists of a translation tool written in GF, the conflict analysis tool CLAN, and some script files used to connect these different modules together. The typical system workflow is as follows:

1. The user starts with a contract (specification, set of requirements, etc.) in plain English, which must be rewritten in CNL. This is primarily a modelling task, (currently) done manually. It requires no technical skills

from the user, but it does demand a knowledge of the CNL syntax and the set of allowed verbs.

2. The CNL version of the contract in AnaCon text format (Fig. 6.8) is then passed to the AnaCon tool, which begins processing the file.

3. The clauses in the contract are translated into their $\mathcal{CL}$ equivalents using GF. This translation is achieved by parsing the CNL clauses into abstract syntax trees, and then re-linearising these trees using the $\mathcal{CL}$ concrete syntax.

4. From the resulting $\mathcal{CL}$ clauses, a dictionary of actions is extracted. Each action is then automatically renamed to improve legibility of the resulting formulae, and a dictionary file is written. The list of mutually exclusive actions from the CNL contract is verified to make sure that each individual action actually does appear in the contract.

5. Using the renamed $\mathcal{CL}$ clauses from the previous step and the list of mutually exclusive actions, an XML representation of the contract is prepared for input into the CLAN tool.

6. This XML contract is then passed for analysis to CLAN via its command-line interface, which checks whether the contract contains normative conflicts. If no such conflicts are found, the user is notified of the success. If CLAN does detect any potential conflicts, the counter-example trace it provides is linearised back into CNL using the GF translator in the opposite direction. The dictionary file is used to re-instate the original action names.

7. The user must then find where the counter-example arises in the original contract. This last step must again be carried out manually, by following the CNL trace and comparing with the original contract.

We have seen the contract language $\mathcal{CL}$ and the conflict detection tool CLAN; in what follows we will briefly describe the two missing components of AnaCon, namely our CNL and GF.

6.5.2.1 Controlled Natural Language

Controlled Natural Languages (CNLs) are artificial languages engineered to be simpler versions of plain natural languages such as English. This simplicity is achieved by carefully selecting the vocabulary and by restricting the language's grammar, and also by not being generic but rather considering a specific application domain. Unlike plain natural languages, the simplifications applied to CNLs usually allow for a formal processing while remaining easy to understand by speakers of the original parent natural language.

As usual in Formal Methods, the more expressive the CNL is, the more complex is its automation of analysis. So, the language designer needs to find a good trade-off between expressiveness and formalisation, which is also affected by the richness of the parent natural language and the specific formalism in which the CNL is defined [WAB+10].

Among other applications, CNLs are useful when considering human-machine interactions which aim for an algorithmic treatment of language. One must answer the following questions when designing a CNL [WAB+10]:

1. Who are the intended users?
2. What is the main purpose of the language?
3. Is the language domain-dependent?

In our concrete setting our answers are:

1. The intended user is any person writing normative texts.
2. The main purpose of the language is that it is close enough to English as to be understood by any person, yet at the same time structured in such a way that its translation into $\mathcal{CL}$ is feasible.
3. The language is not specifically tailored for an application domain, however, it should be easy to parse it in such a way that obligations, permissions and prohibitions are easily identified.

We will not detail the whole CNL used in AnaCon but rather give a brief idea on some of its basic constructs, namely actions and some of the normative modalities.

Since the objective is to translate the CNL into $\mathcal{CL}$ we take *actions* as the most primitive elements in our CNL. In natural language actions correspond to sentences stating who is doing what, roughly following the English sentence structure:

<subject> <verb> <object>

More complex structures as adverbs, prepositional phrases, etc., cannot be expressed in $\mathcal{CL}$, so we omit them from the CNL altogether. Still, since we permit the subject and the object to be free text, the user has the freedom to include more information than just the noun phrase of the subject or the object. It is also possible to have ditransitive verbs, that is, verbs with more than one object. In this case we simply insert both objects in the free text slot for the object. If the verb is intransitive (without objects) then we can just leave the object slot empty. However, the slot for the verb is not free text and must come from a set of predefined verbs. While we do not have to analyse the subject and the object slots, the ability to analyse the verb is important since modal verbs like *must* and *may* have a special meaning in our context.

When analysing an action, we must be able to correctly identify the beginning and the end of each slot, which is difficult when there are free text slots. Our simple solution is to require that the object and subject must be surrounded with curly braces, so the user actually writes:

```
{the  ground  crew}  opens  {the  check−in  desk}
```

Actions can be combined in sequences, conjunctions, disjunctions, etc. using different keywords in the CNL. For instance, in Example 64 we show how the conjunction of two obligations is expressed in CNL.

Example 64: Natural Language versus Controlled Natural Language

Let us consider the following sentence in natural language:

> *The ground crew is obliged to open the check-in desk and request the passenger manifest two hours before the flight leaves.*

Using the CNL defined in AnaCon, this would be re-written as:

```
if  {the  flight}  leaves  {in  two  hours}  then  both
   −  {the  ground  crew}  must  open  {the  check−in  desk}
   −  {the  ground  crew}  must  request  {the  passenger  manifest}
```

We can now build more complex sentences by adding modalities for obligations, permissions and prohibitions. For instance, if we take the action *"the ground crew opens the desk"*, then the different modalities may be written in one of the following ways:

- Obligation:
  ```
  {the ground crew} must open {the desk}
  {the ground crew} shall open {the desk}
  {the ground crew} is required to open {the desk}
  ```
- Permission:
  ```
  {the ground crew} may open {the desk}
  it is optional for {the ground crew} to open {the desk}
  ```
- Prohibition:
  ```
  {the ground crew} must not open {the desk}
  {the ground crew} shall not open {the desk}
  ```

In the case of obligations and prohibitions, the user can specify a reparation clause which must be hold if the contract is violated. In the CNL the reparation is introduced with comma and the keyword *otherwise* after the main action. For example:

```
{the  ground  crew}  must  open  {the  desk},  otherwise
{the  ground  crew}  must  pay  {a  fine}
```

Here we can have an arbitrarily long list of clauses, which are applied in the order in which they are written. The last clause is not followed by *otherwise*, which indicates that its reparation is ⊥.

Even though the structure of the CNL version is noticeably less natural, it is sufficient for our purposes to be merely *close enough* to English as to be understood by any non-technical person.

6.5.2.2 The Grammatical Framework

We have introduced the formal language $\mathcal{CL}$ and a CNL tailored to be translated into (and from) $\mathcal{CL}$, but we have not explained how this bi-directional translation is done. For that we use of the Grammatical Framework (GF) [Ran11] as a grammar formalism and runtime parser/lineariser for converting between CNL and $\mathcal{CL}$.

GF is a logical framework allowing to define logics and languages tailored for specific purposes. It is equipped with mechanisms for mapping abstract logical expressions to a concrete language. While the logical framework encodes the language-independent structure (ontology) of the current domain, all language-specific features can be isolated in the definition of the concrete language. In other words, the definitions in the logical framework comprise the *abstract syntax* of the domain, while the *concrete syntax* is kept clearly separated.

Since GF has both a *parser* and a *lineariser*, the abstract syntax can serve as an interlingua: when a sentence is parsed from the source language, then the meaning of the sentence is extracted as an expression in the abstract syntax. The abstract expression then can be linearised back into some other language and this gives us bi-directional translation between any two concrete languages. In AnaCon, we have two concrete syntaxes – one for English (CNL) and one for the source language of CLAN ($\mathcal{CL}$).

Another important advantage of GF from an engineering point of view is the availability of the *Resource Grammar Library* (RGL) [Ran09]. Since every domain is logically different, it is also necessary to define different concrete syntaxes. When these are natural languages, then it means that a lot of tedious low-level details like word order and agreement have to be implemented again and again for each application. Fortunately, RGL provides general linguistic descriptions for several natural languages which can be reused by using a common language independent API. We implemented the AnaCon syntax for English by using this library, which both simplifies the development and makes it easy to port the system to other languages.

6.5.2.3 Case Studies

AnaCon has been applied to the two of the motivating examples presented in Sect. 6.2.2 (Examples 53 and 54). In both cases the process started by first manually translating the document in natural language into our CNL. AnaCon was then applied and many conflicts were detected. A careful analysis of the first counter-examples gave an idea on whether the problem was at the original document (a real conflict), or at the CNL (due to a wrong modelling). At the beginning most of the conflicts detected were due to a wrong modelling, so after few iterations to get a good CNL, the conflicts

that occurred where due to ambiguities in the way the original documents
were written.

We briefly present in what follows a summary of the result of applying
AnaCon to Example 53. (A full description of the case studies is reported in
[ACS13].) We start by showing the modelling and re-writing process of the
example. Let us consider two clauses from the specification and show their
equivalent CNL representations.

> **Original:** *The ground crew is obliged to open the check-in desk and request the
> passenger manifest from the airline two hours before the flight leaves.*
>
> **CNL:**
>
> ```
> if {the flight} leaves {in two hours} then
>    {the ground crew} must open {the check-in desk} and
>    {the ground crew} must request {the passenger manifest from
>    the airline}
> ```

For this clause, AnaCon gives the following $\mathcal{CL}$ formula as output:

$\mathcal{CL}$: [b3]O(a7&b2)

where from the dictionary file (automatically generated by AnaCon) we see
that:

```
b3 = {the flight} leave {in two hours}
a7 = {the ground crew} open {the check-in desk}
b2 = {the ground crew} request {the passenger manifest from
the airline}
```

We show in what follows an example of a conjunction over clauses.

> **Original:** *Once the check-in desk is closed, the ground crew is prohibited from
> issuing any boarding pass or from reopening the check-in desk.*
>
> **CNL:**
>
> ```
> if {the ground crew} closes {the check-in desk} then both
>  - {the ground crew} must not issue {boarding pass}
>  - {the ground crew} must not reopen {the check-in desk}
> ```

AnaCon gives the following $\mathcal{CL}$ formula as output (again generating the
corresponding action names in the dictionary file):

$\mathcal{CL}$: [b6]((F(a1))∧(F(a4)))

Though we have taken the above two examples individually, in practice
we should consider set of clauses when writing the CNL as clauses often refer
to and depend on each other. When read in natural language, the reader can
easily make the connections between the different clauses, but when it comes
to modelling the contract formally these need to be handled explicitly.

As an example, clause 10 in the example specifies a CTD for violating *any
part* of the contract. Thus combining clauses 1, 8, 9, and 10 from the contract
in Example 53 we obtain:

CNL:

```
if {the flight} leaves {in two hours} then each of
  - {the ground crew} must open {the check−in desk}
      and {the ground crew} must request
      {the passenger manifest from the airline}
  - if {the ground crew} closes {the check−in desk} then
    each of
      - {the ground crew} must send {luggage information to
        airline}
      - {the ground crew} must not issue {boarding pass}
      - {the ground crew} must not reopen {the check−in desk}
```

which results in the following $\mathcal{CL}$ formula:

$\mathcal{CL}$: [b4]((O(b1&a2))∧[b6]((O(b2))∧((F(a1))∧(F(a4))))))

When processed with AnaCon, the first conflicting state reported was reached after a single action:

```
1 counter example found
Clause:
  (((O(a7&b2))_(Oa3))ˆ(((Oa2)_(Ob1))ˆ(([a7]((O(a6.(b4.(a8.a5)
  )))_(Ob7)))ˆ(((F(b5)_(Oa3))ˆ(((Ob6)_(Oa3))ˆ(([b6](Oa9))ˆ((
  [b6](Fa1))ˆ([b6](Fa4)))))))))))
Trace:
  1. the flight leave in two hours
```

The counter-example above contains 2 parts: (i) a $\mathcal{CL}$ formula, and (ii) a trace in CNL. The first part is the formula representing the state of the automaton where the normative conflict happens, which is not particularly interesting to the end user. The second part is a linearisation of the output of CLAN showing what is the sequence of actions leading to the conflict; in this case only one.

A quick analysis of the original contract reveals that the two mutually exclusive actions *opening the check-in desk* and *closing the check-in desk* were erroneously set as mandatory at the same level in the contract. This was a modelling error, and was corrected in a second version of the CNL.

A second version was then rewritten, and the process applied again. After few iterations were more conflicts are found, we arrived to the following rewriting of the final part of the contract:

CNL:

```
if {the airline crew} provides {the passenger manifest to the
ground crew} then each of
  - first {the check−in crew} must initiate {the check−in
  process} ...
  - if {the flight} leaves {in 20 mins} then both
    - {the ground crew} must close {the check−in desk}
    - if {the ground crew} closes {the check−in desk} then
```

```
each of
  - {the ground crew} must send {the luggage information
    to the airline}
  - {the ground crew} must not issue {boarding pass}
  - {the ground crew} must not reopen {the check−in desk}
```

Generated $\mathcal{CL}$: [a6]((O(a9&...))∧
                ([a5]((O(b6))∧[b6]((O(b2))∧((F(a7))∧(F(a4))))))))

Finally, after the iteration process described above we arrived at a final version of the contract without conflicts.

6.5.3 Runtime Verification of Contracts

By definition a contract may be violated, so we should provide means to monitor contracts at runtime. How to obtain a monitor from a given contract? Would it be possible to use the automaton from conflict analysis presented in Sect. 6.5.1? The answer is *yes* and *no*. It is indeed possible for some simple contracts to reuse the automaton as a monitor, though this is not the case in general. The main reason for that is that a monitor needs to: (i) be sufficiently concrete as to refer to the actual 'actions' in the system being monitored (for instance, refer to real methods calls); and (ii) be able to monitor not only actions but also computations (for instance compute averages, percentages, etc.).

So, though in theory it might be possible to extract a monitor from a $\mathcal{CL}$ contract (and indirectly then from a CNL) as done for conflict analysis, there still is a gap to make the approach practical, and more research is needed in order to give a satisfactory solution.

6.5.4 Model Checking Contracts

In order to perform model checking of contracts we have to: (i) get a Kripke structure (or a special kind of automaton, usually a Büchi automaton) of the contract; (ii) write what we want to prove on a property language (usually some kind of temporal logic); and (iii) encode everything on an existing model checker (or develop an *ad hoc* model checker for our specific language, proving that it is indeed possible).

There is no *ad hoc* model checker for $\mathcal{CL}$. However, it is possible to provide a rather involved encoding into existing model checkers, and perform model checking of $\mathcal{CL}$ contracts by following these steps:

1. Model the conventional contract (in natural language) as a $\mathcal{CL}$ expression.
2. Translate the $\mathcal{CL}$ specification into $\mathcal{C}\mu$ (a variant of the μ-calculus).

3. Obtain a Kripke-like model (LTS) from the $\mathcal{C}\mu$ formula.
4. Translate the LTS into the input language of the model checker NuSMV.
5. Perform model checking using NuSMV:

   5.1 check that the model is 'good'; and
   5.2 check some properties about the parties.

6. In case of a counter-example given by NuSMV, interpret it as a $\mathcal{CL}$ clause
   and repeat the model checking process until the property is satisfied.
7. If needed be rephrase the original contract and repeat the process.

The above has been done for the contract shown in this chapter as Example 54 [PPS07].

6.6 Closing Remarks

In this chapter we have presented the formal language $\mathcal{CL}$ for specifying normative texts. We have also shown that $\mathcal{CL}$ has different semantics, and that in particular the trace semantics is useful for monitoring purposes and as the basis for a conflict detection algorithm. We have introduced AnaCon, a framework where contracts are written in a CNL, translated into $\mathcal{CL}$ using GF and analysed for conflict using CLAN. We have finished by briefly mentioning how to perform runtime verification and model checking on $\mathcal{CL}$ contracts.

6.6.1 Annotated Bibliography

The notion of *programming-by-contracts* (or *design by contract*) has been introduced by B. Meyer as early as 1986 [Mey86]. Other notions of contracts as metaphor are JML [BCC+03]), Code Contracts [Log13]), and the notion of service contract specification languages in Service-Oriented Architectures (e.g., ebXML [ebx], WSLA [wsl], and as service level agreements (e.g., WS-Agreement [wsg]).

Pace and Schneider provide an extensive discussion on the semantical challenges in defining a formal language for contracts [PS09a].

The definition of the formal language $\mathcal{CL}$ for specifying (untimed) contracts (cf. Sect. 6.4) was mainly taken from [PS09b] (see also [PS07, PS12] as well as Prisacariu's PhD thesis [Pri10]). A prototype of the conflict analyser for $\mathcal{CL}$, the tool CLAN, is described by Fenech et al. [FPS09c]. A first trace semantics for $\mathcal{CL}$ has been introduced with the main aim of being used for monitoring purposes [KPS08]. The semantics was modified to make it also suitable for conflict analysis [FPS09a] (Sect. 6.4.2.1 is mostly based on the later paper).

Section 6.5 is based on [KPS08, FPS09a]. See [PPS07] for more details about model checking $\mathcal{CL}$ contracts (cf. Sect. 6.5.4).

Díaz et al. introduced the formal graphical language *C-O Diagrams*, extending $\mathcal{CL}$ among other things with real-time constraints (see [DCMS14] for an extended and updated version) [MCDS10]. Camilleri et al. [CPS14] presented a CNL for *C-O Diagrams* following a similar approach as the one presented in this chapter for $\mathcal{CL}$.

This idea of using a CNL as a natural language-like interface for a formal system is not new. In particular, see the Attempto controlled natural language [FKK08], which has played an influential role in the development of the area.

Initial work showing the feasibility to relate $\mathcal{CL}$ and a CNL has been implemented in the tool AnaCon [ACS13, ACS12]. A more detailed description of this framework as well as its application to the two case studies appearing in this chapter (cf. Examples 53 and 54) appears in [ACS13, MRS11].

More recently, Camilleri et al. have defined a new CNL for C-O Diagrams and developed a proof-of-concept web-based tool to transform normative documents in natural language into a formal representation. That way it is possible to perform syntactic and semantic queries, the latter via a translation into UPPAAL timed automata [CGS16, CS17, CHS18, Cam17].

For more details on the Grammatical Framework, see [Ran11].

This chapter has focused on $\mathcal{CL}$ and related tools. The area is, however, quite broad and it is difficult to give an exhaustive list of related work. In what concerns the formalisation of normative concepts in general using deontic logic and other formalisms, see for instance publications appearing in the series of conferences DEON [Deo19], Jurix [AR19] and ICAIL [ICA19], and in the Journal of Artificial Intelligence and Law [Jou19]. For examples of the use of CNL for other formal contracts languages see, for instance, [AGP16, CCP16, ACP18]. See also papers appearing in the CNL workshops [DWK18].

6.6.2 Current Research Directions

The specification and analysis of normative documents (including electronic and legal contracts) is an active research area still at the edge of the state-of-the-art. We could take the contract analysis framework depicted in Fig. 6.1 to guide our discussion on research trends and challenges.

The first challenge is concerned with natural language processing: how to go from a contract written in natural language to a CNL version. Today this is mostly done manually. A first step into this direction is the work by Camilleri et al. that uses the Stanford parser to extract relevant information used to build a CNL [CGS16, CS17, CHS18]. The process is semi-automatic and requires a post-processing that requires knowledge of the grammar of the CNL as well as of the underlying translation process. The aim would be to increase automation, for instance by using machine learning techniques.

The CNL as presented here could be improved to be more expressive, for instance by considering the possibility of describing temporal and timing issues. Also a better treatment of a sentence being able to identify subjects would be a step towards liability analysis. Similarly, having a richer CNL is of no use if the formal language to which this is translated (in this chapter, $\mathcal{CL}$) is not rich enough as to capture real-time issues. C-O Diagrams are definitely an improvement in this sense, but more research is needed in this direction.

Enriching the formal language comes also with the challenge of defining better and richer analysis tools (beyond the conflict detection we have presented). A rich property and query language would be required, as well as algorithmic solutions (and tools) to analyse contracts.

We know that obtaining a runtime monitor directly from the contract is not easy. Current solutions only work for simple contracts; as soon as contracts refer to complex computations, existing techniques are not applicable. We have already mentioned that the automata obtained from $\mathcal{CL}$ contracts are too abstract to be directly used as monitors, and cannot handle complex algorithmic content. This might require the definition of richer libraries with standard computations, and more sophisticated algorithms to plug in such libraries in the monitor extraction process.

Finally, more research needs to be conducted towards the analysis of contracts at runtime. Once a violation is detected it is important to be able to detect liabilities and causalities. Digital forensics is also an interesting direction: how can the logs of the monitors be used as legal evidence in disputes concerning liabilities?

References

[ACP18]   Shaun Azzopardi, Christian Colombo, and Gordon J. Pace. A controlled natural language for financial services compliance checking. In *CNL'18*, volume 304 of *Frontiers in Artificial Intelligence and Applications*, pages 11–20. IOS Press, 2018.

[ACS12]   Krasimir Angelov, John J. Camilleri, and Gerardo Schneider. AnaCon. `http://www.cse.chalmers.se/~gersch/anacon`, Jan 2012.

[ACS13]   Krasimir Angelov, John J. Camilleri, and Gerardo Schneider. A framework for conflict analysis of normative texts written in controlled natural language. *Journal of Logic and Algebraic Programming*, 82(5-7):216–240, July-October 2013.

[AGP16]   Shaun Azzopardi, Albert Gatt, and Gordon J. Pace. Reasoning about partial contracts. In *JURIX'16*, volume 294 of *Frontiers in Artificial Intelligence and Applications*, pages 23–32. IOS Press, 2016.

[AR19]   Michal Araszkiewicz and Víctor Rodríguez-Doncel, editors. *The 32nd Annual Conference on Legal Knowledge and Information Systems*. IOS Press, 2019.

[BCC+03] Lilian Burdy, Yoonsik Cheon, David R. Cok, Michael D. Ernst, Joseph Kiniry, Gary T. Leavens, K. Rustan M. Leino, and Erik Poll. An overview of JML tools and applications. In *FMICS'03*, volume 80 of *ENTCS*, pages 75–91, 2003.

[Cam17]   John J. Camilleri. *Contracts and Computation: Formal Modelling and Analysis for Normative Natural Language*. PhD thesis, Department of Computer Science and Engineering, University of Gothenburg, Gothenburg, Sweden, 2017.

[CCP16]   Aaron Calafato, Christian Colombo, and Gordon J. Pace. A controlled natural language for tax fraud detection. In *CNL'16*, LNCS 9767, pages 1–12. Springer, 2016.

[CGP99]   Edmund M. Clarke, Orna Grumberg, and Doron A. Peled. *Model checking*. The MIT Press, 1999.

[CGS16]   John J. Camilleri, Normunds Gruzitis, and Gerardo Schneider. Extracting Formal Models from Normative Texts. In *NLDB'16*, LNCS 9612, pages 403–408. Springer, 2016.

[CHS18]   John J. Camilleri, Mohammad Reza Haghshenas, and Gerardo Schneider. A Web-Based Tool for Analysing Normative Documents in English. In *SAC-SVT'18*, pages 1865–1872. ACM, 2018.

[CPS14]   John J. Camilleri, Gabrielle Paganelli, and Gerardo Schneider. A CNL for contract-oriented diagrams. In *CNL'14*, volume 8625 of *LNCS*, pages 135–146. Springer, 2014.

[CS17]   John J. Camilleri and Gerardo Schneider. Modelling and analysis of normative documents. *Journal of Logical and Algebraic Methods in Programming*, 91:33–59, October 2017.

[DCMS14]   Gregorio Díaz, M. Emilia Cambronero, Enrique Martínez, and Gerardo Schneider. Specification and Verification of Normative texts using C-O Diagrams. *IEEE Transactions on Software Engineering*, 40(8):795–817, 2014.

[Deo19]   DeonticLogic.org. http://deonticlogic.org/, 2019.

[DWK18]   Brian Davis, Adam Z. Wyner, and Maria Keet, editors. *6th International Workshop on Controlled Natural Language (CNL'16)*, volume 304 of *Frontiers in Artificial Intelligence and Applications*. IOS Press, 2018.

[ebx]   ebXML: Electronic Business using eXtensible Markup Language. www.ebxml.org.

[FKK08]   Norbert E. Fuchs, Kaarel Kaljurand, and Tobias Kuhn. Attempto Controlled English for Knowledge Representation. In *Reasoning Web*, volume 5224 of *Lecture Notes in Computer Science*, pages 104–124. Springer, 2008.

[FPS09a]   Stephen Fenech, Gordon J. Pace, and Gerardo Schneider. Automatic Conflict Detection on Contracts. In *ICTAC'09*, LNCS 5684, pages 200–214. Springer, 2009.

[FPS09c]   Stephen Fenech, Gordon J. Pace, and Gerardo Schneider. Clan: A tool for contract analysis and conflict discovery. In *ATVA'09*, LNCS 5799, pages 90–96. Springer, 2009.

[ICA19]   17th international conference on artificial intelligence and law. http://www.iaail.org/, 2019.

[Jou19]   *Journal of Artificial Intelligence and Law*, 2019.

[KPS08]   Marcel Kyas, Cristian Prisacariu, and Gerardo Schneider. Run-time monitoring of electronic contracts. In *ATVA'08*, volume 5311 of *LNCS*, pages 397–407. Springer, October 2008.

[Log13]   Francesco Logozzo. Practical specification and verification with code contracts. In *HILT'13*, pages 7–8. ACM, 2013.

[MCDS10]   Enrique Martinez, Emilia Cambronero, Gregorio Diaz, and Gerardo Schneider. A Model for Visual Specification of e-Contracts. In *IEEE SCC'10*, pages 1–8. IEEE Computer Society, 2010.

[Mey86]   Bertrand Meyer. Design by contract. Technical Report TR-EI-12/CO, Interactive Software Engineering Inc., 1986.

[MRS11]   Seyed M. Montazeri, Nivir Roy, and Gerardo Schneider. From Contracts in Structured English to CL Specifications. In *FLACOS'11*, volume 68 of *EPTCS*, pages 55–69, 2011.

[PPS07]   Gordon J. Pace, Cristian Prisacariu, and Gerardo Schneider. Model checking contracts –a case study. In *ATVA'07*, LNCS 4762, pages 82–97. Springer, 2007.

[Pri10]   Cristian Prisacariu. *A Dynamic Deontic Logic over Synchronous Actions*. PhD thesis, Department of Informatics, University of Oslo, December 2010.

[PS07]   Cristian Prisacariu and Gerardo Schneider. A formal language for electronic contracts. In *FMOODS'07*, LNCS 4468, pages 174–189. Springer, 2007.

[PS09a]   Gordon J. Pace and Gerardo Schneider. Challenges in the specification of full contracts. In *Integrated Formal Methods (iFM'09)*, LNCS 5423, pages 292–306, 2009.

[PS09b]   Cristian Prisacariu and Gerardo Schneider. CL: An Action-based Logic for Reasoning about Contracts. In *WOLLIC'09*, LNCS 5514, pages 335–349. Springer, 2009.

[PS12]   Cristian Prisacariu and Gerardo Schneider. A dynamic deontic logic for complex contracts. *Journal of Logic and Algebraic Programming*, 81(4):458–490, 2012.

[Ran09]   Aarne Ranta. The GF resource grammar library. *Linguistic Issues in Language Technology*, 2(2), December 2009.

[Ran11]   Aarne Ranta. *Grammatical Framework: Programming with Multilingual Grammars*. CSLI Publications, 2011.

[WAB+10]   Adam Wyner, Krasimir Angelov, Guntis Barzdins, Danica Damljanovic, Brian Davis, Norbert Fuchs, Stefan Hoefler, Ken Jones, Kaarel Kaljurand, Tobias Kuhn, Martin Luts, Jonathan Pool, Mike Rosner, Rolf Schwitter, and John Sowa. On controlled natural languages: properties and prospects. In *CNL'09*, volume 5972 of *LNCS/LNAI*, pages 281–289. Springer, 2010.

[wsg]   Web Services Agreement Specification (WS-Agreement). https://www.ogf.org/documents/GFD.107.pdf.

[wsl]   WSLA: Web Service Level Agreements. https://dominoweb.draco.res.ibm.com/cdedb79080f59ee285256c5900654839.html.

Chapter 7
Formal Methods for Human-Computer Interaction

Antonio Cerone

Abstract Human-computer interaction adds the human component to the operational environment of a system. Furthermore, the unpredictability of human behaviour largely increases the overall system complexity and causes the emergence of errors and failures also in the systems that have been proved correct in isolation. Rather than trying to capture and model human errors that have been observed in the past, as it has been done traditionally in human reliability assessment, we consider cognitive aspects of human behaviour and model them in a formal framework based on the CSP process algebra. We consider two categories of human behaviour, automatic behaviour, mostly representative of a user carrying out everyday activities, and deliberate behaviour, mostly representative of an operator performing tasks driven by specific goals set up within the purpose of a working context. The human cognitive model is then composed with the physical interface/system and with a number of environmental aspects, including available resources, human knowledge and experience. Finally, the overall model is analysed using model checking within the verification framework provided by the Process Analysis Toolkit (PAT). The ATM case study from Chap. 3 and a number of other case studies illustrate the approach.

7.1 Human Errors and Cognition

You are back home from work, tired and hungry. Your partner welcomes you announcing that a nice cake is coming out of the oven soon and, this time, 'properly baked'. You sniff the air and perceive a light burning smell. You then recall that last time the cake did not properly rise, probably because the oven was kept open for too long while inserting the cake and thus the

Antonio Cerone
Nazarbayev University, Nur-Sultan, Kazakhstan

© Springer Nature Switzerland AG 2022, corrected publication 2022
M. Roggenbach et al., *Formal Methods for Software Engineering*,
Texts in Theoretical Computer Science. An EATCS Series,
https://doi.org/10.1007/978-3-030-38800-3_7

initial baking temperature was not high enough. Your partner is announcing that this time there won't be any problems with rising because

1. during the oven pre-heating phase, the temperature was set 20 degrees higher than the temperature indicated in the cake recipe,
2. when such higher temperature was reached, the oven was opened and the cake inserted (supposedly the opening of the oven would have decreased the temperature 20 degrees down, to the one indicated in the recipe), and
3. after closing the oven the temperature setting was immediately lowered to the value indicated in the recipe.

However, the burning smell you perceive is now getting stronger, clearly showing that something went wrong in performing the three-step algorithm above, which supposedly implement our 'baking task'. Your partner swears that the increase of 20 degrees is not too much, because it is a widely tested suggestion from a cooking internet forum and it is confirmed by many positive comments. Can you explain what went wrong? Well, there was some kind of cognitive error during the task execution. But which error exactly?

Normally, cognitive errors occur when a mental process aiming at optimising the execution of a task causes instead the failure of the task itself. The existence of a cognitive cause in human errors started to be understood already at the beginning of last century, when Mach stated: "knowledge and error flow from the same mental sources, only success can tell the one from the other" [Mac05]. But it took till the 1990s to understand that "correct performance and systematic errors are two sides of the same coin" [Rea90].

In our cake baking example, the three-step algorithm that implements the task is in principle correct, but the mental processes used to carry out the task may lead to a cognitive error. In fact, it is the human cognitive processing that does not perform the algorithm correctly, thus causing the error to emerge. Here, the key design point is that we cannot expect human behaviour to adapt to a specific algorithm when performing a task. It is instead the algorithm that must realise the task by taking human performance into account.

In the rest of this section we will briefly review the research trends and milestones in Formal Methods for HCI (Sect. 7.1.1) and state what we mean for *user* (Sect. 7.1.1) and *operator* (Sect. 7.1.1). Section 7.2 introduces the structure of human memory and its main cognitive processes and, in particular, short-term memory (STM), including alternative CSP-based models (Sect. 7.2.1), and long-term memory (LTM) and its further structuring (Sect. 7.2.2). Section 7.3 illustrates how to formally model human behaviour while Sect. 7.4 shows how to combine the model of the human component and the model of the interface to produce the overall model of the interactive system. Finally, Sect. 7.5 addresses the formal verification of the overall interactive system model and delves into the formal analysis of soundness and completeness of cognitive psychology theories; in Example 69.2 of Sect. 7.5.1 we will also reveal what cognitive error caused the cake to burn and why the algorithm used by your partner caused such an error to emerge.

7.1.1 Background

The systematic analysis of human errors in interactive systems has its roots in Human Reliability Assessment (HRA) techniques [Kir90], which mostly emerged in the 1980s. However, these first attempts in the safety assessment of interactive systems were typically based on *ad hoc* techniques [Lev95], with no efforts to incorporate a representation of human cognitive processes. within the model of the interaction.

During the 1980s and 1990s, the increasing use of formal methods led to more objective analysis techniques [Dix91] that resulted, on the one hand, in the notion of *cognitively plausible user behaviour*, based on formal assumptions to bind the way users act driven by cognitive processes [BBD00] and, on the other hand, in the formal description of expected effective operator behaviour [PBP97] and the formal analysis of errors performed by the operator as reported by accident analysis [Joh97]. Thus, research in the formal analysis of interactive systems branched into two separate directions: the analysis of cognitive errors of users involved in everyday-life [Cer11, CB04, CE07, RCB08] and work-related [MRO+15, RCBB14] interactive tasks, and the analysis of skilled operator's behaviour in traditionally critical domains, such as transportation, chemical and nuclear plants, health and defence [CCL08, CLC05, De 15, MPF+16, SBBW09]. The different interaction contexts of a user, who applies attention very selectively and acts mainly under automatic control [Cer11, NS86], and an operator, who deals with high cognitive load and whose attentional mechanisms risk to be overloaded due to coping with Stimulus Rich Reactive Interfaces (SRRIs) [SBBW09], have led to the development of distinct approaches, keeping separate these two research directions. However, users have sometimes to deal with decision points or unexpected situations, which require a 'reactivation' of their attentional mechanisms, and operators must sometime resort to automatisms to reduce attentional and cognitive loads.

In this chapter we propose a modelling approach [Cer16] that unifies these two contexts of human behaviour, which were traditionally considered separately in previous literature, namely

- **user**, i.e., a human who performs everyday activities in a fairly automatic way, and
- **operator**, i.e., a human who performs deliberate tasks making large use of attention explicitly.

User

User refers to ordinary people carrying out everyday activities, such as baking a cake, driving a car, using a smartphone, interacting with an ATM, etc. During such activities, users perform tasks that are initially triggered by specific goals, and then normally proceed in a fairly automatic way until the goal is accomplished.

As an example of everyday activity let us consider the natural language description of the user interaction with an ATM in Example 65 [Cer11, Cer16].

Example 65: ATM Withdrawal Task

The user's *goal* is 'cash withdrawal' and consists of the following basic activities (listed in no specific order).
- When the interface is ready, the user inserts the card and keeps in mind that the card has to be taken back at a later stage.
- When the interface requests a pin, the user enters the pin.
- When the cash has been delivered, the user collects the cash.
- When the card has been returned, the user collects the card and no longer needs to remember to collect it.

The goal 'cash withdrawal' is achieved when the cash is collected.

Notice that there is .no specific order among the basic activities. The user performs a specific basic activity depending on the observed state of the interface associated with that activity. Normally, some ordering is driven by the specific interface with which the user interacts. If we consider the general ATM description in Example 36 from Chap. 3, we notice that all ATMs will deliver the cash only after the user has inserted the card and entered the pin. And, obviously, the card can only be returned after being inserted. Specific ATMs impose further orderings, between card insertion and pin entering as well as between card return and cash delivery. However, if you approach the ATM to start an interaction and notice some cash already delivered, and supposedly forgotten by the previous user, ...you definitely collect it! (independently of whether you give it to the bank or you keep it.) Thus the basic activity of collecting cash may even be the first to occur while performing the task.

Although the task described in Example 65 requires some practice or training, during which the novice user performs deliberate actions, then, after repeated interactions, sufficient knowledge, skill and familiarity will be acquired, thus allowing the user to perform the task in a fairly automatic way. For example, an expert user will automatically insert the card in the right slot when the interface appears in the normal ready state, which the user is familiar with (whatever such a state looks like), and without any need to look for the appropriate slot (which is automatically reached by the hand movement). Such an acquired automatic behaviour allows the user to perform the task efficiently and quickly. However automatic behaviour is also the context in which typical cognitive errors analysed in previous research are most likely to occur as we will see in Sect. 7.5.1.

Furthermore, automatic behaviour is by no means purely reactive, but actually features an implicit, latent form of attention. In this chapter, we will see that, on various occasions during automatic behaviour, deliberate and conscious low-level actions are still required and, when this happens, attention becomes explicit and takes control. We will thoroughly explore the mechanism of attention and we will see that, in some situations, it may also be activated by the failure of those very expectations that the user has developed through experience and training, thus leading to the emergence of cognitive errors in the form of inappropriate deliberate responses.

Operator

Operator refers to a human who performs a task with a general purpose whereby specific goals are set along the way. In this case, failing to achieve the goal is not a task failure, provided the system state is still consistent with the purpose. Examples are operators of an Air Traffic Control (ATC) system, a nuclear power control room, a device to administer a therapy to a patient and a machine of an industrial plant. The operator's task is normally a monitoring one, which requires the performance of deliberate actions when the observed system behaviour is assessed as abnormal.

In Example 66 we consider the natural language description of a task of an operator interacting with a ATC simulator, which shows position, direction and speed of aircraft moving withing a specific sector of the air space [CCL08, CLC05].

Example 66: ATC Task

The operator's *purpose* is to ensure that the aircraft moving through the sector remain horizontally separated by no less than the defined minimum separation distance (5000 m): failure of this requirement is called *separation violation*. Vertical separation is ignored by the simulator. The operator can see position, direction and speed of the aircraft on the screen. The operator's task involves monitoring the movement of aircraft on the screen, searching for pairs of aircraft that are in conflict, that is, which may violate separation. This task comprises the following subtasks:

- **scan the screen** searching for a pair to monitor as possibly being in conflict,
- **classify the pair** as a conflict or a non conflict,
- **prioritise the conflict** by deciding whether there is a need for a plan to resolve an identified conflict,
- **decide an action** to resolve the conflict, possibly defer it or reclassify the conflict as a non conflict while trying to work out the plan of action,

- **perform the action** that has been decided, and
- **new phase subtask** whether to go back to scan the screen or perform an action that was previously deferred or exit the ATC operator role by abandoning the purpose (end of the simulation session, which in real-life would be the end of the operator shift).

Each subtask is driven by a goal, which is set deliberately under the influence of the purpose. For instance, the **scan the screen** task is driven by the deliberately set goal of identifying a part of the air space where there might be a conflicting pair of aircraft. Such a goal has an holistic flavour, since we cannot fully characterise all parameters that the operator considers in order to identify the critical part of air space. Furthermore, not being able to achieve the goal does not represent a task failure, since it is possible that no pair of aircraft violates separation, consistently with the ATC purpose.

Similarly,

- The purpose of the operator of a nuclear plant control room is to ensure the safe functioning of the plant. This purpose results in the monitoring of all system readout, searching for readout configurations that may be indicators of anomalies: goals are deliberately set in order to check specific readout configurations but also, in a more holistic way, by considering configurations which are not normally associated with anomalies and set new subgoals to further investigate them.
- The purpose of the operator of a machine of an industrial plant is to follow standard and specific operating procedures while using the machine. The operator must make deliberate choices depending on the perceived situation and consequently set goals that are consistent with the operating procedures. Furthermore, since operating procedures refer to generic situations and are by no means exhaustive, the operator's choices are not made among a predefined set of possibilities, but normally require a global assessment of the current situation.

We can conclude that an operator cannot automatically act in response to observations, but has to globally assess the observed situation and make informed, deliberate decisions on whether to act and what to do. Goals are thus established throughout the process according to the purpose of the task.

7.2 Human Memory and Memory Processes

Following the *information processing* approach normally used in cognitive psychology, we model human cognitive processes as processing activities that make use of input-output channels, in order to interact with the external environment, and three main kinds of human memory, in order to store information:

- **sensory memory**, where information perceived through the senses persists for a very short time,
- **short-term memory (LTM)**, also called *working memory*, which has a limited capacity and where the information that is needed for processing activities is temporarily stored with rapid access and rapid decay, and
- **long-term memory (LTM)**, which has a virtually unlimited capacity and where information is organised in structured ways, with slow access but little or no decay [DFAB04].

A usual practice to keep information in memory is *rehearsal*. In particular, *maintenance rehearsal* allows us to extend the time during which information is kept in STM, whereas *elaborative rehearsal* allows us to transfer information from STM to LTM.

7.2.1 Short-Term Memory and Closure

The limited capacity of short-term memory has been measured using experiments in which the subjects had to recall items presented in sequence. By presenting sequences of digits, Miller [Mil56] found that the average person can remember 7 ± 2 digits. However, when digits are grouped in *chunks*, as it happens when we memorise phone numbers, it is actually possible to remember larger numbers of digits. Therefore, Miller's 7 ± 2 rule applies to chunks of information and the ability to form chunks can increase people's STM actual capacity.

The limited capacity of short-term memory requires the presence of a mechanism to empty it when the stored information is no longer needed. When we produce a chunk, the information concerning the chunk components is removed from STM. For example, when we chunk digits, only the representation of the chunk stays in STM, while the component digits are removed and can no longer be directly remembered as separate digits. Generally, every time a task is completed, there may be a subconscious removal of information from STM, a process called *closure*: the information used to complete the task is likely to be removed from STM, since it is no longer needed.

We can use CSP to define a general STM model as shown in Example 67.

Example 67: Short-Term Memory: CSP Model

The STM model consists of n states STM i, with $i = 1, \ldots, n$, where
– n is the STM maximum capacity,

- *action* `store` *represents the storage of a piece of information and decreases the available capacity by one unit,*
- *action* `remove` *represents the removal of a piece of information and increases the available capacity by one unit,*
- *action* `closure` *represents the occurrence of a closure, due to the successful completion of the task, and completely clears STM, and*
- *action* `delay` *occurs every time STM is emptied and represents a time delay following the successful or unsuccessful end of the task.*

The empty STM of capacity 7 is modelled by process `STMempty` by defining it as `STM7`:

```
STMempty = STM7;
STM7 = store -> STM6 []
         delay -> STMempty []
         closure -> delay -> STMempty;
STM6 = store -> STM5 [] remove -> STM7 []
         delay -> STMempty []
         closure -> delay -> STMempty;
...
STM2 = store -> STM1 [] remove -> STM3 []
         delay -> STMempty []
         closure -> delay -> STMempty;
STM1 = store -> STM0 [] remove -> STM2 []
         delay -> STMempty []
         closure -> delay -> STMempty;
STM0 = store -> STMmanagement []
         remove -> STM1 []
         delay -> STMempty []
         closure -> delay -> STMempty;
STMmanagement = overloadedSTM ->  delay -> STMempty;
```

The attempt to store information in a full STM is handled by process `STMmanagement`, which in this example is associated with action `overloadedSTM` followed by a `delay`.

Notice that this memory model does not include the representation of the actual pieces of information that can be stored in STM. Information contents need to be represented by further CSP processes, one for every possible piece of information to define the two possible information states, stored and not stored. These further processes must synchronise with the CSP process in Example 67, thus resulting in a complex model, which is not easy to understand and manage and has limited scalability.

In order to develop a more intuitive, manageable and scalable model, we consider the CSP extension implemented in the *Process Analysis Toolkit (PAT)* [PAT19]. In particular, PAT provides integer variables and arrays as syntactic sugar to define system states, without any need to explicitly represent such states as additional synchronising processes. Processes can be then enabled by guards, which check the current values of variables, while events are annotated with performed assignments to variables and, more in general, with any statement block of a sequential program. Notice that the statement block is an atomic action, i.e. it is executed until the end without any interruption or interleaving. However, annotated events cannot synchronise with events of other processes, i.e., the parallel composition operator treats annotated events in the same way as the interleaving operator. PAT also supports the definition of constants, either singulnnand or as part of an enumeration, which associates consecutive integer numbers starting from 0 to the enumerated constants. For example

```
#define low 0;
#define medium 1;
#define high 2;
```

are three declarations of constants, which can be globally introduced in an alternative way as the enumeration

```
enum {low, medium, high};
```

The most obvious array implementation of STM would use each position of the array to store a piece of information. Thus the size of the array would represent the STM maximum capacity. However, the retrieval of information from STM would require to go through all elements of the array. Instead, we consider the implementation in Example 67.1.

Example 67.1: Short-Term Memory: PAT Model

The STM model consists of an array stm *whose capacity is given by the number of possible pieces of information that can be stored. Such a number is defined as a constant* InfoNumber, *which, in this example, equals 10. The various pieces of information (e.g.* Info) *are introduced using an enumeration. The STM maximum capacity is defined as a constant* M, *which, in this example, equals 7.*

```
enum { ... , Info , ... };

#define InfoNumber 10;
#define M 7;

var stmSize = M;
var stm[InfoNumber];
```

By default all positions of the array are initialised to 0. *The storage of information* Info *in the STM is performed by the occurrence of event* store *which is enabled by guard* stmSize < M , *which ensure that the STM is not full, and results in setting the* Info-th *position of array* stm *to* 1 *and incrementing variable* stmSize. *This is achieved with the following construct:*

```
[stmSize < M] store {stm[Info] = 1; stmSize++}
```

The retrieval and removal of information Info *from STM is performed by the occurrence of event* retrieve *which is enabled by guard* stm[Info] == 1, *which ensure that* Info *is in STM, and results in setting the* info-*th position of array* stm *to* 0 *and decrementing variable* stmSize. *This is achieved with the following construct:*

```
[stm[Info] == 1] retrieve {stm[Info] = 0; stmSize--}
```

Closure is achieved by resetting the contents of all positions of the stm array to 0 and assigning 0 to variable stmSize.

All aspects of closure implementation using PAT are explained in details in Sect. 7.3.3.

7.2.2 Long-Term Memory

Long term memory is divided into two types.

- **Declarative** or **explicit** memory refers to our knowledge of the world ('knowing what') and consists of the *events* and *facts* that can be *consciously* recalled:

  - our experiences and specific events in time stored in a serial form (*episodic memory*), and
  - structured record of facts, meanings, concepts and knowledge about the external world, which we have acquired and organised through association and abstraction (*semantic memory*).

- **Procedural** or **implicit** memory refers to our skills ('knowing how') and consists of *rules* and *procedures* that we *unconsciously* use to do things, particularly at the motor level.

Emotions and specific contexts and environments are factors that affect the storage of experiences and events in episodic memory. Information can be transferred from episodic to semantic memory by making abstractions and building associations, whereas *elaborative rehearsal* facilitates the transfer of information from STM to semantic memory in an organised form.

Note that also declarative memory can be used to do things, but in a very inefficient way, which requires a large mental effort in using the short-term memory (*high cognitive load*) and a consequent high energy consumption. In fact, declarative memory is heavily used while learning new skills. For example, while we are learning to drive, ride a bike, play a musical instrument or even when we are learning to do apparently trivial things, such as tying a shoelace, we consciously retrieve a large number of facts from the semantic memory and store a lot of information into STM. Skill acquisition typically occurs through repetition and practice and consists in the creation in procedural memory of rules and procedures (*proceduralisation*), which can be then unconsciously used in an automatic way with limited involvement of declarative memory and STM.

7.3 Human Behaviour and Interaction

In this section we present how to model the human components using PAT.

7.3.1 Input as Perceptions and Output as Actions

Input and output occur in humans through senses and the motor system. In this chapter we give a general representation of input channels in term of *perceptions*, with little or no details about the specific senses involved in the perception, but with a strong emphasis on the semantics of the perception in terms of its potential cognitive effects. For instance, if the user of a vending machine perceives that the requested product has been delivered, the emphasis is on the fact that the perception of the product being delivered induces the user to collect it and not on whether the user has seen or rather heard the product coming out of the machine. We represent output channels in term of *actions*. Actions are performed in response to perceptions.

In Example 65 of Sect. 7.1.1 we can identify a number of perceptions and actions, which we describe in Example 65.1

Example 65.1: Perceptions and Actions

Perceptions:

cardR	the interface is perceived ready,
pinR	the interface is perceived to request a pin,
cashO	the cash is perceived delivered, and
cardO	the card is perceived returned.

Actions:

cardI	the user inserts the card,
pinE.	the user enters the pin,
cashC	the user collects the cash, and
cardC	the user collects the card.

7.3.2 Cognitive Control: Attention and Goals

We have seen in Sect. 7.2.2 that skill acquisition results in the creation in procedural memory of the appropriate rules to automatically perform the task, thus reducing the accesses to declarative memory and the use of the STM, and, as a result, optimising the task performance. Inspired by Norman and Shallice [NS86], we consider two levels of cognitive control:

- **automatic control** is a fast processing activity that requires little or no attention and is carried out outside awareness with no conscious effort implicitly, using rules and procedures stored in the procedural memory, and
- **deliberate control** is a processing activity triggered and focussed by attention and carried out under the intentional control of the individual, who makes explicit use of facts and experiences stored in the declarative memory and is aware and conscious of the effort required in doing so.

For example, let us consider the process of learning to drive

Example 68: Learning to Drive a Car

Automatic control is essential in driving a car and, in such a context, it develops throughout a learning process based on deliberate control: during the learning process the driver has to make a conscious effort to use gear, indicators, etc. in the right way (deliberate control) and would not be able to do this while talking or listening to the radio. Once automaticity in driving is acquired, the driver is aware of the high-level tasks that are carried out, such as driving to office and stopping along the way to buy a newspaper, but is not aware of low-level details that automatically affect the action performance, such as changing gear, using the indicator and the colour of the light, amber or red, while stopping at a traffic light or even turning and whether stopping or not at a traffic light (automatic control).

Let us consider a narrative description of the baking task illustrated at the beginning of Sect. 7.1 in terms of perception, actions and information stored in and retrieved from the STM.

Example 69: Advanced Baking Task

Assuming that we have already put all ingredients in a bowl, the sequence of activities (which may be further decomposed) is as follows.
1. All ingredients are mixed in the bowl.
2. When the mix is perceived having the right consistency, it is poured in a tin.
3. The cake baking temperature is read on a recipe or retrieved from LTM and it is then stored in STM.
4. It is planned to set initially a temperature higher than the baking temperature.
5. The oven is switched on by setting the temperature higher than the cake baking temperature, keeping in mind that the temperature will have to be eventually lowered.
6. After the set temperature is reached, which is perceived through a distinctive warning sound, the oven is opened, the tin is inserted in the oven, the oven is closed and the timer is set, keeping in mind that the cake will have to be eventually taken out of the oven.
7. The temperature setting is lowered to the cake baking temperature.
8. When the cake is baked, which is perceived through a distinctive warning sound associated with the timer, the oven is switched off.
9. The cake is removed from the oven.

Perceptions are briefly stored in sensory memory and only relevant perceptions are transfered to STM using *attention*, a selective processing activity that aims to focus on one aspect of the environment while ignoring others. We can see this focussing activity as the transfer of the selected perception from sensory memory to STM.

For both users and operators the top-level task can be decomposed in a hierarchy of goals and tasks until reaching *basic activities*, which do not require further decomposition and can be performed by executing a single action.

In our model of cognitive behaviour we consider a set Π of perceptions, a set Σ of actions, a set Γ of goals, a set Ξ of purposes, and a set Δ of pieces of cognitive information. The information that can be processed by the human memory is given by the set

$$\Theta = \Pi \cup \Sigma \cup \Gamma \cup \Xi \cup \Delta.$$

In our model, we assume that a piece of information in Θ may belong to one or more of the following categories.

- **Perception transferred to STM** (set Π): a perception transferred from sensory memory to STM as the result of attention.
- **Reference to the future** (set Σ): an action to perform at some point in the future.
- **Cognitive state** (set Δ): a description of the human knowledge about a state of the task or of the system.
- **Received/retrieved information** (set Δ): a piece of information that has been received (i.e., read or heard) or retrieved from LTM.
- **Goal** (set Γ): the outcome of the task, which is initially in STM.
- **Purpose** (set Ξ): the underlying reason for performing the task, which normally influences the goal.

All categories of information apart from purposes may be stored in STM. Therefore $STM \in 2^{\Theta \setminus \Xi}$.

Example 69.1: Categories of information

In the baking task we can distinguish the six possible categories of information.

Perception transfered to STM

> The perceived sound that the oven has reached the right temperature is transferred to STM in Activity 6 and will be then retrieved once another task, which is carried out while waiting for the oven to heat, has been completed or can be interrupted (which will occur in Activity 6).

Reference to the future

> References to the future action of lowering the temperature (to be performed in Activity 7) and to the action of taking the cake out of the oven (to be performed in Activity 9) are stored in STM in Activities 5 and 6.

Cognitive state

> Activities 2, 3 and 6 must store a cognitive state pointing at the next basic activity in order to ensure the correct sequentialisation; in addition Activity 3 must remove its cognitive state, stored by the previous basic activity.

Received/retrieved information

> The read/retrieved baking temperature (Activity 3) is transferred to STM.

Goal and Purpose

> The goal of having the cake baked is initially in STM and is influenced by the purpose of baking the cake in a way that ensure proper raising.

Notice that all categories of information, except for the cognitive state and purpose, are explicit in the narrative description.

A task goal is formally modelled as

$$goal(info)$$

where $info \in 2^{\Theta \backslash \Gamma} \backslash \{\emptyset\}$ is a non-empty set of pieces of information except goals.

Information $info$ characterises the accomplishment of the goal, which results in flashing out STM.

7.3.3 Automatic Control

In automatic control our behaviour is not affected by goals but is driven by perceptions plus pieces of information 'automatically' stored in STM during the top-level task processing. As an example of automatic control let us consider the natural language description of driving a car.

Example 70: Car Driving

Suppose that during working days we always drive to our office, whereas on Saturdays we drive to a supermarket, initially taking the same route as to the office, but then turning into a different road.

It might sometimes happen, especially in a situation of high cognitive load, that we actually drive to our office rather than to the supermarket, as instead we intended. The underlying *cognitive reason (genotype error)* of this *observed error (phenotype error)* is that our automatic control (not driven by the goal to go to the supermarket) may not switch to deliberate control (driven by the goal to go to the supermarket) when we reach the intersection where the two routes diverge.

For each $A \subseteq \Theta$ we define $\bar{A} = \{\bar{\imath} \mid i \in A, \ i \notin \Xi \cup \Gamma\}$ and $\hat{A} = A \cup \bar{A}$. Each element $\bar{\imath} \in \bar{\Theta}$ denotes the absence of the piece of information $i \in \Theta$. Obviously $\hat{\emptyset} = \bar{\emptyset} = \emptyset$.

We model a basic activity under automatic control (*automatic activity*) as a quadruple $(perc, info_1, info_2, act)$, where

- $perc \in \Pi$ is a perception,
- $info_1 \in 2^{\hat{\Theta} \backslash \Xi \backslash \Gamma}$ is the information retrieved and removed from STM,
- $info_2 \in 2^{\Theta \backslash \Xi}$ is the information stored in STM, and
- $act \in \Sigma$ is a human action.

The quadruple $(perc, info_1, info_2, act)$ is subsequently written as

$$info_1 \uparrow perc \Longrightarrow act \downarrow info_2.$$

We formally denote by *none* when a component of a basic activity is absent (perception, action) or is the empty set (information).

Actions may involve an interaction with the system interface or be purely human physical actions with no support from the system. A basic activity whose action is an interaction is called *interactive activity*. A basic activity whose action is a physical action is called *physical activity*. Information is kept promptly available, while it is needed to perform the current top-level task, by storing it in STM. A basic activity is *enabled* (and can be performed) when

- $info_1 \cap \Theta \subseteq STM$,
- there exists $info_3 \subseteq \Theta$ such that $info_1 \cap \bar{\Theta} = \overline{info_3}$ and $info_3 \cap STM = \emptyset$, and
- *perc* is available in the environment.

Thus the basic activity is triggered by the presence of $info_1 \cap \Theta$ in STM, the absence of $info_3 \subseteq \Theta$ from STM, with $\overline{info_3} = info_1 \cap \bar{\Theta}$, and the presence of *perc* in the environment.

The performance of the basic activity results in the removal of $info_1 \cap \Theta$ from STM, the execution of action *act* and the storage of $info_2$ in STM. Therefore, in the absence of closure, the performance of the basic activity changes the value of STM from STM to

$$STM' = (STM \setminus info_1) \cup info_2.$$

When $goal(info) \in STM$, the performance of the basic activity causes closure if

$$info \setminus \Xi \subseteq (STM \setminus info_1) \cup info_2 \cup \{perc, act\}$$

where STM is the content of STM before the performance of the basic activity. In the presence of closure, the performance of the basic activity changes the STM from STM to

$$STM' = (STM \setminus \{info_1, goal(info)\}) \cap \Gamma \cup info_2.$$

Therefore, the closure is determined by the perception, the performance of the action and some pieces of information in STM that make, possibly together with some purposes, the argument of the goal. The closure causes the removal from STM of all information except $info_2$ and the non achieved goals. Note that at least one component of the basic activity on the left of '$\Longrightarrow$' and one on its right have to be distinct from *none*. When the action is *none* and the perception present, the basic activity is an *automatic attentional activity*, in which implicit attention causes the transfer of a perception to STM. When both the action and the perception are *none*, the basic activity is called *cognitive activity*.

The automatic behaviour described in Example 65 is formalised in Example 65.2

Example 65.2: Automatic Behaviour

Let be

Perceptions:	$\Pi = \{cardR, pinR, cashO, cardO\}$,
Actions:	$\Sigma = \{cardI, pinE, cashC, cardC\}$,
Purposes:	$\Xi = \emptyset$,
Cognitive Information:	$\Delta = \emptyset$,
Goals:	$\Gamma = \{goal(cashC)\}$,

Set Ξ is empty since the purpose is not relevant here.

A simple ATM task, in which the user has only the goal to withdraw cash, is modelled by the following four basic activities:

1. $none \uparrow cardR \Longrightarrow cardI \downarrow cardC$
2. $none \uparrow pinR \Longrightarrow pinE \downarrow none$
3. $none \uparrow cashO \Longrightarrow cashC \downarrow none$
4. $cardC \uparrow cardO \Longrightarrow cardC \downarrow none$

The goal ('to withdraw cash') is formally modelled as

$$goal(cashC)$$

Initially the STM only contains the goal:

$$STM = \{goal(cashC)\}$$

All basic activities in this task are automatic interactive activities. A reference to action $cardC$ is stored in STM by Activity 1, which will then be essential in enabling Activity 4. The goal is accomplished when action $cashC$ is performed in Activity 3.

Modelling Automatic Control using PAT

Example 65.3 illustrates how to use PAT to define the infrastructure to model the closure phenomenon for the ATM task described in Example 65.2. The task aims at achieving the goal of withdrawing cash (`getCashGoal`).

Example 65.3: Closure in Automatic Control using PAT

```
enum { getCashGoal };
enum { None,
        CardR, PinR, CashO, StatO, CardO ,
        CardI, PinE, CashC, CardC,
        Interaction }; // 10 items
```

```
#define G 1;      // No. of goal
#define N 10;     // No. of stm array positions
#define M 7;      // STM maximum capacity

var stmGoal = [ 0 ];
var stm[N];
var stmSize;
var perc[N];
```

The storage of goal in STM is modelled by the one position array
stmGoal. The content of this position is initialised to 0. Arrays stm
and perc implement the STM non-goal contents and the perceptions
available in the environment, respectively.

The closure controls the removal of the achieved goal and the removal of
non-goal information in order to free memory space for further processing
towards the achievement of other goals. Example 65.4 illustrates how to use
PAT to model the removal of goal getCashGoal for the ATM task.

Example 65.4: Closure in Automatic Control using PAT

```
Closure() = ba-> (
  [stmGoal[getCashGoal] == 1] cashC ->
    achieveGetCash {stmGoal[getCashGoal] = 0;
                    stmSize--;} -> FlashOut() []
  eact -> Closure() );
```

Event ba marks the beginning of the basic activity, eact marks the end
of the action, and event eba marks the end of the basic activity. Process
Closure is guarded by a condition on the presence of the goal in STM
(stmGoal[getCashGoal] == 1). When the action associated with the
goal is performed (cashC models that the cash is collected) the goal is
achieved (achieveGetCash) and removed from STM by changing to 0
the position of the stmGoal array corresponding to the achieved goal
(getCashGoal) and decrementing stmSize.

Example 65.5 illustrates how to use PAT to model the removal of the non-
goal information for the ATM task.

Example 65.5: Closure in Automatic Control using PAT

```
FlashOut() = closure { var cell = 0;
                       while (cell < M) {
                         if (stm[cell] == 1) {
                             stmSize--;
                         };
                         stm[cell] = 0 ;
```

```
                              cell = cell + 1;
                        }
                  } -> eact -> Closure();
```
Process FlashOut clears the contents of the non-goal part of the STM (array stm). In fact, the storage of goal in STM is separately implemented by array stmGoal to ensure that closure does not remove goals other that the achieved one.

Example 65.6 illustrates how to use PAT to initialise the task with the appropriate goal for the ATM task.

Example 65.6: Closure in Automatic Control using PAT

```
Goals() = [stmSize < M && stmGoal[getCashGoal] == 0]
                    getCash {stmGoal[getCashGoal] = 1;
                                    stmSize++} -> Goals() []
            [stmGoal[getCashGoal] == 1] ba -> eba -> Goals();
```
Process Goals initialises the task by adding the goal (getCashGoal) to the STM by setting to 1 the corresponding position of the stmGoal array and incrementing stmSize, provided that the STM does not exceeds its maximum capacity (stmSize < M).

In general, the storage of goals in STM is modelled by array stmGoal, whose positions are initialised to 0.

Example 65.7 illustrates how to use PAT to model the basic activities for the ATM task described in Example 65.4.

Example 65.7: Automatic Control Task using PAT

```
Task() = ba -> (
  [stmSize < M && perc[CardR] == 1]
    cardI -> eact -> store {stm[CardC] = 1; stmSize++}
    -> eba -> Task() []
  [perc[PinR] == 1]
    pinE -> eact -> eba -> Task() []
  [perc[CashO] == 1]
    cashC -> eact -> eba -> Task() []
  [stmSize > 0 &&
    perc[CardO] == 1 && stm[CardC] == 1]
    retrieve {stm[CardC] = 0; stmSize--}
    -> cardC -> eact -> eba -> Task()
              );

User() = Closure() || Goals() || Task();
```

Each basic activity $info_1 \uparrow perc \implies act \downarrow info_2$ is defined by one choice of the `Task` process. The choice is a process guarded by

- condition `perc[P] == 1`, if $perc \neq none$, where `P` is the position of array `perc` that implements perception $perc$,
- condition `stmSize > n`, if `n+1` is the number of pieces of information in $info_1$, and one condition `stm[I] == 1`, for each piece of information $i \in info_1$, if $info_1 \neq none$, where `I` is the position of array `stm` that implements i, and
- condition `stmSize < M-n`, where `M` is the maximum capacity of STM, if `n` is the cardinality of $info_2 \backslash info_1$.

The first event of the process implements action act. Annotated actions `store` and `retrieve` contain the assignments described in Example 67.1

Process `User` is the parallel composition of the three processes defined in Examples 65.4 and 65.7

Notice that the use of events `eact` and `eba` forces the closure to occur between the action performance and the storage of information in STM. In this way, the same basic activity that causes closure and removal of a goal or subgoal may also store a new goal or information in STM.

In Sect. 7.4 we will see how to combine this user model with an interface model. Then, in Sect. 7.5.1 we will show how to use model checking to formally verify properties of such an overall system.

7.3.4 Deliberate Control

In deliberate control, the role of the goal is not only to determine when closure should occur but also to drive the task: we act deliberately to achieve goals. Thus basic activities are not only driven by perceptions and non-goal information stored in STM, but also by one specific goal stored in STM. A typical case of deliberate behaviour is *problem solving*, in which the task goal is normally reached through a series of steps involving the establishing of subgoals. Achieving the subgoal takes the operator somehow closer to the task goal until this can be achieved directly. This process is illustrated in Example 71.

Example 71: Moving a Box

We need to move a box from point A to point B. The box is full of items. If the box is light enough then we just move it. Otherwise we have first to empty it, then move it and finally fill in it again. Emptying the box is a subgoal that allows us to move a heavy box.

In Example 71 we can identify a number of perceptions and actions as described in Example 71.1

Example 71.1: Perceptions and Actions

Perceptions:
light   the box is perceived light;
heavy the box is perceived heavy.

Actions:

moveBox   the human moves the box;
emptyBox   the human empties the box;
fillBox   the human fills in the box.

Cognitive information:

boxMoved   the fact that the box has been moved with its contents;
boxEmptied   the fact that the box is empty;
boxMovedEmpty   the fact that the box has been moved without
                   its contents.

We model a basic activity under deliberate control (*deliberate activity*) as a quintuple $(goal(info), perc, info_1, info_2, act)$, where

- $goal(info) \in \Gamma$ is the driving goal,
- $perc \in \Pi$ is a perception,
- $info_1 \in 2^{\hat{\Delta} \setminus \Xi \setminus \Gamma}$ is the information retrieved and removed from STM,
- $info_2 \in 2^{\Delta \setminus \Xi}$ is the information stored in STM, and
- $act \in \Sigma$ is a human action.

As above, the tuple is denoted as a rule:

$$goal(info) : info_1 \uparrow perc \Longrightarrow act \downarrow info_2.$$

If $info_1 = none$, the model of the basic activity can be shortened as

$$goal(act) \uparrow perc \Longrightarrow act \downarrow info_2$$

As for automatic activities, also a deliberate activity is

- *interactive* when its action is an interaction,
- *physical* when its action is a purely physical action,
- *attentional* when the action is *none* and the perception is present, and
- *cognitive* when both the action and the perception are *none*.

The basic activity is *enabled* (and can be performed) when

- $\{goal(info')\} \cup (info_1 \cap \Delta) \in STM$, with $info \subseteq info'$ and $info \setminus \Xi = info' \setminus \Xi$,
- $info_1 \cap \bar{\Delta} \notin STM$, and
- $perc$ is available in the environment.

The first condition means that the goal in STM is the same as the one in the basic activity apart from some additional purposes. In fact, on the one hand, a specific purpose $\xi \in \Xi$ does not prevent goals for less specific purposes to be used, since they will still get closer to the goal for purpose ξ, on the other hand, goals for more specific purposes should not be used, since they might take far away from the goal for purpose ξ. The performance of the basic activity and the closure are the same as in the case of automatic control.

The automatic behaviour described in Example 71 is formalised in Example 71.2

Example 71.2: Deliberate Behaviour

Let be
- $\Pi = \{light, heavy\}$,
- $\Sigma = \{moveBox, emptyBox, fillbox\}$,
- $\Xi = \emptyset$,
- $\Gamma = \{goal(boxMoved), goal(boxEmptied)\}$,
- $\Delta = \{boxMoved, boxEmptied, boxMovedEmpty\} \cup \Gamma \cup \Pi \cup \Sigma$.

Set Ξ is empty since the purpose is not relevant here.

The task is modelled by the following seven basic activities:

1. $goal(boxMoved) \uparrow light \Longrightarrow none \downarrow light$
2. $goal(boxMoved) \uparrow heavy \Longrightarrow none \downarrow heavy$
3. $goal(boxMoved) : light \uparrow none \Longrightarrow moveBox \downarrow boxMoved$
4. $goal(boxMoved) : heavy \uparrow none \Longrightarrow none \downarrow goal(boxEmptied)$
5. $goal(boxEmptied) \uparrow none \Longrightarrow emptyBox \downarrow boxEmptied$
6. $goal(boxMoved) :$
   $boxEmptied \uparrow none \Longrightarrow moveBox \downarrow boxMovedEmpty$
7. $goal(boxMoved) :$
   $boxMovedEmpty \uparrow none \Longrightarrow fillBox \downarrow boxMoved$

The task goal is formally modelled as

$$goal(boxMoved)$$

and requires the use of subgoal

$$goal(boxEmptied).$$

Initially the STM only contains the task goal:

$$STM = \{goal(boxMoved)\}$$

Example 71.3 shows the usage of STM while performing the task modelled in Example 71.2 with a heavy box.

Example 71.3: STM usage in Deliberate Behaviour

Initially STM contains the goal of moving the box ($goal(boxMoved)$). The evolution of the content of STM is driven by the deliberate control activities in LTM as described in Example 71.2.

- $STM = \{goal(boxMoved)\}$
  2. $goal(boxMoved) \uparrow heavy \Longrightarrow none \downarrow heavy$
- $STM = \{goal(boxMoved), heavy\}$
  4. $goal(boxMoved) : heavy \uparrow none \Longrightarrow none \downarrow goal(boxEmptied)$
- $STM = \{goal(boxMoved), goal(boxEmptied)\}$
  5. $goal(boxEmptied) \uparrow none \Longrightarrow emptyBox \downarrow boxEmptied$
  (Goal $goal(boxEmptied)$ achieved and removed due to closure)
- $STM = \{goal(boxMoved), boxEmptied\}$
  6. $goal(boxMoved) :$
     $boxEmptied \uparrow none \Longrightarrow moveBox \downarrow boxMovedEmpty$
- $STM = \{goal(boxMoved), boxMovedEmpty\}$
  7. $goal(boxMoved) :$
     $boxMovedEmpty \uparrow fillBox \Longrightarrow moveBox \downarrow boxMoved$
  (Goal $goal(boxMoved)$ achieved and removed due to closure)
- $STM = \{boxMoved\}$

After the fact that the box is heavy ($heavy$) is internalized through Activity 2, the performance of Activity 4 determines the addition of the new goal $goal(boxEmptied)$ to STM and Activity 4 determines the achievement of such a goal. Then Activity 6 determines the moving of the box and, finally, Activity 7 its refilling. The final mental state is the awareness that the box has been moved, which is modelled by the presence of cognitive state $boxMoved$ in STM. All goals have been removed from STM once achieved.

Modelling Deliberate Control using PAT

Example 71.4 illustrates how to use PAT to model the closure phenomenon for the task described in Example 71.3.

Example 71.4: Closure in Deliberate Control using PAT

```
enum { boxMovedGoal, boxEmptiedGoal};
enum { None,
       Heavy, Light,
       moveBox, emptyBox, fillBox,
       BoxMoved, BoxEmptied, BoxMovedEmpty,
       Interaction }; // 10 items
```

```
#define G 2; // No. of goal
#define N 10; // No. of stm array positions
#define M 7; // STM maximum capacity

var stmGoal = [ 0 , 0 ];
var stm[N];
var stmSize;
var perc[N];
var info[N];

Closure() = ba-> (
  [stmGoal[boxMovedGoal] == 1 &&
      (info[boxMoved] == 1 || stm[BoxMoved])]
    achieveBoxMoved {info[BoxMoved] = 0;
                     stmGoal[BoxMoved] = 0;
                     stmSize--;} -> FlashOut() []
  [stmGoal[boxEmptiedGoal] == 1 &&
      (info[boxEmptied] == 1 || stm[BoxEmptied])]
    achieveBoxMoved {info[BoxEmptied] = 0;
                     stmGoal[BoxEmptied] = 0;
                     stmSize--;} -> FlashOut() []
  eact -> Closure() );

 FlashOut() = closure { var cell = 0;
                        while (cell < M) {
                          if (stm[cell] == 1) {
                              stmSize--;
                          };
                          stm[cell] = 0 ;
                          cell = cell + 1;
                        }
                      } -> eact -> Closure();

Goals() =
  [stmSize < M && stmGoal[BoxMovedGoal] == 0]
    move {stmGoal[BoxMovedGoal] = 1;
                                stmSize++} -> Goals() []
  [stmGoal[BoxMovedGoal] == 1] ba -> eba -> Goals() []
  [stmSize < M && stmGoal[BoxEmptiedGoal] == 0]
    move {stmGoal[BoxEmptiedGoal] = 1;
                                stmSize++} -> Goals() []
  [stmGoal[BoxEmptiedGoal] == 1] ba -> eba -> Goals();
```

Array info implements the possibility that the information associated with
the goal achievement is a new piece of information stored in STM by the basic

activity. Thus not only the position of the stmGoal array corresponding to the achieved goal is changed to 0 but also the same position of array info. This is followed by the execution of process FlashOut, which clears the contents of the non-goal part of STM (array stm).

Example 71.5 illustrates how to model Example 71.2 in PAT:

Example 71.5: Deliberate Control Task using PAT

```
Task() = ba -> (
 [stmGoal[BoxMovedGoal] == 1 &&
  stmSize < M && perc[Light] == 1]
 newInfo {info[Light] = 1}
    -> eact -> store {stm[Light] = 1; stmSize++}
    -> eba -> Task() []
 [stmGoal[BoxMovedGoal] == 1 &&
  stmSize < M && perc[Heavy] == 1]
 newInfo {info[Heavy] = 1}
    -> eact -> store {stm[Heavy] = 1; stmSize++}
    -> eba -> Task() []
 [stmGoal[BoxMovedGoal] == 1 &&
  stm[Light] == 1 && stmSize < M]
 moveBox -> newInfo {info[boxMoved] = 1}
    -> eact -> store {stm[boxMoved] = 1; stmSize++}
    -> eba -> Task() []
 [stmGoal[BoxMovedGoal] == 1 &&
  stm[Heavy] == 1 && stmSize < M]
 eact -> store {stmGoal[BoxEmptiedGoal] = 1; stmSize++}
    -> eba -> Task() []
 [stmGoal[BoxEmptiedGoal] == 1 && stmSize < M]
 emptyBox -> newInfo {info[boxEmptied] = 1}
    -> eact -> store {stm[boxEmptied] = 1; stmSize++}
    -> eba -> Task() []
 [stmGoal[BoxMovedGoal] == 1 &&
  stm[boxEmptied] == 1 && stmSize < M]
 moveBox -> newInfo {info[boxMovedEmpty] = 1}
    -> eact -> store {stm[boxMovedEmpty] = 1; stmSize++}
    -> eba -> Task() []
 [stmGoal[BoxMovedGoal] == 1 &&
  stm[boxMovedEmpty] == 1 && stmSize < M]
 fillBox -> newInfo {info[boxMoved] = 1}
    -> eact -> store {stm[boxMoved] = 1; stmSize++}
    -> eba -> Task() []
                   );

User() = Closure() || Goals() || Task();
```

For each choice of the process corresponding to basic activity

$$goal(info) : info_1 \uparrow perc \Longrightarrow act \downarrow info_2$$

for each piece of information $i \in info_2$, the possibility that the information i is associated with the goal achievement is implemented by the assignment `info[I] = 1`, where I is the position of array `info` that implements i.

7.3.5 Operator's Deliberate Behaviour

Operator's behaviour is mainly deliberate. Although there is normally a prefixed sequence of basic activities through which the operator needs to go, each of these activities is driven by a specific goal to be accomplished. However, the operator task does not have a top-level goal. Instead it has a purpose, which influences all goals established (and accomplished) during the task performance.

The 'scan the screen' operator's subtask informally described in Example 66 may be formalised as in Example 66.1.

Example 66.1: 'Scan the Screen' Operator's Subtasks

Let be
- $\Pi = \{globalView, needsFurtherInvestigation, nothingAbnormal\}$,
- $\Sigma = \{moveBox, emptyBox, fillbox\}$,
- $\Xi = \{atcPurpose\}$,
- $\Gamma = \{goal(atcPurpose, identifiedPart),$
  $goal(atcPurpose, assessedPart)\}$
- $\Delta = \{identifiedPart, assessedPart, investigatedPart\} \cup \Gamma \cup \Pi \cup \Sigma$.

The task is modelled by the following four basic tasks:

1. $goal(atcPurpose, identifiedPart) \uparrow globalView$
   $\Longrightarrow identifiedPart \downarrow goal(actPurpose, assessPart)$
2. $goal(atcPurpose, assessedPart) \uparrow needsFurtherInvestigation$
   $\Longrightarrow none \downarrow goal(atcPurpose, investigatedPart)$
3. $goal(atcPurpose, assessedPart) \uparrow nothingAbnormal$
   $\Longrightarrow none \downarrow goal(atcPurpose, identifiedPart)$

Initially
$$STM = \{goal(atcPurpose, identifiedPart)\}$$

Through a global perception of the screen the operator identifies a part of the screen in which there might be a conflict (Activity 1) and sets the subgoal to assess that part $(goal(actPurpose, assessPart))$, while the closure due to the storage of information $identifiedPart$ causes the removal of goal $goal(atcPurpose, identifiedPart)$. If the part of the screen is perceived as in need of further investigation, then subgoal $goal(actPurpose, assessPart)$ is established (Activity 2). If, instead, nothing abnormal is noticed, then subgoal $goal(actPurpose, assessPart)$ is established. In both cases, the closure due to the storage of information $assessPart$ causes the removal of goal $goal(atcPurpose, assessPart)$ (Activity 3).

The purpose is present in STM as argument of some goals, as long as the operator is engaged in the task.

The 'new phase' operator's subtask informally described in Example 66 may be formalised as in Example 66.2.

Example 66.2: 'New Phase' Operator's Substask

Let be
- $\Pi = \{endTask\}$,
- $\Sigma = \emptyset$,
- $\Xi = \{atcPurpose\}$,
- $\Gamma = \{goal(atcPurpose, newPhase),$
  $\quad goal(actPurpose, identifyPart),$
  $\quad goal(atcPurpose, actedOnPair)$
- $\Delta = \{newPhase\} \cup \Gamma \cup \Pi \cup \Sigma$.

  The task is modelled by the following three basic tasks:

1. $goal(atcPurpose, newPhase) \uparrow none$
   $\implies none \downarrow goal(actPurpose, identifyPart)$
2. $goal(atcPurpose, newPhase) \uparrow none$
   $\implies none \downarrow goal(atcPurpose, actedOnPair)$
3. $goal(atcPurpose, newPhase) \uparrow endTask$
   $\implies none \downarrow newPhase$

Initially

$$STM = \{goal(atcPurpose, newPhase)\}$$

In Activity 3 the closure due to the storage of information $newPhase$ causes the removal of goal $goal(atcPurpose, newPhase)$, which is the only goal in STM influenced by purpose $atcPurpose$. Therefore, any trace of the purpose disappears from STM.

7.3.6 Switching Process Control

Familiar perceptions provide a mechanism to switch from deliberate control to automatic control. In an environment, with familiar perception, such as the ones provided by an ATM, the user behaviour proceeds independently of the goal that has triggered it. However, during automatic behaviour there are situations in which the cognitive control must switch back to deliberate control.

This situation is illustrated by Example 65.8, which extends the Example 65.2 by considering two possible goals 'cash withdrawal' and 'statement printing' for the ATM task.

Example 65.8: Automatic and Deliberate Behaviour

Let be
- $\Pi = \{cardR, pinR, selR, cashO, statO, cardO\}$,
- $\Sigma = \{cardI, pinE, cashS, statS, cashC, statC, cardC\}$,
- $\Xi = \emptyset$,
- $\Gamma = \{goal(cashC), goal(statC)\}$.
- $\Delta = \Gamma \cup \Pi \cup \Sigma \cup \{interaction\}$.

Set Ξ is empty since the purpose is not relevant here.

A simple ATM task, in which the user has only the goal to withdraw cash, is modelled by the following four basic tasks:

1. $goal(cashC) : \overline{interaction} \uparrow none \Longrightarrow none \downarrow interaction$
2. $goal(statC) : \overline{interaction} \uparrow none \Longrightarrow none \downarrow interaction$
3. $interaction \uparrow cardR \Longrightarrow cardI \downarrow cardC, interaction$
4. $interaction \uparrow pinR \Longrightarrow pinE \downarrow interaction$
5. $goal(cashC) \uparrow selR \Longrightarrow cashS \downarrow none$
6. $goal(statC) \uparrow selR \Longrightarrow statS \downarrow none$
7. $interaction \uparrow cashO \Longrightarrow cashC \downarrow interaction$
8. $interaction \uparrow statO \Longrightarrow statC \downarrow interaction$
9. $cardC, interaction \uparrow cardO \Longrightarrow cardC \downarrow interaction$

Perception $selR$ denotes that the ATM requests the user to select the transaction between 'cash withdrawal' and 'statement printing'. Perception $statO$ denotes that the statement has been delivered. Actions $cashS$ and $statS$ are the user's selections of 'cash withdrawal' and 'statement printing', respectively. Information $interaction$ models the cognitive state of the user interacting with the ATM; it is initially absent from STM.

The behaviour starts under deliberate control with one of the two possible goals, $goal(cashC)$ ('cash withdrawal') or $goal(cashC)$ ('statement printing') determining the beginning of the interaction (Activities 1 and 2, respectively) by storing $interaction$ in STM. The storage of $interaction$ in STM activates the automatic control driven by perceptions until perception $selR$ requires a decision about which transaction to select. Activities 5 and 6 determine the decision based on the goal in STM, thus under deliberate control. After the decision has been made, automatic control is restored for the rest of the task.

Example 65.9 illustrates how to use PAT to define the infrastructure and model the closure phenomenon for the ATM task described in Example 65.8.

Example 65.9: Closure in Automatic Control using PAT

```
enum { getCashGoal , getStatGoal};
enum { None,
       CardR, PinR, SelR, CashO, StatO, CardO ,
       CardI, PinE, CashS, StatS, CashC, StatC, CardC,
       Interaction }; // 15 items

#define G 1;  // No. of goal
#define N 15; // No. of stm array positions
#define M 7;    // STM maximum capacity

var stmGoal = [ 0 , 0];
var stm[N];
var stmSize;
var perc[N];

Closure() = ba-> (
   [stmGoal[getCashGoal] == 1] cashC ->
     achieveGetCash {stmGoal[getCashGoal] = 0;
                     stmSize--;} -> FlashOut() []
   [stmGoal[getStatGoal] == 1] cashC ->
     achieveGetCash {stmGoal[getStatGoal] = 0;
                     stmSize--;} -> FlashOut() []
   eact -> Closure() );

 FlashOut() = closure { var cell = 0;
                        while (cell < M) {
                          if (stm[cell] == 1) {
                              stmSize--;
                          };
                          stm[cell] = 0 ;
                          cell = cell + 1;
                        }
                       } -> eact -> Closure();]

Goals() =
[stmSize < M && stmGoal[getCashGoal] == 0 &&
  stm[Interaction] == 0] getCash {stmGoal[getCashGoal] = 1;
                                 stmSize++} -> Goals() []
[stmGoal[getCashGoal] == 1] ba -> eba -> Goals() []
[stmSize < M && stmGoal[getStatGoal] == 0 &&
```

```
    stm[Interaction] == 0] getCash {stmGoal[getStatGoal] = 1;
                                        stmSize++} -> Goals() []
[stmGoal[getStatGoal] == 1] ba -> eba -> Goals();
```

With respect to Example 65.4, processes Closure and Goals have the additional choice for the new goal. Moreover, guards in process Goals also include a condition on the absence of interaction from STM: initially there is no goal in STM and the role of process Goals is to establish one goal non deterministically, when the user in not interacting with the ATM.

Example 65.10 illustrates how to use PAT to model the basic activities for the ATM task described in Example 65.8.

Example 65.10: Automatic Control Task using PAT

```
Task() = ba -> (
 [stmGoal[getCashGoal] == 1 && stm[Interaction] == 0] eact
  -> store {stm[Interaction] = 1;
            stmSize++} -> eba-> Task() []
 [stmGoal[getStatGoal] == 1 && stm[Interaction] == 0] eact
  -> store {stm[Interaction] = 1;
            stmSize++} -> eba-> Task() []
 [stm[Interaction] == 1 && stmSize < M &&
 perc[CardR] == 1] cardI
  -> eact -> store {stm[CardC] = 1; stmSize++}
  -> eba -> Task() []
 [stm[Interaction] == 1 &&
 perc[PinR] == 1] pinE
  -> eact -> eba -> Task() []
 [stm[Interaction] == 1 &&
 perc[CashO] == 1] cashC
  -> eact -> eba -> Task() []
 [stm[Interaction] == 1 &&
  perc[StatO] == 1] statC
  -> eact -> eba -> Task() []
 [stm[Interaction] == 1 && stmSize > 0 &&
  perc[CardO] == 1 && stm[CardC] == 1]
  retrieve {stm[CardC] = 0; stmSize--} -> cardC
  -> eact -> eba -> Task()
             );

User() = Closure() || Goals() || Task();
```

7.4 Interface/System Model

In Sect. 7.3.1 we have defined *perceptions* to characterise the human input channel and *actions* to characterise the human output channels, with actions performed in response to perceptions. This describes the human perspective of input/output channels. From a machine perspective, we can also say that a user perception refers to an interface output, which acts as a stimulus for the human. During the interaction, such an output is normally the reaction of the interface to the action performed by the human.

Hence we identify a visible state created by an interface or system with the perception it produces in humans. For example, the interface state created by the action of giving change, performed by the interface of a vending machine, is identified with the perception (sound of falling coins or sight of the coins) produced.

Example 65.11 models one possible ATM interface to support the ATM task described in Example 65.8.

Example 65.11: Old ATM Interface using PAT

```
ATMold() =
  atomic{ [perc[CardR] == 1] cardI ->
    readCard {perc[CardR] = 0 ;
              perc[PinR] = 1} -> ATMold() } []
  atomic{ [perc[PinR] == 1] pinE ->
    readPin {perc[PinR] = 0 ;
             perc[SelR] = 1} -> ATMold() } []
  atomic{ [perc[SelR] == 1] cashS ->
    setCashS {perc[SelR] = 0 ;
              perc[CashO] = 1} -> ATMold() } []
  atomic{ [perc[SelR] == 1] statS ->
    setStatS {perc[SelR] = 0 ;
              perc[StatO] = 1} -> ATMold() } []
  atomic{ [perc[CashO] == 1] ( cashC ->
    detectCashC {perc[CashO] = 0;
                 perc[CardO] = 1} -> ATMold() ) } []
  atomic{ [perc[StatO] == 1] statC ->
    detectStatC {perc[StatO] = 0;
                 perc[CardO] = 1} -> ATMold() } []
  atomic{ [perc[CardO] == 1] cardC ->
    detectCardC {perc[CardO] = 0} ->
    reset {perc[CardR] = 1} -> ATMold() ) };
```

A simple interface may be modelled by a choice between all possible transitions. Each choice is guarded by the source state of the transition, normally represented by the perception provided to the user. The first event of the choice to be performed is the synchronisation event that models the interaction of the human (events CardI, PinE, CashS, StatS, CashC, StatC and CardC in Example 65.11). Each synchronisation event is followed by the 'local' interface event that modifies the interface state by assigning 0 to the source state and 1 to the target state. These local events my be split to increase the readability of the model, as it happens in the last choice of the ATMold process in Example 65.11.

In order to keep the synchronisation event and the associated interface events as an one atomic action, we use the *atomic process* construct available in PAT. The atomic keyword associates higher priority with a process: if the process has an enabled event, the event will execute before any events from non atomic processes. Moreover, the sequence of statements of the atomic process is executed as one single step, with no interleaving with other processes.

The ATM interface modelled in Example 65.11 was very common in the past. However, it was observed that delivering cash or statement before returning the card sometimes caused the user error of forgetting the card. This error is due to the fact that once the goal of collecting the cash or the statement is achieved, STM may be flashed out by the closure phenomenon thus losing some information, possibly including the reference to the action to collect the card. The discovery of this error led to the development of a new ATM interface that returns the card before delivering cash or statement. In terms of interface model this means that the user's selection of a transaction, although it results in the same user perception of seeing the card returned, should determine two distinct state transitions depending on whether the user selects 'cash withdrawal' or 'statement printing'. The new state will then produce the appropriate perception at a later stage.

In general, when dealing with a fairly complex behaviour, possibly resulting from the parallel composition of several subsystems, it is necessary to consider internal system states, which do not present themselves as human perceptions. Therefore, in addition to the perc array to implement perception, we also use an array state to implement internal states.

Example 65.12 models the new ATM interface.

Example 65.12: New ATM Interface using PAT

```
var state[N];

ATMnew() =
  atomic{ [perc[CardR] == 1] cardI ->
    readCard {perc[CardR] = 0;
              perc[PinR] = 1} -> ATMnew() } []
  atomic{ [perc[PinR] == 1] pinE ->
    readPin {perc[PinR] = 0;
```

```
                   perc[SelR] = 1} -> ATMnew() } []
      atomic{ [perc[SelR] == 1] cashS ->
        setCashS {perc[SelR] = 0; perc[CardO] = 1;
                    state[CashO] = 1} -> ATMnew() } []
      atomic{ [perc[SelR] == 1] statS ->
        setStatS {perc[SelR] = 0; perc[CardO] = 1;
                    state[StatO] = 1} -> ATMnew() } []
      atomic{ [state[CashO] == 1 && perc[CardO] == 1] cardC ->
        detectCardC {perc[CardO] = 0; perc[CashO] = 1;
                    state[CashO] = 0} -> ATMnew() } []
      atomic{ [state[StatO] == 1 && perc[CardO] == 1] cardC ->
        detectedCardC {perc[CardO] = 0; perc[StatO] = 1
                    state[StatO] = 0;} -> ATMnew() } []
      atomic{ [perc[CashO] == 1] cashC ->
        detectCashC {perc[CashO] = 0} ->
        reset {perc[CardR] = 1} -> ATMnew() } []
      atomic{ [perc[StatO] == 1] statC ->
        detectCashC {perc[StatO] = 0} ->
        reset {perc[CardR] = 1} -> ATMnew() };  };
```

7.4.1 Experiential Knowledge and Expectations

Section 7.2 illustrated various kinds of memory, which play different roles in processing information. Then in Sects. 7.3.3 and 7.3.4 we described automatic and deliberate behaviour, respectively, and provided a formal notation (and its implementation in PAT) to model basic activities under these two forms of cognitive control. If we wish to associate the location of the rules that model basic activities with distinct parts of the human memory, we can imagine that they are stored in LTM and, more specifically, that automatic basic activities are stored in procedural memory and deliberate basic activities are stored in semantic memory.

We have also mentioned that information may be transferred from sensory memory to STM through attention while facts and knowledge may be transferred from semantic memory to STM. We must add that information may flow from STM to LTM, first to episodic memory, and then produce changes to semantic and procedural memory. In fact, automatic and deliberated basic activities are created through a long-term learning process. In general, users make large use of deliberate activities while learning a task and, during the learning process, they create automatic rules in procedural memory to replace the less efficient rules in semantic memory. However, although automatic control is efficient and requires less STM usage than deliberate control, it may result inappropriate in some situation. In such a case, experiential knowledge already stored in the LTM may be used to solve the situation. Norman and Shallice [NS86] propose the existence of a *Supervisory Attentional System* (SAS), sometimes also called *Supervisory Activating System*, which becomes active whenever none of the automatic tasks are appropriate. The activation of the SAS is triggered by perceptions that are assessed as danger, novelty, requiring decision or are the source of strong feelings such as temptation and anger. The SAS is an additional mechanism to switch from automatic to deliberate control.

In Sect. 7.3.6 we described how to model the switching from automatic to deliberate control due to a required decision. Now we consider how such switching may occur due to the user's assessment of perceptions as the result of acquired experiential knowledge. Example 65.13 extends Example 65.9 with the infrastructure for representing factual and experiential knowledge and the mechanisms to assess perception and produce an appropriate response based on experiential knowledge.

Example 65.13: Closure with Experiential Knowledge

```
enum { safe , danger };      // assessment
enum { normal , abort };     // response
var assessment = safe;
var response = normal;

enum { getCashGoal , getStatGoal};
...

Closure() = ba-> (
  [response == normal && stmGoal[getCashGoal] == 1]
    cashC ->
    achieveGetCash {stmGoal[getCashGoal] = 0;
                    stmSize--;} -> FlashOut() []
  [response == normal && stmGoal[getStatGoal] == 1]
    cashC ->
    achieveGetCash {stmGoal[getStatGoal] = 0;
                    stmSize--;} -> FlashOut() []
  eact -> Closure() );

FlashOut() = ...

Goals() = ...
```

Variable `assessment` records the assessment of the user's perception following the user's action. We enumerate only two possible values: `safe` means that the perception will not affect the user's automatic control, wheras `danger` means that the perception requires a switch to deliberate control and an appropriate response, which will be assigned to variable **response**, whose possible values are **normal** and **abort**. Initially variable **assessment** has value **safe** and variable **response** has value **normal**.

Variable **assessment** is set depending on the user's expectation. We use additional processes to constrain the user's behavior depending on expectations. Such processes are specific to the considered interface/system. Example 65.14 defines the constraints for the ATM.

Example 65.14: Constraints Modelling Expectations

```
ExpectOld() = ba ->
( eba -> ExpectOld() []
  cashS -> eba ->
    ( [perc[CashO] == 1]
          cashExpectMet -> ExpectOld() []
      [perc[CardO] == 1]
          cashExpectFailed {assessment = danger}
              -> ExpectOld() ) []
  statS -> eba ->
    ( [perc[StatO] == 1]
          statExpectMet -> ExpectOld() []
      [perc[CardO] == 1]
          statExpectFailed {assessment = danger}
              -> ExpectOld() )

ExpectNew() = ba ->
( eba -> ExpectNew() []
  cashS -> eba ->
    ( [perc[CardO] == 1]
          cardExpectMet -> ExpectNew() []
      [perc[CashO] == 1]
          cardExpectFailed -> ExpectNew() ) []
    statS -> eba -> ( [perc[CardO] == 1]
          cardExpectMet -> ExpectNew() []
      [perc[StatO] == 1]
          cardExpectFailed -> ExpectNew() )
                    );
);
```

The above model caters for two different user expectations. The first one is **ExpectOld**, where a user used to interact with the old ATM expects to see the cash or statement delivered after selecting 'cash withdrawal' or 'statement printing'; such expectations are not met (**cashExpectMet** or **statExpectMet**, respectively) if the card is returned instead. The second one is **ExpectNew**, where a user used to interact with the new ATM expects to see the card returned after selecting 'cash withdrawal' or 'statement printing'; such expectations is not met (cardExpectMet) if the cash or statement is delivered instead.

Example 65.15 extends Example 65.10 by including the appropriate guards on the assessment of the perception and an abortSession event which assigns abort to the response variable when the assessment of the perception is danger.

Example 65.15: Task with Response to Perception Assessment

```
Task() = ba -> (
 [assessment == safe &&
  stmGoal[getCashGoal] == 1 && stm[Interaction] == 0] eact
   -> store {stm[Interaction] = 1;
             stmSize++} -> eba-> Task() []
 [assessment == safe &&
  stmGoal[getStatGoal] == 1 && stm[Interaction] == 0] eact
   -> store {stm[Interaction] = 1;
             stmSize++} -> eba-> Task() []
 [assessment == safe &&
  stm[Interaction] == 1 && stmSize < M &&
  perc[CardR] == 1] cardI
   -> eact -> store {stm[CardC] = 1; stmSize++}
   -> eba -> Task() []
 [assessment == safe &&
  stm[Interaction] == 1 &&
  perc[PinR] == 1] pinE
   -> eact -> eba -> Task() []
 [assessment == safe &&
  stmGoal[getCashGoal] == 1 && perc[SelR] == 1] cashS
   -> eact -> eba -> Task() []
 [assessment == safe &&
  stmGoal[getStatGoal] == 1 && perc[SelR] == 1] statS
   -> eact -> eba -> Task() []
 [assessment == safe &&
  stm[Interaction] == 1 &&
  perc[CashO] == 1] cashC
   -> eact -> eba -> Task() []
 [assessment == safe &&
  stm[Interaction] == 1 &&
   perc[StatO] == 1] statC
   -> eact -> eba -> Task() []
 [assessment == safe &&
  stm[Interaction] == 1 && stmSize > 0 &&
   perc[CardO] == 1 && stm[CardC] == 1]
```

```
      retrieve {stm[CardC] = 0; stmSize--} -> cardC
      -> eact -> eba -> Task()
   [assessment == danger &&
    stm[Interaction] == 1 && perc[Card0] == 1 ]
      abortSession {assessment = safe;
                    response = abort} -> cardC
      -> eact -> eba -> Task() []
   [assessment == danger &&
    stm[Interaction] == 1 && perc[Card0] == 0 ]
      abortSession {assessment = safe;
                    response = abort}
      -> eact -> eba -> Task()
                 );

   User() = Closure() || Goals() || Task();
```

The response to a danger (guard `assessment == danger`) is to collect the card (event `cardC`), if this is perceived (guard `perc[Card0] == 1`), and abort the interaction session (event `abortSession`, which set `response` to `abort`) or just abort the interaction section if the cards is not perceived (guard `perc[Card0] == 0`), while variable `assessment` is reset to `safe`. Although normally the danger assessment is due to the perception of the card, not considering the assessment of danger possibly due to other reasons would be an overspecification.

7.4.2 Environment and Overall System

Until now we have considered the following components of the overall system:

– the *user's behaviour*

```
      User() = Closure() || Goals() || Task();
```

consisting of the infrastructure for STM (process `Closure`) and goals (process `Goals`) and the human task process `Task` (see Examples 65.7, 71.5, 65.10 and 65.15),

– the *interface or system* (see Examples 65.11 and 65.11), and
– the *user's experiential constraints* (see Example 65.14).

However, as illustrated in Example 65.16 there are further aspects of the environment that influence the interaction and thus ought to be part of the modelled overall system:

• the *initial interface/system state*,
• the *availability of resources*, and
• the *user's knowledge*.

Example 65.16: Closure in Automatic Control using PAT

```
InitState() = initialization {perc[CardR] = 1} -> Skip();

HasCard() = cardI -> NoCard();
NoCard() = cardC -> HasCard();

Resources() = HasCard();

KnowsPin() = pinE -> KnowsPin();

Knowledge() = KnowsPin();

User() = Closure() || Goals() || Task();

Environment() = User() || Resources() || Knowledge() ;

SysOld() = InitState() ; ( Environment() || ATMold() );
SysNew() = InitState() ; ( Environment() || ATMnew() );

UserOld() = Environment() || ExpectOld();
UserNew() = Environment() || ExpectNew();

SysOldUserOld() = InitState() ; ( UserOld() || ATMold() );
SysOldUserNew() = InitState() ; ( UserNew() || ATMold() );
SysNewUserNew() = InitState() ; ( UserNew() || ATMnew() );
SysNewUserOld() = InitState() ; ( UserOld() || ATMnew() );
```

Aspects of the environment are the following.

- *Initial interface state.* We assume that the interface is initially request-
  ing a card. This is expressed by process InitState, which performs
  event initialization to set variable perc[CardR] to 1 and terminate successfully. This process is sequentialised with the main overall
  system process.
- *Availability of resources.* An essential resource for the task is the bank
  card, which has to be available for the user: process Resources consists
  of two states describing the availability (HasCard) and non availability
  (NoCard) of the card.
- *User's knowledge.* The user must know the pin in order to perform the
  task. In our example we implicitly assumed that the user knows the
  pin, thus process KnowsPin models only the correct pin in terms of
  event pinE. However, we might want to consider also the case of using
  a wrong pin, in order to explore its impact on the interaction and the
  emergent errors This would require a more sophisticated version of
  process KnowsPin.

Finally, the overall system is modelled by processes `SysOld` and `SysNew`, corresponding to the two possible interfaces, but with no assumptions on the user's experiential knowledge, and by processes `SysOldUserOld` and `SysOldUserNew`, `SysNewUserNew` and `SysNewUserOld`, which include constraints on user expectations.

7.5 Model Checking Analyses

From an analytical point of view we focus on two aspect: overall system verification and task failure analysis. First, in Sect. 7.5.1 we illustrate how to verify whether the design of the interface and the other environment components addresses cognitive aspects of human behaviour such as closure phenomena and user expectations that trigger the SAS to activate attention (overal system verification). Then, in Sect. 7.5.2, we consider patterns of behaviour featuring persistent operator errors may lead to a task failure.

7.5.1 Overall System Verification

Model checking techniques provide an effective analytical tool to exhaustively explore the system state space and capture the behaviour that emerges from the combination of several system components. Closure, automatic behaviour, expectancy and attention are phenomena that represent distinct components of human cognition and action, and their combination results in an apparently holistic ways of performing tasks. In this context model checking can be used to capture errors that emerge when environment design, which includes physical devices, interfaces and their operational environment, cannot deal with the closure phenomena, or when the outcome of the interaction between automatic behaviour and environment does not meet human expectations. We use Linear Temporal Logic (LTL), as described in Sect. 2.5.3, to specify system properties and then we use PAT model checking capabilities to verify such properties on the CSP model. PAT support the definition of *assertions* of the form

$$\text{\#assert } system \models property$$

where *system* is the model we aim to verify and *property* is a property expressed in LTL. The PAT model checker verifies whether the property is valid on the model and, if not, provides a counterexample. The counterexample provided by the model checking analysis can then be exploited to improve the environment design.

We consider three kinds of properties:

- *Functional correctness* the user or operator can always complete the task by successfully accomplishing the goal,
- *Non-functional correctness* in spite of successfully accomplishing the goal, the system may violate some non-functional properties (e.g. the user of an ATM forgets the card after collecting the cash), and
- *Cognitive Overload* the STM is overloaded above a considered upper limit, which may lead to a failure in accomplishing the goal when the STM is loaded by additional uncompleted tasks.

We illustrate the functional correctness in Example 65.17

Example 65.17: Functional Property Verification using PAT

Since there are two possible goals, to get cash and to get a statement, **functional correctness** aims to verify for each interface design, whether there are cognitive errors that may prevent the user from collecting the cash and from collecting the statement. The property that the user is always able to collect the cash can be expressed by formalising that "a user who selects 'cash' will collect the cash before the end of the interaction section". Since the end of the interaction section may be characterised as the beginning of a new interaction section, which occurs when a card is inserted again, we can say that "a user who selects 'cash' will collect the cash before a card is inserted". Furthermore "the user collects the cash before a card is inserted" can also be expressed as "no card is inserted until the user collects the cash". Finally, our original property can be expressed as "if a user selects 'cash' then no card is inserted until the user collects the cash", which can be immediately translated into LTL. Similarly, the properties that the user is always able to collect the statement can be expressed by formalising that "a user who selects 'statement' will collect the statement before the end of the interaction section" or equivalently as "if a user selects 'statement' then no card is inserted until the user collects the statement", which again can be immediately translated into LTL.

```
#assert SystemNewUserNew() |=
    [] ( cashS -> ( ! cardI U cashC ) );
#assert SystemNewUserNew() |=
    [] ( statS -> ( ! cardI U statC ) );

#assert SystemOldUserNew() |=
    [] ( cashS -> ( ! cardI U cashC ) );
#assert SystemOldUserNew() |=
    [] ( statS -> ( ! cardI U statC ) );
```

```
#assert SystemNewUserOld() |=
    [] ( cashS -> ( ! cardI U cashC ) );
#assert SystemNewUserOld() |=
    [] ( statS -> ( ! cardI U statC ) );

#assert SystemOldUserOld() |=
    [] ( cashS -> ( ! cardI U cashC ) );
#assert SystemOldUserOld() |=
    [] ( statS -> ( ! cardI U statC ) );
```

The PAT model checker shows all systems except `SystemNewUserOld` to be functionally correct. In fact, although the new design of the ATM works in an ideal world where all ATMs are designed according to the new criterion, there are still some ATMs, especially in the developing world, that are designed according to the old criterion. Thus we can imagine that a user from one of such countries, while visiting a country where all ATMs are designed according to the new criterion, is likely to assess the early return of the card as a danger and is prone to abandon the interaction forgetting to collect the cash. This situation is formalised by the counterexample returned in the verification of `SystemNewUserOld`.

In general, when the system behaviour consists of a loop of user sessions each characterised by a *begin* event and there are two events *choose* and *accomplish* which characterise the choice and accomplishment of the goal, then functional correctness can be expressed as

   `#assert` *system* `|=` `[]` *choose* `-> (` `!` *begin* `U` *accomplish* `))`.

In some cases, also non-functional properties may be characterised in this way. For example, in the case of *safety* properties, there might be a system internal event *internal*, rather than a user's choice, as a precondition for the user not to lose some owned resource currently used by the system. If *return* is the event characterising the return of the resource to the user, then the safety property can be expressed as

   `#assert` *system* `|=` `[]` *internal* `-> (` `!` *begin* `U` *return* `))`;

We illustrate the verification of safety in Example 65.18

Example 65.18: Safety Property Verification using PAT

As an example of **nonfunctional correctness** we consider the **safety** property that aims to verify, for each interface design, whether there are cognitive errors that may prevent the user from collecting the returned card.

```
#assert SystemNewUserNew() |=
    [] ( readCard -> ( ! cardI U cardC ) );
```

```
#assert SystemOldUserNew() |=
    [] ( readCard -> ( ! cardI U cardC ) );
#assert SystemNewUserOld() |=
    [] ( readCard -> ( ! cardI U cardC ) );
#assert SystemOldUserOld() |=
    [] ( readCard -> ( ! cardI U cardC ) );
```

The PAT model checker shows that the safety property is valid with the new ATM (SystemNewUserNew and SystemNewUserOld) and not with the old ATM (SystemOldUserOld and SystemOldUserOld), independently of the user experience. The counterexample captures possible post-completion errors in using the old design of the ATM and shows that such errors cannot occur in the new design of the ATM.

We can now understand what cognitive error caused the cake of Example 65 to burn and why the algorithm used by your partner caused such an error to emerge. This is illustrated in Example 69.2.

Example 69.2: Why Cakes and Engines Burn

The baking tasks is divided in two separate parts, with a long period in between that is likely to be devoted to many other tasks. Each part is actually a task in itself with a specific goal. The goal of the first task is achieved when the cake is inserted in the oven and the oven is closed (Activity 6), thus causing STM closure. Therefore, the subsidiary task of lowering the temperature setting may be forgotten (Activity 7), with the result that the cake burns.

The obvious solution to this problem is to swap Activity 6 and Activity 7, thus preventing the occurrence of a post-completion error. The problem here is not in the interface, but in the algorithm, that is, the protocol that is used to carry out the task. This subtle form of post-completion error is difficult to eliminate in practice, since the solution count on the human to strictly adhere to a given protocol.

For example, on 24 May 2013, the fan cowl doors of an aircraft were left unlatched on both engines after completing scheduled maintenance (forgetting this subsidiary task after the achievement of the maintenance goal). As the aircraft departed London Heathrow Airport, the fan cowl doors from both engines detached, puncturing a fuel pipe on the right engine and damaging the airframe and some aircraft systems. While the flight returned to Heathrow an external fire developed on the right engine, which was then shut down. The aircraft managed to safely land using the left engine. All the passengers and crew evacuated the aircraft via the escape slides.

Cognitive load expresses the amount of information stored in STM at a given time. Thus cognitive overload occurs when the amount of information stored in STM is above a considered upper limit. We illustrate the analysis of cognitive overload in Example 65.19

Example 65.19: Cognitive Overload Analysis using PAT

As an example of **cognitive overload** we consider an upper limit of 5 piece of information stored in STM.,

```
#define cognitiveOverload (stmSize > 5);

#assert SystemNewUserNew()  |= [] ! cognitiveOverload;
#assert SystemOldUserNew()  |= [] ! cognitiveOverload;
#assert SystemNewUserOld()  |= [] ! cognitiveOverload;
#assert SystemOldUserOld()  |= [] ! cognitiveOverload;
```

PAT provides a `define` construct to define constants. This can be used to define boolean constants to be used as proposition within assertions, as in Example 65.19. This way, the model checker can determine the mental capabilities an operator has to possess to avoid cognitive overload.

7.5.2 Task Failures Analysis

The purpose of the operator's behaviour is to prevent the system from reaching a failure state. In this case the unwanted result of the interaction is the task failure. Although it is acceptable that the operator makes errors, since recovery from errors is always possible, if this recovery does not occur and the operator persists in making errors, then the system will eventually reach a failure state.

Applied psychology uses experiments, natural observation and other data gathering instruments to identify and categorise the operator's patterns of behaviour that may lead to a task failure. The goal of this kind of studies is to capture all possible patterns of behaviour that may lead to a task failure in order to design system controls, support tools, environment settings and working schedules that prevent operators from entering such dangerous patterns of behaviour.

Formal methods can support applied psychology by verifying whether the decomposition of a task failure into patterns of behaviour is sound and complete. The task failure F and its empirically defined decomposition $\mathcal{D} = \{P_1, \ldots P_n\}$ into patterns of behaviour can be formalised in LTL. The decomposition $\mathcal{D}$ is

- *sound* if each of the P_i is sufficient to cause the task failure F, and
- *complete* if one of the P_i is necessary to cause the task failure F.

Then model checking can be used to verify the soundness of the decomposition

$$\bigwedge_{P \in \mathcal{D}} (P \Rightarrow F)$$

and the completeness of the decomposition

$$F \Rightarrow \bigvee_{P \in \mathcal{D}} P$$

We informally illustrate this methodology in Example 66.3.

Example 66.3: Task Failure Analysis using PAT

We can characterise a separation violation as an operator who persistently misses the intention to carry out a specific action to solve the conflict [CCL08, CLC05]. We distinguish between intention and action to be able to model an unintended action that does not match the intention [Rea90]. Although this is not part of our analysis, such a mismatch would be relevant in the analysis of errors induced by a specific interface design, which could be carried out on this case study by introducing alternative interface designs and using our formal cognitive framework as in the ATM case study.

The formalisation of the ATC task failure decomposition suggested by Lindsay and Connelly [LC02] is

1. **Failure of scanning** when the operator fails to monitor a specific part of the interface, thus missing possible conflicts,
2. **Persistent mis-classification** when the operator persistently classifies as a non conflict what is actually a conflict,
3. **Persistent mis-prioritisation** when the operator persistently gives a low priority to a conflict, thus missing to solve it, and
4. **Defer action for too long** when the operator persistently delays to implement an already developed plan to solve a conflict.

Model checking analysis using PAT shows that decomposition of the task failure is sound but not complete. The counterexample shows that the definition of persistent mis-classification by Lindsay and Connelly mixes two different kinds of behaviour, one fully characterising persistent misclassification and the other being a part of another property which was not captured through empirical analysis. This property, which we call **Contrary decision process**, occurs when a conflict is persistently reclassified as a non conflict. Once such a property is added to the decomposition and the notion of persistent misclassification is redefined in a way that does not overlap with it, the decomposition becomes complete.

7.6 Closing Remarks

In this chapter, we presented a formal approach to the specification, modelling and analysis of interactive systems in general and, more specifically, of human-computer interaction. Systems are modelled using the CSP extension implemented in the Process Analyzer Toolkit (PAT), properties are specified using temporal logic formulae, either on states or events, and analysis is carried out by exploiting the model checking capabilities of PAT.

The approach is illustrated through two classical examples. The Automated Teller Machine (ATM) example was already introduced in Chap. 3 and is used in this chapter to illustrate the automatic behaviour of a user who carries out everyday activities with just implicit attention, but who may resort to explicit attention when in need of making a decision, driven by the task goal, or when realising the occurrence of an anomalous situation, such as a danger. Both functional properties, such as being enabled to achieve the goal (withdrawing cash or printing a statement, in the case of the ATM), and safety properties (remembering to collect the card, in the case of the ATM) are analysed. The Air Traffic Control (ATC) system example is introduced in this chapter to illustrate the deliberate behaviour of an operator who performs a task with a general purpose, whereby specific goals are set along the way. Although, in general, failing to achieve the goal is not a task failure, provided the system state is still consistent with the purpose, a pattern of behaviour featuring persistent operator errors may indeed lead to a task failure. In this context, model checking is used to support applied psychology by analysing an empirically defined decomposition of the task failure into patterns of behaviour, in order to verify whether the decomposition is sound and complete.

7.7 Annotated Bibliography

There is a large number of textbooks on human-computer interaction. The most comprehensive and appropriate to provide an accessible introduction to the concepts used in this chapter are by Dix et al. [DFAB04], by Preece, Rogers and Sharp [PRS17] and by Thimbleby [Thi07]. The first has an emphasis on modelling. It provides an extensive introduction to human behaviour and interaction from a cognitive science perspective and also presents, mostly at an intuitive level, a variety of formal approaches for dealing with some aspects of HCI and tackling specific challenges. The second has an emphasis on designing for user experience. It is intended as a book for practitioners and has a broader scope of issues, topics and methods than traditional human-computer interaction textbooks, with a focus on diversity of design and evaluation process involved. However, it is less concerned with cognition than the book by Dix et al. The third draws on sound computer science principles, with a strong formal methods flavour. It uses state machines and graph theory as a powerful and insightful way to analyse and design better interfaces and examines specific designs and creative solutions to design problems

Looking more specifically at modelling cognition, historical but still actual works are by Newell and Simon [NS72], Card et al. [CEB78]. In a later work Card, Moran and Newel [CMN83] introduced a somehow formal notation, which inspired, on the one hand, the development of a plethora of cognitive architectures over the last 40 years and, on the other hand, the use of formal methods in HCI.

Kotseruba and Tsotsos published a broad overview of these last 40 years of cognitive architectures [KT18], featuring 84 cognitive architectures and mapping them according to perception modality, implemented mechanisms of attention, memory organisation, types of learning, action selection and practical applications. A similar, but less comprehensive survey by Samsonovich [Sam10] collects the descriptions of 26 cognitive architectures submitted by the respective authors. Finally, Laird et al. [LLR17] focus on three among the most known cognitive architectures, ACT-R, Soar and Sigma, and compare them based on their structural organisation and approaches to model core cognitive abilities.

In 1991 two nice surveys on the first formal approaches in HCI were compiled by Haan, van der Veer and van Vliet [GdHvV91], based on a psychology perspective, and by Dix [Dix91], based on a computer science perspective. Although the scientific community working on the use of formal methods in HCI is quite small, there have been a number of significant results over the last 20 years. Such results mainly appear in the proceedings of the international workshops on on Formal Methods for Interactive Systems (FMIS), which run from 2006, though not every year, and in journal special issues associated with such workshop. Some of these special issues and other papers in the area appeared in the journal Formal Aspects of Computing. Two important collection of works on formal methods approaches to HCI have been recently edited by Weyers, Bowen, Dix and Palanque [WBDP17] and by Oulasvirta, Kristensson, Bi and Howes [OKBH18].

7.7.1 Current Research Directions

The way the validity of both functional and nonfunctional properties is affected by user behaviour is quite intricate. It may seem obvious for functional properties that an interactive system can deploy its functionalities only if it is highly usable. However, usability may actually be in conflict with functional correctness, especially in applications developed for learning or entertainment purpose. More in general, high usability may be in conflict with user experience, whereby the user expects some challenges in order to test personal skills and knowledge, enjoy the interaction and avoid boredom. Usability is also strictly related to critical nonfunctional properties such as safety [CCL08] and security [CE07]. Such relationship is actually two ways. On one side improving usability increases safety and/or security [CCL08]. On the other side introducing mechanisms to increase safety and/or security may reduce usability, and as a result, may lead to an unexpected global decrease in safety [IBCB91] and/or security [CE07]. Investigating such complex relationships is an important research direction.

In the real world humans frequently have to deal with operating environments that

1. continuously produce, change and invalidate human expectations as part of an evolutionary learning process [Cer16, IBCB91],
2. deploy constraining social contexts [IBCB91] and cultural differences [Hei07], and
3. provide a large amount of stimuli, which are perceived through several modalities at the same time and interpreted and combined according to temporal and contextual constraints (multimodal interaction) [CFG07].

The formal approach proposed in this chapter as well as all approaches that aim at applying formal methods to generic HCI problems presuppose that

1. expectation are a priori constraints rather than part of a learning process,
2. cognitive behavior depends on a specific social and cultural context, and
3. human cognition and actions are directly triggered by isolated perceptions.

Therefore, it is important to

1. define cognitive mechanisms that build

   - expectations in semantic memory out of experience stored in episodic memory, and
   - procedures in procedural memory out of knowledge stored in semantic memory,

   thus mimicking the information flow from STM first to episodic memory and then to LTM (see Sect. 7.4.1),
2. enable multiple, interacting instantiations of cognitive architectures as part of a complex sociotechnical system, and
3. define, at the cognitive architecture level, mechanisms for the fusion of multiple modalities.

These objectives may not be easily accomplished using formal notations with limited data structures such as CSP, even in the extended form provided by PAT. A more sophisticated modelling language with extensive data structures, possibly featuring an object-oriented paradigm, is needed. With this aim in mind the Maude rewrite system [Ö17] has been proposed as a possible candidate [Cer18, BMO19]. Furthermore, the definition of the Behavioural and Reasoning Description Language (BRDL) [Cer20] and its implementation using Real-Time Maude [CO20] have recently paved the way for the insilico simulation of experiments carried out in cognitive psychology with human subjects [CM21] as well as the simulation of long-term human learning processes [CP21].

Furthermore, the intrinsic unpredictability of human behaviour requires the validation of any a priori model on real data. Using text mining techniques and appropriate ontologies, abstract event logs that match the representation used in the cognitive model could be extracted from the dataset and used to constrain the model before performing formal verification. This could be done at different levels, from a correspondence one-to-one between real interaction history and constraints to the representation of a set of real interaction histories with a single constraint.

Finally, the use of formal methods for system modelling and analysis requires high expertise in mathematics and logic, which is not common among typical users, such as interaction design and usability experts as well as psychologists and other social scientists. Therefore, the development of tools that address the need and skills of such typical users is essential for the acceptance and diffusion of formal methods in the HCI area.

References

[BBD00]   R. Butterworth, Ann E. Blandford, and D. Duke. Demonstrating the cognitive plausability of interactive systems. *Form. Asp. of Comput.*, 12:237–259, 2000.

[BMO19]   Giovanna Broccia, Paolo Milazzo, and Peter Csaba Ölveczky. Formal modeling and analysis of safety-critical human multitasking. *Innovations in Systems and Software Engineering*, 2019.

[CB04]   Paul Curzon and Ann E. Blandford. Formally justifying user-centred design rules: a case study on post-completion errors. In *Integrated Formal Methods*, LNCS 2999, pages 461–480. Springer, 2004.

[CCL08]   A. Cerone, S. Connelly, and P. Lindsay. Formal analysis of human operator behavioural patterns in interactive surveillance systems. *Softw. Syst. Model.*, 7(3):273–286, 2008.

[CE07]   Antonio Cerone and Norzima Elbegbayan. Model-checking driven design of interactive systems. In *Proceedings of FMIS 2006*, volume 183 of *Electronic Notes in Theoretical Computer Science*, pages 3–20. Elevier, 2007.

[CEB78]   S.K. Card, W.K. English, and B.J. Burr. Evaluation of mouse, rate-controlled isometric joystick, step keys, and text keys for text selection on a CRTl. *Ergonomics*, 21:601–613, 1978.

[Cer11]   Antonio Cerone. Closure and attention activation in human automatic behaviour: A framework for the formal analysis of interactive systems. In *Proc. of FMIS 2011*, volume 45 of *Electronic Communications of the EASST*, 2011.

[Cer16]   Antonio Cerone. A cognitive framework based on rewriting logic for the analysis of interactive systems. In *Software Engineering and Formal Methods*, LNCS 9763, pages 287–303. Springer, 2016.

[Cer18]   Antonio Cerone. Towards a cognitive architecture for the formal analysis of human behaviour and learning. In *STAF collocated workshops*, LNCS 11176, pages 216–232. Springer, 2018.

[Cer20]   Antonio Cerone. Behaviour and reasoning description language (BRDL). In
            Javier Camara and Martin Steffen, editors, *SEFM 2019 Collocated Workshops
            (CIFMA)*, LNCS 12226, pages 137–153. Springer, 2020.
[CFG07]   M.C. Caschera, F. Ferri, and P. Grifoni. Multimodal interaction systems: infor-
            mation and time features. *International Journal of Web and Grid Services*,
            3(1):82–99, 2007.
[CLC05]   Antonio Cerone, Peter Lindsay, and Simon Connelly. Formal analysis of
            human-computer interaction using model-checking. In *Proc. of SEFM 2005*,
            pages 352–361. IEEE, 2005.
[CM21]   Antonio Cerone and Diana Murzagaliyeva. Information retrieval from seman-
            tic memory: BRDL-based knowledge representation and Maude-based com-
            puter emulation. In *SEFM 2020 Collocated Workshops (CIFMA)*, LNCS
            12524, pages 159–175. Springer, 2021.
[CMN83]   S.K. Card, T.P Moran, and A. Newell. *The Psychology of Human-Computer
            Interaction*. Laurence Erlbaum, 1983.
[CO20]   Antonio Cerone and Peter Csaba Ölveczky. Modelling human reasoning in
            practical behavioural contexts using Real-Time Maude. In *FM'19 Collocated
            Workshops (FMIS)*, LNCS 12232, pages 424–442. Springer, 2020.
[CP21]   Antonio Cerone and Graham Pluck. A formal model for emulating the gener-
            ation of human knowledge in semantic memory. In *From Data to Models and
            Back (DataMod 2020)*, LNCS 12611, pages 104–122. Springer, 2021.
[De 15]   Raquel Araujo De Oliveira. *Formal Specification and Verification of Interac-
            tive Systems with Plasticity : Applications to Nuclear-Plant Supervision*. PhD
            thesis, University of Grenoble, 2015.
[DFAB04]   Alan Dix, John Finlay, Gregory Abowd, and Russel Beale. *Human-Computer
            Interaction*. Pearson Education, 3rd edition, 2004.
[Dix91]   Alan John Dix. *Formal Methods for Interactive Systems*. Academic Press,
            1991.
[GdHvV91]   G.C van der Veer G. de Haan and J.C. van Vliet. Formal modelling techniques
            in human-computer interaction. *Acta Psychologica*, 78:27–67, 1991.
[Hei07]   Rüdiger Heimgärtner. Cultural differences in human computer interaction:
            Results from two online surveys. In *Open Innovation. Proc. of the 10th Sym-
            posium for Information Science*, pages 145–157, 2007.
[IBCB91]   Ioanna Iacovides, Ann Blandford, Ann Cox, and Jonathan Back. How exter-
            nal and internal resources influence user action: the case of infusion device.
            Cognition, Technology and Work, 18(4):793–805, 1991.
[Joh97]   C. Johnson. Reasoning about human error and system failure for accident
            analysis. In *Proc. of INTERACT 1997*, pages 331–338. Chapman and Hall,
            1997.
[Kir90]   B. Kirwan. Human reliability assessment. In *Evaluation of Human Work*,
            chapter 28. Taylor and Francis, 1990.
[KT18]   Iuliia Kotseruba and John K. Tsotsos. 40 years of cognitive architectures: core
            cognitive abilities and practical applications. *Artificial Intelligence Review*,
            2018.
[LC02]   Peter Lindsay and Simon Connelly. Modelling erroneous operator behaviours
            for an air-traffic control task. In *Third Australasian User Interfaces Confer-
            ence (AUIC2002)*, pages 43–54. Australian Computer Society, 2002.
[Lev95]   N. G. Leveson. *Safeware: System Safety and Computers*. Addison-Wesley,
            1995.
[LLR17]   J.E. Lairs, C. Lebiere, and P.S. Rosembloom. A standard model for the mind:
            towards a common computational framework across artificial intelligence, cog-
            nitive science, neuroscience, and robotics. *AI Magazine*, 38:13–26, 2017.
[Mac05]   C. Mach. *Knowledge and Error*. Reidel, 1905. English translation, 1976.
[Mil56]   G. A. Miller. The magical number seven, plus or minus two: Some limits on
            our capacity to process information. *Psychological Review*, 63(2):81–97, 1956.

[MPF+16]   C. Martinie, P. Palanque, R. Fahssi, J. P. Blanquart, C. Fayollas, and C. Seguin. Task model-based systematic analysis of both system failures and human errors. *IEEE Trans. Human-Mach. Syst.*, 46(2):243–254, 2016.

[MRO+15]   Paolo Masci, Rimvydas Rukšėnas, Patrick Oladimeji, Abigail Cauchi, Andy Gimblett, Yunqiu Li, Paul Curzon, and Harold Thimbleby. The benefits of formalising design guidelines: a case study on the predictability of drug infusion pumps. *Innovations Syst. Softw. Eng.*, 11(2):73–93, 2015.

[NS72]   A. Newel and H.A. Simon. *Human Problem Solving*. Prentice-Hall, 1972.

[NS86]   Donald A. Norman and Tim Shallice. Attention to action: Willed and automatic control of behavior. In *Consciousness and Self-Regulation*, volume 4 of *Advances in Research and Theory*. Plenum Press, 1986.

[Ö17]   Peter Csaba Ölveczky. *Designing Reliable Distributed Systems — A Formal Methods Approach Based on Executable Modeling in Maude*. Springer, 2017.

[OKBH18]   Antti Oulasvirta, Per Ola Kristensson, Xiaojun Bi, and Andrew Howes, editors. *Computational Interaction*. Oxford University Press, 2018.

[PAT19]   PAT: Process Analysis Toolkit. User manual (online version). `http://pat.comp.nus.edu.sg/wp-source/resources/OnlineHelp/htm/index.htm`, 1 Dec 2019.

[PBP97]   P. Palanque, R. Bastide, and F. Paterno. Formal specification as a tool for objective assessment of safety-critical interactive systems. In *Proc. of INTERACT 1997*, pages 323–330. Chapman and Hall, 1997.

[PRS17]   Jennifer Preece, Yvonne Rogers, and Helen Sharp. *Interaction Design — beyond human-computer interaction*. Wiley, 5th edition, 2017.

[RCB08]   Rimvydas Rukšėnas, Paul Curzon, and Ann E. Blandford. Modelling rational user behaviour as games between an angel and a demon. In *Sixth IEEE International Conference on Software Engineering and Formal Methods*, pages 355–364. IEEE Computer Society, 2008.

[RCBB14]   R. Rukšėnas, P. Curzon, A. E. Blandford, and J Back. Combining human error verification and timing analysis: A case study on an infusion pump. *Form. Asp. of Comput.*, 26:1033–1076, 2014.

[Rea90]   James Reason. *Human Error*. Cambridge University Press, 1990.

[Sam10]   Alexei V. Samsonovic. Toward a unified catalog of implemented cognitive architectures. In *Proceedings of the 1st Annual Meeting on Biologically Inspired Cognitive Architectures (BICA 2010)*, pages 195–244. IOS Press, 2010.

[SBBW09]   Li. Su, Howard Bowman, Philip Barnard, and Brad Wyble. Process algebraic model of attentional capture and human electrophysiology in interactive systems. *Form. Asp. of Comput.*, 21(6):513–539, 2009.

[Thi07]   Harold Thimbleby. *Press On*. MIT Press, 2007.

[WBDP17]   Benjamin Weyers, Judy Bowen, Alan Dix, and Philippe Palanque, editors. *The Handbook of Formal Methods in Human-Computer Interaction*. Springer, 2017.

Chapter 8
Formal Verification of Security Protocols

Markus Roggenbach, Siraj Ahmed Shaikh, and Hoang Nga Nguyen

Abstract Security protocols address the question of how one communicates 'securely' in an untrusted 'hostile' environment. This chapter introduces the general notions of cryptography, communication protocols, security goals, and security protocols. Taking the Needham-Schroeder authentication protocol as an example, the chapter demonstrates that it is difficult to get the design of security protocols 'right'. This raises the need for a rigorous approach to analysing security protocols at a design level. To this end, the chapter discusses the CSP approach of modelling security protocols, security properties, and when a protocol satisfies a desired property. It then presents two different approaches for protocol analysis: (1) an automated approach via model checking, where the challenge lies in finding the right coding tricks in order to 'tame' state space explosion; and (2) a manual approach using rank functions, where the challenge lies in finding a suitable protocol invariant.

8.1 Introduction

Let's start with a fairy tale. "In days of yore and in times and tides long gone before there dwelt in a certain town of Persia two brothers one named Kasim and the other Ali Baba."[1]

Markus Roggenbach
Swansea University, Wales, United Kingdom

Siraj Ahmed Shaikh
Coventry University, Coventry, United Kingdom

Hoang Nga Nguyen
Coventry University, Coventry, United Kingdom

[1] See https://www.pitt.edu/~dash/alibaba.html#burton for the full text of this story.

© Springer Nature Switzerland AG 2022, corrected publication 2022
M. Roggenbach et al., *Formal Methods for Software Engineering*,
Texts in Theoretical Computer Science. An EATCS Series,
https://doi.org/10.1007/978-3-030-38800-3_8

"Open Sesame!", shouts Ali Baba in this famous tale of *Ali Baba and the Forty Thieves* from the *Arabian Nights*, where these are the magic words that open the treasure cave. As the tale goes, Ali Baba follows a group of forty thieves to the cave where they hide their loot, and he overhears the magic words, which are *key* to opening the cave. Later, Kasim, Ali Baba's brother, also gets the magic words as he overhears him to say "Open Sesame" to get into the cave. In his haste and excitement over the riches, Kasim ends up forgetting the magic words to get out of the cave once he is in.

Kasim meets his fate and gets killed in the cave by the thieves once they find him. Ali Baba ends up getting rich and happy eventually. The moral of the story is that effective key management is *key* to security and safety! And of course greed does not serve one well.

Three important security principles emerge from the story:

– First, challenge and response is used to open the cave door. Only the one who can say the magic words can enter or leave the cave. This is a simple example of a challenge response protocol, where the challenge is the door only opens in response to a magic word (password), and the valid response is the password "Open Sesame!".
– Secondly, eavesdropping is always a possibility, as it was for Ali Baba and Kasim.
– Thirdly, the perfect cryptography assumption is made, i.e., the cave can't be fooled. There is no other way to open the door but to say "Open Sesame!".

This chapter introduces an important application of Formal Methods: the design and verification of security protocols. Such protocols are used to provide properties such as authentication, confidentiality, integrity or non-repudiation.

Security protocols underpin a number of common technologies in everyday use. An electronic purse, cf. Example 9, is one application where all financial transactions are conducted over a protocol with desired security properties. Similarly, electronic voting, cf. Example 14, relies on security protocols to ensure, e.g., secrecy of the vote, i.e., only the voter knows who they voted, and privacy of voters, i.e., if they have participated in the election or not.

Among the many formalisms suitable to address the challenge of security protocol verification, we choose the process algebra CSP as the underlying formalism for this chapter, see Chap. 3 for a thorough introduction to CSP; an understanding of the syntax as provided in Sect. 3.2.1 and the traces model as provided in Sect. 3.2.2 would suffice for this chapter. As distributed systems, security protocols belong to the natural remit of process algebra. Over the years, in particular the CSP community has made significant contributions to the verification of security protocols and has developed a rich set of analysis methods, cf. Sect. 8.7.1.

Section 8.2 introduces the reader to the basic principles of security. Section 8.3 introduces the Needham-Schroeder authentication protocol, discusses why it fails to provide authentication, and presents the Needham-Schroeder-

Lowe protocol, which corrects the discussed flaw. In Sect. 8.4, we introduce the CSP approach of how to formally model authentication and security protocols in general. Such models open the possibility to rigorously reason and prove protocols with respect to desired properties. Section 8.5 applies model checking to the Needham-Schroeder authentication protocol and shows how a counter example trace is automatically found. It also provides a correctness result for the Needham-Schroeder-Lowe protocol under the assumption that an attacker uses a limited number of reasoning steps only. In contrast, Sect. 8.6 utilises the so-called Rank-Function Theorem to give a manual correctness proof of the Needham-Schroeder-Lowe protocol, without making any assumptions w.r.t. the number of an attacker's reasoning steps.

8.2 Basic Principles

This section lays out some fundamental principles: these are basica notations of cryptography, cf. Sect. 8.2.1, security principles, cf. Sect. 8.2.2, and security protocols, cf. Sect. 8.2.3.

8.2.1 Cryptography

The term *cryptography* is derived from the Greek word *kryptos*, meaning hidden. A related notion *cryptanalysis* is to do with analysing hidden information with intent to uncover it. Both cryptography and cryptanalysis are branches of *cryptology*, which is the science of hiding. The purpose of cryptography is to hide the content of a message so that it is not revealed to everyone. This is made possible by converting the message from its original form, known as *plaintext*, to a distorted incomprehensible form known as *ciphertext*. The process of converting plaintext to ciphertext is called *encryption* and the process of converting ciphertext back to plaintext is called *decryption*. Encryption and decryption operations are implemented in the form of algorithms known as *cryptographic algorithms* or *cryptosystems*, where plaintext (or ciphertext) and encryption (or decryption) keys are parameters. Most cryptosystems are based on assumptions that certain problems such as, e.g., factorisation of numbers, are hard problems in terms of complexity theory. We denote a ciphertext produced as a result of encrypting a plaintext message m using an encryption key k as $\{m\}_k$.

A ciphertext $\{m\}_k$ shall ensure that only those possessing key k are able to decrypt the ciphertext and access m in plaintext form. In practice ciphertexts may be open to manipulation, as the legacy of the *Data Encryption Standard*, or *DES*, reminds us. DES was introduced in the 1970s as an encryption algorithm; as of late 1990s it has been considered to be vulnerable to brute-force attacks; consequently it has been replaced by the *Advanced Encryption Standard*, or *AES*. For the purposes of analysis of security protocols, assumptions are often made on cryptography. This helps abstract away from cryptographic weaknesses. Throughout this chapter, we take the *perfect encryption assumption*, which says all cryptographic methods perform as they are described to do so. Cryptography is therefore assumed to be perfect, that is to say, no plaintext, partly or wholly, can be derived from a ciphertext unless the appropriate decryption key for the ciphertext is known. Taking the perfect encryption assumption helps to separate concerns: the focus is only on those attacks which manipulate the protocol; attacks on messages are not considered.

Not only does the possession of the key, therefore, becomes essential to the retrieval of plaintext from a ciphertext, it also provides us with the means to dictate who can, or cannot, retrieve it; equally, the possession of a key enables the generation of a ciphertext using that key. Both concepts are exercised in the design of security protocols to implement the principles discussed previously. Two forms of cryptography are commonly distinguished:

- *Shared-key cryptography* provides a system where the encryption key and the corresponding decryption key are the same. If a message m is encrypted with key k then the same key k is used to decrypt $\{m\}_k$ to retrieve m in plaintext form. This is a conventional form of cryptography and is in use where two communicating entities already share a secret key to communicate secret messages. A typical example in use these days is the AES, see above.

- *Public-key cryptography* makes use of a pair of corresponding encryption and decryption keys that are different, such that the possession of the encryption key does not allow any knowledge of the decryption key. The system allows an entity to possess a pair of such keys where the encryption key is made public, called the *public key*, while the decryption key is kept private to the entity, called the *private key*. For such a system to work, it is essential that the private key remains private to the entity.

For an entity A, we denote public and private kEys as pk_A and sk_A. A message m encrypted with A's public key as $\{m\}_{pk_A}$ could only be decrypted using A's private key sk_A, where

$$\{\{m\}_{pk_A}\}_{sk_A} = m \tag{8.1}$$

This allows any entity to send private messages to A without the need for sharing a key between them.

Another useful application of public key cryptography is to allow for signing messages, i.e., a message m is encrypted with A's private key as $\{m\}_{sk_A}$. This could only be decrypted using A's public key pk_A, where

$$\{\{m\}_{sk_A}\}_{pk_A} = m \tag{8.2}$$

This allows any entity A to sign a message, where anyone can verify that the message originated from A.

Closely following [CLRS09], we provide an example of a public key cryptosystem.

Example 72: RSA Public Key Cryptosystem

RSA (Rivest-Shamir-Adleman) is a public key cryptosystem, for which a participant creates the pair of public and secret keys as follows:
1. Select at random two large prime numbers p and q with $p \neq q$.
2. Let $n \triangleq pq$.
3. Select a small odd number e which is relatively prime to $(p-1)(q-1)$, i.e., the only natural number that divides both of them is 1.
4. Compute d as the the multiplicative inverse of e modulo $(p-1)(q-1)$, i.e., $de = 1 \ mod \ (p-1)(q-1)$.
5. The pair (e, n) is the public key.
6. The pair (d, n) is the secret key.
   Given a message $m \in \{0, \ldots, n-1\}$, we define

- $\{m\}_{e,n} \triangleq m^e \ mod \ n$ and
- $\{m\}_{d,n} \triangleq m^d \ mod \ n$.

With these definitions, both Eqs. (8.1) and (8.2) hold [CLRS09].

We mention in passing that in cryptography the above discussed 'textbook RSA' is perceived not to be secure. It suffers from the following attack: given an encrypted message $\{m\}_{pk}$, an attacker can 'guess' a message m' and check if $\{m'\}_{pk} = \{m\}_{pk}$. To prevent such and other attacks, in practice RSA is often complemented with a so-called Optimal Asymmetric Encryption Padding (OAEP). OAEP adds an element of randomness and turns a deterministic encryption scheme into a probabilistic one; further it prevents partial decryption of cipher texts.

Often, shared-key and public-key cryptography are used in tandem. For instance, the network protocol suite IPSec and the security protocol TLS use public-key cryptography when establishing a fresh session key for subsequent shared-key encrypted communications to leverage the advantages of both, while avoiding the drawbacks of each.

8.2.2 Principles of Security

We introduce a few core principles on which security properties are based [SC14].

- *Confidentiality* is an assurance on the secrecy of data. Data is said to be confidential to a set of entities if it is only available to those entities, and not disclosed to any other outside of the set. This is an important principle as it allows for some information to be exclusively shared on the basis of which further security principles can be achieved.
- *Integrity* is an assurance that data is not modified or manipulated in any way from inception. Any such modification may be intentional by an unauthorised entity attempting to achieve a malicious purpose, or accidental, for example, due to the medium of communication.
- *Availability* is an assurance that data is available when needed. An intruder can't prevent an authorised user from accessing or modifying data.

The three principles above provide a coarse classification for a range of security properties. In this chapter, we concern ourselves with authentication, which is a property that stems from integrity [SC14].

Authentication is used to refer to *data* or *entity authentication*. Data authentication ensures that some data seemingly coming from an entity A, has certainly come from A and no other entity. Entity authentication provides a similar guarantee on the identity of an entity, such that if B believes that it is communicating with some entity A, then it is indeed A that B is communicating with. Data integrity together with entity authentication guarantees data authentication.

Example 73: Online Auction Site

Some of the security principles can be illustrated in online auction sites such as eBay.

During an auction, the identities of bidders have to be kept *confidential* to ensure the seller or other bidders are not able to determine who is interested or who they are up against. Disclosure of bidder identities could lead to violation of bidder privacy or coercion.

The amount of each bid needs to preserve *integrity* in line with the expectations of the bidders. Each bid is a commitment from the bidder to the price they are willing to pay. Malicious modification of the amount of the bid could either lead the bidder to lose the auction if it is reduced, or force the bidder to pay more than committed if it is increased.

The seller needs the guarantee that each bid is *authenticated*, that is it is actually coming from the rightful bidder. False bids injected into the auction will not lead to a sale given the bidder never committed.

The online auction provider needs to ensure the website remain equally *available* for all bidders to bid at any time before the auction closes.

Of course there are other critical properties for online auctions but we have mentioned only the ones relevant to the chapter.

8.2.3 Security Protocols

A *communication protocol* is an agreed sequence of actions performed by two or more communicating entities in order to accomplish a purpose, for example fault tolerance over a noisy communication medium. Such protocols are critical to connectivity at every level, from cellular devices to Internet of Things (IoT). One example of a protocol is the well-known Internet Protocol (IP), which serves to address device interfaces, packages data into acceptable chunks for transfer, and routes data packets across the globe. In this chapter protocols are informally described using the notation given

$$(i) \ A \rightarrow B : m$$

This simply denotes that in the ith step of a protocol entity A sends a message m destined for entity B. We use A, B and C throughout the chapter as participant entities. Message can consist of several parts which are concatenated. We write message concatenation with a '.', e.g., $m_1.m_2$ is a message consisting of two messages m_1 and m_2. In order to avoid brackets, we use the convention that concatenation is a left-associative operator.

A *security protocol* is a communication protocol that provides assurance on security. To this end, such a protocol may use cryptography as a building block towards achieving one or more security principles; note however that cryptography alone may not suffice to achieve a principle.

Example 74: A Simple Security Protocol

Going back to Example 11 from Chap. 1, we can formulate the protocol description given there in natural language in more formal terms:

"The aim of the protocol is that A and B share a secret key s at the end. Participant A generates a fresh session key k, signs it with his secret key $sk(A)$ and encrypts it using B's public key $pk(B)$. Upon receipt B decrypts this message using the private secret key, verifies the digital signature and extracts the session key k. B uses this key to symmetrically encrypt the secret s." [CK11].

Using the notation introduced above, we can write this protocol as follows:

$$(1) \ A \rightarrow B : \{\{k\}_{sk(A)}\}_{pk(B)}$$
$$(2) \ B \rightarrow A : \{s\}_k$$

Note, that this notation does not take care of actions that a participant is performing after receiving a message, e.g., the activity "upon receipt B decrypts this message" is abstracted from.

It is important to introduce the notion of an *intruder* here. Security protocols operate in environments where intruders also operate. Essentially, intruders are legitimate users who may operate to achieve malicious goals. These include spoofing (appearing as some other user) or capturing confidential data (which they are not authorised to access), ultimately to achieve some advantage. A successful realisation of such an attempt is an *attack*.

The capabilities of such an intruder are important to comprehend to ensure that protocols are designed to withstand attacks that such intruders can launch. One model of an intruder is presented by Dolev and Yao [DY83], which allows it to

- block messages, where a message is withheld from recipients;
- replay messages, where an old message could be retransmitted to a recipient of choice;
- spoof messages, where messages are constructed to falsely appear to come from a different source;
- manipulate messages, where multiple messages could be assembled into one or deassembled into fragments of choice; and
- encrypt or decrypt messages, however only where the intruder is in possession of the relevant keys (perfect encryption assumption).

Assuming such an intruder, the protocol shown in Example 74 above is not secure:

Example 74.1: Man in the Middle Attack

The protocol is vulnerable to a so-called "man in the middle" attack where the intruder breaches security by intercepting messages and manipulating them:

$$(1.1)\ A \rightarrow I : \{\{k\}_{sk(A)}\}_{pk(I)}$$
$$(2.1)\ I(A) \rightarrow B : \{\{k\}_{sk(A)}\}_{pk(B)}$$
$$(2.2)\ B \rightarrow I(A) : \{s\}_k$$

The attack goes as follows: participant A wants to talk to participant I. However, I is an intruder. Rather than responding to A, participant I is initiating a conversation with another participant B. To this end, participant I decrypts the message it receives and obtains $\{k\}_{sk(A)}$—that is possible as I knows its own private key and thus can apply equation (8.1)—and encrypts $\{k\}_{sk(A)}$ with the public key of a participant B whom I would like to speak with, pretending to be A. I sends this message to B, pretending to be A. This is expressed by the notion $I(A)$. B verifies that the message came from A: first by decrypting it with $sk(B)$ and then using $pk(A)$ and Eq. (8.2) to check that the message k 'makes sense', i.e., is a legitimate session key. Then B sends the secret key s to I, in the belief to speak with A. This is expressed in step (2.2) as $I(A)$.

I can decrypt this message, as it can obtain k by applying Eq. 8.2: it can decrypt $\{k\}_{sk(A)}$ using A's public key.

In this example, the intruder I was using the ability to encrypt or decrypt messages as any other protocol participant.

Some authors consider more powerful intruder models, where the intruder has more capabilities, in particular w.r.t. the perfect encryption assumption. For instance, Schneider [Sch02] demonstrates how the intruder could be allowed to manipulate composition of messages through the exclusive-or function. Compared to the Dolev-Yao model, the intruder is therefore given further capability to interfere with protocol messages.

8.3 Needham-Schroeder Protocol for Authentication

This chapter introduces the original Needham-Schroeder protocol [NS78] (short: N-S protocol) as one of the earliest authentication protocols. The protocol was designed to provide mutual entity authentication to a pair of participants. This is followed by Gavin Lowe's attack [Low95] on the protocol, which demonstrates the challenge of getting the design of such protocols correct with respect to their authentication goals. Then the amended version, often known as the Needham-Schroder-Lowe protocol (short: N-S-L protocol), is described.

The design of authentication protocols is often based on a *challenge-response* mechanism where an authenticating entity A sends out a challenge to some entity B being authenticated and expects a response, which is some manipulated form of the challenge value that only B can generate. This design principle is often referred to as *authentication by correspondence* [Gol03]. Such manipulation between the challenge and response forms of the messages could be achieved using cryptography. Some renowned examples of authentication protocols based on the correspondence principle include Kerberos[2] (derived from the Needham-Schroeder protocol), RADIUS (described by a number of documents issued by the Internet Engineering Task Force IETF) and IPSec (also standardized by IETF). Such protocols can offer different guarantees on authentication. Lowe [Low97b] lays out a hierarchy of such guarantees, out of which we introduce two, closely following Lowe.

Definition 1 *(Aliveness)* We say that a protocol guarantees to an initiator A *aliveness* of another entity B if, whenever A (acting as initiator) completes a run of the protocol, apparently with responder B, then B has previously been running the protocol.

Aliveness therefore serves the simple purpose of confirming to an initiator that a responder is alive and responding in some protocol run. However, there

---

[2] http://web.mit.edu/kerberos/.

are no guarantees to the initiator with whom the responder is running the protocol with. This is the weakest of the authentication guarantees suggested by Lowe. Some protocols may serve this guarantee from a responder to an initiator, in which case we speak of aliveness as well.

On the other hand the strongest of Lowe's guarantees is presented below.

Definition 2 *(Injective agreement)* We say that a protocol guarantees to an initiator A *injective agreement* with a responder B on a set of data items ds if, whenever A (acting as initiator) completes a run of the protocol, apparently with responder B, then B has previously been running the protocol, apparently with A, and B was acting as responder in his run, and the two agents agreed on the data values corresponding to all the variables in ds, and each such run of A corresponds to a *unique* run of B.

Injective agreement ensures that both protocol participants are engaged in the same run with each other, every time the protocol is run. Additionally, every run provides explicit agreement on any parameters of the protocol. Essentially, this serves for mutual authentication between A and B.

Strong guarantees such as injective agreement above require some mechanism to ensure uniqueness of protocol runs. To ensure that a response is fresh, and not a replay from a previous run of the protocol between some A and B, a random value is often used, referred to as a *nonce*. At the start of every run, A would generate a fresh nonce N_A and include it in the challenge it sends to B. Upon receipt of a response from B, A would check for the value of the nonce to ensure it is the same. This provides A with a guarantee that B has engaged in this fresh run of the protocol. This is important where protocols are designed to run several times; in such a situation the attacker could replay an older message and use it to deceive a participant.

For the purposes of this chapter, we consider nonces as arbitrary values for single use. We assume that nonces are

- *fresh* every time they are generated;
- *unpredictable* such that no participant can determine the value of a nonce yet to appear; and
- *not able to reveal the identity* of the participant that produced the nonce.

For protocol implementation nonces are typically realised as randomly generated numbers. How to realise random number generators, and how to prove that they match the criteria listed above, is beyond the scope of this chapter.

One of the first authentication protocols was presented by Needham and Schroeder [NS78] in 1978. It became the basis for the well-known Kerberos protocol widely used in a range of applications including imap, POP, SMTP, FTP, SSH and Microsoft applications.

Example 75: Needham-Schroeder Protocol

The protocol provides two participants A and B to authenticate to each other using public-key cryptography. A shortened three-message version of the protocol is presented below. The original seven-message version of the protocol [NS78] includes additional messages where A and B simply communicate with a trusted server to obtain public keys for each other; the heart of the protocol remains the same in the shortened version.

$$(1)\ A \to B : \{N_A.A\}_{pk_B}$$
$$(2)\ B \to A : \{N_A.N_B\}_{pk_A}$$
$$(3)\ A \to B : \{N_B\}_{pk_B}$$

A initiates the protocol by sending out an encrypted message to B including its own identity concatenated with a freshly generated nonce N_A. The message is encrypted with public key pk_B of B so only B can read it (using the corresponding private key). B responds by generating a fresh nonce N_B and sends it back to A along with N_A. The message is encrypted with public key pk_A of A so only A can read it. The second message allows A to authenticate B as it confirms N_A to be the nonce that it originally sent out to B. A responds with the final message of the protocol including only N_B and encrypting it with pk_B, back to B, which allows B to authenticate A.

The protocol aims to satisfy injective agreement; the two nonces by A and B are the respective data items that both participants agree on in each run.

In 1995, Lowe presented a man in the middle attack on the N-S protocol [Low95]. Such an attack allows an attacker to impersonate one participant in a session with another and serves to demonstrate the subtleties of designing such protocols.

Example 75.1: Lowe's attack on the N-S protocol

Lowe's attack involves two runs of the protocol involving an intruder I. As in Example 74.1, we denote I pretending to be A as $I(A)$.

$$(1.1)\ A \to I : \{N_A.A\}_{pk_I}$$
$$(2.1)\ I(A) \to B : \{N_A.A\}_{pk_B}$$
$$(2.2)\ B \to I(A) : \{N_A.N_B\}_{pk_A}$$
$$(1.2)\ I \to A : \{N_A.N_B\}_{pk_A}$$
$$(1.3)\ A \to I : \{N_B\}_{pk_I}$$
$$(2.3)\ I(A) \to B : \{N_B\}_{pk_B}$$

The first run of the protocol shows A initiating an honest run with I with a fresh nonce N_A. Before I responds, it initiates a second run of the protocol with B, pretending to be A, and forwards N_A to B by encrypting it with pk_B. B responds to the request by encrypting N_A and a fresh nonce N_B, and encrypting it with A's public key pk_A. At this stage, B believes that it is taking part in a run with A, whereas A believes that it is taking part in a run with I; note that N_B is of the form that does not convey to A the identity of the entity that generated the nonce.

For B to be assured that it communicates with A, it waits for the final message of the protocol. To mislead B, I relays the response from B onwards to A in response to the earlier run of the protocol. A finds the message as expected and returns N_B encrypted with pk_I, which I is able to decrypt. I encrypts it with pk_B to send it onwards to B as part of the final message of the second run. As a result of this, B is misled to believe that it has just completed a protocol run with A.

Lowe's attack shows that the protocol fails on injective agreement between A and B. Injective agreement insists that the responder B has to be running the protocol with A, which has shown to be not the case.

For the intruder to complete the last step (2.3) of the manipulated run, it needs A to decrypt B's nonce. Hence the attack cannot be fulfilled without A's engagement. Therefore, despite of the attack, A's aliveness is guaranteed to B.

The attack demonstrates how the protocol fails to achieve the goal of mutual authentication between the two participants.

Lowe [Low95] presents a modification to the original protocol, aptly known as the *Needham-Schroeder-Lowe protocol* (short: N-S-L protocol), to address the flaw demonstrated in Example 75.1

Example 76: Needham-Schroeder-Lowe Protocol

The modification involves including the identity of the responding participant in the second message of the protocol, which allows the identities of both participants to be made explicit in encrypted messages. This prevents I from simply forwarding the message received from B in the second run of the protocol attack as a response back to A in the first run as shown in Example 75.1; ultimately, this prevents A to send out a response in the first run that I uses to manipulate B and successfully complete the attack. The modified version of the protocol is as follows.

$$(1)\ A \rightarrow B : \{N_A.A\}_{pk_B}$$
$$(2)\ B \rightarrow A : \{B.N_A.N_B\}_{pk_A}$$
$$(3)\ A \rightarrow B : \{N_B\}_{pk_B}$$

> Injective agreement ought to be guaranteed given that B explicitly
> states its identity in Step (2) of the protocol, to avoid manipulation
> of the message in other interleaving runs.

Clearly, designing such protocols is a challenge. Lowe's attack on the N-S protocol appeared some seventeen years later from the publication of its design. This begs the question therefore, how does one check for correctness of protocols with respect to the desired security properties? Indeed, this is the central motivation underlying this chapter. The approach presented in the rest of the chapter allows us to formally model security protocols, and effectively verify their properties with respect to desired guarantees.

8.4 Formal Specification of Protocols and Properties

The design of authentication protocols is difficult to get correct as Sect. 8.3 demonstrates. This raises the need for a rigorous approach to analysing the design of such protocols. This section presents a framework to express authentication properties as formal specifications for protocols in CSP. Following sections present approaches to verify such specifications. Section 8.4.1 discusses the underlying abstraction principles, Section 8.4.2 defines the message space for protocol descriptions, Sect. 8.4.3 describes how to model protocol participants in the context of a reliable network. Naturally, to assume a reliable network is unrealistic. Consequently, Sect. 8.4.4 provides a formalisation of the Dolev-Yao model in CSP. Section 8.4.5 presents a formal specification of authentication in terms of correspondence.

8.4.1 Protocol Abstraction Through Use of Symbolic Data

Up to now, we studied protocols presented in an informal protocol notation. Now we want to provide formal models in CSP. This will allow us to actually prove properties.

The formal models in CSP will capture the message exchange between the protocol participants on some level of abstraction. Note that exchange of messages between computers on the technical side is quite a complex task in itself: the Open Systems Interconnection model[3] defines seven different layers of message exchange. The new abstraction here concerns data. Up to now, one could think of concrete data being exchanged in between participants, e.g., in the context of RSA a key could be a sequence of 2048 bits. Now we will

---

[3] https://www.iso.org/standard/20269.html.

abstract from these concrete values by defining symbols representing data in order to simplify protocol analysis.

Whenever one describes a system on two different levels of abstraction, say a concrete model CM and and abstract model AM, one can link these two levels in the following way. An abstraction is called **sound** w.r.t. a property φ if

$$AM \models \varphi \implies CM \models \varphi.$$

An abstraction is called **complete** w.r.t. a property φ if

$$CM \models \varphi \implies AM \models \varphi.$$

Sound abstractions allow one to establish system properties by carrying out system analysis on the abstract, i.e., simpler system.

In order to ease protocol analysis, we apply a data abstraction as follows:

1. We represent the set of all participants by two users and one intruder.
2. We represent the set of all keys by the keys associated with these three participants; we encode the perfect encryption assumption by ensuring that between theses values only Eqs. (8.1) and (8.2) hold.
3. We represent the set of nonces by one nonce for each participant, which is used only once in the protocol run.

Using this data abstraction, it is assumed that the abstract Csp models are sound and complete w.r.t. the protocols presented in the informal protocol notation. As the protocol notation is an informal one, we can't connect the concrete and the abstract level by a proof. In principle it would be possible to also provide formal models of protocols on the concrete level. However, in the literature the assumption is that the above presented abstraction is a trivial one, i.e., it would not be worth to provide a formal proof for it.

In passing, we would like to mention that Ranko Lazić has worked on data abstraction in Csp [Laz99]. With his techniques, one could provide a non-symbolic model of a security protocol in Csp and try to prove that—w.r.t. selected security properties—a symbolic Csp model is sound and correct.

8.4.2 Message Space

The message space of a protocol is the set of all messages that can appear in any run of the protocol. This space usually has a structure. In general, one considers the following sets: $\mathcal{U}$ denotes a set of protocol participants that may run it, $\mathcal{N}$ is a set of all nonces and $\mathcal{K}$ is a set of all cryptographic keys that can be used by participants. Thus the set of all atoms $\mathcal{A}$ is defined as

$$\mathcal{A} ::= \mathcal{U} \mid \mathcal{N} \mid \mathcal{K}$$

In the context of cryptophic protocols, it is usual to construct the message space with notation borrowed from grammars; alternatively, one could write a disjoint union rather than "|".

Given these atoms $\mathcal{A}$, the message space $\mathcal{M}$ containing all messages that may ever appear in a protocol run is given by the grammar

$$\mathcal{M} ::= \mathcal{A} \mid \{\mathcal{M}\}_{\mathcal{K}} \mid \mathcal{M}.\mathcal{M}$$

There are three cases: messages can be atoms, messages can be encrypted under some key, and messages can be paired.

Example 75.2: Message Space of the N-S Protocol

For the N-S protocol the atoms of the message space are given as follows:
$\mathcal{U} \triangleq \{A, B, I\}$
$\mathcal{N} \triangleq \{N_A, N_B, N_I\}$
$\mathcal{K} \triangleq \{pk_A, pk_B, pk_I, sk_A, sk_B, sk_I\}$
Examples of messages that can be formed include A, $\{N_A.A\}_{pk_B}$, and $A.B$. Note that the message space is infinite thanks to both, encryption and pairing.

The pairing operator is binary. In order to avoid writing brackets, we assume that it is left-associative, i.e., its operands are grouped from the left. This allows us to write messages such as $B.N_A.N_B$ without brackets, knowing it stands for $(B.N_A).N_B$.

8.4.3 Protocol Participants in a Reliable Network

When analysing security protocols in CSP, the behaviour of the participants is usually represented by processes. Participants can communicate over channels, over which they can send messages (denoted as '!') or on which they can receive messages (denoted by '?'). Channels have a name, e.g., *send*, and consist of several components, which are separated by '.' (for selecting a specific channel), '!' indicates sending a value and '?' indicates receiving a value and binding it to a variable. We will apply the following naming conventions: on a *send* channel, the first component is the sender, the second component the receiver; on a *receive* channel, the first component is the receiver, the second component the sender; on both channels, the third component is for the actual message.

In our example below we use three operators for describing protocol participants:

- The *Stop* operator, which denotes the process that does not do anything,

- The *action prefix* operator $a \to P$, which describes a process that first engages in the event a and then behaves like P, and, finally,
- The *external choice* operator

$$\Box_{x \in X} P(x)$$

which denotes a collection of processes $P(x)$ for $x \in X$, one of which will be executed.

Note that as we will be working over the traces model, external and internal choice are the same. In the literature on modelling security protocols in CSP usually the external choice operator is used, even when conceptually the choice is an internal one.

In order to model communication networks and the composition of the overall system, we use two operators:

- The *generalised parallel* operator $P \, [| \, Channels \, |] \, Q$ combines two processes P and Q to a new process representing a system, in which P and Q exchange messages over communication channels listed in *Channels*.
- The *interleaving* operator $|||_{U \in \mathcal{X}} U$ combines all processes listed in the set $\mathcal{X}$ to a new process representing a system, in which the processes of $\mathcal{X}$ run independently of each other.

For a more thorough introduction to CSP see Chap. 3.

In order to demonstrate the attack on the N-S protocol, it suffices that the participants run only one instance of the protocol. With the channels and the above listed operators, we can model the participants of the N-S protocol and a reliable network as follows:

Example 75.3: Modelling the participants of the N-S protocol

The initiator A can choose to run the protocol with any other protocol participant $b \in \mathcal{U}$:

$$A = \Box_{b \in \mathcal{U}, b \neq A} \; send.A.b!\{N_A.A\}_{pk_b} \to$$
$$receive.A.b?\{N_A.n\}_{pk_A} \to$$
$$send.A.b!\{n\}_{pk_b} \to Stop$$

Given a chosen participant b, the process follows the specified protocol conforming to the sequence of messages required. b and n are variables: b is bound to a specific value at the beginning of the process, n is bound to a specific value when A receives a message in the second step. After running the protocol once A stops. Similarly for process B :

$$B = \; receive.B?a?\{n.a\}_{pk_B} \to$$
$$send.B.a!\{n.N_B\}_{pk_a} \to$$
$$receive.B.a?\{N_B\}_{pk_B} \to Stop$$

Here, n and a are variables. Both a and n are bound when B receives its first message. The question mark before the a indicates that a is a variable. The CSP process B is only prepared to receive messages if both occurrences of a in $receive.B?a?\{n.a\}_{pk_B}$ have the same value. In $receive.B.a?\{N_B\}_{pk_B}$ we write a '?' in front of $\{N_B\}_{pk_B}$, though the process B is only willing to receive this specific value and no binding to a variable takes place.

Note that each message in the protocol corresponds in CSP to one send and one receive action in different processes.

We model a 'secure' *Network* as a process that receives messages, passes them on to the intended receiver, and then starts over again:

$$Network = send?i?j?m \rightarrow receive!j!i!m \rightarrow Network$$

Protocol participants and network then can be composed together to a system, e.g, to the system being composed from A and B :

$$System = (\|\|_{U \in \{A,B\}} U) \, [\| \, send, receive \, \|] \, Network$$

All participants run independently of each other $(\|\|_{U \in \{A,B\}} U)$, however they communicate over the *Network* using the *send* and *receive* channels.

Figure 8.1 provides a visualisation of the communication structure within this *System*. Participants A and B have each two *send* and *receive* channels to and from the *Network*. As there is no participant I, the *Network* is free to use the channels to or from I, though they are not connected with any process.

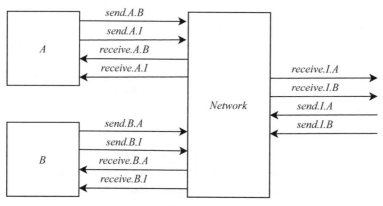

Fig. 8.1 Communication structure of $(\|\|_{U \in \{A,B\}} U) \, [\| \, send, receive \, \|] \, Network$

This modelling makes use of CSP pattern matching: a process is only willing to receive those messages that match the specified pattern. This abstracts from implementation details such as first receiving a message, then decrypting and analysing it, before possibly rejecting it.

8.4.4 Modelling the Intruder and the Network

Recall that the Dolev-Yao model allows an intruder to block, replay, spoof and manipulate messages that appear on any public communication channel. A *generates* $\vdash$ relation can be used to characterise the message manipulation that the intruder is capable of. Figure 8.2 presents the rules that define this relation.

Given a set $S \subseteq \mathcal{M}$ of messages, the generates relation $\vdash \subseteq \mathcal{P}(\mathcal{M}) \times \mathcal{M}$ is the smallest relation closed under the following rules:

1. if $m \in S$ then $S \vdash m$
2. if $S \vdash m$ and $S \vdash k$ then $S \vdash \{m\}_k$
3. if $S \vdash \{m\}_{pk_b}$ and $S \vdash sk_b$ then $S \vdash m$
4. if $S \vdash m_1.m_2$ then $S \vdash m_1$ and $S \vdash m_2$
5. if $S \vdash m_1$ and $S \vdash m_2$ then $S \vdash m_1.m_2$

Fig. 8.2 Rules for generating messages using the $\vdash \subseteq \mathcal{P}(\mathcal{M}) \times \mathcal{M}$ relation

The relation characterises the intruder's message manipulation. For a given set S of intruder's knowledge, the intruder can generate any message it is aware of (Rule 1). The intruder can encrypt (Rule 2) or decrypt (Rule 3), can pair (Rule 4) or unpair messages (Rule 5) it is aware of.

Note that we realise only Eq. (8.1) in the intruder. Equation (8.2) does not play a role in Lowe's attack nor is it used anywhere in the protocol. Schneider proves that the N-S-L protocol is also secure if the intruder has the added capability of using Eq. (8.2) [Sch98b]. For demonstrating the principles of protocol analysis, it suffices to consider Eq. 8.1 only.

The literature may include an additional property:

$$\text{if } S \vdash m \text{ and } S \subseteq S' \text{ then } S' \vdash m.$$

This rule says that any growth in knowledge from S to S' allows the intruder to generate from S' at least as many messages as from S. As this property is a consequence of other stated rules it is excluded.

Example 75.4: Working with the generates relation

In the N-S protocol, initially the intruder knows all participants as well as the public keys of all participants and its own secret key sk_I. Thus, its initial set of knowledge is

$$Init \triangleq \{A, B, I, pk_A, pk_B, pk_I, sk_I\}.$$

In step (1.1) of the protocol run in Eample 75.1, the intruder receives the message $\{N_A.A\}_{pk_I}$. This increases the intruder's knowledge to a set denoted as *IK* as follows:

$$IK = Init \cup \{\{N_A.A\}_{pk_I}\}.$$

With *IK* the intruder can decrypt the message $\{N_A.A\}_{pk_I}$:
- we have $\{N_A.A\}_{pk_I} \in IK$ and thus $IK \vdash \{N_A.A\}_{pk_I}$ using Rule 1,
- we have $sk_I \in IK$ and thus $IK \vdash sk_I$ using Rule 1, and
- thus we obtain $IK \vdash N_A.A$ using Rule 3.

Furthermore, the intruder can encrypt the message $N_A.A$ with B's public key, as $pk_B \in IK$ and Rule 2 allows for encryption.

An *Intruder* process is introduced to model the capabilities of a Dolev-Yao intruder. A set of messages *IK* constitutes as *intruder knowledge* that the process is parameterised with:

$$Intruder(IK) = (send?i?j?m \rightarrow Intruder(IK \cup m)) \qquad (8.3)$$

$$\square$$

$$(\square_{i,j \in \mathcal{U}, IK \vdash m} \; receive!i!j!m \rightarrow Intruder(IK))$$

The process models what the intruder is capable of and as such

- receive m, sent by i to j on *send*, and then behave as an intruder with m added to its knowledge, or
- send any message m, generated under $\vdash$ from set *IK*, to any i pretending to come from any j along *receive*.

IK builds up as the intruder observes traffic passing through the network, hence simply

$$Intruder = Intruder(Init)$$

Here, *Init* is the intruder's initial knowledge. In the context of the N-S and the N-S-L protocols, *Init* refers to the set given in Example 75.4.

We compose a network *NET* containing all protocol participants communicating with each other over a medium that is completely under the control of the Dolev-Yao intruder:

$$NET = (|||_{U \in \mathcal{U}, U \neq I} U) \, [|\, send, receive \,|] \, Intruder$$

NET is modelled to force all participants to synchronise with the *Intruder* over the *send* and *receive* channels. Such a model of the network allows the intruder to have Dolev-Yao capabilities to:

- block messages, through simply allowing the intruder a choice to forward any message it receives on the *send* channel on to the *receive* channel, which it may not do so,
- replay messages, where any message the intruder holds in *IK* can be retransmitted over the *receive* channel to any participant of choice apparently coming from any other, and
- spoof, manipulate, encrypt or decrypt messages from *IK* under the generates relation $\vdash$ to forward on to the receive channel to any participant of choice apparently coming from any other.

Furthermore, the intruder process is capable to act as protocol participant I:

Example 75.5: The intruder as a protocol participant

We consider all messages from a 'normal' protocol run, where the intruder using its identity I would like to initiate authentication, say with participant B:

- In the first step, the intruder would send the message $\{N_I.I\}_{pk_B}$ to B. As I, N_I, pk_B are initially known to the intruder process, it is capable of forming this message and communicating it to B via the channel message *receive.B.I*.
- In the second step, the intruder ought to receive a message $\{N_I.n\}_{pk_I}$ from B. The intruder listens to the channel *send.B.I* and is willing to receive any message. According to the protocol, B sends $\{N_I.N_B\}_{pk_I}$. The intruder can decrypt this message using sk_I, and adds N_B to its knowledge base.
- Having N_B in its knowledge base from the second step, the intruder can form the message $\{N_B\}_{pk_B}$ and send it to B via the channel *receive.B.I*.

Similarly, the intruder can take on the role of the follower process, e.g., can participate in a run of the N-S protocol for A initiating authentication with I.

We use trace semantics to formally express the relation between authenticatee and authenticator. The underlying idea is to make explicit a participant's involvement in their run of the protocol with respect to a corresponding participant.

Example 75.6: The N-S protocol attack in CSP trace semantics

We can now formally study the N-S attack:

$$NET = (A \;|||\; B) \;[|\; send, receive \;|]\; Intruder$$

Compared with the setting of Example 75.3, now the intruder process replaces the network component.

The process NET has many traces including the one from the man in the middle attack demonstrated in Example 75.1:

$$\langle\; send.A.I.\{N_A.A\}_{pk_I},$$
$$recv.B.A.\{N_A.A\}_{pk_B},$$
$$send.B.A.\{N_A.N_B\}_{pk_A},$$
$$recv.A.I.\{N_A.N_B\}_{pk_A},$$
$$send.A.I.\{N_B\}_{pk_I},$$
$$recv.B.A.\{N_B\}_{pk_B} \;\rangle$$

Analysing this trace, the above described deception can no be detected easily. The reason is that it does not include the beliefs of the participants, namely whom A and B are believing to run the protocol with.

Example 75.6 shows that taking the traces of a security protocol as they are can make it difficult to analyse the protocol for authentication. In such circumstances, one often 'enriches' such model of a system in order to indicate particular behaviours of interest. In the next section we will use 'signal events' to indicate what stage of the protocol a participant has reached through a protocol run.

8.4.5 Formalising Authentication

To make explicit the role of protocol participants, signal events are defined [SBS09] as a structure to indicate the status of a participant. This technique is known as instrumentation. In instrumentation, signal events are added to a process in order to allow for more fine grained observations. When one makes use of instrumentation, it is important to check that the chosen instrumentation does not alter the original behaviour, i.e., after instrumentation, all original behaviours must remain possible, and no new behaviours are added.

For the purpose of analysis in this chapter, we use a set $\mathcal{S}$ of *signals*:

$$\mathcal{S} ::= Running.\mathcal{U}.\mathcal{U}.\mathcal{N} \;|\; Commit.\mathcal{U}.\mathcal{U}.\mathcal{N}$$

Each of the two event values is meant to indicate a particular stage in a protocol run: a *Running* signal indicates that an authenticatee (being authenticated) is engaged in some protocol run; a *Commit* signal acknowledges successful authentication on behalf of an authenticator in a protocol run. For a pair of participants engaged in a protocol run, it follows that *Running* is always expected to precede *Commit*.

We illustrate the use of signals in the example below.

Example 75.7: Instrumenting the N-S protocol

We instrument the N-S protocol from Example 75. We first extend the protocol notation by including the relevant signals for each step:

(1) $A \rightarrow B : \{N_A.A\}_{pk_B}$ $\qquad\qquad\qquad$ $Running.B.A.N_A$

(2) $B \rightarrow A : \{N_A.N_B\}_{pk_A}$ $\qquad$ $Commit.A.B.N_A, Running.A.B.N_B$

(3) $A \rightarrow B : \{N_B\}_{pk_B}$ $\qquad\qquad\qquad$ $Commit.B.A.N_B$

The signal $Running.B.A.N_A$ indicates that B believes it is being authenticated to A with nonce N_A serving as the unique challenge. The corresponding $Commit.A.B.N_A$ signal indicates A's belief that it has authenticated B in this run. Conversely A's authentication to B is captured using the second pair of signals. This instrumentation can be realised as follows:

$$A = \Box_{b \in \mathcal{U}, b \neq A} \; send.A.b!\{N_A.A\}_{pk_b} \rightarrow$$
$$receive.A.b?\{N_A.n\}_{pk_A} \rightarrow$$
$$Commit.A.b.N_A \rightarrow$$
$$Running.A.b.n \rightarrow$$
$$send.A.b!\{n\}_{pk_b} \rightarrow Stop$$

$$B = \; receive.B?a?\{n.a\}_{pk_B} \rightarrow$$
$$Running.B.a.n \rightarrow$$
$$send.B.a!\{n.N_B\}_{pk_a} \rightarrow$$
$$receive.B.a.\{N_B\}_{pk_B} \rightarrow$$
$$Commit.B.a.N_B \rightarrow Stop$$

The N-S protocol shall provide mutual authentication, therefore there are two authentication attempts in one protocol run.

The first of these attempts is:
- B's signal $Running.B.A.N_A$ says that B believes it is running the protocol with A,
- this ought to correspond with A's subsequent signal $Commit.A.B.N_A$, which says that A believes it has authenticated B.

The second of these attempts is:

- A's signal *Running.A.B.N_B* says that A believes it is running the protocol with B,
- this ought to correspond with B's subsequent signal *Commit.B.A.N_B*, which says that B believes is has authenticated A.

We distinguish the two attempts by annotating the signal data using nonces.

Naturally, authentication properties would always be expressed over the beliefs of honest participants. Therefore, our use of instrumentation focusses on honest participants. It does not allow us to express any guarantees on behalf of the intruder.

We formalise authentication in terms of signals using trace specifications.

Definition 3 *(a precedes b)* Let a and b be some events, let tr be a trace. We say that a **precedes** b in tr, if any occurrence of event b is preceded by an occurrence of event a, i.e.,

$$\forall tr'.tr' \frown \langle b \rangle \leq tr \implies a \in tr'$$

Here, $\frown$ stands for concatenation of traces, $\langle \_ \rangle$ is a constructor that turns a single event into a one element trace, $s \leq t$ holds if s is a prefix of t, and $a \in tr$ holds if event a occurs in the trace tr.

Once a predicate is defined on a single trace, we can lift it to a predicate on processes:

Definition 4 *(The **sat** relation)* A process P satisfies a **precedes** b if this predicate holds for every trace tr of P:

$$P \text{ sat } a \text{ precedes } b \iff \forall tr \in traces(P). \, a \text{ precedes } b$$

Without explicitly giving procedures of encoding protocols in CSP and then instrumenting them with signals, we provide formal definitions of different levels of authentication.

Definition 5 *(Formal definition of aliveness)* For A authenticating some B, a protocol P satisfies aliveness if for P extended with signals the following holds:

$$P \text{ sat } Running.B.X.n \text{ precedes } Commit.A.B.m$$

where X is an arbitrary participant, and n, m are some nonces.

Note that in Definition 5, the nonces n and m may not be the same. This definition only guarantees to A that B has run the protocol at some stage, which is what aliveness stipulates in Definition 1.

Definition 6 *(Formal definition of injective agreement)* For A authenticating some B using nonce N_A as a challenge, a protocol P satisfies injective agreement if for P extended with signals, where A will only commit once with N_A, the following holds:

$$P \text{ sat } Running.B.A.N_A \text{ precedes } Commit.A.B.N_A$$

Corresponding to Definition 2, this formal specification insists that both A and B are engaged in the same run and aware of each other. The insistence on the agreement on data values is met through the use of the same nonce. The uniqueness assumption on nonces implies uniqueness of runs.

Definition 3 defines a relatively weak relationship between a and b, namely, it insists only that there is an a event happening before a b event can occur. However, it allows for many b events to occur afterwards. In the application to authentication, this means that one *Running* event can be followed by many *Commit* events.

To guarantee injective agreement, as is the objective of Definition 6, having several commit events following one run event needs to be excluded. We exclude this by saying that the instrumentation of participant A is allowed only once to send a *Commit* event with a particular nonce N_A.

8.5 Protocol Analysis by Model Checking

In the following, we model security protools in the machine readable version $Csp_M{}^4$ of Csp. This means especially that the alphabet of communications needs to be constructed as well. The tool FDR (Failures Divergence Refinement) provides a number of automated analysis methods. Here, we will use a check on trace refinement, i.e., we utilise the tool in order to automatically check if all traces of a formal model of a security protocol are traces of a property specification. Concretely, we will discuss the encoding of the N-S protocol and of the aliveness and injective agreement properties in Csp_M. Our encoding is inspired by the principles underlying the tool Casper[5] [Low97a].

The modelling of security protocols in Csp_M involves quite a number of intricate coding tricks, both on the encoding of data in the functional programming language that is part of Csp_M, and on the encoding of processes in Csp. Thus, this section requires good knowledge of Csp and Csp_M.

These coding tricks address a number of challenges: how to obtain a finite set of messages, how to ensure that there are only finitely many messages that the intruder can derive, and, finally, how to deal with the intruder's huge state space. We will discuss this to some detail.

---

[4] https://cocotec.io/fdr/.

[5] https://www.cs.ox.ac.uk/gavin.lowe/Security/Casper/.

8.5.1 Encoding the Message Space in CSP_M

The first step in making our CSP specification of a security protocol machine readable is to represent its message space (cf. Sect. 8.4.2). To this end, CSP_M offers syntax to describe the CSP alphabet in form of a functional programming language. This includes (among many other features) the possibility to define (recursive) data types, provides a built-in generic type for forming sequences, supports set comprehension, and allows one to define functions using pattern matching.

Using CSP_M, we can represent the message space of Example 75.2 as follows:

Example 75.8: Message space of the N-S protocol in CSP_M, first attempt

We first describe data used in the protocol:

```
datatype USER = A | B | I
datatype NONCE = NA | NB | NI
datatype KEY = PKA | PKB | PKI | SKA | SKB | SKI
PKEY = { PKA, PKB, PKI }
SKEY = { SKA, SKB, SKI }

pk(A) = PKA pk(B) = PKB pk(I) = PKI
sk(PKA) = SKA sk(PKB) = SKB sk(PKI) = SKI
```

The datatypes USER, NONCE, and KEY consist of the constants listed. For instance, USER comprises of the three values A, B and I. There are two sets of keys, the set PKEY collects all public keys, the set SKEY collects all private keys. Finally, there are functions pk and sk. pk associate participants with their public keys. sk associate public keys with their secret counterparts. Based on this data, we compile the message space:

```
datatype MESSAGE = User.USER | Nonce.NONCE | Key.KEY |
                   Enc.KEY.MESSAGE | Sq.Seq(MESSAGE)
```

The datatype MESSAGE comprises the atomic messages, where constructors User, Nonce, and Key indicate from which set they are coming. It also comprises compound messages, where the constructor Enc indicates encrypting a message with a key, and the constructor Sq indicates pairing of messages. For pairing, we are using CSP_M's predefined constructor Seq, i.e., we implement the paring operator with an associative one. This fits with our use of the pairing operator as a left-associative one. The datatype MESSAGE is recursive and infinite.

Infinite data types such as MESSAGE are legitimate CSP_M specifications. However, when simulating or model checking a process, FDR diverges over infinite data enumerating all possible events. For this reason, we define a finite set of messages those that can be exchanged in one run of the N-S protocol with three participants.

Example 75.9: Message space of the N-S protocol in CSP_M, second attempt

---

Building upon the datatypes from Example 75.8 up to MESSAGE as a prefix, we define:

```
datatype MESSAGE_LABEL = Msg1 | Msg2 | Msg3

MSG1_CONT = { Enc.pkb.(Sq.<Nonce.na, User.a>) |
                pkb <- PKEY, na <- NONCE, a <- USER }
MSG1_BODY = { (Msg1, c) | c <- MSG1_CONT }

MSG2_CONT = { Enc.pka.(Sq.<Nonce.na, Nonce.nb>) |
                pka <- PKEY, na <- NONCE, nb <- NONCE }
MSG2_BODY = { (Msg2, c) | c <- MSG2_CONT }

MSG3_CONT = { Enc.pkb.(Sq.<Nonce.nb>) |
                pkb <- PKEY, nb <- NONCE }
MSG3_BODY = { (Msg3, c) | c <- MSG3_CONT }

MSG_CONT = Union({MSG1_CONT, MSG2_CONT, MSG3_CONT})
MSG_BODY = Union({MSG1_BODY, MSG2_BODY, MSG3_BODY})
```

In order to obtain a finite message space, for each of the three message types exchanged in the protocol, we construct a set of messages that includes all messages that could possibly be exchanged by that stage of the protocol run. We label these messages with Msg1, Msg1, or Msg3 respectively.

We discuss our encoding for the first message only. In the protocol as presented in Example 75 this message is

$$\{N_A, A\}_{pk_B}$$

i.e., it is an encrypted message, where the key is pk_B, and the content consists of pairing up the nonce N_A and the entity name A. The structure of the message is given by

```
Enc.pkb.(Sq.<Nonce.na, User.a>)
```

Note the symbols < and >, which are used as sequence constructors in CSP_M. The variables pkb, na, and a are bound to the following specific values

$$\text{pkb} \leftarrow \text{PKEY, na} \leftarrow \text{NONCE, a} \leftarrow \text{USER}$$

i.e., pkb can be any public key, na can be any nonce, and a can be any user. Finally, this set of messages is wrapped up as

$$\text{MSG1\_BODY} = \{ \ (\text{Msg1, c}) \ | \ \text{c} \leftarrow \text{MSG1\_CONT} \ \}$$

Similar constructions are applied to the messages in the second and third steps. Finally, we construct sets MSG_CONT and MSG_BODY, that contain all possible message contents and message bodies respectively.

This message space is a finite one. This is due to the finiteness of the domains we start with, and due to using only constructions that, given finite sets as parameters, result in a finite set.

8.5.2 Protocol Encoding

Encoding (instrumented) protocols in CSP_M is more or less straight forward:

Example 75.10: The N-S protocol in CSP_M

```
channel send : USER . USER . MSG_BODY
channel recv : USER . USER . MSG_BODY

channel running, commit : USER . USER . NONCE

User_A =
  [] b : diff(USER,{A}) @
  send.A.b.(Msg1, Enc.pk(b).(Sq.<Nonce.NA, User.A>))
  -> [] nb : NONCE @
  recv.A.b.(Msg2, Enc.PKA.(Sq.<Nonce.NA, Nonce.nb>))
  -> send.A.b.(Msg3, Enc.pk(b).(Sq.<Nonce.nb>))
  -> STOP
...
Network = send?a?b?m -> recv.b.a.m -> Network

System =
  (User_A ||| User_B) [| {|send, recv|} |] Network
```

> The process `User_A` directly encodes the three steps in which partici-
> pant A is engaged in, where the choice operators are restricted to finite
> sets, ensuring finite branching.
>
> The system composition with parallel operators uses the CSP_M con-
> struct $\{|\texttt{send, recv}|\}$, which denotes the set of all events that can be
> communicated over the channels `send` and `recv`.

8.5.3 Encoding the Intruder in CSP_M

The challenge in encoding the intruder as a CSP process is how to represent
the intruder's growing knowledge. This knowledge ranges over all the user
names, the nonces, the keys, as well as all the messages that can be formed
from these. For the case of the N-S protocol, this is a finite set because of our
construction in Example 75.9. The size of this set, still substantially large,
provides a challenge for model checking; managing its size is therefore key.

In the N-S protocol, the potential knowledge of the intruder is a subset of
all facts:

Example 75.11: Potential intruder knowledge

```
Fact = Union(
  { {|User|}, {|Nonce|}, {|Key|},
  {Enc.k.(Sq.<Nonce.n, User.u>) |
      k <- PKEY, n <- NONCE, u <- USER },
  {Enc.k.(Sq.<Nonce.n, Nonce.n'>) |
      k <- PKEY, n <- NONCE, n' <- NONCE},
  {Enc.k.(Sq.<Nonce.n>) |
      k <- PKEY, n <-NONCE }}
```

In our first encoding of the intruder, see Eq. 8.3, the intruder process was
parameterised over the set of facts known in a particular state. Thus, there
is an exponential number of states, as there are $2^{|\texttt{Fact}|}$ subsets. Inspired by
the Casper tool, we represent the facts by having $|\texttt{Fact}|$ many processes that
are running in parallel. Each of these can be in two states, i.e., we still have
an exponentiell size of states. However, by using parallel composition, we
play to the strength of model checking in FDR. The intruder has some initial
knowledge `IK`.

**Example 75.12: Knowledge representation by parallel pro-
cesses**

In the beginning the intruder knows all public keys, its own secret key
and nonce, as well as the names of all users (cf. Example 75.1):
```
Init = {Key.PKA, Key.PKB, Key.PKI,
```

```
                  Key.SKI, Nonce.NI,
                  User.A, User.B, User.I}

      Intruder(K) =
        || F : Fact @ [KnownAlpha(F)]
        if member(F,K) then Known(F) else Unknown(F)
```

The replicated alphabetised parallel || F:Fact @ [KnownAlpha(F)] creates |Fact| many processes. These processes take either the form Known(F) or Unknown(F). This state depends whether the fact F belongs to the set K. The processes are synchronised over sets KnownAlpha(F). Section B.1 provides a concise definition of the replicated operator in particular of the synchronisation sets between the processes.

```
      KnownAlpha(F) =
        Union({
          { send.x.y.(Msg1,F) | x <- USER, y <- USER,
                                       member(F, MSG1_CONT)},
          ...
          { recv.x.y.(Msg3,F) | x <- USER, y <- USER,
                                       member(F, MSG3_CONT)},
          { infer.F.r |
              r <- rules_match_head(F,Deduction_Rules) },
          { infer.rule_head(r).r |
              r <- rules_contain_body(F,Deduction_Rules) }
        })
```

The synchronisation sets KnownAlpha(F) allows the F process to engage in

- any event on the **send** channel, where F is part of the message content; or
- any event on the **recv** channel, where F is part of the message content.

The process can further engage in

- any event on the **infer** channel, where F is deduced; or
- any event on the **infer** channel, where F is part of the body of a rule.

The first two cases allow 'listening' to communication, the latter two participation on the inference of facts.

The intruder has capabilities as described in the Dolev-Yao model. Figure 8.2 describes the intruder's deduction capabilities. Rules 2 and 5 of the of the Dolev-Yao model allow for infinitely many derivations and thus infinitely many different facts in the message space as defined in Sect. 8.4.2. In our encoding in CSP_M, we are working with finitely many facts only. Thus, we will limit these derivations to finitely many, taking into account that participants can run the protocol at most once.

In a nutshell, the intruder can combine or encrypt, and decompose or decrypt, messages based on known facts. This can be encoded by sets of rules in CSP_M for the the N-S protocol as follows:

Example 75.13: Encoding the intruder's deduction capabilities

The intruder's deduction rules are encoded as pairs, where the first component is the conclusion of the rule, and the second component is its condition. The operational interpretation of a rule is: if only the facts stated in the condition are known, then it is possible to deduce the conclusion.

```
Msg1_Encryption_Deduction_Rules =
  { (Enc.k.(Sq.<Nonce.n, User.u>),
    {Key.k, Nonce.n, User.u}) |
     k <- PKEY, n <- NONCE, u <- USER }


...


Encryption_Deduction_Rules =
  Union({Msg1_Encryption_Deduction_Rules,
         Msg2_Encryption_Deduction_Rules,
         Msg3_Encryption_Deduction_Rules})

Msg1_Decryption_Deduction_Rules =
  { (Nonce.n ,
    {Enc.k.(Sq.<Nonce.n, User.u>), Key.sk(k)}) |
     k<-PKEY, n <- NONCE, u <- USER }


...


Decryption_Deduction_Rules =
  Union({Msg1_Decryption_Deduction_Rules,
         Msg2_Decryption_Deduction_Rules,
         Msg3_Decryption_Deduction_Rules})

Deduction_Rules =
  Union({Encryption_Deduction_Rules,
         Decryption_Deduction_Rules})
```

The rules are formed along the three message types. The set of Deduction_Rules is the union of all these rules.

Note that we only encode encryption with a public key and decryption with a secret key, i.e., we realise Eq. 8.1, but ignore Eq. 8.2.

In the chosen encoding approach, there are |Fact| many processes. These processes are either in the Known state or in the Unknown state, representing the current intruder's knowledge. Every message that the intruder listens to on the network potentially increases this knowledge, i.e., the state of a process changes from Unknown to Known or remains as Known. Furthermore, the intruder can apply inference rules. This is encoded via a synchronisation between processes in state Known: provided the intruder knows all facts in the condition, the process encoding the fact stated in the rule's conclusion can change from Unknown to Known. The intruder can send those messages which are known.

Example 75.14: Inference modelled by synchronization

We declare a channel infer to make learning observable as a communication.

```
channel infer : Fact . Deduction_Rules

rule_head((f,_))  = f
rule_body((_,bs)) = bs

rules_match_head(h,Rs) = {r | r <- Rs, rule_head(r) == h}
rules_contain_body(b,Rs) = { r | r <- Rs,
                                member(b, rule_body(r))}
```

There are four functions that decompose rules: to return a rule's head or body, and to obtain rules matching a given head or body. Here, the "_" is a wildcard symbol, that acts as a placeholder for any value in the appropriate domain.

```
Unknown(F) =
  (member(F, MSG1_CONT) &
   [] x : diff(USER, {I}),
      y : USER @ send.x.y.(Msg1,F) -> Known(F))
  [] ... []
  ([] r : rules_match_head(F,Deduction_Rules) @
   infer.F.r -> Known(F))
```

An F-process in state Unknown can listen to any message containing F that is communicated via the send channel in the network and switch to state Known. Provided F can be inferred, an F-process can turn into state Known.

```
Known(F) =
  (member(F, MSG1_CONT) &
    (([] x : USER, y : USER @
```

```
        send.x.y.(Msg1,F) -> Known(F)) []
    ([] x : USER, y : USER @
      recv.x.y.(Msg1,F) -> Known(F))))
  []...[]
  ([] r : rules_contain_body(F,Deduction_Rules) @
    infer.rule_head(r).r -> Known(F))
```

An F-process in state `Known` can listen to any message containing `F` that is communicated via the `send` channel or send any message containing `F` involving arbitrary users as senders and receivers. As the intruder keeps their knowledge, an F-process in state `Known` always remains in state `Known`.

An F-process in state `Known` can synchronize with other processes in state `Known` to infer knowledge.

Building over the message space, the protocol participants and the intruder encoded in CSP_M, it becomes straightforward to compose the overall system.

Example 75.15: N-S protocol in CSP_M

We abstract the `Intruder` process in two ways: we hide all events on the `infer` channel; we optimise the resulting process with a CSP_M specific function `chase`, which does not effect the traces of the system but reduces computational overhead.

```
    IntruderHideInfer = chase(Intruder(Init) \ {| infer |})

    Unsecure_System = (User_A ||| User_B)
                       [| {|send, recv|} |]
                       IntruderHideInfer
```

The N-S protocol in CSP_M is then the parallel composition of the two users together with the abstracted `Intruder` process—like in Example 75.6. We call it `Unsecure_System`, as we know that the protocol fails to provide authentication.

8.5.4 Encoding and Verifying the Security Properties

Having encoded the system under consideration, the next step is to encode the security property we are interested in.

Example 75.16: Aliveness and Injective Agreement in Csp_M

We define a generic process `Precedes` over a two element alphabet $\{e, d\}$ such that

$$\texttt{Precedes(e,d) sat } e \textbf{ precedes } d$$

We use the predefined Csp_M `RUN` process that in each step offers all the events which are element of its parameter set.

```
Precedes(e,d) = e -> RUN({e,d})
```

Note that over an alphabet with two elements $\{e, d\}$, any trace that fulfills e **precedes** d needs to begin with e: suppose a trace t would start with d, then the property would not hold for the first element of t. When a trace has e as its first element, it will fulfill the property e **precedes** d regardless of what follows.

The encoding of aliveness and injective agreement in Csp_M follows Definitions 5 and 6. The external choice operator in the process encoding aliveness allows to choose any participant as a partner.

```
Aliveness(a,b) = [] x : USER, n : NONCE, m : NONCE @
                    Precedes(running.b.x.n, commit.a.b.m)

InjectiveAgreement(a,b,na) =
    Precedes(running.b.a.na, commit.a.b.na)
```

Having now encoded both the system and the security properties in Csp_M, we can set up the model checking:

Example 75.17: Model Checking the N-S protocol

```
AlphaAliveness(a,b) = { running.b.x.n, commit.a.b.m |
                    x <- USER, n <- NONCE, m <- NONCE }

assert Aliveness(A,B) |||
        RUN(diff(Events,AlphaAliveness(A,B)))
        [T= Unsecure_System
```

The `assert` keyword declares a refinement statement, in our case using the `[T=` keyword for traces refinement. Here we investigate if the `Unsecure_System` satisfies the aliveness property for `A` authenticating `B`. We check if the `Unsecure_System` refines the `Aliveness(A,B)` process. For the refinement check, the alphabet of the processes on both sides of the refinement sign needs to be identical. We let `Aliveness(A,B)` interleave with a `RUN` process, where the alphabet of the `RUN` process consists of `Events`—all events possible—except those which belong to the alphabet of the `Aliveness(A,B)` process, i.e., `AlphaAliveness`.

The model checker confirms that both refinements hold, i.e., with `Aliveness(A,B)` and `Aliveness(B,A)`. Thus, we have an automated proof for the claim that the N-S protocol offers aliveness.

However, setting up model checking for injective agreement leads to a violation of the refinement:

```
assert InjectiveAgreement(B,A,NB) |||
       RUN(diff(Events,AlphaInjectiveAgreement(B,A,NB)))
       [T= Unsecure_System
```

The model checker FDR provides a counter example to this assertion:

```
send.A.I.(Msg1, Enc.PKI.Sq.<Nonce.NA, User.A>)
recv.B.A.(Msg1, Enc.PKB.Sq.<Nonce.NA, User.A>)
running.B.A.NA
send.B.A.(Msg2, Enc.PKA.Sq.<Nonce.NA, Nonce.NB>)
recv.A.I.(Msg2, Enc.PKA.Sq.<Nonce.NA, Nonce.NB>)
commit.A.I.NA
running.A.I.NB
send.A.I.(Msg3, Enc.PKI.Sq.<Nonce.NB>)
recv.B.A.(Msg3, Enc.PKB.Sq.<Nonce.NB>)
commit.B.A.NB
```

This counter example is exactly the attack as discussed in Example 75.1, provided we disregard the signal events due to instrumentation. Note that by looking at the messages, we still can't be sure who was actually initiating them. For instance, message exchange (2.1) $I \rightarrow B : \{N_A, A\}_{pk_B}$ from Example 75.1 appears in the counter example as `recv.B.A.(Msg1, Enc.PKB.Sq.` $< $ `Nonce.NA, User.A` $>$ `)`, i.e., the intruder impersonates participant A. This is possible as, in the Dolev-Yao model, the intruder has control over the network.

Concerning the property that we want to check, this counter example ends with the signal `commit.B.A.NB`, though the signal `running.A.B.NB` has not appeared. In other words, participant B believes it has communicated with A, while A has not engaged with B.

The model checker successfully finds the man in the middle attack from Example 75.1.

Applying the same encoding techniques to the N-S-L protocol allows to analyse N-S-L protocol as well. Overall we obtain:

Theorem 1 (Model checking results)

1. *The N-S protocol does not provide injective agreement.*
2. *The N-S-L protocol provides the injective agreement property to initiator A and responder B and vice versa, under the assumption that the implemented inference relation is complete w.r.t. derivations for* `Fact`.

Proof By model checking—see above.   ∎

In the above encoding in CSP_M, we implemented an inference mechanism mimicking the generates relation $\vdash$. It is obvious that the CSP_M inference is sound in the following sense:

Let $m \in \text{Fact}$ and $K \subseteq \text{Fact}$, then it holds:
If the CSP_M inference can derive m from K, then $K \vdash m$.

However, it is unclear if the converse direction holds, i.e., if the inference mechanism implemented in CSP_M is complete. The CSP_M inference can only make Fact many inference steps: there are only Fact many processes, each of which can only once turn from unknown to known—made visible by the infer event. However, a derivation $K \vdash m$ can use any number of inference steps. If one assumes completeness of CSP_M inference, then the intruder implemented in CSP_M has the same capabilities as the Dolev-Yao intruder and Theorem 1 shows that there is no attack possible on the N-S-L protocol.

8.6 Protocol Analysis by Theorem Proving

This section presents an alternative to the previously discussed model checking approach to verification. When using model checking, it is a challenge to encode security protocols in such a way that their state spaces does not become too large. Though the Casper tool automates the translation of protocols into CSP_M, there is no guarantee that the generated CSP_M model will be small enough for successful verification with FDR. An alternative that does not suffer from the state space explosion problem is Schneider's rank function approach [Sch98b]. The approach has been realised within the interactive theorem prover PVS [ES05]. This allows, in principle, to prove a protocol to be correct using two encodings in two different tools, one of these encoding is via model checking in FDR, the other encoding is via theorem proving in PVS. Use of multiple tools to demonstrate correctness helps to strenghten trust in the result, cf. section "Tool qualification" in Chap. 1.

8.6.1 Rank Functions

A rank function is a function from the message space $\mathcal{M}$ extended by signals $\mathcal{S}$ to the set $\{0, 1\}$, i.e.,

$$\rho : \mathcal{M} \cup \mathcal{S} \to \{0, 1\}$$

Rank functions are used to partition the message space, namely into messages of rank 0 and messages of rank 1. Rank 1 is assigned to messages that are 'allowed' to be seen by the intruder, and rank 0 to those that are 'forbidden' to be seen by the intruder. The idea is that message of rank 1 can be 'freely circulated' and won't give the intruder any knowledge that could be used to compromise security. In a protocol run the intruder's knowledge shall remain of rank 1.

So, for example, the identities of protocol participants are always assumed to be known by all participants. Hence we can assign them rank 1: $\rho(\mathcal{U}) \triangleq 1$. Public keys are always known and hence are assigned rank 1. Private keys of honest users however are assumed to be never compromised, hence they have rank 0. Naturally, the intruder knows its own private key, thus it gets rank 1. For the N-S-L protocol, this would mean:

$$\rho(k) \triangleq \begin{cases} 0 \; ; \; k = sk_A \vee \; k = sk_B \\ 1 \; ; \; otherwise \end{cases}$$

Also the signals that we introduced in order to analyse protocols will be assigned a rank. We will give the full details of a rank function later in the chapter.

Note that the original theory was introduced using the integers as the domain of rank functions. This might help in developing rank function. However, there is a general result stating that the existence of a rank function into $\{0, 1\}$ is a necessary and sufficient condition for existence of a suitable rank function into the integers [HS05].

A given rank function can be lifted from atomic events to composed ones. The rank of a composed event $c.m$, where c is a channel and m is a message, is the rank of message m passed on channel c:

$$\rho(c.m) \triangleq \rho(m).$$

The rank of a set S is the rank of the element in the set with the lowest rank:

$$\rho(S) \triangleq \min\{\rho(s) \mid s \in S\}.$$

The rank of a sequence tr is the rank of the element in the sequence with the lowest rank:

$$\rho(tr) \triangleq \min\{\rho(s) \mid s \text{ in } tr\}.$$

We define $\min\{\} \triangleq 1$, i.e., it is the maximum value that a rank function can take.

A process P **maintains** ρ if it does not introduce a message of rank 0 after receiving messages of rank 1 only, i.e., it never 'leaks a secret'. More formally:

Definition 7 *(Maintains rank)*

$$P \textbf{ maintains } \rho$$
$$\Leftrightarrow$$
$$\forall tr \in traces(P).\rho(tr \upharpoonright receive) = 1 \implies \rho(tr \upharpoonright (send \cup \mathcal{S})) = 1$$

where P is a process and ρ is a rank function. The channels *receive* and *send* are seen here as sets of events that can be communicated over them. The function *restrict* denoted by $\upharpoonright$ projects an arbitrary trace to a trace containing only elements that belong to a given set A. It is defined as follows:

$$\langle\rangle \upharpoonright A \triangleq \langle\rangle; \quad (\langle a\rangle \frown tr) \upharpoonright A \triangleq \begin{cases} \langle a\rangle \frown (tr \upharpoonright A) & \text{if } a \in A \\ tr \upharpoonright A & \text{if } a \notin A \end{cases}$$

8.6.2 The Rank Function Theorem

In order to state the main result of this chapter, we generalise the precedes predicate from pairs of events to pairs of event sets.

Definition 8 *(R precedes T)* Let R and T be some disjoint sets of events.

1. Let tr be a trace. We say that R **precedes** T in tr, if any occurrence of an event $b \in T$ is preceded by an occurrence of an event $a \in R$, i.e.,

$$\forall b \in T, tr'.tr' \frown \langle b\rangle \le tr \implies \exists a \in R.a \in tr'$$

2. A process P satisfies R **precedes** T if this predicate holds for every trace tr of P:

$$P \text{ sat } R \text{ precedes } T \iff \forall tr \in traces(P)\,.\,R \text{ precedes } T$$

If we want to study the behaviour of a process P up to the first appearance of an event from R, this can be expressed in CSP as

$$P \,[|\, R \,|]\, Stop$$

The process *Stop* does not engage in any event. The synchronous parallel operator $[|\,R\,|]$ says that events in the set R need to be synchronised, i.e., they can happen only if both processes are willing to engage in them. Therefore as long as events are not from R any event in which $P\,[|\,R\,|]\,Stop$ engages is coming from P. If P wants to engage in an event from R, it is blocked by *Stop*. The above construction 'preserves' all traces of P *up until* the first appearance of an event from R.

Network restriction and precedes predicates are closely related:

Theorem 2 (Linking "**precedes**" with network restriction) *With the notations as above it holds that*

$$P \text{ sat } R \text{ precedes } T \Leftrightarrow \forall tr \in traces(P\,[|\,R\,|]\,Stop)).tr \upharpoonright T = \langle\rangle.$$

Proof "$\Rightarrow$" Let $tr \in traces(P\,[|\,R\,|]\,Stop))$. Then $tr \upharpoonright R = \langle\rangle$, as *Stop* does not engage in any event and thus blocks the execution of events in the set

R. We have to show that tr does not contain an event from T. Assume, that tr contains some event b from T. Then there exists $tr' \frown \langle b \rangle \leq tr$. Then, by assumption, there must occur some $a \in R$ in tr'. Contradiction to $tr \upharpoonright R = \langle \rangle$.

"$\Leftarrow$" Let $tr \in traces(P)$. Assume that R **precedes** T does not hold for tr. Then there exists some $tr' \frown \langle b \rangle \leq tr$ with $b \in T$ such that for all $a \in R$ it holds that a does not occur in tr'. As $tr' \frown \langle b \rangle \leq tr$, we have $tr' \frown \langle b \rangle \in traces(P)$—this holds as the CSP traces semantics is prefix closed. Then $tr' \frown \langle b \rangle \in traces(P \,[\![\, R \,]\!]\, Stop)$. However, $tr' \frown \langle b \rangle \upharpoonright T \neq \langle \rangle$, as $b \in T$. Contradiction. ■

Theorem 2 provides a general strategy for security properties that can be characterised by a **precedes** predicate. Consider the network NET restricted by a particular set R, i.e., the process $NET \,[\![\, R \,]\!]\, Stop$; if this process has no traces in which events from T occur, the protocol is correct; otherwise it is not correct.

The following theorem, originally stated in [Sch98a], turns this idea into an effective proof method for security protocol analysis. It essentially reduces the analysis for the precedes predicate to the analysis of the maintains rank property.

For the formulation of this theorem, we—like in our protocol encoding in CSP_M for the sake of model checking—distinguish between the user name, e.g., $A, B, I \in \mathcal{U}$, and the CSP process $User_i$ that says how user i behaves in the context of the protocol.

Theorem 3 (Rank Function Theorem) *Consider the following network:*

$$NET = (|||_{i \in \mathcal{U}} \, User_i \,[\![\, send, receive \,]\!]\, Intruder(Init),$$

where

- $\mathcal{U}$ *is a finite set of user names,*
- $\bar{S}$ *is some finite set of channel names for signal events,*
- $User_i$ *is a process that describes how participant i behaves, where*

  - $User_i$ *can send messages on channel send.i,*
  - $User_i$ *can receive messages on channel receive.i,*
  - $User_i$ *can report progress to the environment by sending signals on channel signal.i,*
  - *the alphabet of $User_i$ is a subset of*

    $$\{send.i.u.m, receive.i.u.m, signal.i.u.m \mid u \in \mathcal{U}, m \in \mathcal{M}, signal \in \bar{S}\},$$

    and

- *the Intruder process is given by*

$$Intruder(IK) = (send?i?j?m \rightarrow Intruder(IK \cup \{m\}))$$
$$\Box$$
$$(\Box_{i,j \in \mathcal{U}, IK \vdash m} \; receive.i.j!m \rightarrow Intruder(IK))$$

where $Init \subseteq \mathcal{M}$ is the initial knowledge of the intruder.

If for some disjoint sets of events R and T there is a rank function

$$\rho : \mathcal{M} \cup \{signal.i.j.m \mid i, j \in \mathcal{U}, m \in \mathcal{M}, signal \in \bar{\mathcal{S}}\} \rightarrow \{0, 1\}$$

satisfying

(R1) $\forall m \in \mathbf{Init}.\rho(m) = 1$
(R2) $\forall S \subseteq \mathcal{M}, m \in \mathcal{M}.(\forall m' \in S.\rho(m') = 1 \wedge \mathbf{S} \vdash m) \implies \rho(m) = 1$
(R3) $\forall t \in T.\rho(t) = 0$
(R4) $\forall i \in \mathcal{U}.User_i \, [\![\, R \,]\!] \, Stop \; \mathbf{maintains} \; \rho$

then NET **sat** R **precedes** T.

Proof Assume for constructing a contradiction that (R1), (R2), (R3) and (R4) hold but not NET **sat** R **precedes** T. Then we know by Theorem 2 that there exists a trace $tr \in traces(NET [\![\, R \,]\!] \, Stop))$ such that $tr \upharpoonright T \neq \langle \rangle$, i.e., tr includes some events from T. Let tr' be the prefix of tr that ends with the first event of rank 0 in tr. We know that tr includes events of rank 0, as tr includes some events from T and events from T have rank 0 thanks to (R3). By construction, the trace tr' has the form $tr_0 \frown \langle e \rangle$, where $\rho(tr_0) = 1$ and $\rho(e) = 0$. As the CSP traces semantics is prefix closed, $tr' \in traces(NET [\![\, R \,]\!] \, Stop)$.

Now consider the last event e of tr'. Considering the alphabet of the network, this event is of the form

- $receive.i.j.x$,
- $send.i.j.x$, or
- $signal.i.j.x$

for some i, j, and x.

Case 1: tr' ends with an event $e = receive.i.j.x$

By construction of the network, such an event can only happen as a synchronisation in which the $Intruder$ process participates. For the trace tr', the knowledge of the $Intruder$ is given by $Init \cup (tr_0 \upharpoonright send)$, i.e., its initial knowledge enlarged by the traffic the $Intruder$ process was listening to on the $send$ channel. Thanks to (R1), the messages in $Init$ have rank 1. We know $\rho(tr_0) = 1$. Therefore, $\rho(tr_0 \upharpoonright send) = 1$. Thus, thanks to (R2), $\rho(receive.i.j.x) = 1$. Contradiction to $\rho(e) = 0$.

Case 2: tr' ends with an event $e = send.i.j.x$ or $e = signal.i.j.x$.

By construction of the network, any occurrence of an event in the set $\{send_i, receive_i, signal_i\}$ requires interaction of the process $User_i$. Thus,

$tr_i = tr' \restriction \{send_i, receive_i, signal_i\} \in traces(User_i \,[|\, R \,|]\, Stop)$. Thanks to (R4), $User_i \,[|\, R \,|]\, Stop$ **maintains** ρ. Thus, as in tr_i we have only messages of rank 1 on the *receive* channel, $\rho(e) = 1$. Contradiction to $\rho(e) = 0$. ∎

In the context of security protocol analysis, Theorem 3 has the following interpretation: the condition **R1** of the theorem ensures the intruder should not be in possession of any messages of rank 0 to start with; **R2** ensures that given a set of messages of positive rank the intruder should only be able to generate messages (under the generates relation $\vdash$) of rank 1; **R3** ensures events in T are of rank 0; and **R4** checks for every protocol participant that its corresponding process **maintains** ρ given restriction on R.

Note that condition R4 requires us to look at individual processes only: while the result holds for the network NET, i.e., for protocols like N-S or N-S-L, condition R4 deals with sequential processes only that represent the protocol participants. Overall, we do not have to consider processes running in parallel. This eases the analysis considerably.

8.6.3 Applying the Rank Function Theorem

In this section we focus on proving that the N-S-L protocol provides injective agreement. We will also briefly discuss where the proof fails in the case of the N-S protocol.

First, let's take a look at the instrumented N-S-L protocol. Here, we consider only the CSP specifications for participants A and B, as instrumentation does neither change the intruder process nor the network composition. We instrument the protocol for proving B authenticating A using nonce N_B as a challenge. This means by Definition 6 that we aim to prove that the protocol satisfies the two properties:

- *Running.A.B.N_B* **precedes** *Commit.B.A.N_B* and
- nonce N_B is used only once.

The Rank Function Theorem can be utilised to check the **precedes** property. That nonce N_B is used only once is a consequence of our assumptions on nonces, cf. Sect. 8.3.

Example 76.1: The N-S-L protocol instrumented

A *Running* signal indicates that an authenticatee (being authenticated)—in this case A—is engaged in some protocol run. A *Commit* signal acknowledges successful authentication on behalf of an authenticator—in this case B—in a protocol run.

For A this results in the following placement of signals:

$$A = \Box_{b \in \mathcal{U}, b \neq A} \; send.A.b!\{N_A.A\}_{pk_b} \rightarrow$$
$$receive.A.b?\{b.N_A.n\}_{pk_A} \rightarrow$$
$$Running.A.b.n \rightarrow$$
$$send.A.b!\{n\}_{pk_b} \rightarrow Stop$$

Instrumenting B requires to add a *Commit* signal:

$$B = \; receive.B?a?\{n.a\}_{pk_B} \rightarrow$$
$$send.B.a!\{B.n.N_B\}_{pk_a} \rightarrow$$
$$receive.B.a?\{N_B\}_{pk_B} \rightarrow$$
$$Commit.B.a.N_B \rightarrow Stop$$

Rather than using the instrumented processes A and B from Example 76.1 directly for our analysis, we are replacing them with different processes that are better suited for the analysis. In Example 76.2 for process A we are using a semantically equivalent process, and for process B we are using a refined process $\hat{B}$, i.e., $\hat{B}$ has less behaviours.

Example 76.2: Protocol participants for rank function analysis

The CSP trace semantics is a linear time semantics. Thus, the point where a process is branching does not matter. This allows us to move the choice which nonce we are receiving to the beginning of the process. We equivalently rewrite the process A into $A = \Box_{b \in \mathcal{U}, b \neq A, n \in \mathcal{N}} \bar{A}(b, n)$ where

$$\bar{A}(b,n) \triangleq \; send.A.b!\{N_A.A\}_{pk_b} \rightarrow$$
$$receive.A.b?\{b.N_A.n\}_{pk_A} \rightarrow$$
$$Running.A.b.n \rightarrow$$
$$send.A.b!\{n\}_{pk_b} \rightarrow Stop$$

For protocol runs, in which the event $Commit.B.A.N_B$ does not occur, the property $Running.A.B.N_B$ **precedes** $Commit.B.A.N_B$ holds. Thus, to prove that the N-S-L protocol provides the injective agreement property to initiator A and responder B, it suffices to analyse only those protocol runs in which the event $Commit.B.A.N_B$ occurs. Only process B engages in *Commit* events. B engages in the event $Commit.B.A.N_B$ if $a = A$. Thus, it suffices to analyse

$$(A \,|||\, \hat{B}) \, [|\, send, receive \,|] \, Intruder(Init)$$

where

$$\hat{B} \triangleq receive.B.A?\{n.A\}_{pk_B} \rightarrow$$
$$send.B.A!\{B.n.N_B\}_{pk_A} \rightarrow$$
$$receive.B.A?\{N_B\}_{pk_B} \rightarrow$$
$$Commit.B.A.N_B \rightarrow Stop$$

Constructing a Rank Function

The first step in applying Theorem 3 is to provide a rank function.

Theorem 2 serves as a guiding principle to assign ranks to the message space: messages that are allowed to appear in the restricted network have rank 1, while messages that ought not to appear get rank 0. Another principle is to have as few 'secrets' as possible, i.e., messages of rank 0. This eases dealing with the proof obligations arising from Theorem 3.

Finding a rank function that satisfies the theorem's conditions is a challenge. However, heuristic approaches to constructing rank functions can assist [SBS06]. Rank functions are not unique: there can be more than one rank function that fulfills the theorem's condition. Note that rank functions are constructed to support the analysis of a given property. Should one wish to analyse the same protocol for a different property, it follows that a different rank function is probably needed.

As discussed above, we want to prove that the instrumented N-S-L protocol from Example 76.1 has the property

$$B \text{ authenticates } A \text{ using nonce } N_B,$$

i.e., formally, that the following relation holds:

$$NET \text{ sat } Running.A.B.N_B \text{ precedes } Commit.B.A.N_B.$$

In our proof we need to restrict the network by the set $R \triangleq \{Running.A.B.N_B\}$ and to choose the set $T \triangleq \{Commit.B.A.N_B\}$. Specifically to this guarantee, we define a rank function, closely following [HS05].

Example 76.3: A rank function for the N-S-L protocol

$$\rho(\mathcal{U}) \triangleq 1$$

$$\rho(\mathcal{K}) \triangleq \begin{cases} 0 \text{ ; if } k \in \{sk_A, sk_B\} \\ 1 \text{ ; otherwise} \end{cases}$$

$$\rho(\mathcal{N}) \triangleq \begin{cases} 0 \text{ ; if } n = N_B \\ 1 \text{ ; otherwise} \end{cases}$$

$$\rho(\{m\}_{pk_b}) \triangleq \begin{cases} 1 & ; \text{if } b = A, m \in B.\mathcal{N}.N_B \\ \rho(m) & ; \text{otherwise} \end{cases}$$

$$\rho(\{m\}_{sk_b}) \triangleq \begin{cases} 0 & ; \text{if } b = A, m \in \{B.\mathcal{N}.N_B\}_{pk_A} \\ \rho(m) & ; \text{otherwise} \end{cases}$$

$$\rho(\mathcal{S}) \triangleq \begin{cases} 0 & ; \text{if } s = Commit.B.A.N_B; \\ 1 & ; \text{otherwise} \end{cases}$$

$$\rho(m_1.m_2) = min\{\rho(m_1), \rho(m_2)\}$$

The specific rationale behind this definition is as follows:

- The protocol participants are all expected to be known, so that any participant can initiate communication with any other. Hence rank 1 is assigned to all values in $\mathcal{U}$.
- We assume that all secret keys of honest users remain private. Hence rank 0 is assigned to them. The intruder knows its own secret key, thus $\rho(sk_A) = 1$. All public keys have rank 1.
- As nonce N_B is used for authentication, it is assigned rank 0. For the chosen guarantee, all other nonces can be assigned rank 1.
- When in dialog with participant A, in its second step participant B circulates the message $\{B.N_A.N_B\}_{pk_A}$, thus we assign such messages rank 1 and explicitly set the decryption of these messages to rank 0. The rank of all other messages depends on their contents only.
- Theorem 2 requires signals in set T to be of rank 0: consequently we assign $\rho(Commit.B.A.N_B) = 0$; all other signals can be freely circulated and hence assigned rank 1.
- In order to deal with the pairing function, we say that a pair of messages has rank 0 when one of the messages involved has rank 0.

One can prove by induction on the message structure that the target domain of ρ is indeed $\{0, 1\}$.

Given a rank function, applying Theorem 3 means to check if the theorem's four conditions are all fulfilled. We will do this in turn for each of the conditions.

Checking for R1

Example 76.4: Condition R1 holds for the N-S-L protocol

According to Example 75.4 for both the N-S and the N-S-L protocol the set *Init* is given by

$$Init \triangleq \{A, B, I, pk_A, pk_B, pk_I, sk_I\}.$$

All elements of *Init* are of rank 1. Thus, the condition is satisfied.

Checking for R2

Let us recall the rules as stated in Fig. 8.2:

1. $m \in S$ then $S \vdash m$
2. $S \vdash m$ and $S \vdash k$ then $S \vdash \{m\}_k$
3. $S \vdash \{m\}_{pk_b}$ and $S \vdash sk_b$ then $S \vdash m$
4. $S \vdash m_1.m_2$ then $S \vdash m_1$ and $S \vdash m_2$
5. $S \vdash m_1$ and $S \vdash m_2$ then $S \vdash m_1.m_2$

When deriving messages using these rules, we count the application of one rule as one step. This allows us to check property R2 by induction on the length of the derivation.

Example 76.5: Condition R2 holds for the N-S-L protocol

Let $S \subseteq \mathcal{M}$ with $\rho(S) = 1$. Let $S \vdash m$. We want to show that $\rho(m) = 1$.

Induction base: Only rule (1) is applicable. If $m \in S$, then by assumption $\rho(m) = 1$.

Induction hypothesis (I.H.): We have $\rho(m) = 1$ for all messages m derived with up to n rule applications, $n \geq 1$.

Induction step: The message m has been derived with $n+1$ rule applications, $n \geq 1$.

Case $S \vdash \{m\}_k$ by rule (2):
Then there exist shorter derivations $S \vdash m$ and $S \vdash k$. By I.H. we know $\rho(m) = \rho(k) = 1$. By definition of ρ, we know that $k \in \{pk_A, pk_B, pk_I, sk_I\}$. We consider two cases:
- $k \in \{pk_A, pk_B, pk_I\}$. If $m \in B.\mathcal{N}.N_B$ and $k = pk_A$, we have $\rho(\{m\}_k) = 1$. Otherwise, $\rho(\{m\}_k) = \rho(m) = 1$, by I.H.
- $k = sk_I$: Then $\rho(\{m\}_{sk_I}) = \rho(m) = 1$, by I.H.

Case $S \vdash m$ by rule (3):
Then there exist shorter derivations $S \vdash \{m\}_{pk_b}$ and $S \vdash sk_b$ for some $b \in \mathcal{U}$. By I.H. we know $\rho(\{m\}_{pk_b}) = \rho(sk_b) = 1$. Thus, $sk_b = sk_I$ and $b = I$. According to the definition of ρ, for $b = I$ we have $\rho(\{m\}_{pk_I}) = \rho(m)$. We know by I.H. that $\rho(\{m\}_{pk_I}) = 1$. Therefore, $\rho(m) = 1$.

Case $S \vdash m_1$ or $S \vdash m_2$ by rule (4):
Then there exists a shorter derivation with $S \vdash m_1.m_2$. By I.H. it holds that $\rho(m_1.m_2) = 1$. By definition of ρ, we have $\rho(m_1.m_2) = min\{\rho(m_1), \rho(m_2)\}$. It follows that $\rho(m_1) = \rho(m_2) = 1$.

Case $S \vdash m_1.m_2$ by rule (5):

Then there exist shorter derivations with $S \vdash m_1$ and $S \vdash m_2$. By I.H. it holds that $\rho(m_1) = \rho(m_2) = 1$. Thus, by definition of ρ, we have $\rho(m_1.m_2) = min\{\rho(m_1), \rho(m_2)\} = 1$.

Checking for R3

The only event in set T is the signal event $Commit.B.A.N_B$, which has rank 0. Thus, the condition is satisfied.

(stop-law)
$$Stop \,[|\,R\,|]\, Stop = Stop$$

(input-law)
$$(c?x : T \rightarrow P(x)) \,[|\,R\,|]\, Stop = c?x : U \rightarrow (P(x) \,[|\,R\,|]\, Stop),$$
where $U = T\backslash\{t \,|\, c.t \in R\}$

(output-law)
$$send.x.y.m \rightarrow P \,[|\,R\,|]\, Stop$$
$$= \begin{cases} Stop & ; \text{ if } send.x.y.m \in R \\ send.x.y.m \rightarrow (P \,[|\,R\,|]\, Stop) & ; \text{ otherwise} \end{cases}$$

(signal-law)
$$(s \rightarrow P) \,[|\,R\,|]\, Stop = \begin{cases} Stop & ; \text{ if } s \in R \\ s \rightarrow (P \,[|\,R\,|]\, Stop) & ; \text{ otherwise} \end{cases}$$

Here, s is an arbitrary signal event.

(distr-law)
$$(\Box_{i \in I} P(i)) \,[|\,R\,|]\, Stop = \Box_{i \in I}(P(i)) \,[|\,R\,|]\, Stop)$$

Fig. 8.3 CSP laws concerning restriction

Checking for R4

In order to check for condition R4, there are two calculi that support discharging this proof obligation. The first calculus deals with network restriction: given a process involving a restriction operator, this calculus allows us to derive an equivalent CSP process without restriction operator. The second calculus supports us in checking if a CSP process maintains rank. It reduces the proof obligation "maintains rank" over a 'complex' process to a proof obligation over a simpler process and the checking of some side conditions. Thanks to these two calculi, it is often possible to discharge proof obligation R4 in an, admittedly lengthy, however rather mechanical way.

Checking for R4—Dealing with the Network Restriction

Figure 8.3 collects a number of algebraic laws that hold for restriction (over the CSP traces model $\mathcal{T}$). Chapter 3 discusses how to develop and prove such CSP laws. Furthermore, the following theorem holds:

Theorem 4 (Restriction with disjoint event set) *If* $R \cap \alpha(P) = \emptyset$ *then* $P \, [\![\, R \,]\!] \, Stop = P$, *where* $\alpha(P)$ *denotes the alphabet of process* P.

Applying the calculus provided by these laws, and also taking Theorem 4 into account, we compute restriction free processes associated with the processes describing the behaviour of participants A and B under restriction:

Example 76.6: Restricting participants A and $\hat{B}$

The processes $\bar{A}(b, n)$ are generic in two variables, namely in the user name $b \in \{B, I\}$ and in the nonce $n \in \{N_A, N_B, N_I\}$. We consider two cases:

<u>Case 1</u>: $b = B$ and $n = N_B$.

$$\bar{A}(B, N_B) \, [\![\, Running.A.B.N_B \,]\!] \, Stop$$

$$= (send.A.B!\{N_A.A\}_{pk_B} \rightarrow$$
$$receive.A.B?\{B.N_A.N_B\}_{pk_A} \rightarrow$$
$$Running.A.B.N_B \rightarrow$$
$$send.A.B!\{N_B\}_{pk_B} \rightarrow Stop) \, [\![\, Running.A.B.N_B \,]\!] \, Stop$$

$$= send.A.B!\{N_A.A\}_{pk_B} \rightarrow$$
$$receive.A.B?\{B.N_A.N_B\}_{pk_A} \rightarrow Stop$$

Here, we first apply the (output-law), followed by the (input-law) (note that the process is willing to engage in only one specific event here), and finally the (signal-law).

<u>Case 2</u>: $b \neq B$ or $n \neq N_B$.
We address this case by applying Theorem 4:

$$\bar{A}(b, n) \, [\![\, Running.A.B.N_B \,]\!] \, Stop = \bar{A}(b, n)$$

<u>Combining the two cases</u> we obtain:

$$A \, [\! | \, Running.A.B.N_B \, |\!] \, Stop$$

$$= (\bar{A}(B, N_A) \, \square \, \ldots \, \square \, \bar{A}(I, N_I)) \, [\! | \, Running.A.B.N_B \, |\!] \, Stop$$

$$= (\bar{A}(B, N_A) \, [\! | \, Running.A.B.N_B \, |\!] \, Stop) \\ \square \, \ldots \, \square \\ (\bar{A}(I, N_I) \, [\! | \, Running.A.B.N_B \, |\!] \, Stop)$$

$$= (send.A.B! \{N_A.A\}_{pk_B} \, \to \\ receive.A.B? \{B.N_A.N_B\}_{pk_A} \, \to \, Stop) \\ \square \\ (\square_{b \neq B \vee n \neq N_B} \bar{A}(b, n))$$

We first rewrite process A into six processes combined by external choice, one process for each possible behaviour of A; then we apply the (distr-law) and bring the restriction into the scope of each of the subprocesses; these subprocesses can be analysed according to the two cases, which gives the overall result.

It follows from Theorem 4 that $\hat{B} \, [\! | \, Running.A.B.N_B \, |\!] \, Stop = \hat{B}$ as the event $Running.A.B.N_B$ does not occur in process $\hat{B}$.

Checking for R4—Checking for Maintains Rank

There is a calculus that, for sequential processes, checks if a process preserves rank or not. This provides an easy method to analyse the behaviour of protocol participants. Theorem 5 presents some selected rules.

Theorem 5 (Proof rules) *The proof rules stated in Fig. 8.4 are correct.*

Proof By definition, a process P maintains ρ if

$$\forall tr \in traces(P).\rho(tr \upharpoonright receive) = 1 \implies \rho(tr \upharpoonright (send \cup S)) = 1.$$

(stop-rule) We have $traces(Stop) = \{\langle \rangle\}$. For the trace $\langle \rangle$ it holds:

$$\rho(\langle \rangle \upharpoonright receive) = 1 \Rightarrow \rho(\langle \rangle \upharpoonright (send \cup S)) = 1 \\ \Leftrightarrow \rho(\langle \rangle) = 1 \Rightarrow \rho(\langle \rangle) = 1 \\ \Leftrightarrow \top \Rightarrow \top \\ \Leftrightarrow \top$$

(output-rule) Let $tr \in traces(send.x.y.m \to P)$.
Case 1: If $tr = \langle \rangle$, then the claim holds, see the proof for the **stop** rule.
Case 2: If $tr \neq \langle \rangle$, then there exists some $tr' \in traces(P)$ such that $tr = send.x.y.m \frown tr'$. Let P **maintains** ρ and $\rho(m) = 1$. With this we obtain:

(stop-rule) $$\frac{}{Stop \textbf{ maintains } \rho}$$

(output-rule) $$\frac{P \textbf{ maintains } \rho, \ \rho(m) = 1}{send.x.y.m \to P \textbf{ maintains } \rho}$$

(signal-rule) $$\frac{P \textbf{ maintains } \rho, \ \rho(data) = 1}{event.data \to P \textbf{ maintains } \rho}$$

Here, $event \in Running.\mathcal{U}.\mathcal{U} \cup Commit.\mathcal{U}.\mathcal{U}$.

(input-rule) $$\frac{\forall y, m \in M.\rho(m) = 1 \Rightarrow (P(y, m) \textbf{ maintains } \rho)}{receive.x?y?p \to P(y, p) \textbf{ maintains } \rho}$$

Here, M is a set of messages, that the process is willing to receive: CSP uses pattern matching to determine this set. We write p for such a pattern and define $M = \{m \in \mathcal{M} \,|\, m \text{ matches } p\}$.

(choice-rule) $$\frac{\forall j \in J.P_j \textbf{ maintains } \rho}{\Box_{j \in J} P_j \textbf{ maintains } \rho}$$

Fig. 8.4 Proof rules for maintaining rank

- For the left side of the implication it holds:
  $(send.x.y.m \smallfrown tr') \restriction receive) = tr' \restriction receive$ and thus
  $\rho(send.x.y.m \smallfrown tr') \restriction receive)) = \rho(tr' \restriction receive)$.
- For the right side of the implication it holds that,

$$\rho(send.x.y.m \smallfrown tr') \restriction (send \cup \mathcal{S}))$$
$$= \min(\rho(send.x.y.m), \rho(tr' \restriction (send \cup \mathcal{S})))$$
$$= \min(1, \rho(tr' \restriction (send \cup \mathcal{S})))$$
$$= \rho(tr' \restriction (send \cup \mathcal{S})).$$

Thus,
$$\rho(tr \restriction receive) = 1 \Rightarrow \rho(tr \restriction (send \cup \mathcal{S})) = 1$$
$$\Leftrightarrow \rho(tr' \restriction receive) = 1 \Rightarrow \rho(tr' \restriction (send \cup \mathcal{S})) = 1$$
$$\Leftrightarrow P \textbf{ maintains } \rho.$$

P **maintains** ρ holds as it is a premise of the rule.

(signal-rule) Analogously to the **(output-rule)**.

(input rule) Let $tr \in traces(receive.x?y?p \to P(y, p))$,
Case 1: If $tr = \langle\rangle$, then the claim holds, see the proof for the ***stop*** rule.
Case 2: If $tr \neq \langle\rangle$, then there exists some $tr' \in traces(P)$ such that $tr = receive.x.y.m \smallfrown tr'$ for some y and m. Let $\forall y, m \in M.\rho(m) = 1 \Rightarrow (P(y, m)$ **maintains** ρ.

Case 2.1: If $\rho(m) = 0$, then $\rho(receive.x.y.m \smallfrown tr') \restriction receive = 0$ and the implication holds as the premise is wrong.

Case 2.2: If $\rho(m) = 1$, we obtain that $\rho(receive.x.y.m) \frown tr' \upharpoonright receive =$ $min(1, \rho(tr' \upharpoonright receive)) = \rho(tr' \upharpoonright receive)$. We also have $receive.x.y.m \frown tr' \upharpoonright$ $(send \cup \mathcal{S}) = tr' \upharpoonright (send \cup \mathcal{S})$. With this we compute:

$$\rho(receive.x.y.m \frown tr') = 1 \Rightarrow \rho(receive.x.y.m \frown tr' \upharpoonright (send \cup \mathcal{S})) = 1$$
$$\Leftrightarrow \rho(tr' \upharpoonright receive) = 1 \Rightarrow \rho(tr' \upharpoonright (send \cup \mathcal{S})) = 1$$
$$\Leftrightarrow P(y, m) \text{ maintains } \rho.$$

According to the rule's premise, $P(y, m)$ **maintains** ρ holds for all y and m with $\rho(m) = 1$.

(choice-rule) Let $tr \in traces(\Box_{j \in J} P_j)$. Then there exists a $k \in J$ such that $tr \in traces(P_k)$, as $traces(\Box_{j \in J} P_j) = \bigcup_{j \in J} traces(P_j)$. We know that $\forall j \in J . P_j$ **maintains** ρ. Thus, in particular for k we have P_k **maintains** ρ. ∎

Now we apply the proof rules given in Theorem 5 to check if the restricted processes maintain rank.

Example 76.7: The processes for A maintain rank

Based upon the results of Example 76.6, for participant A we distinguish three cases.

<u>Case 1:</u> $b = B, n = N_B$.
   The process to consider is:

$$\bar{A}(B, N_B) \, [\!| \, Running.A.B.N_B \, |\!] \, Stop$$
$$= send.A.B!\{N_A.A\}_{pk_B} \rightarrow$$
$$receive.A.B?\{B.N_A.N_B\}_{pk_A} \rightarrow Stop$$

$\rho(\{B.N_A.N_B\}_{pk_A}) = 1$. As $Stop$ maintains ρ (stop-rule), we know that

$$receive.A.B?\{B.N_A.N_B\}_{pk_A} \rightarrow Stop$$

maintains ρ (input-rule). We compute

$$\rho(\{N_A.A\}_{pk_B}) = \rho(N_A.A) = min\{\rho(N_A), \rho(A)\} = min\{1, 1\} = 1.$$

Thus, applying the (output-rule) we conclude that $\bar{A}(B, N_B)$ maintains ρ.

<u>Case 2:</u> $b = I, n = N_B$.
   The process to consider is:

$$\bar{A}(I, N_B) \, [\!| \, Running.A.B.N_B \, |\!] \, Stop$$
$$= send.A.I!\{N_A.A\}_{pk_I} \rightarrow$$
$$receive.A.I?\{I.N_A.N_B\}_{pk_A} \rightarrow$$
$$Running.A.I.N_B \rightarrow$$
$$send.A.I!\{N_B\}_{pk_I} \rightarrow Stop$$

1. $\rho(\{I.N_A.N_B\}_{pk_A}) = min\{\rho(I), \rho(N_A), \rho(N_B)\} = min\{1, 1, 0\} = 0.$
   Thus,

   $$P_1 \triangleq receive.A.I?\{I.N_A.N_B\}_{pk_A} \rightarrow$$
   $$Running.A.I.N_B \rightarrow$$
   $$send.A.b!\{N_B\}_{pk_I} \rightarrow Stop$$

   maintains ρ (input-rule).
2. $\rho(\{N_A.I\}_{pk_I}) = 1.$ As P_1 maintains ρ, $send.A.b!\{N_A.A\}_{pk_b} \rightarrow P_1$
   maintains ρ (output-rule).

<u>Case 3:</u> $b \in \{B, I\}$, $n \in \{N_A, N_I\}$.
The processes to consider are

$$\bar{A}(b, n) \,[|\, Running.A.B.N_B \,|]\, Stop$$
$$= send.A.b!\{N_A.A\}_{pk_b} \rightarrow$$
$$receive.A.b?\{b.N_A.n\}_{pk_A} \rightarrow$$
$$Running.A.b.n \rightarrow$$
$$send.A.b!\{n\}_{pk_b} \rightarrow Stop$$

1. *Stop* maintains ρ according to the (stop-rule).
2. For $n \in \{N_A, N_I\}$, we have $\rho(\{n\}_{pk_b}) = 1.$ As *Stop* maintains ρ, we
   have that $P_1(b, n) \triangleq send.A.b!\{n\}_{pk_b} \rightarrow Stop$ preserves ρ (output-rule).
3. $\rho(Running.A.b.n) = 1.$ As $P_1(b, n)$ maintains ρ, we also have that
   $P_2(b, n) \triangleq Running.A.b.n \rightarrow P_1(b, n)$ maintains ρ (signal-rule).
4. As $n \neq N_B$, we have that $\{b.N_A.n\}_{pk_A} \notin B.\mathcal{N}.N_B.$ Thus, the rank
   of the received message is

   $$\rho(\{b.N_A.n\}_{pk_A})$$
   $$= min\{\rho(b), \rho(N_A), \rho(n)\}$$
   $$= min\{1, 1, 1\}$$
   $$= 1.$$

   As $P_2(b, n)$ maintains ρ, also $P_3(b, n) \triangleq receive.A.b?\{b.N_A.n\}_{pk_A} \rightarrow$
   $P_2(b, n)$ maintains ρ (input-rule).
5. $\rho(\{N_A, A\}_{pk_b}) = min\{\rho(N_A), \rho(A)\} = min\{1, 1\} = 1.$
   As $P_3(b, n)$ maintains ρ, also $send.A.b!\{N_A, A\}_{pk_b} \rightarrow P_3(b, n)$ maintains ρ (output-rule).

<u>Bringing the cases together:</u>
We have:

- $\bar{A}(B, N_B)$ maintains ρ,
- $\bar{A}(I, N_B)$ maintains ρ, and
- $\bar{A}(b, n)$ maintains ρ for all $b \in \{B, I\}$ and $n \in \{N_A, \mathcal{N}_I\}$.

Therefore we know that

$$A \, [| \, Running.A.B.N_B \, |] \, Stop$$

maintains ρ (choice-rule).

After the analysis of A, we now analyse the relevant part of B :

Example 76.8: The process $\hat{B}$ maintains rank

Recall the process under consideration:

$$\begin{aligned} \hat{B} = \ & receive.B.A?\{n.A\}_{pk_B} \ \rightarrow \\ & send.B.A!\{B.n.N_B\}_{pk_A} \ \rightarrow \\ & receive.B.A?\{N_B\}_{pk_B} \ \rightarrow \\ & Commit.B.A.N_B \rightarrow Stop \end{aligned}$$

1. $\rho(\{N_B\}) = \rho(N_B) = 0$.
   Thus, $Q_1 \triangleq receive.B.A?\{N_B\}_{pk_B} \ \rightarrow \ Commit.B.A.N_B \ \rightarrow \ Stop$
   maintains ρ (input-rule).
2. $\rho(\{B.n.N_B\}_{pk_A}) = 1$ for all $n \in \mathcal{N}$.
   As Q_1 maintains ρ, also $Q_2(n) \triangleq send.B.A!\{B.n.N_B\}_{pk_A} \ \rightarrow \ Q_1$
   maintains ρ for all $n \in \mathcal{N}$ (output-rule).
3. As $Q_2(n)$ maintains ρ for all $n \in \mathcal{N}$, $receive.B.A?\{n.A\}_{pk_B} \rightarrow Q_3(n)$
   maintains ρ (input-rule).

Overall, this analysis shows that $\hat{B}$ maintains ρ.

As conditions R1 to R4 hold we conclude that the N-S-L protocol satisfies that B authenticates A using nonce N_B. Similarly, one can establish that A authenticates B using nonce N_A.

Theorem 6 (Theorem proving result) *The N-S-L protocol provides the injective agreement property to initiator A and responder B.*

Proof By manual proof—see above. ∎

Note that—as was to be expected—the proof fails for the N-S protocol when using the above function ρ : in the second step of the N-S protocol, participant B sends the message $\{n.N_B\}_{pk_a}$. For this message it holds that $\rho(\{n.N_B\}_{pk_a}) = \rho(n.N_B) = min\{\rho(n), 0\} = 0$. Therefore, B does not maintain ρ and Theorem 3 does not apply. Of course, this result only shows that the proof fails with this specific rank function. However, as we know from the previous section on model checking, the N-S protocol fails to provide injective agreement. Therefore, there does not exist a rank function that allows to prove the N-S protocol to be correct.

8.7 Closing Remarks

We have introduced the notion of a security protocol as a means to establish security properties such as authentication. Examples demonstrated that even without 'breaking' the cryptosystem it is possible to successfully attack security protocols, i.e., even when assuming perfect encryption it matters which messages are exchanged. While security protocols utilise cryptographic primitives, their analysis is a scientific area different from cryptanalysis.

As Formal Methods we used the process algebra CSP with model checking and theorem proving. Thanks to CSP's primitives to send and receive values over communication channels, it is straightforward to give a formal model of the participants and their communication. CSP's parallel operators allow to compose a network of participants. Thanks to CSP's choice operator, a Dolev-Yao intruder can easily be expressed.

As for authentication properties, we demonstrated two ways to formalise them. We could encode them as properties of traces with the help of a **precedes** predicate, i.e., we were expressing them on the semantic level of CSP, cf. Definitions 5 and 6. But we also could characterise them as a refinement statement with respect to specific CSP processes, i.e., on the syntactic level of CSP, cf. Example 75.16.

The chapter discussed two proof methods: model checking, Sect. 8.5, and theorem proving, Sect. 8.6. These methods can be considered to be 'complementary' to each other:

- With *model checking*, it is 'easy' to find counter examples, i.e., to prove that a property does not hold. However, it can be a challenge to find an encoding that circumvents the state space explosion problem. It is often the case that the model checker times out when trying to prove the correctness of a protocol, i.e., it is 'hard' to prove that a property holds.
  The encoding of the N-S protocol in CSP_M was far more complex than our original encoding of the protocol in CSP. One reason for the additional complexity arose from the need to provide a formal model for the data used: messages were restricted to those which actually could appear in a single run. Yet another challenge was to encode the intruder. As the model checker FDR is optimised to analysing processes running in parallel, we encoded the intruder's reasoning on facts through process synchronisation. The result in Theorem 1 that the N-S-L protocol is correct with regards to a Dolev-Yao intruder holds only under the assumption that the implemented inference relation is complete with respect to derivations for Fact.
- With *theorem proving*, it is 'easy' to prove a security property, provided one has 'guessed' a rank function for which the four conditions of Theorem 3 hold. The correctness result in Theorem 6 for the N-S-L protocol holds for a Dolev-Yao intruder without making any further assumptions.

While establishing R1 to R4 for a rank function requires 'manually intensive' and 'lengthy' proofs, thanks to the two presented calculi these proofs
require 'little' ingenuity and are therefore 'well suited' for automation
through proof tactics in an interactive theorem prover.

However, when the property does not hold, one has to argue that all possible rank functions fail one of the conditions R1–R4. To establish such a
result would be a real challenge, i.e., it is 'hard' to prove that a security
property does not hold.

Note that Theorem 3 provides a sound, however, not a complete proof
method, i.e., it is not necessarily the case that there exists a rank function
if a property holds for a security protocol.

As the size of the state space does not play a role in the rank function
approach, it actually has been extended to deal with multiple runs [Sch98b].
An intruder might collect more information listening to several protocol runs
compared to listening to only one as in our analyses. Possibly, this larger
knowledge could be exploited for 'more informed' attacks. With model checking, considering multiple runs would increase the state space, and could possibly lead to a timeout. Consequently, many protocol analyses with model
checking are restricted to the single run scenario. The question under which
circumstances 'single run security' implies 'multiple run security' is still open
for research.

In a nutshell, the choice of the method depends on the hypothesis one
wants to establish: if the hypothesis is "the protocol is flawed", model checking is well suited to establish it; if the hypothesis is "the protocol is correct",
theorem proving is suitable.

8.7.1 Annotated Bibliography

Concerning the process algebra CSP, the book "The Modelling and Analysis of Security Protocols: The CSP Approach" [RSG+00] provides a good
overview on the techniques that one can apply to protocol analysis in the
context of CSP.

When model checking security protocols with FDR, the informal specification needs to be converted into an appropriate script that FDR takes as an
input. In order to assist in the generation of such machine-readable script,
Gavin Lowe [Low97a] has developed a specialist tool, *Compiler for the Analysis of Security Protocols*, better known as *Casper*, to convert informal protocol
specification into a script that can be processed by FDR.

The insight gained into the design of a protocol due to rank functions
analysis is invaluable. Constructing a rank function, however, is not always
trivial and remains arduous. Heuristic approaches to constructing rank functions assist in the manual process of verification [SBS06]. This still leaves the

problem of verifying arbitrarily large networks where assignment of ranks to messages is not straightforward; a decision procedure exists that attempts to address the problem by reducing it to the verification of a network with limited number of protocol runs, and as a result, a smaller message space [HS05]. Handling arbitrarily large networks is still a challenge [HS02, HS06]. Rank functions have been extended to handle a biometric-based authentication protocol for mobile nodes operating over a 3G telecommunication network [SD06, DS07b]. A different line of enquiry deals with group authentication protocols involving human 'channels' (relying on manual interaction) [NR06].

The tasks of modelling security protocols and the manual checking of proof conditions to prove the formal correctness of such protocols remain labour intensive and error-prone. Over the years, this has encouraged the use of machine assistance for such analyses in an attempt to reduce the possibility of human-error, mechanising repetitive procedures and, consequently, adding speed to the overall process. Rank functions analysis has benefited from a bespoke tool known as *RankAnalyzer* [Hea00] and the use of the general-purpose theorem-proving Proof Verification System (PVS) tool [ES05].

The formal analysis of security protocols has developed into a comprehensive body of knowledge, building on a wide variety of formalisms and treating a diverse range of security properties. The book "Formal Models and Techniques for Analyzing Security Protocols" [CK11] provides an overview covering various modelling and verification techniques. The paper "Comparing State Spaces in Automatic Security Protocol Analysis" [CLN09] provides some comparison from a tool perspective. There are a number of tools available for symbolic security protocol analysis, of which we want to mention a few:

- *Scyther* is a tool for the automatic verification of security protocols. It allows for standard analysis with respect to various adversary models.
- *Tamarin* is a theorem prover for the symbolic modeling and analysis of security protocols. Protocols and adversaries are specified by giving rewrite rules. Security properties are modelled as trace properties and are checked against the traces of the transition system.
- *ProVerif* is an automatic cryptographic protocol verifier based on a representation of the protocol by Horn clauses. It can be used to establish properties such as secrecy (the adversary cannot obtain the secret), strong secrecy (the adversary does not see the difference when the value of the secret changes), and authentication.

8.7.2 Current Research Directions

This chapter has demonstrated the use of the Dolev-Yao intruder model. An area of further research aims to extend the boundaries of this model and to incorporate further properties of existing cryptographic primitives and handle new primitives. This is critical if protocols with real-world complexity are to be analysed for even more sophisticated attacks. The rank functions approach has been extended to analyse protocols relying on algebraic properties of exclusive-or (XOR) primitives used to combine messages [Sch02], and for protocols making use of the Diffie-Hellman scheme [DS07a].

Handling security properties other than authentication is also of interest. Some work in this area has dealt with anonymity [SS96], perfect forward secrecy [DS05], non-repudiation [Sch98a] and temporal aspects of security properties [ES00]; as new properties permanently emerge there is an ongoing need to address these new properties.

Another area of research belongs to emerging scenarios for security protocols operating in pervasive environments, where location and attribute authentication properties, along with nodes with limited processing abilities, are modelled for.

One example of this can be found in RFID technology as is used, e.g., in supply chains. RFID tags have limited computational capabilities and power supply. Thus, RFID technology relies on specific protocols. For many of these protocols it is not known if they achieve the desired security services, while for many of them attacks have been demonstrated [VDR08]. Manik Lal Das et al. give an analysis of an RFID authentication scheme in the context of the internet of things [DKM20].

Another example are Industrial Internet of Things (IIoT) edge security challenges, e.g., when it comes to data integrity. An attacker might alter data from sensors, thus let automation and decision making come to wrong conclusions or degrade network performance. As IIoT networking and devices are heterogenous and resource-constrained, deploying standard security protocols is challenging. Thus, many IIoT networks are operating unprotected. Developing and analysing suitable protocols is an ongoing research topic, see, e.g., [CGBK20]. Yet another topic is how to guarantee security in heterogenous settings, when IIoT networks consists of edge nodes running various different protocols.

References

[CGBK20] Karanjeet Choudhary, Gurjot Singh Gaba, Ismail Butun, and Pardeep Kumar. MAKE-IT - A lightweight mutual authentication and key exchange protocol for industrial internet of things. *Sensors*, 20(18):5166, 2020.

[CK11] Véronique Cortier and Steve Kremer, editors. *Formal Models and Techniques for Analyzing Security Protocols*. IOS Press, 2011.

[CLN09] Cas J. F. Cremers, Pascal Lafourcade, and Philippe Nadeau. Comparing state spaces in automatic security protocol analysis. In *Formal to Practical Security*, LNCS 5458. Springer, 2009.

[CLRS09] Thomas H. Cormen, Charles E. Leiserson, Ronald L. Rivest, and Clifford Stein. *Introduction to Algorithms*. The MIT Press, 2009.

[DKM20] Manik Lal Das, Pardeep Kumar, and Andrew Martin. Secure and privacy-preserving RFID authentication scheme for internet of things applications. *Wirel. Pers. Commun.*, 110(1):339–353, 2020.

[DS05] Rob Delicata and Steve A. Schneider. Temporal rank functions for forward secrecy. In *18th IEEE Computer Security Foundations Workshop*. IEEE Computer Society, 2005.

[DS07a] Rob Delicata and Steve Schneider. An algebraic approach to the verification of a class of Diffie-Hellman protocols. *International Journal of Information Security*, 6(2-3):183–196, 2007.

[DS07b] Christos K. Dimitriadis and Siraj A. Shaikh. A biometric authentication protocol for 3G mobile systems: Modelled and validated using CSP and rank functions. *International Journal of Network Security*, 5(1), 2007.

[DY83] D. Dolev and A. Yao. On the security of public key protocols. *IEEE Trans. Inf. Theor.*, 29(2):198–208, 1983.

[ES00] Neil Evans and Steve Schneider. Analysing time dependent security properties in CSP using PVS. In *ESORICS*, LNCS 1895. Springer, 2000.

[ES05] Neil Evans and Steve A. Schneider. Verifying security protocols with PVS: widening the rank function approach. *J. Log. Algebr. Program.*, 64(2):253–284, 2005.

[Gol03] D. Gollmann. Authentication by correspondence. *IEEE Journal on Selected Areas in Communications*, 21(1):88–95, 2003.

[Hea00] James Heather. *Using rank functions to verify authentication protocols*. PhD thesis, Royal Holloway University of London, 2000.

[HS02] James Heather and Steve Schneider. Equal to the task? In *ESORICS*, LNCS 2502. Springer, 2002.

[HS05] James Heather and Steve Schneider. A decision procedure for the existence of a rank function. *Journal of Computer Security*, 13(2):317–344, 2005.

[HS06] James Heather and Steve Schneider. To infinity and beyond or, avoiding the infinite in security protocol analysis. In *ACM Symposium on Applied Computing*, pages 346–353. ACM, 2006.

[Laz99] Ranko S. Lazić. A semantic study of data independence with applications to model checking, 1999. PhD thesis. University of Oxford.

[Low95] Gavin Lowe. An attack on the Needham-Schroeder public-key authentication protocol. *Information Processing Letters*, 56(3):131–133, 1995.

[Low97a] Gavin Lowe. Casper: A compiler for the analysis of security protocols. In *Computer Security Foundations Workshop*. IEEE, 1997.

[Low97b] Gavin Lowe. A hierarchy of authentication specifications. In *Computer Security Foundations Workshop*. IEEE, 1997.

[NR06] Long H. Nguyen and Andrew W. Roscoe. Efficient group authentication protocol based on human interaction. *IACR Cryptol. ePrint Arch.*, 2006.

[NS78]   Roger M. Needham and Michael D. Schroeder. Using encryption for authentication in large networks of computers. *Communications of the ACM*,
            21(12):993–999, 1978.

[RSG+00]   P.Y.A. Ryan, S.A. Schneider, M.H. Goldsmith, G. Lowe, and A.W. Roscoe.
            The Modelling and Analysis of Security Protocols: The CSP Approach.
            Addison-Wesley, 2000.

[SBS06]   Siraj A. Shaikh, Vicky J. Bush, and Steve A. Schneider. A heuristic for constructing rank functions to verify authentication protocols. In *Preproceedings
            of the 3rd International Verification Workshop at IJCAR 2006*, pages 112–
            127, 2006.

[SBS09]   Siraj A. Shaikh, Vicky J. Bush, and Steve A. Schneider. Specifying authentication using signal events in CSP. *Computers & Security*, 28(5):310–324,
            2009.

[SC14]   Spyridon Samonas and David Coss. The CIA strikes back: Redefining confidentiality, integrity and availability in security. *Journal of Information System
            Security*, 2014.

[Sch98a]   Steve A. Schneider. Formal analysis of a non-repudiation protocol. In *CSFW*,
            pages 54–65, 1998.

[Sch98b]   Steve A. Schneider. Verifying authentication protocols in CSP. *IEEE Trans.
            Software Eng.*, 24(9):741–758, 1998.

[Sch02]   Steve Schneider. Verifying authentication protocol implementations. In *Formal
            Methods for Open Object-Based Distributed Systems*. Kluwer, 2002.

[SD06]   Siraj A. Shaikh and Christos K. Dimitriadis. Analysing a biometric authentication protocol for 3G mobile systems using CSP and rank functions. In
            Security in Pervasive Computing, LNCS 3934, pages 211–226. Springer, 2006.

[SS96]   Steve Schneider and Abraham Sidiropoulos. CSP and anonymity. In
            ESORICS, LNCS 1146. Springer, 1996.

[VDR08]   Ton Van Deursen and Sasa Radomirovic. Attacks on RFID protocols. *Cryptology ePrint Archive*, 2008.

Part IV
Wrapping up

Chapter 9
Origins and Development of Formal Methods

John V. Tucker

Abstract This chapter offers an historical perspective on the development of Formal Methods for software engineering. It surveys some of the problems and solution methods that have shaped and become our theoretical understanding and practical capability for making software. Starting in the 1950s, the history is organised by the topics of programming, data, reasoning, and concurrency, and concludes with a selection of notes on application areas relevant to the book. Although the account emphasizes some contributions and neglects others, it provides a starting point for studying the development of the challenging and ongoing enterprise that is software engineering.

9.1 Where do Formal Methods for Software Engineering Come From?

Let us look at early software and ask, how it was made and who for? Two domains are well known: scientific software and business software—both were pioneering, large scale and critically important to their users. Science and business had a profound effect on the early development and adoption of computing technologies, though *what* was computed was already computed long before electronic computers and software emerged and changed the scale, speed and cost of computation.

The initial development of programming was largely shaped by the need to make computations for scientific, engineering and business applications. An important feature of applications in numerical calculations and simulations in science and engineering that is easily taken for granted is that the problems, theoretical models, algorithms and data are mathematically precise and

John V. Tucker
Swansea University, Wales, United Kingdom

© Springer Nature Switzerland AG 2022, corrected publication 2022
M. Roggenbach et al., *Formal Methods for Software Engineering,*
Texts in Theoretical Computer Science. An EATCS Series,
https://doi.org/10.1007/978-3-030-38800-3_9

well-studied. This means that programming is based on a firm understanding of phenomena, its mathematical description in equations, approximation algorithms for solving equations, and the nature of errors. To a large extent the same can be said of business applications. In contrast, for the early applications of computers that were *non*-numerical there was little or no rigorous understanding to build upon. In particular, there is a third domain for pioneering software, that of computer science itself, where the creation of high-level programming languages and operating systems were truly new and even more challenging! Whilst the physical and commercial worlds had models that were known for centuries, the systems that managed and programmed computers were unknown territory. Indeed for the 1970s and 1980s, Fred Brooks' reflections [Bro75] on software engineering after making the operating system OS/360 for the IBM 360 series was required reading in university software engineering courses (see also [Bro87]). Formal Methods for software engineering begins in making software to use computers, with programming languages and operating systems.

Formal Methods owe much to the failure of programmers to keep up with the growth in scale and ambition of software development. A milestone in the creation of the subject of software engineering were the discussions and reports at a NATO Summer School at Garmisch, Germany, on the "software crisis" in 1968; the dramatic term "software crisis" was coined by Fritz Bauer (1924–2015) [NR69].[1] One early use of the term is in the 1972 ACM Turing Award Lecture by Edsger Dijkstra (1930–2002) [Dij72]:

> "The major cause of the software crisis is that the machines have become several orders of magnitude more powerful! To put it quite bluntly: as long as there were no machines, programming was no problem at all; when we had a few weak computers, programming became a mild problem, and now we have gigantic computers, programming has become an equally gigantic problem."

Among the responses to the crisis was the idea of making the whole process of software development more "scientific", governed by theoretically well-founded concepts and methods. A metaphor and an aspiration was the contemporary standards of engineering design, with its mathematical models, experimental discipline and professional regulation, as in civil engineering. Enter a new conception of software engineering whose Formal Methods were to provide new standards of understanding, rigour and accountability in design and implementation. Today, we can organise Formal Methods through their systematic methodologies for design and validation, techniques for formally modelling systems, software tools for exploring the models, and mathematical theories about the models. In addition to this technical organisation, Formal Methods can also be organised by their use in different application domains. Here my emphasis is on original formal modelling techniques and mathematical theories.

---

[1] An account of the conference and copies of the proceedings are available at http://homepages.cs.ncl.ac.uk/brian.randell/NATO.

Certainly, Formal Methods based on fledgling theories about programming existed before this conception of software engineering emerged. Questions about what a program is supposed to be doing, and to what extent it is doing what it is supposed to do, are timeless. Thinking scientifically about programming is an activity much older than software development—thanks to Charles Babbage (1791–1871) and Ada Lovelace (1815–1852). We will look at how and when some of the technical ideas in this book entered software engineering. Technically, they can be grouped around *programming*; *specifications of data*; *reasoning* and *proof*; and *concurrency*. Necessarily, my historical observations are impressionistic and highly selective. However, they should provide a useful foundation upon which understanding and experience will grow.

9.2 Logic

A simple and profound observation is that programs are made by creating data representations and equipping them with basic operations and tests on the data—a programming concept called a *data type*. To understand a program involves understanding

(i) how these representations and their operations and tests work;
(ii) how the operations and tests are scheduled by the control constructs to make individual programs; and
(iii) how programs are organised by constructs that compose and modularise programs to make software.

The issues that arise are fundamentally *logical issues* and they are addressed by seeking better *logical understanding* of the program and its behaviour. Most Formal Methods adapt and use logical and algebraic concepts, results, and methods to provide better understanding of program behaviour. Thus, it is in Formal Methods for reasoning—logic—are to be found the origins of Formal Methods.

Computer science drew on many of the technical ideas in logic, especially for formalisms for describing algorithms: early examples are *syntax and semantics of first-order languages, type systems, decision problems, λ-calculus, recursion, rewriting systems, Turing machines*, all of which were established in the 1930s, if not before. Later, after World War II, many more logical theories and calculi were developed, especially in philosophy, where subtle forms of reasoning—occurring in philosophical arguments rather than mathematical proofs—were analysed formally: examples are *modal* and *temporal logics*, which found applications in computer science much later.

Whilst it is true that many of Formal Methods come from mathematical and philosophical logic, in virtually each case the logical concepts and tools needed adaptation, extension and generalisation. Indeed, new mathematical theories were created around the new problems in computing: excellent examples are the theories of *abstract data types* and *process algebra*.

The case for mathematical logic as a theoretical science that is fundamental to the future of computing was made eloquently by John McCarthy (1927–2011) in a 1963 paper that has elements of a manifesto [McC63]:

> "It is reasonable to hope that the relationship between computation and mathematical logic will be as fruitful in the next century as that between analysis and physics in the last. The development of this relationship demands a concern for both applications and for mathematical elegance."

Over fifty years later the fruitful relationship is thriving and is recognised in computer science and beyond. Very advanced theories about data, programming, specification, verification have been created—clearly establishing the connections envisioned by McCarthy. So, today, logical methods are advanced and commonplace. Their origins and nature require explaining to young computer scientists who encounter them as tools. In science, there are few more dramatic examples of the fundamental importance of research guided by the curiosity of individuals—rather than by the directed programmes of companies, organisations and funding bodies—than the legacy of logic to computer science.

9.3 Specifying Programming Languages and Programs

What is a program? What does a program do? Formal Methods for developing programs begin with the problem of defining programming languages. This requires methods for defining

(i) the syntax of the language, i.e., spelling out the properties of texts that qualify as legal programs of the language; and

(ii) the semantics of the language, i.e., giving a description of what constructs mean or what constructs do.

Formal Methods often make precise informal methods but in the case of programming and programming languages there were few informal methods.

Early languages of the 1950s, like Fortran for scientific computation, and the later Cobol for commercial data processing, established the practicality and financial value of high-level machine-independent languages.[2] But their features were simple and, being close to machine architectures, their informal descriptions were adequate for users and implementors. The need for more expressive high-level languages presented problems. An early success was the definition of the syntax of Algol 60 using mathematical models of grammars [Nau+60, Nau62]. The method used was to become known as *BNF notation*,

---

[2] By 1954 the cost of programming was becoming comparable with the cost of computer installations, which was a prime motivation for IBM's development of Fortran [Bac98].

sometimes named after its creators John Backus (1924–2007) and Peter Naur (1928–2016).

The definition of syntax took up some important ideas from linguistics from the 1950s, where the search for a mathematical analysis of what is common to natural languages led to the formal grammars of Noam Chomsky [Cho56, Cho59] and his four-level hierarchy. The BNF notation corresponded exactly with Chomsky's *context-free grammars*. The mathematical analysis of languages defined by grammars was taken over by computer scientists, motivated by parsing and translating programming languages. Its generality enabled its applications to grow widely. The resulting *formal language theory* was one of the first new scientific theories to be made by computer scientists. It found its place in the computer science curriculum in the 1960s, symbolised by the celebrated classic by Ullman and Hopcroft [UH69].[3] The technical origins of Formal Methods for the syntax of natural and computer languages lie in mathematical logic, especially computability and automata theory where decision problems were a central topic and rewriting rules for strings of symbols were well known as models of computation. Thus, the works of Alan Turing and Emil Post (1897–1954) are an influence on theory-making from the very beginning: see Greibach [Gre89] for an account of the development of formal language theory.

The definition of the semantics of programming languages has proved to be a much harder problem than that of syntax. One needs definitive explanations for what programming constructs do in order to understand the implications of choices in the design of languages, and the consequences for the programs that may be written in them. A semantics is needed as a reference standard, to guarantee the portability of programs and to reason about what programs do.

Most programming languages are large and accommodate lots of features that are thought to be useful in some way. This criterion of 'usefulness' varies a great deal. Variants of features to allow programmers lots of choice add to the size, and the interaction between features add to the semantic complexity. Thus, semantic definitions of whole languages are awkward and are rarely achieved completely in a formal way. However, the semantical analysis of languages that are focussed on a *small* number of programming constructs has proved to be very useful—modelling constructs and their interaction in a controlled environment, as it were. For example, simple languages containing just a few constructs can be perpetual sources of insights.[4] Over many years, these studies have led to ambitious attempts to find systematic methods for cataloging and predicting the semantic consequences of choosing constructs for programming languages. However, the basic approaches to defining

---

[3] The following decade saw a rich harvest of textbooks on processing syntax, six by Ullman, Hopcroft and Aho.

[4] Imperative programming has at its heart a language containing only assignments, sequencing, conditional branching and conditional iteration.

formally the meaning of a programming language have been settled since the 1970s.

First, there are *operational semantics*, where the constructs are explained using mathematical models of their execution. A natural form of operational semantics defines an abstract machine and explains the behaviour of constructs in terms of changes of states of the machine. Operational semantics aligns with interpreters. An important historical example are the Formal Methods developed to define the semantics of the language PL/1 at the IBM Vienna Laboratories. The PL/1 language was an important commercial development for IBM, complementing the convergence of IBM users' software and machines represented by OS/360. PL/1—like OS/360—was a huge challenge to the state of the art of its day. Starting in 1963, the language was developed in New York and Hursley, UK. The task of providing a complete and precise specification of the new language was given to Vienna in 1967, and led to remarkable advances in our knowledge of programming languages, through the work of many first-rate computer scientists (e.g., the contributions of Hans Bekić (1936–1982) and Peter Lucas (1935–2015), see [BJ84]). Their methods resulted in the Vienna Definition Language for the specification of languages [Luc81].

Secondly, there are *denotational semantics*, where the constructs are interpreted abstractly as so-called *denotations*, normally using mathematical objects of some kind. In a first attempt at this denotational approach, Christoper Strachey (1916–1975) and Robert Milne made a huge effort to develop such a mathematically styled semantics for languages. An early important example of the application of the approach is Peter Mosses's semantics for Algol 60 [Mos74]. The mathematical ideas needed led to a new semantic framework for computation called *domain theory*. This was based upon modelling the approximation of information using orderings on sets; it was proposed by Dana Scott (1932-) and developed by him and many others into a large, comprehensive and technically deep mathematical subject. Domains of many kinds were created and proved to be suited for defining the semantics of functional languages where recursion is pre-eminent. Denotational semantics also involve abstract meta-languages for the purpose of description and translation between languages.

Thirdly, there are *axiomatic semantics*, where the behaviour of constructs are specified by axioms. Axiomatic semantics defines the meaning of constructs by means of the logical formulae that correctly describe input-output behaviours, or even the logical formulae that can be proven in some logic designed around the language. Axiomatic semantics focus on what a programmer can know and reason about the behaviour of his or her programs.

An important development was the attempt by Robert Floyd (1936–2001) to provide rules for reasoning on the input/output behaviour of flow charts and Algol fragments [Flo67]. In these early investigations, the behaviour of a program was described in terms of expressions of the form

$$\{P\}S\{Q\},$$

where property P is called a *pre-condition*, S is a program, and property Q is called a *post-condition*; the expressions came to be called *Floyd-Hoare triples*. This means, roughly, if P is true then after the execution of S, Q is true. There are different interpretations depending upon whether the pre-condition P implies the termination of program S. If P implies the termination of program S then the interpretation is called *total correctness*; and if P fails to imply the termination of program S then the interpretation is called *partial correctness*.

This approach to program specification was developed by Tony Hoare in 1969, in an enormously influential article on axiomatic methods. He proposed to use proof systems tailored to the programming syntax as a way of specifying programming languages for which program verification is a primary goal [Hoa69]. At the time, this so-called Floyd-Hoare approach to verification was seen as a high-level method of defining the semantics of a whole language for the benefit of programmers. Called *axiomatic semantics*, it was applied to the established language Pascal [HW73].

The theoretical study of Floyd-Hoare logics that followed was also influential as it raised the standard of analysis of these emerging semantic methods. To use the current semantics of programs to prove soundness for logics proved to be difficult and error prone. Stephen Cook (1939-) offered soundness and completeness theorems for a Floyd-Hoare logic for a simple imperative language based on first-order pre- and post-conditions in Floyd-Hoare triples; these demonstrated that the known rules were correct and, indeed, were "enough" to prove all those $\{P\}S\{Q\}$ that were valid [Coo78].

Unfortunately, the completeness theorems required a host of special assumptions, essentially restricting them to programs on the data type of natural numbers, with its very special computability and definability properties. Indeed, the completeness theorems were difficult to generalise to any other data type. The applicability of Floyd-Hoare logics attracted a great deal of theoretical attention, as did their technical problems. The development of Floyd-Hoare logics for new programming constructs grew [Bak80, Apt81, Apt83, RBH+01]. However, the deficiencies in the completeness theorems widened. The assertion language in which the pre- and post-conditions were formalised was that of first-order logic, which was not expressive of essential computational properties (such as weakest pre-conditions and strongest post-conditions) for data types other than the natural numbers. One gaping hole was the need to have access to the truth of *all* first-order statements about the natural numbers—a set *infinitely* more uncomputable than the halting problem (thanks to a 1948 theorem of Emil Post [Pos94]). Another problem was a multiplicity of non-standard models of the data [BT82b], and the failure of the methods applied to a data type with two or more base types [BT84b].

A completeness theorem is much more than a confirmation that there are enough rules in a proof system; it establishes precisely what semantics the proof system is actually talking about—something immensely valuable, if not essential, for a method for defining programming languages. In the case of Floyd-Hoare logic for **while** programs, the semantics the proof system is *actually* talking about was surprising [BT84a], for example, it was non-deterministic.

The relationship between the three different methods of defining semantics was addressed early on—e.g., by the former Vienna Lab computer scientist Peter Lauer in [Lau71]—but the links between the semantics of programs and the formal systems for reasoning were weak and error prone. For example, a decade later, Jaco de Bakker (1939–2012) made a monumental study of operational and denotational programming language semantics and their soundness and completeness with respect to Floyd-Hoare logics in [Bak80]. Again the theory was limited to computation on the natural numbers. Later the theory was generalised in [TZ88] to include abstract data types using an assertion language that was a weak second-order language, and a lot of new computability theory for abstract algebraic structures [TZ02].

The late 1960s saw the beginnings of an intense period of thinking about the nature of programming and programs that sought concepts and techniques that were independent of particular programming languages. New methods for developing data representations and developing algorithms focussed on a rigorous understanding of program structure and properties, and became a loosely defined paradigm called *structured programming*. For a classic example, in 1968, Edsger Dijkstra pointed out that the use of the **goto** statement in programs complicated massively their logic, was a barrier to their comprehension and should be avoided [Dij68c]. Throughout the 1970s, increasingly sophisticated views of programming and programs grew into the new field of *programming methodology*, which was perfect for encouraging the growth of formal modelling and design methods. For example, in the method of *stepwise refinement* an abstractly formulated specification and algorithm are transformed via many steps into a concrete specification and program, each transformation step preserving the correctness of the new specification and program. This method of developing provably correct programs was promoted by Edsger Dijkstra [Dij76]. The abstract formulations used novel computational concepts such as *non-determinism* in control and assignments, *concurrency*, and *abstract data type specifications* to make high-level descriptions of programs, and turned to the languages of mathematical logic to formalise them. For example, a formal theory of program refinement employing infinitary language and logic was worked out by Back [Bac80].

Many interesting new developments and breakthroughs in Formal Methods have their roots in the scientific, curiosity-driven research and development we have mentioned. For example, the specification languages and their associated methodologies are intended to describe and analyse systems independently of—and at a higher level of abstraction than is possible with—programming languages. An early example is the *Vienna Development Method (VDM)*, which originates in the IBM Vienna Laboratory work on Formal Methods and the exercise of developing a compiler—the Vienna Definition Language.

The general applicability of VDM in software engineering was established by Cliff Jones (1944-) and Dines Bjørner (1937-) [Jon80, Jon90, BJ82]. A second example is the method Z, based on set theory and first-order logic, created by Jean-Raymond Abrial (1938-); an early proposal is [ASM80]. Bjørner went on to develop the influential RAISE (Rigorous Approach to Industrial Software Engineering) Formal Method with tool support; his reflections on his experiences with these enterprises are informative [BH92].

Ways of visualising large complex software documentation—whether requirements or specifications—and relating them to programming were developed: to the venerable flowcharts were added: *Parnas tables* [Par01]; *statecharts* [Har87]; and the *Unified Modeling Language (UML)* family of diagrams [BJR96].

In hindsight, the influence of these semantic methods has been to establish the problem of specifying and reasoning about programs as a central problem of computer Science, one best tackled using Formal Methods based upon algebra and logic. The mathematical theories and tools that were developed were capable of analysing and solving problems that arose in programming languages and programming. Moreover, they also offered the prospect of working on a large scale in practical software engineering, on realtime, reactive and hybrid systems.

9.4 Specifications of Data

The purpose of computing is to create and process data. Of all the concepts and theories to be found in Formal Methods to date, perhaps the simplest and most widely applicable is the theory of *abstract data types*. The informal programming idea of an abstract data type is based upon this:

Principle. *Data—all data, now and in the future—consists of a set of objects and a set of operations and tests on those objects. In particular, the operations and tests provide the only way to access and use the objects.*

This informal notion can be found in Barbara Liskov and Steve Zilles 1974 article [LZ74].[5] Liskov saw in abstract data types a fundamental abstraction that could be applied pretty much anywhere and would contribute to the methodologies emerging in structured programming; more importantly the abstraction could be implemented and a bridge formed between computational structures and operations. Liskov designed CLU to be the first working programming language to provide such support for data abstraction [LSR+77, LAT+78]. Liskov's thinking about abstract data types is focussed by the construct of the *cluster* which, in turn, is inspired by the concepts of

---

[5] Along with suggestions about encapsulation, polymorphism, static type checking and exception handling!

modularity + encapsulation.

These two concepts derive from David Parnas' hugely influential ideas of information hiding and encapsulation, with their emphasis on interfaces separating modules, and specifications and their implementations [Par72a, Par72b, Par01].

Encapsulation means that the programmer in CLU can access data only via the operations listed in the header of a cluster, and is ignorant of the choices involved in data representations. This raises the question what, and how, does the programmer know about the operations? The answer is by giving axioms that specify the properties of the operations. Steve Zilles had presented this idea using axioms that were equations in a workshop organised by Liskov in 1973, where he defined a data type of sets. This is the start of the emphasis on the algebraic properties of data types [Zil74, LZ75]. The notion was designed to improve the design of programming languages—as in Liskov's CLU. It helped shape the development of modular constructs such as objects and classes, e.g., in the languages C++ [Str80] and Eiffel [Mey91, Mey88]. But it did much more. It led to a deep mathematical theory, new methods for specification and verification, and contributed spinouts seemingly removed form abstract data types.

Soon abstract data types became a new field of research in programming theory, as the Workshop in Abstract Data Types (WADT), begun and initially sustained in Germany from 1982, and the 1983 bibliography [KL83] and bear witness.

The formal theory of data types developed quickly but rather messily. The idea of abstract data type was taken up more formally by John Guttag in his 1975 PhD (e.g., in [Gut75, Gut77]), and by others who we will meet later. Guttag studied under Jim J Horning (1942–2013) and took some initial and independent steps toward a making a theory out of Zillies's simple idea [GH78]. As it developed it introduced a number of mathematical ideas into software engineering: *universal algebra, initial algebras, final algebras, axiomatic specifications based on equations, term rewriting*, and *algebraic categories*. Most of these ideas needed considerable adaption or extension: an important example is the use of *many sorted structures*—a topic barely known in algebra. Experienced computer scientists, mainly at IBM Yorktown Heights, began an ambitious research programme on Formal Methods for data: Joseph Goguen (1941–2006), Jim Thatcher, Eric Wagner, Jesse Wright formed what they called the ADJ Group and wrote about many of the basic ideas needed for a fully formal theory of data in a long series of some 18 papers [Gog89, GTW78, Wag01], though Goguen left the group to pursue other collaborations. Most of their work is unified by the fundamental notion of initiality, and the problems of specifying abstract data types using axioms made of equations. The theory was elegant and very robust, and encouraged the emergence of specification as an independent subject in software engineering.

Mathematically, the theory of abstract data types grew in scope and sophistication. For example, Guttag and others had noted the relevance of the connection between computability and equations. Equations were used by Kurt Gödel to define the computable functions in 1934; the technique was suggested by Jacques Herbrand (1908–1931) and became a standard method called *Gödel-Herbrand computability*. But the connection was too removed from the semantic subtleties of specifications, e.g., as pointed out by Sam Kamin in 1977 [Kam77]. A major classification programme on the scope and limits of modelling abstract data types, and of making axiomatic specifications for them, was created by Jan A Bergstra (1951-) and John V Tucker (1952-), who discovered intimate connections between specification problems and the computability of the algebraic models of the data [BT87, BT82a, BT95]. Begun in 1979, some of the original problems were settled relatively recently [KM14]. They exploited deeply the theory of computable sets and functions to make a mathematical theory about digital objects in general. For example, one of their results established that any data type that could be implemented on a computer can be specified uniquely by some small set of equations using a small number of auxiliary functions [BT82a]: *Let A be a data type with n subtypes. Then A is computable if, and only if, A possesses an equational specification, involving at most $3(n + 1)$ hidden operators and $2(n + 1)$ axioms, which defines it under initial and final algebra semantics simultaneously.*

The theory also spawned algebraic specification languages and tools that could be used on practical problems. For example, Guttag and Horning collaborated fruitfully on the development of the LARCH specification languages, based upon ideas in [GH82]. The LARCH specification languages had a single common language for the algebraic specification of abstract data types (called LSL, the Larch Shared Language), and various interface languages customised to different programming languages; there were also tools such as the Larch Prover for verification. This was work of the 1980s, culminating in the monograph [GH93]. Important systems tightly bound to the mathematical theory are the OBJ family, which began early with OBJ (1976) and led to CafeObj (1998), and Maude (1999); and the programming environment generator ASF+SDF(1989) [BHK89]. Such software projects were major undertakings: the *Common Algebraic Specification Language CASL* used in this book began in 1995 and was completed in 2004 [Mos04]!

The development of specification languages demanded further extensions and generalisations of the mathematical foundations; examples include new forms of rewriting systems, the logic and algebra of partial functions, and the theory of institutions. Some of these ingredients have led to substantial theoretical textbooks, such as for term rewriting [Ter03], for abstract data types [LEW96], and for institutions [ST12].

Partial functions arise naturally in computation when an algorithm fails to terminate on an input; they have been at the heart of computability theory since Turing's 1936 paper. They also arise in basic data types of which the most important examples are division $1/x$, which is not defined for $x = 0$, and the *pop* operation of the stack, which does not return data from an empty stack and so is not defined. Such partial functions cause difficulties when working with equations and term rewriting; they are especially disliked by theoreticians captivated by beauty of the algebraic methods applied to total functions. Making partial operations total is an option, e.g., the idea of introducing data to flag errors, but one that is not always attractive mathematically. As abstract data types and specification methods expanded in the 1980s, issues of partiality could not be ignored. A good impression of the variety of treatments of partiality that were becoming available can be gained from Peter Mosses's [Mos93], who was later to take on the task of managing the definition of CASL where decisions on partial semantics were needed. Monographs [Bur86] and [Rei87] on partiality appeared in the decade. New treatments continue to be developed such as [HW99], and the iconoclastic but surprisingly practical and algebraically sound $1/0 = 0$ of [BT07].

The theory of institutions was created by Joseph Goguen and Rod Burstall with the aim of capturing the essence of the idea of a logical system and its role in Formal Methods. The use of logical calculi was burgeoning and the concept aimed to abstract away and become independent of the underlying logical system; it did this by focussing on axioms for satisfaction. Institutions could also describe the structuring of specifications, their parameterization and refinement, and proof calculi. Institutions see the light of day through the algebraic specification language CLEAR [BG80] and more independently in [GB84]; Goguen and Burstall's polished account appears only 12 years later in [GB92]. A interesting reflection/celebration of institutions and related attempts is [Dia12]. Institutions offer a general form of template for language design, comparison and translation, albeit one that is very abstract. They have been applied to modelling languages like UML and ontology languages like OWL, and to create new languages for both such as the distributed ontology, modelling and specification Language DOL [Mos17]. Specification as a fundamental concept and as a subject in its own right was advanced by work involving abstract data types. Unsurprisingly in view of the universal importance of data types, several areas in computer science first tasted Formal Methods through abstract data types or benefitted from ideas and methods spun out of its research. It was out of this research community came the first use of Formal Methods for testing by Marie-Claude Gaudel (1946-) and her coworkers in e.g., [BCF+86, Gau95].

The design of better language constructs was a motivation for abstract data types, and the initial concerns with abstraction, modularity, reuse, and verification have proved to be timeless and very general. Inherent in thinking about data abstraction are ideas of genericity. With a background in abstract data types, David Musser (who had worked with John Guttag), Deepak Kapur (who had worked with Barbara Liskov) and Alex Stepanov proposed a language Tecton [KMS82] for generic programming in 1982. Stepanov went on to design of the standard template library (STL) for C++, i.e., the C++ standard collection classes, which has been influential in the evolution of C++ and other languages. Generic programming is a programming paradigm based that focuses on finding the most abstract formulations of algorithms and then implementing efficient generic representations of them; it leads to libraries of re-usable domain-specific software. Much of these language developments took place in industrial labs, starting with the General Electric Research Center, New York.

Like grammars and automata, abstract data types are timeless scientific ideas.

9.5 Reasoning and Proof

Despite the fact that logic is fundamentally about reasoning and logic was so influential in computing, reasoning about programs was slow to gather momentum and remained remote from practical software development. The mathematical logician and computer pioneer Alan Turing applied logic in his theoretical and practical work and, in particular, addressed the logical nature of program correctness in an interesting report on checking a large routine in 1949, see [MJ84]. In 1960, John McCarthy drew attention to proving correctness [McC62]: "Primarily, we would like to be able to prove that given procedures solve given problems" and, indeed:

> "Instead of debugging a program, one should prove that it meets its specification, and this proof should be checked by a computer program. For this to be possible, formal systems are required in which it is easy to write proofs."

Earlier, we noted the rise of such formal systems for program correctness after Floyd and Hoare in the late 1960s, motivated by the needs of users of programs.

The development of reasoning about programs has followed three paths. First, there was the development of logics to model and specify computational properties, such as program equivalence and correctness. To add to the selection of first-order and second-order logics mentioned, temporal logics were proposed by Burstall [Bur74] and Kröger [Krö77, Krö87]; and, earlier, in a particularly influential 1977 article about properties arising in sequential and, especially, concurrent programming, Amir Pnueli (1941–2009) applied *linear temporal logic (LTL)* [Pnu77].

The second path is the formulation of special methodologies and languages for constructing correct programs. One example is *design-by-contract*, associated with Bertrand Meyer's language Eiffel. In this object-oriented language, software components are given verifiable interface specifications, which are styled *contracts*; these specifications augment abstract data types with preconditions, postconditions and invariants. Another example is Cliff Jones' *rely guarantee methods* for designing concurrent programs, originating in [Jon81]. Rely guarantee methods are designed to augment Floyd-Hoare triples to control information about the environment of a parallel program. The method is discussed in Willem Paul de Roever's substantial work on concurrency [RBH+01]. Building in annotations such as pre and post conditions into languages is gaining interest when developing and maintaining high-quality software. The Microsoft programming language Spec# is an extension of the existing object-oriented .NET programming language C# that adds this feature to methods, together with relevant tools [BLS05]. The pre and post condition annotations support the concept of APIs in programming, though algebraic specifications seem more appropriate [BH14].

The third path is the development of software to support verification such as theorem provers—both general and specialised—and model checkers. Theorem provers are systems that can prove statements based upon the language and rules of a formal logic. Experiments with making theorem provers for mathematical purposes started early and continues to be a driving force in their development, but already in the 1960s their potential for use in software development was recognised. An excellent introduction to their development is [HUW14].

One family tree of theorem provers with verifying computer systems in mind begins with Robin Milner (1934–2010) and his development of the Logic for Computable Functions (LCF) for computer assisted reasoning; see [Gor00] for a detailed account. Milner was initiated into theorem proving through working on David Cooper's 1960s programme to make software for reasoning about programs (e.g., equivalence, correctness) using first-order predicate logic and first-order theories, programmed in the language POP-2 [Coo71]. Cooper experimented with programming decision procedures and was the first to implement Presburger's Theorem on arithmetic with only addition [Coo72]. In a telling reflection in 2003, Milner observed

> "I wrote an automatic theorem prover in Swansea for myself and became shattered with the difficulty of doing anything interesting in that direction and I still am. ... the amount of stuff you can prove with fully automatic theorem proving is still very small. So I was always more interested in amplifying human intelligence than I am in artificial intelligence." [Mil03].

Milner's LCF is the source of the functional programming language ML—for Meta Language—which plays a pivotal role in many subsequent approaches to reasoning, as well as being a functional language of great interest and influence in its own right. LCF is a source for several major theorem proving tools that qualify as breakthroughs in software verification, such as HOL and Isabelle. Mike Gordon (1948–2017) created and developed HOL over decades and demonstrated early on the value of theorem provers in designing hardware at the register transfer level, where errors are costly for manufacturers and users. The long road from HOL verifications of experimental to commercial hardware, and its roots in scientific curiosity, is described in the unpublished lecture [Gor18].

Other contemporary theorem provers that delivered significant milestones in applications are Robert S Boyer and J Strother Moore's theorem prover [BKM95] and John Rushby's PVS [ORS92]. The Boyer-Moore system, officially known as Nqthm, was begun in 1971 and over four decades accomplished a number of important verifications including a microprocessor [Hun85] and a stack of different levels of software abstraction [Moo89]; the later industrial strength version ACL2 provided verifications of floating point for AMD processor implementations [MLK96]. PVS appeared in 1992 but is one of a long line of theorem provers built and/or developed at SRI, starting with Jovial in the 1970s [EGM+79]. The aim of PVS is provide a general working environment for system development using Formal Methods, in which large formalizations and proofs are at home. Thus, PVS combines a strong specification language and proof checker, supported by all sorts of tools and libraries relevant to different application areas. The progress of PVS was influenced by work for NASA and is now very widely used [ORS+95, Cal98].

Theorem provers based on quite different logics have also proved successful. An intuitionistic logic created in 1972 to model mathematical statements and constructive reasoning by Per Martin-Löf (1942-) based on types is the basis for many theorem provers and programming languages. Robert Constable developed Nuprl, first released in 1984 [Con86], and others based upon dependent type theories and functional programming have followed, such as Coq [BC04] and Agda [BDN09]. The use of types goes back to logic and Bertrand Russell (1872–1970)—see [Con10].

Model checkers seek to find when a formula φ is satisfiable in a model. The verification technique is particularly suited to concurrency where formulae in temporal logic can express properties such as mutual exclusion, absence of deadlock, and absence of starvation, and their validity tested in a state transition graph, called a *Kripke structure*. For such concurrency problems, *linear temporal logic (LTL)* was applied by Amir Pnueli; the logic contained operators F (sometimes) and G (always), augmented with X (next-time) and U (until) and the program proofs were deductions in the logic. In 1981 the value and efficiency of satisfiability was established by Edmund M Clarke and Allen Emerson [CE81] and, independently, by Jean-Pierre Queille and Joseph Sifakis [QS82] who showed how to use model checking to verify finite state concurrent systems using temporal logic specifications. For example, Clarke and Emerson used *computation tree logic (CTL)* with temporal operators A (for all futures) or E (for some future) followed by one of F (sometimes), G (always), X (next-time), and U (until). Personal accounts of the beginnings of model checking are [Cla08] by Clarke and [Eme08] by Emerson. For an introduction to temporal logic methods see the monograph [DGL16].

The role of logic is to express properties of programs in logical languages and to establish rules for deducing new properties from old. To make a program logic, such as a Floyd-Hoare logic, typically there are two languages and sets of rules—the language of specifications and the language of programs. It is possible to combine specifications and programs into a single formal calculus, and there are plenty of formal systems that seem to possess such unity. Given that the calculi are intended for reasoning about programs, the complicating factor is what is meant by programs. In such calculi using abstract models of algorithms is practical, but using programming languages with their syntactic richness is not. This approach to verification impinges on the distinction between algorithm and program. Verifying algorithms is distinct from verifying programs. A significant example of this view of reasoning is Leslie Lamport's work on concurrent and distributed algorithms that culminates in a calculus called the Temporal Logic of Actions (TLA) in which formulae contain temporal logic operators with actions to model algorithms [Lam94].

In recalling these Formal Methods and tools we have neglected to track their progress in application. A useful early account of their path toward breakthroughs in theorem proving and model checking is [CW96].

9.6 Concurrency

Concurrency in computing refers to the idea that two or more processes exist, that they are taking place at the same time and communicating with one another. The phenomenon is ubiquitous in modern software, but it can take on many forms and leads to very complicated behaviour. Analysing, and to some extent taming, the logical complexity of concurrency has been another significant achievement of Formal Methods over the past 50 years.

Early on concurrency was found to be fundamental in the design of operating systems, where in the simplest of machines many processes need to be running at the same time, monitoring and managing the machine's resources, computations, input-output, and peripherals. Until quite recently, there was one processor that had to schedule all instructions so as to create and maintain an approximation to simultaneous operation. The solution was to break up the different processes and interleave their instructions—the processor speed being so great that for all practical purposes the effect would be simultaneous operation. This technique later became known as the *arbitrary interleaving* of processes.

The problems that arose from this necessary concurrency in operating systems required computer scientists to isolate the phenomenon and to create special constructs such as Dijkstra's semaphore for the THE multiprogramming system [Dij63, Dij68a, Dij68b]. The topic was soon central to research in programming methodology. Programming concurrency led to all sorts of special algorithmic constructs and reasoning techniques initially to extend the Formal Methods that had bedded down for sequential languages. An important paper extending Floyd-Hoare style verification to parallel programs is [OG76]. But parallelism also demanded a substantial rethink of how we specify semantics. Gordon Plotkin's introduction of the method of what became called *structural operational semantics* (SOS) in 1980 [Plo04a, Plo04b] is something of a milestone, evident in his elegant semantics for the concurrent language CSP [Plo83]. But more radical approaches to semantic modelling were needed to understand the fabulously complicated behaviours.

An important insight of work on concurrency was this:

Principle. *For logical purposes, concurrency as implemented by the interleaving of processes could be defined by reducing it to non-determinism.*

Specifically, the instructions were split up and groups of instructions from each sequence were processed but one could not know which groups would be scheduled when, only that the order within each sequence would be preserved.

Later, special small *concurrent languages* were developed, such as Tony Hoare's first formulation of CSP in 1978 [Hoa78]. The semantics was difficult to define, and program verification was even more problematic but achievements were (and continue to be) made. Parallel programs are *much* more complicated than sequential. Difficulties arise because of a global memory that is shared between parallel programs, or because programs have local memories and have to pass messages between them when executed in parallel; communications can be synchronous or asynchronous. All sorts of new general computational phenomena arise, such as deadlock and livelock. A valuable guide is [RBH+01], which also contains a substantial gallery of photographs of contributors to the verification of concurrent programs; and textbooks such as [AO91, ABO09].

For the theoretician, a radical and influential departure from the down to earth methods of Floyd-Hoare triples was needed. To raise the level of abstraction of thought from concrete languages to purely semantic models of the amazingly varied and complex behaviour possible in the execution of independent programs that can and do communicate. This change of thinking can be found in the attempt by Hans Bekić to create an abstract theory of processes in the IBM Vienna Laboratory in 1971 [Bek71], work that has become more widely known thanks to [BJ84]. The key point is that thinking about processes replaces the focus on input and output that dominates earlier semantic modelling and is needed in Floyd-Hoare specifications.

A search began for an analogous calculus for concurrent processes. Influenced by the purity of lambda calculus for the definition of functions, a major development were the early attempts of Robin Milner, who essentially launched the search with his *Calculus of Communicating Systems* (CCS) of 1980 [Mil80]. Among a number of innovations:

(i) Milner, like Bekić, thought about the notion of process in an abstract way; a process is a sequence of atomic actions put together by operations of some kind in a calculus—rather like the notion of string as a concatenated sequence of primitive symbols equipped with various operations.

(ii) Milner solved the problem of finding operators to make a calculus that focussed on the troublesome problem of communication between processes.

The idea of a process calculus led Tony Hoare to re-analyse the ideas of his CSP language [Hoa78] and create a new calculus called (for a period) *Theoretical* CSP.

These calculus approaches took off with new energy and in all sorts of new directions. The sharpness of the mathematical tools uncovered a wide spectrum of semantics for concurrent processes. The relationship between processes, especially their equivalence, emerged as a fundamental but very complex topic. There were *many* ways of viewing processes and their equivalence in formal calculi, for the world has many systems. Milner's longstanding interest [Mil70, Mil71a, Mil71b] in the idea of a process simulating another was a basic idea of CCS. In contrast, Hoare's CSP compared processes by notions of refinement.

In the emerging process theory, notions of system equivalence soon multiplied taking many subtly different forms [Gla96]. David Park (1935–1990) introduced a technical notion into process theory called *bisimulation* [Par81]. Ideas of bisimulation focus on when two systems can each simulate the operation of the other. Bisimulation in concurrency also took on many forms and, indeed unsurprisingly, bisimulation notions suitably generalised were found to have wide relevance [Rog00]. Bisimulation stimulated interest in applying and developing new semantic frameworks for computation, such as *game semantics* [Cur03] and *coalgebraic methods* [San11, SR11].[6]

De Bakker and Zucker took the process notion and created a new theory of process specification based on metric space methods for the solution of equations [BZ82]. They were inspired technically by Maurice Nivat's lectures on formal languages based on infinite strings [Niv79] where the languages were defined using fixed points provided by the Banach contraction theorem. The metric space theory of processes expanded providing an alternate theory of nondetermisitic processes [BR92].

An important advancement of the nascent theory was to refine further the fundamental issues that the principle demanded, non-determinsim and sequencing. A pure form of process theory called *Algebra of Communicating Processes* (ACP) was created by Jan Bergstra and Jan Willem Klop in 1982. They viewed processes algebraically and axiomatically: a process algebra was a structure that satisfied the axioms of ACP, and a process was simply an element of a process algebra! In particular, the axioms of ACP were equations that defined how operators made new processes from old. The equations made ACP subject to all sorts of algebraic constructions such as initial algebras, inverse limits etc. Thus, ACP took an independent direction, inspired by the world of abstract data types and rewriting. Interestingly, ACP was developed along the way of solving a problem in de Bakker-Zucker process theory (on fixed points of so called non-guarded equations). It was Bergstra and Klop who first coined the term *process algebra* in this first publication [BK82]. The term later came to cover all work at this level of abstraction. Their theory was

---

[6] Just as studies of recursive definitions of higher types in programming languages led to the semantic framework of domain theory.

extended with communication and provided a third effective way of working with concurrency [BK84].

These theories were not without software tools. The basic science of model checking was the basis of a range of useful tools. An early example is the *Concurrency workbench* of 1989, which was able to define behaviours in an extended version of CCS, or in its synchronous cousin SCCS, analyse games to understand why a process does or does not satisfy a formula, and derive automatically logical formulae which distinguish non-equivalent processes. Model checking technologies lie behind tools for mature concurrent process algebras. Another early influential tool is SPIN by Gerard Holtzman, which has been extended significantly and has become well-known [Hol97, Hol04]. For Csp, the refinement checker FDR is also such a tool [Ros94]. For the process algebra ACP, μCRL and its successor mCRL2 [GM14] offers simulation, analysis and visualization of behaviour modelled by ACP; its equational techniques also include abstract data types.

Within ten years of Milner's CCS, substantial textbooks and monographs became available, many of which have had revisions: for the CCS family [Mil89, Hen88], for the Csp family [Hoa85, Ros97, Ros10], and for the ACP family [BW90, Fok00, BBR10]; and a major *Handbook of Process Algebra* [BPS01] was created.

Semantic modelling often leads to simplifications that are elegant and long lasting and reveal connections with other subjects that are unexpected. The operational semantics of processes revealed the very simple and invaluable idea of the *labelled transition system*. The axiomatic algebraic approach led to stripped down systems of axioms that capture the essence of concurrent phenomena. However, algebraic methods are so exact and sensitive that many viable formulations of primitive computational actions and operations on processes were discovered and developed—we have mentioned just CCS, Csp and ACP families of theories. Each of these families have dozens of useful theories that extend or simplify their main set of axioms in order to model new phenomena or case studies. For instance, in different ways process algebras were extended with basic quantitive information such as time (e.g., [MT90b, Low95, BB96]), and probabilities (e.g., [GJS90, BBS92, MMS+96]), often starting with Milner's CCS family of processes. The addition of a concept of mobility was quite complicated, this being first attempted by Robin Milner et al. in the π calculus in 1992 [MPW92].

The diversity of theories of concurrent processes soon became evident, it took years to come to terms that this diversity is inherent. A useful overview of the history of these three process algebras is [Bae05]. The semantic tools that were created or renovated by concurrency research and simplified by use are sufficiently well understood to have found their way into basic courses in computer science (e.g., first year undergraduate [MS13]).

9.7 Formal Methods Enter Specialist Areas

The early development of Formal Methods focussed on general problems of programming and as they matured they were applied in and influenced specialist areas of software and hardware engineering, especially where the formal tools were discovered to be effective, or the problems to be in need of deep understanding or radical ideas.

Integrated circuit design. The development of Very Large Scale Integration (VLSI) technologies enabled chips to be designed using software tools, fabricated in large or small numbers, and so deployed with low cost. Exploring application specific hardware to transform performance—e.g., in signal and graphics processing—led to widespread interest in the customisation of chips. This opened up hardware design to the ideas, methods and theories in algorithm design and structured programming of the 1970s. The interest of computer scientists were aroused by an influential text-book by Carver A. Mead and Lynn Conway [MC80], which discussed algorithms and a modular design methodology. Formal Methods were particularly relevant to structured VLSI design because correctness issues loomed large: (i) hardware once made cannot be easily changed and so errors are costly to repair; (ii) customised hardware is needed to control physical processes and so human safety is an explicit concern; (iii) architectures of hardware are often more regular and simpler logically than those of software and are more amenable to the application of formal modelling and reasoning techniques. Using the theorem provers Boyer-Moore and Gordon's HOL to model and verify CPUs were breakthroughs in theorem proving. A survey that emphasises the direct influence of Formal Methods on progress made in hardware design in the decade is [MT90a].

Safety critical systems. The essential role of software engineering in automation is another example. The automation of industrial plant in the 1960s, such as in steel making, has expanded to a wide range of machines and systems, such as aeroplanes, railways, cars, and medical equipment, where the safety of people is—and very much remains—an important worry. The use of Formal Methods in the development of such systems is now well established in an area called safety-critical software engineering. An important event in Formal Methods for safety-critical computing was the introduction of new software engineering standards for military equipment and weapons. In 1986, the UK's Ministry of Defence circulated its *Defence Standard 00-55*. Its strong requirements made it controversial and it was not until 1989 that the Ministry published as Interim standards *Defence Standard 00-55. The Procurement of Safety Critical Software in Defence Equipment*, see [Tie92].[7] Relevant for this application domain of automation and

---

[7] Along with Defence Standard 00-55 there was an umbrella standard for identifying and reducing risks and so to determine when 00-55 would apply. The standards have been

safety critical computing in general are the Formal Methods for hybrid systems.

Safety critical software engineering needs to grow as automation deepens its hold on our work places, infrastructure, homes and environment, and software is desired that is smart in making anticipations. But human safety is not merely a matter of exact specifications that are correctly implemented. Human safety is dependent on good design that understands the human in the context.

Human-computer interaction. The field of human-computer interaction (HCI) has also experimented with Formal Methods to explore and improve design of systems. HCI developed in the 1970s influenced by an assortment of emerging display and text processing systems and cognitive psychology e.g., [CEB77, CMN83]. The first formal techniques that were tried were state diagrams [Par69] and grammars [Rei81, Mor81, Shn82], which were applied to user behaviour, e.g., to model sequences of actions on a keyboard and other input devices. The important exemplar of the text editor had been treated as a case study in Formal Methods research [Suf82]. HCI interest in formal methods begins in earnest in the mid 1980s with Alan Dix, Harold Thimbleby, Colin Runciman, and Michael Harrison [DR85, DH86, Thi86, Dix87], and the formal approach is evident in Thimbleby's influential text [Thi90]. These beginnings are brought together in some early books [TH90, Dix91], and the growth and present state of Formal Methods in HCI is charted in the substantial *Handbook of Formal Methods in Human-Computer Interaction* [WBD+17], e.g., in expository chapters such as [OPW+17]. A particularly interesting and growing area is HCI for safe technologies for healthcare. There is a great deal of software and hardware involved in the treatment of patients in hospital, and at home, and their effectiveness and safety are an serious issue because of the low quality of their design and user experience [Thi19].

Security. Lastly, with our capabilities and our appetite to connect together software devices come deep worries about security. These worries are affecting much software engineering as the need to identify and work on vulnerabilities on legacy and new software becomes commonplace. Access control, broadly conceived, is an important area for security models that codify security policies. For a system or network they specify who or what are allowed to access the system and which objects they are allowed to access. Access problems are encountered in the design of early operating systems, of course. The 1973 mathematical model that Bell and Padula designed for military applications was particularly influential that was developed and deployed in many security applications [BLaP73, BLaP76]. However, John McLeans's formal analysis [McL87], some 14 years later, revealed technical problems with the

---

revised several times subsequently and the explicit requirement for Formal Methods has been removed.

model that were controversial. Formal Methods were attracting attention in what was a small but growing computer security community.

Another example of early pioneering formal work is Dorothy Denning's formal studies of information flow [Den76, DD77]. Rather abstractly, data is assumed to be classified and that the classification is hierarchical. The relationship between two types of data in the hierarchy is represented by an ordering, and so the classification forms an ordered structure that is a lattice. Information can only flow in one direction, from lower to higher, or between equal, classifications. Later, Goguen and Meseguer also made a telling contribution with their simple criterion for confidentiality based on classifying the input-output behaviour of an automaton, called the *non-interference model* [GM82, GM84]. An impression of the early use of Formal Methods in tackling security problems can be gained from [Lan81], and Bell's reflections [Bel05].

A natural source of vulnerabilities is communication between processes. Communication protocols were a primary source of case studies for developing process algebras from the beginning. Early security applications of Formal Methods to such problems can be found in the mid 1990s: in [Low96], Gavin Lowe uses the concurrent process algebra CSP and its model refinement checker FDR to break and repair the then 17-year old Needham-Schroeder authentication protocol [NS78] that aims to check on the identities of processes before they exchange messages. There is so much more on all these issues, of course.

Although Formal Methods have been applied in many domains of programming, there are some where they have found few applications. One striking example is scientific computation. This is because the various scientific and engineering fields are firmly based on rigorous mathematical models and techniques well studied in Analysis, Geometry and Probability, and programmers are necessarily scientists and engineers with focussed on data and what it might indicate. However, growth in the appetite for detail in software simulation, in the complexity and longevity of software, and the logical challenges of programming the parallel architectures of supercomputers, is stimulating interest in software engineering for science and engineering domains. A pioneering example of the application of Formal Methods to numerics is [Hav00].

9.8 In Conclusion

So where do Formal Methods for software engineering come from? Although Formal Methods are tools for software developers to solve problems set by users in many domains, they largely arose in solving problems of computer science. The problems were recognised and explored theoretically. The early theory-makers collected and adapted tools from logic and algebra, and from them they forged new mathematics, new theories and new tools. Often they found what they needed in small neglected corners of logic and algebra. The theory-makers were driven to speculate and experiment with ideas, sometimes behind and sometimes in front of the technologies of the day. It started with the specification of programming languages—syntax and semantics. As our understanding grew, languages developed alongside Formal Methods. Software tools of all kinds demand languages for descriptions. That digital computation is in its nature logical and algebraic was understood early on—it is clear in Turing's view of computation. That logical and algebraic theories could be so expanded and refined to embrace practical large scale hardware and software design, and the immense and diverse world of users, is a remarkable scientific achievement, one that is ongoing and is at the heart of research and development of Formal Methods.

However, from the beginning, the speed of innovation in software and hardware has been remarkable—as any history of computer science since the 1950s makes clear. This relentless development generates productivity for users, profit for innovators, and challenges for regulators. It has certainly outstripped the complex and patient development of the underlying science of software engineering, e.g., in safety and especially security. Formal Methods have come a long way and have mastered many theoretical and practical problems of enormous complexity and significance. They are the foundations for an enduring science of computing.

On a personal note, I thank Markus Roggenbach for his invitation and encouragement to write this account of the origins and early development of formal Formal Methods for software engineering. I have benefitted from information and advice from Antonio Cerone, Magne Haveraaen, Faron Moller, Bernd-Holger Schlingloff, Harold Thimbleby, and Henry Tucker. I find myself a witness to many of the technical innovations that make up the story so far. I know that this first hand experience has led to bias toward some achievements and to neglect of others, but hopefully I—and certainly others—will have opportunities to correct my shortcomings. This survey has been shaped by the themes of this textbook, and the extensive Formal Methods archives in Swansea University's *History of Computing Collection*. As my efforts here suggests, deeper histories will be needed as the subject matures, our understanding grows, and breakthroughs mount up.

References

[ABO09] Krzysztof Apt, Frank S de Boer and Ernst-Rüdiger Olderog, *Verification of Sequential and Concurrent Programs*, Springer, 2009.

[All81] Frances E. Allen, The history of language processor technology in IBM, *IBM Journal of Research and Development*, 25 (5) (1981), 535–548.

[AO91] Krzysztof Apt and Ernst-Rüdiger Olderog, *Verification of Sequential and Concurrent Programs*, Springer, 1991.

[Apt81]   Krzysztof Apt, Ten years of Hoare's logic: A survey – Part I, *ACM Transactions on Programming Languages and Systems*, 3 (4) (1981), 431–483.

[Apt83]   Krzysztof Apt, Ten years of Hoare's logic: A survey – Part II: Nondeterminism, *Theoretical Computer Science*, 28 (1-2) 1983, 83–109.

[ASM80]   Jean-Raymond Abrial, Stephen A Schuman, and Bertrand Meyer, A specification language, in A M Macnaghten and R M McKeag (editors), *On the Construction of Programs*, Cambridge University Press, 1980.

[Bac80]   Ralph-Johan Back, *Correctness Preserving Program Refinements: Proof Theory and Applications*, Mathematical Centre Tracts 131, Mathematical Centre, Amsterdam, 1980.

[Bac98]   John Backus, The history of Fortran I, II, and III, *IEEE Annals of the History of Computing*, 20 (1998) (4), 68–78.

[Bae05]   Jos C.M. Baeten, A brief history of process algebra, *Theoretical Computer Science*, 335 (2-3) (2005), 131–146.

[Bak80]   Jaco de Bakker, *Mathematical Theory of Program Correctness*, Prentice-Hall International Series in Computer Science, 1980.

[BB96]   Jos C.M. Baeten and Jan A. Bergstra, Discrete time process algebra, *Formal Aspects of Computing*, 8 (1996) (2), 188–208.

[BBR10]   Jos C M Baeten, T. Basten, and M.A. Reniers, *Process Algebra: Equational Theories of Communicating Processes*, Cambridge Tracts in Theoretical Computer Science 50, Cambridge University Press, 2010.

[BBS92]   Jos C M Baeten, Jan A Bergstra, and Scott A. Smolka, Axiomatising probabilistic processes: ACP with generative probabilities, in *CONCUR 92*, Lecture Notes in Computer Science 630, Springer, 1992, 472–485.

[BBS95]   Jos C M Baeten, Jan A Bergstra, and Scott A. Smolka, Axiomatizing probabilistic processes: ACP with generative probabilities, *Information and Computation*, 121(1995) (2), 234–254.

[BC04]   Yves Bertot and Pierre Castéran, *Interactive Theorem Proving and Program Development. Coq art: The Calculus of Inductive Constructions*, Texts in Theoretical Computer Science: an EATCS series, Springer, 2004.

[BCF+86]   L. Bouge, N.Choquet, L. Fribourg, and M.-C. Gaudel. Test set generation from algebraic specifications using logic programming. *Journal of Systems and Software*, 6 (4) (1986) 343–360.

[BD02]   Manfred Broy and Ernst Denert (editors), *Software Pioneers: Contributions to Software Engineering*, Springer-Verlag, 2002.

[BDN09]   Ana Bove, Peter Dybjer and Ulf Norell, A Brief Overview of Agda – A Functional Language with Dependent Types, in Stefan Berghofer, Tobias Nipkow, Christian Urban and Makarius Wenzel (editors), *Theorem Proving in Higher Order Logics*, Lecture Notes in Computer Science 5674, Springer-Verlag, 2009, 73–78.

[Bek71]   Hans Bekić, Towards a mathematical theory of processes, Technical Report TR 25.125, IBM Laboratory Vienna, 1971. Reprinted in [BJ84].

[Bel05]   D. Elliott Bell, Looking back at the Bell-La Padula model, *ACSAC '05 Proceedings of the 21st Annual Computer Security Applications Conference*, IEEE Computer Society, 2005, 337–351.

[BG80]   Rod Burstall and Joseph Goguen. The semantics of Clear, a specification language. In Dines Bjørner (editor), *Abstract Software Specification – 1979 Copenhagen Winter School*, Lecture Notes in Computer Science 86, Springer, 1980, 292–332.

[BH92]   Dines Bjørner, Anne Elisabeth Haxthausen, Klaus Havelund, Formal, model-oriented software development methods: From VDM to ProCoS and from RAISE to LaCoS, *Future Generation Computer Systems* 7 (2-3) (1992), 111–138.

[BH14]   Anya Helene Bagge and Magne Haveraaen, Specification of generic APIs, or: why algebraic may be better than pre/post, in *High integrity language technology 2014*, ACM, 2014, 71–80.

[BHK89]   Jan A Bergstra, Jan Heering, and Paul Klint (editors), *Algebraic Specification*, ACM Press/Addison-Wesley, 1989.

[BJ82]   Dines Bjørner and Cliff Jones, *Formal Specification and Software Development*, Prentice Hall International, 1982.

[BJ84]   Hans Bekić and Cliff Jones (editors), *Programming Languages and Their Definition: H. Bekić (1936–1982)*, Lecture Notes in Computer Science 177, Springer, 1984.

[BJR96]   Grady Booch, Ivar Jacobson and James Rumbaugh, *The Unified Modeling Language for Object-Oriented Development Documentation Set Version 0.91. Addendum UML Update*, Rational Software Corporation, 1996.

[BK82]   Jan A Bergstra and Jan Willem Klop, Fixed point semantics in process algebra. Technical Report IW 208, Mathematical Centre, Amsterdam, 1982.

[BK84]   Jan A Bergstra and Jan Willem Klop, Process algebra for synchronous communication. *Information and Control* 60 (1-3) (1984), 109–137.

[BKM95]   Robert S.Boyer, Matt Kaufmann, Joseph S Moore, The Boyer-Moore theorem prover and its interactive enhancement, *Computers and Mathematics with Applications*, 29 (2) (1995), 27–62.

[BLaP73] D. Elliott Bell and Leonard J. LaPadula, Secure computer systems: Vol. I mathematical foundations, Vol. II mathematical model, Vol III refinement of the mathematical model. Technical Report MTR-2547 (three volumes), Mitre Corporation, Bedford, MA, 1973.

[BLaP76] D. Elliott Bell and Leonard J. LaPadula, Secure computer system: Unified exposition and Multics interpretation. Technical Report ESD-TR-75-306, Mitre Corporation, Bedford, MA, 1976.

[BLS05]   Mike Barnett, K. Rustan M. Leino, and Wolfram Schulte, The Spec# programming system: an overview, in Gilles Barthe, Lilian Burdy, Marieke Huisman, Jean-Louis Lanet, and Traian Muntean (editors), *Construction and Analysis of Safe, Secure and Interoperable Smart Devices*, Lecture Notes in Computer Science 3362, Springer-Verlag, 2005, 49–69.

[BPS01]   Jan A Bergstra, Alban Ponse and Scott A Smolka, *Handbook of Process Algebra*, Elsevier 2001.

[BR92]   Jaco de Bakker and Jan Rutten (editors), *Ten Years of Concurrency Semantics. Selected Papers of the Amsterdam Concurrency Group*, World Scientific, 1992.

[Bro75]   Fred Brooks, *The Mythical Man-Month: Essays on Software Engineering*, Addison Wesley, 1975. Republished 1995.

[Bro87]   Fred Brooks, No Silver Bullet – Essence and Accidents of Software Engineering, *IEEE Computer*, 20 (1987) (4), 10–19.

[BT82a]   Jan A Bergstra and John V Tucker, The completeness of the algebraic specification methods for data types, *Information and Control*, 54 (1982), 186–200.

[BT82b]   Jan A Bergstra and John V Tucker, Expressiveness and the completeness of Hoare's logic, *Journal of Computer and System Sciences*, 25 (3) (1982), 267–284.

[BT84a]   J A Bergstra and John V Tucker, The axiomatic semantics of programs based on Hoare's logic, *Acta Informatica*, 21 (1984), 293–320.

[BT84b]   J A Bergstra and John V Tucker, Hoare's logic for programming languages with two data types, *Theoretical Computer Science*, 28 (1984) 215–221.

[BT87]   Jan A Bergstra and John V Tucker, Algebraic specifications of computable and semicomputable data types, *Theoretical Computer Science*, 50 (1987), 137–181.

[BT95]   Jan A Bergstra and John V Tucker, Equational specifications, complete term rewriting systems, and computable and semicomputable algebras, *Journal of ACM*, 42 (1995), 1194–1230.

[BT07]   Jan A Bergstra and John V Tucker, The rational numbers as an abstract data type, *Journal of the ACM*, 54 (2), (2007), Article 7.

[Bur74]   Rod Burstall, Program proving as hand simulation with a little induction, in *Information Processing '74*, 308-312. North-Holland, 1974.

[Bur86]   Peter Burmeister, *A Model Theoretic Oriented Approach to Partial Algebras*, Akademie-Verlag, 1986.

[BW90]   Jos C.M. Baeten and W. Weijland. *Process Algebra*, Cambridge Tracts in Theoretical Computer Science 18, Cambridge University Press, 1990.

[BZ82]   Jaco de Bakker and Jeffrey I Zucker, Processes and the denotational semantics of concurrency, *Information and Control*, 54 (1/2) (1982), 70–120.

[Cam03]   Martin Campbell-Kelly, *From Airline Reservations to Sonic the Hedgehog: A History of the Software Industry*, MIT Press, 2003.

[Cal98]   James L Caldwell, Formal methods technology transfer: A view from NASA, *Formal Methods in System Design*, 12 (1998), 125–137.

[CE81]   Edmund M Clarke and E Allen Emerson, Design and synthesis of synchronization skeletons using branching time temporal logic, in Dexter Kozen (editor), *Logic of Programs: Workshop, Yorktown Heights, NY, May 1981*, Lecture Notes in Computer Science 131, Springer-Verlag, 1981.

[CEB77]   Stuart K Card, William K. English and Betty J Burr, Evaluation of mouse, rate-controlled isometric joystick, step keys and text keys on a CRT, *Ergonomics*, 21:8 (1977) 601–613.

[Cho56]   Noam Chomsky, Three models for the description of language, *IRE Transactions on Information Theory*, 2 (1956), 113–124.

[Cho58]   Noam Chomsky, and G. A. Miller, Finite state languages, *Information and Control*, 1 (1958), 91–112.

[Cho59]   Noam Chomsky, On certain formal properties of grammars, *Information and Control*, 2 (1959), 137–167.

[Cla08]   Edmund M. Clarke, The birth of model checking, in Orna Grumberg and Helmut Veith (editors), *25 Years of Model Checking: History, Achievements, Perspectives*, Lecture Notes in Computer Science 5000, Springer-Verlag, 1–26, 2008.

[CMN83]   Stuart K. Card, Thomas P. Moran, Allen Newell, *The Psychology of Human-Computer Interaction*, L. Erlbaum Associates, 1983.

[Con86]   Robert L Constable, *Implementing Mathematics with the NUPRL Proof Development System*, Prentice Hall, 1986.

[Con10]   Robert L Constable, The triumph of types: *Principia Mathematica*'s impact on computer science, Unpublished: available, e.g., at https://sefm-book.github.io.

[Coo71]   David Cooper, Programs for mechanical program verification, in B. Melzer and D. Michie, editors, *Machine Intelligence 6*, Edinburgh University Press, 1971, 43–59.

[Coo72]   David Cooper, Theorem proving in arithmetic without multiplication, in B. Melzer and D. Michie (editors), *Machine Intelligence 7*, Edinburgh University Press, 1972, 91–99.

[Coo78]   Stephen A Cook, Soundness and completeness of an axiom system for program verification. *SIAM J. Computing* 7 (1) (1978), 70–90.

[Cur03]   Pierre-Louis Curien. Symmetry and interactivity in programming. *Bulletin of Symbolic Logic*, 9: 2 (2003), 169–180.

[CW96]   Edmund M Clarke and Jeannette Wing, Formal methods: state of the art and future directions, *ACM Computing Surveys*, 28 (4) (1996), 626–643.

[Day12]   Edgar Daylight, *The Dawn of Software Engineering: From Turing to Dijkstra*, Lonely Scholar, 2012.

[Den76]   Dorothy Denning, A lattice model of secure information flow, *Communications of the ACM* 19 (5) (1976) , 236–243.

[DD77]   Dorothy Denning and Peter Denning, Certification of programs for secure information flow,*Communications of the ACM* , 20 (7) (1977) , 504–513.

[DGL16]   Stéphane Demi, Valentin Goranko and Martin Lange, *Temporal Logics in Computer Science and Finite State Systems*, Cambridge Tracts in Theoretical Computer Science 58, Cambridge UP, 2016.

[Dia12]   Răzvan Diaconescu, Three decades of institution theory, in Jean-Yves Béziau (editor), *Universal Logic: An Anthology*, Springer Basel, 2012, 309–322.
[Dij63]   Edsger W Dijkstra, Over de sequentialiteit van procesbeschrijvingen, E.W. Dijkstra Archive. Center for American History, University of Texas at Austin, undated, 1962 or 1963.
[Dij68a]   Edsger W Dijkstra, Cooperating sequential processes, in F. Genuys (editor) *Programming Languages*, 1968, Academic Press, 43–112.
[Dij68b]   Edsger W Dijkstra, The structure of the THE multiprogramming system, *Communications of the ACM*, 11 (5) (1968), 341–346.
[Dij68c]   Edsger W Dijkstra, Go to statement considered harmful, *Communications of the ACM*, 11 (3) (1968), 147–148.
[Dij72]   Edsger W Dijkstra, The humble programmer, *Communications of the ACM*, 11 (10) (1972), 859–866.
[Dij76]   Edsger W Dijkstra, *A Discipline of Programming*, Prentice-Hall, 1976.
[Dij82]   Edsger W Dijkstra, *Selected Writings on Computing: A Personal Perspective*. Springer-Verlag, 1982.
[Dix87]   Alan J Dix, *Formal Methods and Interactive Systems: Principles And Practice*, D.Phil. Thesis, Department of Computer Science, University of York, 1987.
[Dix91]   Alan J Dix, *Formal Methods for Interactive Systems*, Cambridge UP, 1991.
[DH86]   Alan J Dix and M D Harrison, Principles and interaction models for window managers, in M D Harrison A F Monk (editors) *People and computers: designing for usability*, Cambridge UP, 1986, 352–366.
[DR85]   Alan J Dix and Colin Runciman, Abstract models of interactive systems, P J and S Cook (editors), *People and computers: designing the interface*, Cambridge UP, 1985, 13–22.
[EGM+79]   B. Elspas, M. Green, M. Moriconi, and R. Shostak, *A JOVIAL verifier*. Technical report, Computer Science Laboratory, SRI International, January 1979.
[Eme08]   E. Allen Emerson, The beginning of model checking: A personal perspective, in Orna Grumberg and Helmut Veith (editors), *25 Years of Model Checking: History, Achievements, Perspectives*, Lecture Notes in Computer Science 5000, Springer-Verlag, 27–45, 2008.
[Flo62]   Robert W Floyd, On the nonexistence of a phrase structure grammar for ALGOL 60. *Communications of the ACM*, 5 (9) (1962), 483–484.
[Flo67]   Robert W Floyd, Assigning meanings to programs, in Jacob Schwartz (editor) *Proceedings of Symposia in Applied Mathematics* 19, American Mathematical Society, 1967, pp. 19–32.
[Fok00]   Willem Jan Fokkink, *Introduction to Process Algebra*, Texts in Theoretical Computer Science, An EATCS Series. Springer, January 2000.
[Fox66]   Leslie Fox (editor), *Advances in Programming and Non-numerical Computation*, Pergamon Press, 1966.
[Gau95]   Marie-Claude Gaudel, Testing can be formal, too, in Peter D. Mosses, M. Nielsen and M. I. Schwartzbach, (editors), *TAPSOFT*, Lecture Notes in Computer Science 915, Springer-Verlag, 1995, 82–96.
[Gla96]   Rob van Glabbeek, *Comparative concurrency semantics and refinement of actions*, CWI Tract 109, CWI Amsterdam, 1996.
[GH78]   John V Guttag and James J. Horning, The algebraic specification of abstract data types, *Acta Informatica* 10 (1) (1978) 27–52.
[GH82]   John V Guttag, James J. Horning, and Janette.M. Wing, Some notes on putting formal specifications to productive use, *Science of Computer Programming*, 2 (1) (1982), 53–68.
[GH93]   John V. Guttag and James J. Horning (editors), *Larch: Languages and Tools for Formal Specification*, Springer-Verlag, 1993.
[Gut75]   John V Guttag, *The Specification and Application to Programming of Abstract Data Types*, PhD Thesis, Department of Computer Science, University of Toronto, 1975.

[Gut77]   John V Guttag, Abstract data types and the development of data structures, *Communications of the ACM*, 20 (6) (1977), 396–404.

[Har87]   David Harel, Statecharts: a visual formalism for complex systems, *Science of Computer Programming*, 8 (3) (1987), 231–274.

[Hav00]   Magne Haveraaen, Case study on algebraic software methodologies for scientific computing, *Scientific Programming* 8 (4) (2000), 261–273.

[Hen88]   Matthew Hennessy, *Algebraic Theory of Processes*, MIT Press, 1988.

[HJ89]   C A R Hoare and Cliff B. Jones, *Essays in Computing Science*, Prentice Hall International, 1989.

[Hoa69]   C A R Hoare, An axiomatic basis for computer programming, *Communications of the ACM*, 12 (10) (1969), 576–580, 583.

[Hoa78]   C A R Hoare, Communicating Sequential Processes, *Communications of the ACM*, 21 (8) (1978), 666–677.

[Hoa85]   C A R Hoare, *Communicating Sequential Processes*, Prentice-Hall, 1985.

[Hol97]   Gerard Holzmann, The model checker SPIN, *IEEE Transactions on Software Engineering* 23 (5) (1997), 279–295.

[Hol04]   Gerard Holzmann, *The SPIN Model Checker: Primer and Reference Manual*, Addison-Wesley, 2004.

[Hun85]   Warren A Hunt, Jr, *FM8501: A Verified Microprocessor*, PhD Thesis, University of Texas at Austin, 1985.

[HW73]   C A R Hoare and N Wirth, An axiomatic definition of the programming language Pascal, *Acta Informatica*, 2 (4) (1973), 335–355.

[HW99]   Magne Haveraaen and Eric G. Wagner, Guarded algebras: disguising partiality so you won't know whether it's there, in (editors), in Didier Bert, Christine Choppy and Peter D. Mosses, *Recent Trends in Algebraic Development Techniques*, Lecture Notes in Computer Science 1827, Springer-Verlag, 1999, 182–200.

[Jon80]   Cliff Jones, *Software Development: A Rigorous Approach*, Prentice Hall International, 1980.

[Jon81]   Cliff Jones, *Development Methods for Computer Programs including a Notion of Interference*. PhD Thesis, Oxford University, 1981. Published as: Programming Research Group, Technical Monograph 25.

[Jon90]   Cliff Jones, *Systematic Software Development using VDM*, Prentice Hall 1990.

[Jon01]   Cliff Jones, The Transition from VDL to VDM, *Journal of Universal Computer Science*, 7 (8) (2001), 631–640.

[Jon03]   Cliff Jones, The early search for tractable ways of reasoning about programs, *IEEE Annals of the History of Computing*, 25 (2) (2003), 26–49.

[Jon13]   Capers Jones, *The Technical and Social History of Software Engineering*, Addison Wesley, 2013.

[GB84]   Joseph A Goguen and Rod Burstall, Introducing institutions, in Edward Clarke and Dexter Kozen (editors), *Logics of Programming Workshop*, Lecture Notes in Computer Science 164, Springer, 1984, 221–256.

[GB92]   Joseph A Goguen and Rod Burstall, Institutions: Abstract model theory for specification and programming, *Journal of the Association for Computing Machinery*, 39 (1) (1992), 95–146.

[GJS90]   Alessandro Giacalone, Chi-chang Jou, and Scott A Smolka, Algebraic reasoning for probabilistic concurrent systems, in Manfred Broy and Cliff Jones (editors), *Proceedings of IFIP Technical Committee 2 Working Conference on Programming Concepts and Methods*, North Holland, 1990, 443–458.

[GM82]   Joseph A Goguen and Jose Meseguer, Security policies and security models, in *Proceedings 1982 Symposium on Security and Privacy, Oakland, CA*, IEEE Computer Society, 1982, 11–20.

[GM84]   Joseph A Goguen and Jose Meseguer, Inference control and unwinding, in *Proceedings 1984 Symposium on Security and Privacy, Oakland, CA*, IEEE Computer Society, 1984, 75–86.

[GM14]   Jan F Groote and M.R.Mousavi, *Modeling and analysis of communicating systems*, The MIT Press, 2014.

[Gog89]   Joseph A Goguen, Memories of ADJ, *Bulletin of the EATCS*, No. 36, October 1989, 96–102. Also *Current Trends in Theoretical Computer Science: Essays and Tutorials*, World Scientific (1993), 76–81.

[Gor00]   Michael Gordon, From LCF to HOL: a short history, in Gordon Plotkin, Colin P Stirling, and Mads Tofte (editors), *Proof, Language, and Interaction*, MIT Press, 2000, 169–185.

[Gor18]   Michael Gordon, The unforeseen evolution of theorem proving in ARM processor verification. Talk at Swansea University, 28th April 2015. http://www.cl.cam.ac.uk/archive/mjcg/SwanseaTalk. Retrieved February 2018.

[Gre89]   Sheila A Greibach, Formal languages: origins and directions, *IEEE Annals of the History of Computing*, 3 (1) (1998), 14–41.

[Gri78]   David Gries (editor), *Programming Methodology. A Collection of Articles by Members of IFIP WG 2.3*, Springer, 1978.

[GTW78]   Joseph A Goguen, Jim W Thatcher, and Eric G Wagner. An initial algebra approach to the specification, correctness, and implementation of abstract data types. In R T Yeh (editor) *Current Trends in Programming Methodology. IV: Data Structuring*, Prentice-Hall, 1978, 80–149.

[HUW14]   John Harrison, Josef Urban and Freek Wiedijk, History of interactive theorem proving, in Dov M. Gabbay, Jörg H Siekmann, and John Woods *Handbook of the History of Logic. Volume 9: Computational Logic*, North-Holland, 2014, 135–214.

[Kam77]   Sam Kamin, Limits of the "algebraic" specification of abstract data types, *ACM SIGPLAN Notices*, 12 (10) (1977), 37–42.

[KL83]   B Kutzler and F Lichtenberger, *Bibliography on Abstract Data Types*, Lecture Notes in Computer Science 68, Springer, 1983.

[KM14]   Bakhadyr Khoussainov and Alexei Miasnikov, Finitely presented expansions of groups, semigroups, and algebras, Transactions American Mathematical Society, 366 (2014), 1455–1474.

[KMS82]   Deepak Kapur, David R Musser, and Alex A Stepanov, Tecton: A language for manipulating generic objects, in J. Staunstrup (editor), *Program Specification*, Lecture Notes in Computer Science 134, Springer-Verlag, 1982, 402–414.

[Krö77]   Fred Kröger, LAR: A logic of algorithmic reasoning, *Acta Informatica* 8 (3) (1977), 243–266.

[Krö87]   Fred Kröger, *Temporal Logic of Programs*, EATCS Monographs in Theoretical Computer Science 8, Springer-Verlag, 1987.

[Lam94]   Leslie Lamport, The temporal logic of actions, ACM Transactions on Programming Languages and Systems 16 (3) (1994), 872–923.

[Lan81]   Carl Landwehr, Formal models for computer security, *ACM Computing Surveys*, 13 (3) (1981), 247–278.

[LAT+78] Barbara Liskov, Russell Atkinson, Toby Bloom, J. Eliot Moss, J. Craig Schaffert, Robert Scheifler, and Alan Snyder, *CLU Reference Manual*, Computation Structures Group Memo 161, MIT Laboratory for Computer Science, Cambridge, MA, July 1978.

[Lau71]   Peter Lauer, *Consistent Formal Theories of the Semantics of Programming Languages*, PhD Thesis, Queen's University of Belfast, 1971. Published as TR 25.121, IBM Lab. Vienna.

[LEW96]   Jacques Loeckx, Hans-Dieter Ehrich, Markus Wolf, *Specification of Abstract Data Types: Mathematical Foundations and Practical Applications*, John Wiley and Sons, 1996.

[LG86]   Barbara Liskov and John V Guttag, *Abstraction and Specification in Program Development*, MIT Press and McGraw Hill, 1986.

[Low95]   Gavin Lowe, Probabilistic and prioritized models of timed CSP. *Theoretical Computer Science*, 138 (1995), 315–352.

[Low96]   Gavin Lowe, Breaking and fixing the Needham-Schroeder public-key protocol using FDR. *Software - Concepts and Tools*, 17:93–102, 1996.

[LSR+77]   Barbara Liskov, Alan Snyder, Russell Atkinson, and J. Craig Schaffert, Abstraction mechanisms in CLU, *Communications of the ACM*, 20:8, 1977, 564–576.

[Luc70]   Peter Lucas, On the semantics of programming languages and software devices, in Randall Rustin (editor), *Formal Semantics of Programming Languages, Courant Computer Science Symposium 2*, Prentice Hall, 1970, 52–57.

[Luc81]   Peter Lucas, Formal Semantics of Programming Languages: VDL, *IBM Journal of Research and Development* 25 (5) (1981), 549–561.

[LZ74]   Barbara Liskov and Stephen Zilles, Programming with abstract data types, *ACM Sigplan Conference on Very High Level Languages*, April 1974, 50–59.

[LZ75]   Barbara Liskov and Stephen Zilles, Specification techniques for data abstractions, in *IEEE Transactions on Software Engineering*, 1 (1975), 7–19.

[MC80]   Carver A. Mead and Lynn Conway, *Introduction to VLSI Systems*, Addison-Wesley, 1980.

[McC62]   John McCarthy, Towards a mathematical science of computation, in Cicely M Popplewell (editor), *Information Processing 1962*, Proceedings of IFIP Congress 62, Munich, Germany, August 27 - September 1, 1962. North-Holland, 1962, 21–28.

[McC63]   John McCarthy, A basis for a mathematical theory of computation, in P Braffort and D Hirshberg (editors), *Computer Programming and Formal Systems*, North-Holland, 1963, 33–69.

[McL87]   John Mclean, Reasoning about security models, *1987 IEEE Symposium on Security and Privacy, Oakland, CA*, IEEE Computer Society, 1987, 123–131.

[Mey88]   Bertrand Meyer, *Object-Oriented Software Construction*, Prentice Hall, 1988.

[Mey91]   Bertrand Meyer, *Eiffel: The language*, Prentice Hall, 1991.

[Mey92]   Bertrand Meyer, Applying "Design by Contract", *IEEE Computer*, 25 (10) 1992, 40–51.

[Mil70]   Robin Milner, A Formal Notion of Simulation Between Programs, Memo 14, Computers and Logic Research Group, University College of Swansea, UK, 1970.

[Mil71a]   Robin Milner, Program Simulation: An Extended Formal Notion, Memo 15, Computers and Logic Research Group, University College of Swansea, UK, 1971.

[Mil71b]   Robin Milner, An Algebraic Definition of Simulation Between Programs, Stanford Computer Science Report No. STAN-CS-71-205, 1971.

[Mil80]   Robin Milner, *A Calculus of Communicating Systems*, Lecture Notes in Computer Science 92, Springer-Verlag, 1980.

[Mil89]   Robin Milner, *Communication and Concurrency*, Prentice Hall, 1989.

[Mil03]   Robin Milner, Interview with Martin Berger at the University of Sussex, `http://users.sussex.ac.uk/~mfb21/interviews/milner`. Retrieved March 2019.

[MJ84]   F L Morris and Cliff B Jones, An early program proof by Alan Turing, *Annals of the History of Computing*, 6 (2) (1984), 139–143.

[MLK96]   J Strother Moore, Tom Lynch and Matt Kaufmann, A mechanically checked proof of the correctness of the kernel of the $AMD5_K86$ floating-point division algorithm, *IEEE Transactions on Computers* 47 (9) (1998), 913–926.

[MLP79]   R A Millo, R J Lipton, and A J Perlis, Social processes and proofs of theorems and programs, *Communications of the ACM*, 22 (5) (1979), 271–280.

[MMS+96]   Carroll Morgan, Annabelle McIver, Karen Seidel and J. W. Sanders, Refinement-oriented probability for CSP, *Formal Aspects of Computing*, 8 (6) (1996) 617–647.

[Moo89]   J Strother Moore, A mechanically verified language implementation, Journal of Automated Reasoning, 5 (4) (1989), 461–492.

[Mos74]   Peter D Mosses, *The mathematical semantics of Algol 60*, Oxford University Programming Research Group Technical Report 12, 1974.

[Mos93] Peter D. Mosses. The use of sorts in algebraic data type specification, in Michel Bidoit and Christine Choppy (editors), *Recent Trends in Data Type Specification*, Lecture Notes in Computer Science 655, Springer-Verlag, 1993, 66–91.

[Mos04] Peter D Mosses (editor), *CASL Reference Manual. The Complete Documentation of the Common Algebraic Specification Language*, Lecture Notes in Computer Science 2960, Springer-Verlag, 2004.

[Mos17] Till Mossakowski, The Distributed Ontology, Model and Specification Language DOL, in Phillip James and Markus Roggenbach (editors), *Recent Trends in Algebraic Development Techniques*, Lecture Notes in Computer Science 10644, Springer-Verlag, 2017.

[Mor81] Thomas P Moran, The Command Language Grammar: A representation for the user interface of interactive computer systems, *International Journal of Man-Machine Studies* 15 (1) (1981), 3–50.

[MP67] John McCarthy and J Painter, Correctness of a compiler for arithmetic expressions, Jacob T Schwartz (editor), *Proceedings of Symposia in Applied Mathematics 19. Mathematical Aspects of Computer Science*, American Mathematical Society, 1967, 33–41.

[MPW92] Robin Milner, Joachim Parrow and David Walker, A calculus of mobile processes, *Information and Computation*, 100 (1) (1992), 1–40.

[MS76a] Robert Milne and Christopher Strachey, *A Theory of Programming Language Semantics. Part A: Indices and Appendices, Fundamental Concepts and Mathematical Foundations*, Chapman and Hall, 1976.

[MS76b] Robert Milne and Christopher Strachey, *A Theory of Programming Language Semantics. Part B: Standard Semantics, Store Semantics and Stack Semantics*, Chapman and Hall, 1976.

[MS13] Faron Moller and Georg Struth, *Modelling Computing Systems: Mathematics for Computer Science*, Springer, 2013.

[MT90a] Kevin McEvoy and John V Tucker, Theoretical foundations of hardware design, in Kevin McEvoy and John V Tucker (editors), *Theoretical Foundations of VLSI Design*, Cambridge Tracts in Theoretical Computer Science 10, Cambridge UP, 1990, 1–62.

[MT90b] Faron Moller and Chris Tofts, A temporal calculus of communicating systems, In Jos C. M. Baeten and Jan Willem Klop (editors), *CONCUR 90, Theories of Concurrency: Unification and Extension*, Lecture Notes in Computer Science 458. Springer-Verlag, 1990, 401–415.

[MT92] Karl Meinke and John V Tucker, Universal algebra, in S. Abramsky, D. Gabbay and T Maibaum (editors) *Handbook of Logic in Computer Science. Volume I: Mathematical Structures*, Oxford University Press, 1992, 189–411.

[Nau+60] Peter Naur et al. Report on the Algorithmic Language ALGOL60, *Communications of the ACM*, 3 (5) (1960), 299–314.

[Nau62] Peter Naur (editor), Revised Report on Algorithmic Language ALGOL 60, *Communications of the ACM*, i, No. I, (1962), 1–23.

[Nau66] Peter Naur, Proof of algorithms by general snapshots, *BIT* 6 (1966), 310–316.

[Niv79] Maurice Nivat, Infinite words, infinite trees, infinite computations, in J W de Bakker and J van Leeuwen (editors), *Foundations of Computer Science III*, Mathematical Centre Tracts 109, 1979, 3–52.

[NR69] Peter Naur and Brian Randell (editors), *Software Engineering: Report of a conference sponsored by the NATO Science Committee, Garmisch, Germany, 7-11 October, 1968*, Brussels, Scientific Affairs Division, NATO (1969), 231pp.

[NR85] Maurice Nivat and John Reynolds (editors), *Algebraic Methods in Semantics*, Cambridge University Press, 1985.

[NS78] Roger Needham and Michael Schroeder, Using encryption for authentication in large networks of computers, *Communications of the ACM*, 21(12) 1978, 993–999.

[OG76]   Susan Owicki and David Gries, An axiomatic proof technique for parallel programs I, *Acta Informatica* 6 (4) (1976), 319–340.

[OPW+17] R Oliveira, P Palanque, B Weyers, J Bowen, A Dix, State of the art on formal methods for interactive systems, in B Weyers, J Bowen, A Dix, P Palanque (editors) *The Handbook of Formal Methods in Human-Computer Interaction*, Human–Computer Interaction Series. Springer, Cham, 2017.

[ORS92]   Sam Owre, John Rushby and Natarajan Shankar, PVS: A Prototype Verification System, in Deepak Kapur (editor), *11th International Conference on Automated Deduction (CADE)* Lecture Notes in Artificial Intelligence 607, Springer-Verlag, 1992, 748–752.

[ORS+95] S. Owre, J.Rushby, N. Shankar, and F.von Henke. Formal verification for fault-tolerant architectures: Prolegomena to the design of PVS, *IEEE Transactions on Software Engineering*, 21(2) (1995), 107–125.

[Par69]   David Parnas, On the use of transition diagrams in the design of a user Interface for an interactive computer system, *Proceedings 24th National ACM Conference* (1969), 379–385.

[Par72a]   David Parnas, A technique for software module specification with examples, *Communications of the ACM*, 15 (5) (1972), 330–336.

[Par72b]   David Parnas, On the criteria to be used in decomposing systems into modules, *Communications of the ACM*, 15 (12)(1972), 1053–58.

[Par81]   David Park, Concurrency and automata on infinite sequences, in Peter Deussen (editor), *Theoretical Computer science*, Lecture Notes in Computer Science 104, Springer, 1981, 167–183.

[Par01]   David Parnas, *Software Fundamentals – Collected Papers by David L Parnas*, Daniel M.Hoffman and David M Weiss (editors), Addison-Wesley, 2001.

[Plo83]   Gordon D. Plotkin, An operational semantics for CSP, in D Bjørner (editor), *Proceedings IFIP TC2 Working Conference: Formal Description of Programming Concepts II*, North-Holland, 1983, 199–223.

[Plo04a]   Gordon D. Plotkin, The origins of structural operational semantics, *Journal of Logic and Algebraic Programming*, 60-61, (2004), 3–15.

[Plo04b]   Gordon D. Plotkin, A structural approach to operational semantics, DAIMI FN-19, Computer Science Department, Aarhus University, 1981. Also: *Journal of Logic and Algebraic Programming*, 60-61, (2004), 17–139.

[Pnu77]   Amir Pnueli, The temporal logic of programs, Proceedings of the 18th Annual Symposium on Foundations of Computer Science, IEEE 1977, 46–57.

[Pos94]   Emil L. Post, Degrees of recursive unsolvability. Preliminary report, in Martin Davis (editor), *Solvability, Provability, Definability: The Collected Works of Emil L. Post*. Birkhäuser, Boston, Basel, Berlin, 1994, 549–550.

[QS82]   J.-P. Queille and Josef Sifakis, Specification and verification of concurrent systems in CESAR. In: Symposium on Programming. Lecture Notes in Computer Science, vol. 137, Springer, 1982, 337–351.

[RB69]   Brian Randell and John N. Buxton (editors), *Software Engineering Techniques: Report of a conference sponsored by the NATO Science Committee, Rome, Italy, 27-31 October, 1969*, Brussels, Scientific Affairs Division, NATO (1970), 164pp.

[RBH+01] Willem-Paul de Roever, Frank de Boer, Ulrich Hannemann, Jozef Hooman, Yassine Lakhnech, Mannes Poel, and Job Zwiers, *Concurrency Verification: Introduction to Compositional and Noncompositional Methods*, Cambridge University Press, 2001.

[Rei81]   Phyllis Reisner, Formal grammar and human factors design of an interactive graphics system, *IEEE Transactions on Software Engineering*, 7 (2) (1981), 229–240.

[Rei87]   Horst Reichel, *Initial Computability Algebraic Specifications, and Partial Algebras*, Clarendon Press, 1987.

[Rog00]   Markus Roggenbach, Mila Majster-Cederbaum, Towards a unified view of bisimulation: a comparative study. *Theoretical Computer Science*, 238 (2000), 81–130.

[Ros94]   A W Roscoe, Model checking CSP, in *A Classical Mind: Essays in Honour of C A R Hoare*, Prentice Hall, 1994, 353–2378.

[Ros97]   A W Roscoe, *Theory And Practice Of Concurrency*, Prentice Hall, 1997.

[Ros10]   A W Roscoe, *Understanding Concurrent Systems*, Springer, 2010.

[San11]   Davide Sangiorgi, *Introduction to Bisimulation and Coinduction*, Cambridge UP, 2011.

[Shn82]   Ben Shneiderman, Multi-party grammars and related features for defining interactive systems. *IEEE Transactions on Systems, Man, and Cybernetics*, 12 (2) (1982), 148–154.

[SR11]   Davide Sangiorgi and Jan Rutten (editors), *Advanced Topics in Bisimulation and Coinduction*, Cambridge Tracts in Theoretical Computer Science 52, Cambridge UP, 2011.

[ST95]   Viggo Stoltenberg-Hansen and John V Tucker, Effective algebras, in S Abramsky, D Gabbay and T Maibaum (editors) *Handbook of Logic in Computer Science. Volume IV: Semantic Modelling*, Oxford University Press, 1995, 357–526.

[ST12]   Donald Sannella and Andrzej Tarlecki, *Foundations of Algebraic Specification and Formal Software Development*, EATCS Monographs in Theoretical Computer Science, Springer, 2012.

[Ste66]   T B Steel (editor), *Formal Language Description Languages for Computer Programming*, North-Holland, 1966.

[Str80]   Bjarne Stroustrup, Classes: An abstract data type facility for the C language. *Bell Laboratories Computer Science Technical Report*, CSTR 84. April 1980.

[Suf82]   Bernard Sufrin, Formal specification of a display editor, *Science of Computer Programming*, 1(1982), 157–202.

[Ter03]   Terese, *Term Rewriting Systems*, Cambridge Tracts in Theoretical Computer Science 55, Cambridge University Press, 2003.

[Tie92]   Margaret Tierney, Software engineering standards: the 'formal methods debate' in the UK, *Technology Analysis and Strategic Management* 4 (3), (1992), 245–278

[TH90]   Harold W Thimbleby and Michael Harrison, *Formal Methods in Human-Computer Interaction*, Cambridge UP, 1990.

[Thi86]   Harold W Thimbleby, User interface design and formal methods, *Computer Bulletin*, series III, 2 (3) (1986) 13–15, 18.

[Thi90]   Harold W Thimbleby, *User Interface Design*, ACM Press, Addison-Wesley, 1990.

[Thi19]   Harold W Thimbleby, *Death by Design: Stories of digital healthcare*, in preparation.

[TZ88]   John V Tucker and Jeffrey I Zucker, *Program Correctness over Abstract Data Types with Error-state Semantics*, North-Holland, Amsterdam, 1988.

[TZ00]   John V Tucker and Jeffrey I Zucker, Computable functions and semicomputable sets on many sorted algebras, in S. Abramsky, D. Gabbay and T Maibaum (editors), *Handbook of Logic for Computer Science. Volume V: Logical and Algebraic Methods*, Oxford University Press, 2000, 317–523.

[TZ02]   John V Tucker and Jeffrey I Zucker, Origins of our theory of computation on abstract data types at the Mathematical Centre, Amsterdam, 1979-80, in F de Boor, M van der Heijden, P Klint, J Rutten (eds), *Liber Amicorum: Jaco de Bakker*, CWI Amsterdam, 2002, 197–221.

[UH69]   Jeffrey Ullman and John E. Hopcroft, *Formal Languages and Their Relation to Automata*, Addison Wesley, 1969.

[Wag01]   Eric G. Wagner, Algebraic specifications: some old history, and new thoughts, Unpublished, 2001.

[WBD+17]   B Weyers, J Bowen, A Dix, P Palanque (editors) *The Handbook of Formal Methods in Human-Computer Interaction*, Human Computer Interaction Series, Springer, Cham, 2017.

[You82]   Edward Yourdon, *Writings of the Revolution: Selected Readings on Software Engineering*, Yourdon Press, 1982.

[Zil74]   Steve Zilles, *Algebraic specifications of data types*, Project MAC Progress Report 11, Massachusetts Institute of Technology, Cambridge, MA, 1974, 52–58.

Correction to: Formal Methods for Software Engineering

Markus Roggenbach ⓘ, **Antonio Cerone, Bernd-Holger Schlingloff,
Gerardo Schneider, and Siraj Ahmed Shaikh**

Correction to:
M. Roggenbach et al., *Formal Methods for Software*
Engineering, **Texts in Theoretical Computer Science.**
An EATCS Series, https://doi.org/10.1007/978-3-030-38800-3

In the original version of the book, the following corrections have been incorporated:
The Author names have been updated on Springer Link for all chapters.

The updated version of the book can be found at
https://doi.org/10.1007/978-3-030-38800-3

Authors' Conclusion

A brief postponement
Brings the most distant goal within reach.

B. Brecht

Challenge

The challenge of software engineering has been acknowledged for decades. The latter half of the last century saw the costs of software development start to rise as the cost of computer hardware was falling. Over the years, software development projects have overrun on time and resources, running risks of non-delivery and unmanageability; cost overruns remain considerably high even by conservative estimates [JMO06]. The quality of software produced is also of concern. Found to be error-prone, non-compliant, low quality and non maintainable, credible efforts to survey software problems over the years show that the situation persists [Neu06]. Languages and tools may have changed with new application domains, but practically all modern software still contains subtle errors.

The situation has contributed to a sense of software crisis, noted as such early on by community and industry leaders [NR69, Dij72]. The term has since been somewhat associated to the debate on software industry. It is reasonable to presume the state of software engineering will continue this course unless methods and practices are addressed. The question is how long is the crisis to last? This book is motivated by the very challenge and has set out to address the fundamental need for clear and coherent reasoning. The premise however is beyond past debate. A new era has dawned with breakthroughs in electronics and communications, and a wider acceptance of software technology. As modern applications evolve, the demands on software to comply with increasing expectations grow. As such three trends have emerged over the recent years.

© Springer Nature Switzerland AG 2022
M. Roggenbach et al., *Formal Methods for Software Engineering*,
Texts in Theoretical Computer Science. An EATCS Series,
https://doi.org/10.1007/978-3-030-38800-3

- Modern computing is becoming pervasive [Sat01]. No longer is software sitting comfortably on desktops, but seamlessly dispersed across urban and household installations, needing to be open to cross-layer configuration and communication. This is truly a reflection of the service-oriented world that software has to serve.
- Software is increasingly designed to support critical systems, which require safety and reliability guarantees to be predicated over execution. Admittedly, margins of errors on such systems, from jet engine controllers to programmable syringe pumps, are intolerable.
- User centricity is vital. As systems become more interactive, how they appear to and engage end-users has received more attention than ever before [Dix10]. More so, user interface design is subject to cognitive reasoning to allow for better and safer human-computer interaction, with the ultimate goal of being error-free. This has implications for software design and development; users have become insiders to the process in essence.

Readers would recognise these influences over the many examples of applications spread across the preceding chapters. Such trends only add to the challenge of software engineering, as they bring increased complexity in requirements and specification, added layers of abstraction for design, and the potential for subtle errors at the implementation stage. The message of this book is clear: Formal Methods are a step in the direction of addressing the software engineering challenge. Evidence suggests that they offer a clear beneficial impact on quality of the software produced, and time and cost incurred in the process [WLBF09]. A mathematical underpinning of the engineering process allows for explicit requirements to be expressed, precise specifications to be derived, critical properties proved using some form of theorem-proving or model checking, and the final outcome to be comprehensively tested. At each stage, Formal Methods allow for ambiguity to be addressed and rigour applied to provide some assurance in the final product. Undoubtedly, successful adoption depends on the application of Formal Methods only at a selected stage of the engineering process, so as to provide assurance where it is most needed and avoid additional cost where possible.

Contribution

We intended to present Formal Methods to the reader with a careful balance of depth and breadth.

Part I took the reader through two formal languages. Logic is an obvious choice to lay down the basics of formal reasoning as it underpins rational thought. The process algebraic language of CSP allows one to study typical phenomena of parallel systems, including non-determinism, livelock, and deadlock.

Part II moved the attention on to practical methods employed. The purpose has been to show how Formal Methods address aspects of the development lifecycle. Chapter 4 presented the reader with an algebraic specification language developed in response to a need for a common framework by the research community. Chapter 5 demonstrated the utility of Formal Methods for software testing, to show that systematic treatment of testing can be used to demonstrate the correctness of software.

Part III of the book was intended to show the application of Formal Methods to a diverse set of domains. Chapter 6 has explored the notion of contracts and components as used in distributed systems. Chapter 7 tackled interactive systems which are increasingly subject to rigorous treatment given their relevance to safety-critical applications from air traffic control to automated teller machines. Chapter 8 presented the challenge of designing security protocols. As they increasingly underpin electronic communication infrastructure to provide authentication, confidentiality and other important properties, security protocols have historically been difficult to design and prove for correctness.

Chapter 9 finally provided a historical perspective for Formal Methods, to allow the reader to develop a wider mindset about the interrelations between different schools of thought.

Takeaways for the Reader

The 2020 white paper "Rooting Formal Methods within Higher Education Curricula for Computer Science and Software Engineering" [CRD+20] makes the following propositions:

- Current software engineering practices fail to deliver dependable software.
- Formal methods are capable of improving this situation, and are beneficial and cost-effective for mainstream software development.
- Education in formal methods is key to progress things.
- Education in formal methods needs to be transformed.

With this textbook we hope to have helped improving this situation. In particular, we argue that Formal Methods have come out of the niche of safety critical applications, and that it is reasonable to apply them also in mainstream software development.

For the academic teacher, our book offers an example driven approach for teaching the subject. It puts an emphasis on the application of Formal Methods, while still preserving mathematical rigour. The book describes a selected set of Formal Methods in one integrated setting.

For students of software engineering, the book offers an accessible account to understand what Formal Methods are about. Studying this book should enable them to apply various Formal Methods to concrete software engineering challenges. As different Formal Methods share foundations, we hope that the material of this book will allow the reader to comprehend the essence of newly encountered Formal Methods. In the same spirit, we hope that studying our book may encourage the reader to undertake own research in Formal Methods for software engineering.

References

[CRD⁺20] Antonio Cerone, Markus Roggenbach, James Davenport, Casey Denner, Marie Farrell, Magne Haveraaen, Faron Moller, Philipp Koerner, Sebastian Krings, Peter Ölveczky, Bernd-Holger Schlingloff, Nikolay Shilov, and Rustam Zhumagambetov. Rooting formal methods within higher education curricula for computer science and software engineering—A White Paper, 2020. https://arxiv.org/abs/2010.05708.

[Dij72] Edsger Dijkstra. The humble programmer. *Communications of the ACM*, 15(10):859–866, 1972.

[Dix10] Alan Dix. Human-computer interaction: A stable discipline, a nascent science, and the growth of the long tail. *Interacting with Computers*, 22(1):13–27, 2010.

[JMO06] Magne Jorgensen and Kjetil Molokken-Ostvold. How large are software cost overruns? A review of the 1994 chaos report. *Information and Software Technology*, 48(4):297–301, 2006.

[Neu06] Peter Neumann. Risks of untrustworthiness. In *22nd Annual Computer Security Applications Conference*. IEEE Computer Society, 2006.

[NR69] Peter Naur and Brian Randell, editors. *Software Engineering: Report of a conference sponsored by the NATO Science Committee*. NATO Scientific Affairs Division, Brussels, 1969.

[Sat01] M. Satyanarayanan. Pervasive computing: vision and challenges. *IEEE Personal Communications*, 8(4):10–17, 2001.

[WLBF09] Jim Woodcock, Peter Gorm Larsen, Juan Bicarregui, and John S. Fitzgerald. Formal methods: Practice and experience. *ACM Computing Surveys*, 41(4), 2009.

Appendix A
Syntax of the Logics in this Book

Here, we give an overview of the syntax of the various logics and formalisms used in this book. Moreover, we provide examples for each formalism.

The symbol "$\triangleq$" stands for "equal by definition" or "is defined as".

A.1 Regular Expressions

Syntax

Let $\mathcal{A} = \{a_1, \ldots, a_n\}$ be a finite *alphabet*, i.e., a set of *letters*.

$$Regexp_{\mathcal{A}} \quad ::= \quad \mathcal{A} \quad | \quad \emptyset \quad | \quad (Regexp_{\mathcal{A}} \; Regexp_{\mathcal{A}}) \quad |$$
$$(Regexp_{\mathcal{A}} + Regexp_{\mathcal{A}}) \quad | \quad Regexp_{\mathcal{A}}^*$$

Abbreviations

- $\varepsilon \triangleq \emptyset^*$ (the *empty word*)
- $\varphi^+ \triangleq (\varphi \, \varphi^*)$ (one or more *repetitions* of φ)
- $. \triangleq (((a_1 + a_2) + \ldots) + a_n)$ (*any* letter)
- $* \triangleq .^*$ (*any word*)
- $\varphi? \triangleq (\varepsilon + \varphi)$ (*maybe* one φ)
- $\varphi^0 \triangleq \varepsilon$ and $\varphi^n \triangleq (\varphi \, \varphi^{n-1})$ for any $n > 0$ (*exactly n times* φ)
- $\varphi_m^n \triangleq (\varphi^m \varphi?^{n-m})$ for $0 \leq m \leq n$ (*at least m and at most n* φ)

Examples

- (a^*b) (a sequence of a followed by b)
- $(\varepsilon + a^+)$ (arbitrary many a, same as a^*)
- $._3^3$ (any three-letter word)

© Springer Nature Switzerland AG 2022
M. Roggenbach et al., *Formal Methods for Software Engineering*,
Texts in Theoretical Computer Science. An EATCS Series,
https://doi.org/10.1007/978-3-030-38800-3

A.2 Propositional Logic

Syntax

Let $\mathcal{P} = \{p_1, \ldots, p_n\}$ be a finite set of proposition symbols. The symbol '$\bot$' is called *falsum*, and '$\Rightarrow$' is called *implication*.
$$\mathrm{PL}_\mathcal{P} ::= \quad \mathcal{P} \quad | \quad \bot \quad | \quad (\mathrm{PL}_\mathcal{P} \Rightarrow \mathrm{PL}_\mathcal{P})$$

Abbreviations

- $\neg\varphi \triangleq (\varphi \Rightarrow \bot)$ (*negation*)
- $\top \triangleq \neg\bot$ (*verum*)
- $(\varphi \vee \psi) \triangleq (\neg\varphi \Rightarrow \psi)$ (*disjunction*)
- $(\varphi \wedge \psi) \triangleq \neg(\neg\varphi \vee \neg\psi)$ (*conjunction*)
- $(\varphi \Leftrightarrow \psi) \triangleq ((\varphi \Rightarrow \psi) \wedge (\psi \Rightarrow \varphi))$ (*equivalence*)
- $(\varphi \oplus \psi) \triangleq (\varphi \Leftrightarrow \neg\psi)$ (*exclusive-or*)
- $\bigoplus(\varphi_1, \ldots, \varphi_n) \triangleq \bigvee_{i \leq n}(\varphi_i \wedge \bigwedge_{j \leq n, j \neq i} \neg\varphi_j)$ (*choice*)
- $\bigvee_{i \leq n} \varphi_i \triangleq (\bigvee_{i \leq n-1} \varphi_i \vee \varphi_n)$, if $n > 0$, and $\bigvee_{i \leq 0} \varphi_i \triangleq \bot$
- $\bigwedge_{j \leq n, j \neq i} \varphi_j \triangleq (\bigwedge_{j \leq n-1, j \neq i} \varphi_j \wedge \varphi_n)$, if $i \neq n$, and $\bigwedge_{j \neq j} \varphi_j \triangleq \top$

Example formulae

- $((\bot \Rightarrow \bot) \Rightarrow \bot)$
- $(p \Rightarrow (p \vee q))$
- $(\texttt{motor\_59kW} \Rightarrow \neg\texttt{gearshift\_automatic})$

A.3 First- and Second-Order Logic

Basic First-Order logic

Syntax

Let $\Sigma = (\mathcal{F}, \mathcal{R}, \mathcal{V})$ be a first-order signature (function symbols, relation symbols, variables). The symbol '$\exists$' is called the *existential quantifier*.

$Terms:$ $\mathcal{T}_\Sigma ::= \quad \mathcal{V} \quad | \quad \mathcal{F}(\mathcal{T}_\Sigma, \ldots, \mathcal{T}_\Sigma)$
$FOL_\Sigma ::= \quad \mathcal{R}(\mathcal{T}_\Sigma, \ldots, \mathcal{T}_\Sigma) \quad | \quad \bot \quad | \quad (FOL_\Sigma \Rightarrow FOL_\Sigma) \quad | \quad \exists \mathcal{V} \, FOL_\Sigma$

Abbreviation

- $\forall x \, \varphi \triangleq \neg \exists x \, \neg \varphi$ (*universal* quantifier)

Example formulae

- $\forall x \exists y (p(x) \Rightarrow q(x, y))$
- $(\exists x \forall y \, p(x, y) \Rightarrow \forall y \exists x \, p(x, y))$
- $\forall x \exists y \, eq(y, f(x))$
- $(Hugo \prec (Erna + Hugo))$

First-order logic with equality

Syntax

Assume the symbol "=" (*equals*) is not part of the signature.
$$FOL_\Sigma^= ::= \mathcal{R}(\mathcal{T}_\Sigma, \ldots, \mathcal{T}_\Sigma) \mid (\mathcal{T}_\Sigma = \mathcal{T}_\Sigma) \mid \bot \mid (FOL_\Sigma^= \Rightarrow FOL_\Sigma^=) \mid \exists \mathcal{V} \, FOL_\Sigma^=$$

Example formulae

- $((a = b) \wedge (b = c) \Rightarrow (a = c))$
- $f(x) = x^2 + 2 * x + 1$
- $\forall x \exists y \, (y = f(x))$
- $\exists x \exists y \, (\neg(x = y) \wedge \forall z (z = x \vee z = y))$

Many-sorted FOL$^=$

Syntax

In many-sorted logic, the universe is structured into different *sorts* S, each function symbol and variable is of a dedicated sort, and arguments of functions and relations must respect the sort constraint.

Example formulae

- $\exists x : String. \, (length(x) = 0)$ (where $0 : Nat$, and $length : String \rightarrow Nat$.)
- $\forall x : Person. \exists y : Year. \, (y = birth(x))$ (where $birth : Person \rightarrow Year$.)
- $\forall x : \texttt{float}. \, \exists y : \texttt{int}. \, (\texttt{Math.round}(x) = y)$

FOL with Partiality

Syntax

A first-order signature with partiality $\Sigma = (S, \mathcal{F}_t, \mathcal{F}_p, \mathcal{R}, \mathcal{V})$ distinguishes between *total* function symbols $\mathcal{F}_t$ and *partial* function symbols $\mathcal{F}_p$. The unary predicate *def* indicates whether a function result is defined or not.

Example formulae

- $\forall x : real.\ \sqrt{x} > 0$ (where $\sqrt{\cdot} : real \rightarrow?\ real$)
- $\forall x : real.\ (x > 0 \ \Rightarrow\ \sqrt{x} > 0)$
- $\forall x : real.\ (def\ \sqrt{x} \ \Leftrightarrow\ x \geq 0)$
- $\forall s : String.\ (\neg isEmpty(rest(s)) \ \Rightarrow\ def\ first(rest(s)))$
- $\forall s : String.\ \forall c : Char.\ (isEmpty(s) \Rightarrow \neg(first(s) = c))$

Monadic second-order logic

Syntax

Monadic second order logic allows quantification both of individual variables $x \in \mathcal{V}^0$ and of (unary) predicate variables $P \in \mathcal{V}^1$.
$$\mathcal{T}_\Sigma ::= \quad \mathcal{V}^0 \ \mid \ \mathcal{F}(\mathcal{T}_\Sigma, \ldots, \mathcal{T}_\Sigma)$$
$$MSO_\Sigma ::= \quad \mathcal{R}(\mathcal{T}_\Sigma, \ldots, \mathcal{T}_\Sigma) \ \mid \ \mathcal{V}^1(\mathcal{T}_\Sigma) \ \mid \ \bot \ \mid \ (MSO_\Sigma \Rightarrow MSO_\Sigma)$$
$$\mid \ \exists \mathcal{V}^0\ MSO_\Sigma \ \mid \ \exists \mathcal{V}^1\ MSO_\Sigma$$

Abbreviation

- $\forall P\, \varphi \triangleq \neg \exists P\, \neg \varphi$ (*universal* second-order quantifier)

Example formulae

- $\forall x \exists P\ (P(x) \wedge \forall y(P(y) \Rightarrow R(x,y)))$
- $\forall x \forall y\ (eq(x,y) \Leftrightarrow \forall P\ (P(x) \Leftrightarrow P(y)))$
- $\forall P(\forall x(\forall y(y < x \Rightarrow P(y)) \Rightarrow P(x)) \Rightarrow \forall x P(x))$

A.4 Non-Classical Logics

Modal Logic

Syntax

The symbol '$\diamond$' is called *modal possibility* or *diamond* operator.

$$ML_{\mathcal{P}} ::= \quad \mathcal{P} \quad | \quad \bot \quad | \quad (ML_{\mathcal{P}} \Rightarrow ML_{\mathcal{P}}) \quad | \quad \diamond ML_{\mathcal{P}}$$

Abbreviation

- $\Box \varphi \triangleq \neg \diamond \neg \varphi$ (modal *necessity* or *box* operator)

Example formulae

- $\diamond \diamond \Box p$
- $(p \Rightarrow \Box \Box p)$
- $(\Box \diamond \; cloudy \Rightarrow \diamond \Box \; cloudy)$

Multimodal logic

Syntax

In multimodal logic, there is one diamond operator for each accessibility relation $R \in \mathcal{R}$.

$$\mathrm{MML}_{\Sigma} ::= \quad \mathcal{P} \quad | \quad \bot \quad | \quad (\mathrm{MML}_{\Sigma} \Rightarrow \mathrm{MML}_{\Sigma}) \quad | \quad \langle \mathcal{R} \rangle \mathrm{MML}_{\Sigma}$$

Abbreviation

- $[R]\varphi \triangleq \neg \langle R \rangle \neg \varphi$ (multimodal box operator)

Example formulae

- $\langle a \rangle \langle b \rangle [a] \, p$
- $(\langle a \rangle p \Rightarrow \langle b \rangle \, p)$
- $(isHobby \Rightarrow (\langle int \rangle \; isHome \wedge \langle ext \rangle \; isClub))$

Deontic logic

Syntax

In deontic logic, the modal possibility operator is interpreted as *permission* 'ℙ'. Nesting of modalities is not allowed.

$$\text{SDL}_\mathcal{P} ::= \quad \text{PL}_\mathcal{P} \quad | \quad (\text{SDL}_\mathcal{P} \Rightarrow \text{SDL}_\mathcal{P}) \quad | \; \mathbb{P} \, \text{PL}_\mathcal{P}$$

Abbreviations

- $\mathbb{O} \, p \triangleq \neg \mathbb{P} \neg p$ (*obligation* operator)
- $\mathbb{F} \, p \triangleq \neg \mathbb{P} \, p$ (*prohibition* operator)

Example formulae

- $(buy \Rightarrow (\mathbb{P} \, use \land \mathbb{O} \, pay))$
- $(\mathbb{O} \, q \land \mathbb{O} \neg q)$
- $(flight\_leaves \Rightarrow (\mathbb{O} \, desk\_opens \land \mathbb{O} \, request\_man))$

Linear temporal logic

Syntax

LTL has besides the modal *next*-operator '$\bigcirc$' the binary *until*-operator '$\mathcal{U}$'.

$$\text{LTL}_\mathcal{P} ::= \quad PL_\mathcal{P} \quad | \quad (\text{LTL}_\mathcal{P} \Rightarrow \text{LTL}_\mathcal{P}) \quad | \quad \bigcirc \text{LTL}_\mathcal{P} \quad | \quad (\text{LTL}_\mathcal{P} \, \mathcal{U} \, \text{LTL}_\mathcal{P})$$

Abbreviations

- $\diamondsuit \varphi \triangleq (\top \, \mathcal{U} \, \varphi)$ (*eventually*- or *sometime*-operator)
- $\square \varphi \triangleq \neg \diamondsuit \neg \varphi$ (*globally*- or *always*-operator)

Example formulae

- $(\square \, sleeping \Rightarrow \bigcirc \, sleeping)$
- $(\square (\varphi \Rightarrow \bigcirc \varphi) \Rightarrow (\varphi \Rightarrow \square \varphi))$
- $((\psi \lor \varphi \land \bigcirc (\varphi \, \mathcal{U} \, \psi)) \Rightarrow (\varphi \, \mathcal{U} \, \psi))$
- $\square \diamondsuit \, phil_0 \, eating$

Appendix B
Language Definition of CSP

B.1 Syntax

B.1.1 Processes

The CSP syntax is given by the following grammar, which is defined relatively to an alphabet of events Σ and a set of process names PN. After the %% we list the name of the process and give its representation in CSP_M.

$P, Q ::=$	$Stop$	%% deadlock process	`STOP`
	$\mid Skip$	%% terminating process	`SKIP`
	$\mid Div$	%% diverging process	
	$\mid N$	%% process name	`N`
	$\mid a \to P$	%% action prefix	`a -> P`
	$\mid ?x : A \to P$	%% prefix choice	
	$\mid c.a \to P$	%% channel communication	`c.a -> P`
	$\mid c?x \to P$	%% channel input	`c?x -> P`
	$\mid c!a \to P$	%% channel output	`c!a -> P`
	$\mid P \square Q$	%% external choice	`P [] Q`
	$\mid P \sqcap Q$	%% internal choice	`P |~| Q`
	$\mid$ **if** $cond$ **then** P **else** Q	%% conditional	`if cond then P` `else Q`
	$\mid cond \,\&\, P$	%% guarded process	`cond & P`
	$\mid P \,\fatsemi\, Q$	%% sequential composition	`P; Q`
	$\mid P \triangle Q$	%% interrupt	`P /\ Q`
	$\mid P \,[\|\, A \,\|]\, Q$	%% general parallel	`P [|A|] Q`
	$\mid P \,[A \parallel B]\, Q$	%% alphabetised parallel	`P [A||B] Q`
	$\mid P \,\|\|\|\, Q$	%% interleaving	`P ||| Q`
	$\mid P \parallel Q$	%% synchronous parallel	`P [| Events |] Q`
	$\mid P \setminus A$	%% hiding	`P \ A`
	$\mid P[\![R]\!]$	%% renaming	
	$\mid \square_{i \in I} P_i$	%% replicated external choice	`[]i:I@P(i)`
	$\mid \sqcap_{j \in J} P_j$	%% replicated internal choice	`|~|j:J@P(j)`
	$\mid \|\|\|_{i \in I} P_i$	%% replicated interleaving	`|||i:I@P(i)`
	$\mid [\|\, A \,\|]_{i \in I} P_i$	%% replicated general parallel	`[|A|]i:I@P(i)`
	$\mid \|_{i \in I} (A_i, P_i)$	%% replicated alphabetised	`||i:I@[A(i)]P(i)`

© Springer Nature Switzerland AG 2022
M. Roggenbach et al., *Formal Methods for Software Engineering*,
Texts in Theoretical Computer Science. An EATCS Series,
https://doi.org/10.1007/978-3-030-38800-3

where

- $a \in \Sigma$,
- $N \in PN$,
- c is a channel, $a \in T(c)$—the type of c—and $events(c) \subseteq \Sigma$ (cf. Sect. 3.2.1),
- $cond$ is a condition in a logic of choice (not determined by CSP),
- $A, B \subseteq \Sigma$,
- $R \subseteq \Sigma \times \Sigma$ such that for all $a \in \Sigma$ there exists $a' \in \Sigma$ with $(a, a') \in R$,
- I is a finite index set, $(P_i)_{i \in I}$ is a family of processes, $(A_i)_{i \in I}$ is a family of sets with $A_i \subseteq \Sigma$ for all $i \in I$,
- J is a non-empty index set, $(P_j)_{j \in J}$ is a family of processes.

The CSP$_M$ column has been left blank in several cases:

- To the best of our knowledge, the prefix choice operator $?x : A \to P$ has no counterpart in CSP$_M$ (though in case of a finite set A it can be simulated via replicated external choice).
- Renaming is written in CSP$_M$ as

  ```
  P [[a <- b]]
  ```

  where a and b are events. CSP$_M$ offers a rich comprehension syntax for expressing complex renaming—see, e.g., the FDR user manual or th book A.W. Roscoe, *The theory and practice of concurrency*, Prentice Hall, 1998, for a documentation of this syntax.
- The CSP$_M$ process CHAOS is similar to Div, however, can't terminate as it is defined as

$$CHAOS(A) = (\sqcap_{a \in A} a \to CHAOS(A)) \sqcap Stop$$

Note that—in order to keep the language manageable in the context of a brief introduction—we refrain from introducing a number of CSP constructs such as multi channels (allowing multiple data transfers in a single event), linked parallel, untimed timeout etc.

B.1.2 Operator Precedences

In order to reduce the number of brackets needed, the following operator precedence is generally assumed for CSP operators:

- Renaming binds tighter than
- prefix, guard, and sequential composition bind tighter than
- interrupt, external choice, internal choice bind tighter than
- the parallel operators bind tighter than
- conditional.

B.1.3 Process Equations

A process equation takes the form

$$N = P$$

where $N \in PN$ and P is as described by the above grammar.

B.2 Semantics

B.2.1 Static Semantics

CSP static semantics concerns aspects of the language such as the scope of variables, the relation between synchronisation sets and process alphabets, and that process names have a unique definition.

CSP distinguishes between two kinds of variables, which we call *static* and *dynamic*. Static variables are bound by replicated operators, and by process names occurring on the lhs of an equation; they are not affected by process termination. Dynamic variables are bound by prefix choice and channel input; their binding lasts up to process termination.

Synchroniation sets and processes in alphabetised parallel are related as follows:

- The process $P\,[\,A \parallel B\,]\,Q$ is wellformed if the alphabet of P is a subset of A and the alphabet of Q is a subset of B.
- The process $\parallel_{i \in I}(A_i, P_i)$ is wellformed, if for all $i \in I$, the alphabet of P_i is a subset of A_i.

A system of equations is wellformed, if there is exactly one equation for each occurring process name.

B.2.2 Syntactic Sugar

Semantically, several operators introduced above are syntactic sugar. Some of them expand directly:

- $c?x \to P \triangleq ?y : events(c) \to P[value(y)/x]$ where $value(c.a) = a$
- $c!a \to P \triangleq c.a \to P$
- $cond \,\&\, P \triangleq \textbf{if } cond \textbf{ then } P \textbf{ else } Stop$
- $P\,[\,A \parallel B\,]\,Q \triangleq P\,[\!|\,A \cap B\,|\!]\,Q$
- $P\,[\!|\!|\,Q \triangleq P\,[\!|\,\emptyset\,|\!]\,Q$
- $P \parallel Q \triangleq P\,[\!|\,\Sigma\,|\!]\,Q$

Note that channels are syntactic sugar.

Some replicated operators are defined inductively over the number of elements in the finite index set I:

- $\square_{i \in I} P_i \triangleq P_k \square \left(\square_{i \in I \setminus \{k\}} P_i \right)$ (for $k \in I$)
  where $\square_{i \in \{\}} P_i \triangleq Stop$

- $|||_{i \in I} P_i \triangleq P_k ||| \left(|||_{i \in I \setminus \{k\}} P_i \right)$ (for $k \in I$)
  where $|||_{i \in \{\}} P_i \triangleq Skip$

- $[|A|]_{i \in I} P_i \triangleq P_k [|A|] \left([|A|]_{i \in I \setminus \{k\}} P_i \right)$ (for $k \in I$)
  where $[|A|]_{i \in \{\}} P_i \triangleq Skip$

- $\|_{i \in I} (A_i, P_i) \triangleq P_k [A_k \| \bigcup_{i \in I \setminus \{k\}} A_i] \left(\|_{i \in I \setminus \{k\}} (A_i, P_i) \right)$ (for $k \in I$)
  where $\|_{i \in \{\}} (A_i, P_i) \triangleq Skip$.

  The choice to set $\|_{i \in \{\}} (A_i, P_i) \triangleq Skip$ is in accordance with the FDR documentation and also agrees with CSP algebraic laws. However, FDR operation is different. It appears to be $\|_{i \in \{\}} (A_i, P_i) \triangleq RUN\_ext\_Skip(A)$, where

$$RUN\_ext\_Skip(A) = (?x : A \to RUN\_ext\_Skip(A)) \square Skip$$

Note that prefix choice and replicated internal choice are part of the core language rather than syntactic sugar. Both these operators allow for infinite branching in CSP.

B.2.3 Core Language

$$
\begin{array}{lll}
P, Q ::= & Stop & \%\% \text{ deadlock process} \\
 & |\ Skip & \%\% \text{ terminating process} \\
 & |\ Div & \%\% \text{ diverging process} \\
 & |\ N & \%\% \text{ process name} \\
 & |\ a \to P & \%\% \text{ action prefix} \\
 & |\ ?x : A \to P & \%\% \text{ prefix choice} \\
 & |\ P \square Q & \%\% \text{ external choice} \\
 & |\ P \sqcap Q & \%\% \text{ internal choice} \\
 & |\ \textbf{if } cond \textbf{ then } P \textbf{ else } Q & \%\% \text{ conditional} \\
 & |\ P \,\mathbin{\raise0.3ex{\scriptstyle\circ}}_9\, Q & \%\% \text{ sequential composition} \\
 & |\ P \triangle Q & \%\% \text{ interrupt} \\
 & |\ P [|A|] Q & \%\% \text{ general parallel} \\
 & |\ P \setminus A & \%\% \text{ hiding} \\
 & |\ P[\![R]\!] & \%\% \text{ renaming} \\
 & |\ \sqcap_{j \in J} P_j & \%\% \text{ replicated internal choice}
\end{array}
$$

B.3 Operational Semantics

The definitions presented subsequently closely follow the book A.W. Roscoe, *The theory and practice of concurrency*, Prentice Hall, 1998.

For ease of presentation, we assume the following implicit typing for the labels of transitions:

- $a \in \Sigma$—i.e., a is an element of the alphabet.
- $b \in \Sigma \cup \{\checkmark\}$—i.e., b stands for an observable event.
- $x \in \Sigma \cup \{\tau\}$—i.e., x stands for a non-terminating event.

Furthermore, we add the state Ω representing a process after termination. This special state helps to treat termination in the case of the parallel operator. Note that the semantics has the property that the state Ω can only be reached by transitions labelled with $\checkmark$.

- *Stop*—deadlock process.

$$\text{This process has no firing rule.}$$

- *Skip*—successfully terminating process.

$$\overline{Skip \xrightarrow{\checkmark} \Omega}$$

- *Div*—diverging process.

$$\overline{Div \xrightarrow{\tau} Div}$$

- *N*—process name.

$$\frac{}{N \xrightarrow{\tau} P} \quad \text{if there is an equation } N = P$$

- $a \to P$—action prefix.

$$\overline{(a \to P) \xrightarrow{a} P}$$

- $?x : A \to P$—prefix choice.

$$\frac{}{(?x : A \to P) \xrightarrow{a} P[a/x]} \quad a \in A$$

- $P \,\square\, Q$—external choice.

$$\frac{P \xrightarrow{b} P'}{P \,\square\, Q \xrightarrow{b} P'} \qquad \frac{Q \xrightarrow{b} Q'}{P \,\square\, Q \xrightarrow{b} Q'}$$

$$\frac{P \xrightarrow{\tau} P'}{P \,\square\, Q \xrightarrow{\tau} P' \,\square\, Q} \qquad \frac{Q \xrightarrow{\tau} Q'}{P \,\square\, Q \xrightarrow{\tau} P \,\square\, Q'}$$

- $P \sqcap Q$—internal choice.

$$\overline{P \sqcap Q \xrightarrow{\tau} P} \qquad \overline{P \sqcap Q \xrightarrow{\tau} Q}$$

- **if** *cond* **then** P **else** Q—conditional.

$$\frac{}{\textbf{if } cond \textbf{ then } P \textbf{ else } Q \xrightarrow{\tau} P} \qquad \textit{if } cond \textit{ evaluates to true}$$

$$\frac{}{\textbf{if } cond \textbf{ then } P \textbf{ else } Q \xrightarrow{\tau} Q} \qquad \textit{if } cond \textit{ evaluates to false}$$

- $P \,\talloblong\, Q$—sequential composition.

$$\frac{P \xrightarrow{x} P'}{P \,\talloblong\, Q \xrightarrow{x} P' \,\talloblong\, Q} \qquad \frac{P \xrightarrow{\checkmark} P'}{P \,\talloblong\, Q \xrightarrow{\tau} Q}$$

- $P \triangle Q$—interrupt.

$$\frac{P \xrightarrow{x} P'}{P \triangle Q \xrightarrow{x} P' \triangle Q} \qquad \frac{Q \xrightarrow{b} Q'}{P \triangle Q \xrightarrow{b} Q'}$$

$$\frac{P \xrightarrow{\checkmark} P'}{P \triangle Q \xrightarrow{\checkmark} \Omega} \qquad \frac{Q \xrightarrow{\tau} Q'}{P \triangle Q \xrightarrow{\tau} P \triangle Q'}$$

- $P \,[\![\, A \,]\!]\, Q$—general parallel.

$$\frac{P \xrightarrow{a} P', Q \xrightarrow{a} Q'}{P \,[\![\, A \,]\!]\, Q \xrightarrow{a} P' \,[\![\, A \,]\!]\, Q'} a \in A$$

$$\frac{P \xrightarrow{a} P'}{P \,[\![\, A \,]\!]\, Q \xrightarrow{a} P' \,[\![\, A \,]\!]\, Q} a \in \Sigma \backslash A \qquad \frac{Q \xrightarrow{a} Q'}{P \,[\![\, A \,]\!]\, Q \xrightarrow{a} P \,[\![\, A \,]\!]\, Q'} a \in \Sigma \backslash A$$

$$\frac{P \xrightarrow{\tau} P'}{P \,[\![\, A \,]\!]\, Q \xrightarrow{\tau} P' \,[\![\, A \,]\!]\, Q} \qquad \frac{Q \xrightarrow{\tau} Q'}{P \,[\![\, A \,]\!]\, Q \xrightarrow{\tau} P \,[\![\, A \,]\!]\, Q'}$$

$$\frac{P \xrightarrow{\checkmark} \Omega}{P \,[\![\, A \,]\!]\, Q \xrightarrow{\tau} \Omega \,[\![\, A \,]\!]\, Q} \qquad \frac{Q \xrightarrow{\checkmark} \Omega}{P \,[\![\, A \,]\!]\, Q \xrightarrow{\tau} P \,[\![\, A \,]\!]\, \Omega}$$

$$\frac{}{\Omega \,[\![\, A \,]\!]\, \Omega \xrightarrow{\checkmark} \Omega}$$

- $P \setminus A$—hiding.

$$\frac{P \xrightarrow{a} P'}{P \setminus A \xrightarrow{\tau} P' \setminus A} a \in A \qquad \frac{P \xrightarrow{x} P'}{P \setminus A \xrightarrow{x} P' \setminus A} x \notin A$$

$$\frac{P \xrightarrow{\checkmark} P'}{P \setminus A \xrightarrow{\checkmark} \Omega}$$

- $P[\![R]\!]$—relational renaming.

$$\frac{P \xrightarrow{a} P'}{P[\![R]\!] \xrightarrow{c} P'[\![R]\!]}\text{if}aRc \qquad \frac{P \xrightarrow{\tau} P'}{P[\![R]\!] \xrightarrow{\tau} P'[\![R]\!]}$$

$$\frac{P \xrightarrow{\checkmark} P'}{P[\![R]\!] \xrightarrow{\checkmark} \Omega}$$

- $\sqcap_{j\in J}P_j$—replicated internal choice.

$$\frac{}{\sqcap_{j\in J}P_j \xrightarrow{\tau} P_k}k \in J$$

B.4 Denotational Semantics

In the context of CSP denotational semantics the following notations are standard:

- Σ : alphabet of events.
- $\Sigma^{\checkmark} \triangleq \Sigma \cup \{\checkmark\}$: alphabet of events extended by the termination event $\checkmark$.
- Σ^* : all non-terminating traces over Σ.
- $\Sigma^{*\checkmark} \triangleq \Sigma^* \cup \{s \frown \langle\checkmark\rangle \mid s \in \Sigma^*\}$: all 'interesting' traces over Σ, i.e., the non-terminating and the terminating ones.

B.4.1 The Traces Model $\mathcal{T}$

In the traces model $\mathcal{T}$ the denotation of a process P over an alphabet Σ is a set of traces $T = traces(P)$ with

- $T \subseteq \Sigma^{*\checkmark}$.

The Domain $\mathcal{T}$ and Its Ordering

- Healthiness condition on process denotations T:

T1 $T \neq \emptyset$ and T is prefix-closed, i.e., $\forall s \in \Sigma^{*\checkmark}, t \in T . s \leq t \implies s \in T$

- Domain: $\mathcal{T} \triangleq \{T \subseteq \Sigma^{*\checkmark} \mid T\text{fulfills T1}\}$.
- $\mathcal{T}$ is a complete partial order with bottom element: $(\mathcal{T}, \subseteq, traces(Stop))$.
- Traces refinement:

  - $P \sqsubseteq_T Q \iff [\![Q]\!]_{\mathcal{T}} \subseteq [\![P]\!]_{\mathcal{T}}$.
  - The process $Stop$ refines all processes.

– The process $RUN\_With\_Skip_{\Sigma} = (?x : \Sigma \to RUN\_With\_Skip_{\Sigma}) \sqcap Skip$ is the least refined process.

Semantic Clauses

The function $traces(\_)$ maps CSP processes to observations in the traces domain $\mathcal{T}$. $traces(\_)$ is defined relatively to a function $M : PN \to \mathcal{T}$ that gives interpretations to process names. While we write $traces_M(\_)$ in the main text, it is common practice to omit the index M when writing semantical clauses.

$$
\begin{aligned}
traces(Stop) &\triangleq \{\langle\rangle\} \\
traces(Skip) &\triangleq \{\langle\rangle, \langle\checkmark\rangle\} \\
traces(Div) &\triangleq \{\langle\rangle\} \\
traces(N) &\triangleq M(N) \\
traces(a \to P) &\triangleq \{\langle\rangle\} \\
&\quad \cup \{\langle a\rangle \frown s \mid s \in traces(P)\} \\
traces(?x : A \to P) &\triangleq \{\langle\rangle\} \\
&\quad \cup \{\langle a\rangle \frown s \mid s \in traces(P[a/x]) \wedge a \in A\} \\
traces(P \Box Q) &\triangleq traces(P) \cup traces(Q) \\
traces(P \sqcap Q) &\triangleq traces(P) \cup traces(Q) \\
traces(\textbf{if } cond \textbf{ then } P \textbf{ else } Q) &\triangleq \begin{cases} traces(P) & ; cond \text{ evaluates to true} \\ traces(Q) & ; \text{otherwise} \end{cases} \\
traces(P \mathbin{\raisebox{0.2ex}{$\scriptstyle\text{\textbardbl}$}}_9 Q) &\triangleq (traces(P) \cap \Sigma^*) \\
&\quad \cup \{s \frown t \mid s \frown \langle\checkmark\rangle \in traces(P) \wedge \\
&\qquad\qquad t \in traces(Q)\} \\
traces(P \triangle Q) &\triangleq traces(P) \\
&\quad \cup \{s \frown t \mid s \in traces(P) \cap \Sigma^* \wedge \\
&\qquad\qquad t \in traces(Q)\} \\
traces(P \mathbin{[\![} A \mathbin{]\!]} Q) &\triangleq \bigcup \{s \mathbin{[\![} A \mathbin{]\!]} t \mid s \in traces(P) \wedge \\
&\qquad\qquad t \in traces(Q)\} \\
traces(P \setminus A) &\triangleq \{s \setminus A \mid s \in traces(P)\} \\
traces(P[\![R]\!]) &\triangleq \{t \mid \exists s \in traces(P). \, sR^*t\} \\
traces(\sqcap_{j \in J} P_j) &\triangleq \bigcup_{j \in J} traces(P_j)
\end{aligned}
$$

The auxiliary notations $t_1 \mathbin{[\![} A \mathbin{]\!]} t_2$, $t \setminus A$, and R^* are defined as follows:

- Given traces $t_1, t_2 \in \Sigma^{*\checkmark}$ and a set $A \subseteq \Sigma$, $t_1 \mathbin{[\![} A \mathbin{]\!]} t_2$ is inductively defined by:

$$\langle\rangle \;[\![\,A\,]\!]\; \langle\rangle \qquad\quad \triangleq \{\langle\rangle\}$$
$$\langle\rangle \;[\![\,A\,]\!]\; \langle x\rangle \frown t_2 \triangleq \emptyset$$
$$\langle\rangle \;[\![\,A\,]\!]\; \langle y\rangle \frown t_2 \triangleq \{\langle y\rangle \frown u \mid u \in \langle\rangle \;[\![\,A\,]\!]\; t_2\}$$
$$\langle x\rangle \frown t_1 \;[\![\,A\,]\!]\; \langle\rangle \qquad \triangleq \emptyset$$
$$\langle x\rangle \frown t_1 \;[\![\,A\,]\!]\; \langle x\rangle \frown t_2 \triangleq \{\langle x\rangle \frown u \mid u \in t_1 \;[\![\,A\,]\!]\; t_2\}$$
$$\langle x\rangle \frown t_1 \;[\![\,A\,]\!]\; \langle x'\rangle \frown t_2 \triangleq \emptyset$$
$$\langle x\rangle \frown t_1 \;[\![\,A\,]\!]\; \langle y\rangle \frown t_2 \triangleq \{\langle y\rangle \frown u \mid u \in \langle x\rangle \frown t_1 \;[\![\,A\,]\!]\; t_2\}$$
$$\langle y\rangle \frown t_1 \;[\![\,A\,]\!]\; \langle\rangle \qquad \triangleq \{\langle y\rangle \frown u \mid u \in t_1 \;[\![\,A\,]\!]\; \langle\rangle\}$$
$$\langle y\rangle \frown t_1 \;[\![\,A\,]\!]\; \langle x\rangle \frown t_2 \triangleq \{\langle y\rangle \frown u \mid u \in t_1 \;[\![\,A\,]\!]\; \langle x\rangle \frown t_2\}$$
$$\langle y\rangle \frown t_1 \;[\![\,A\,]\!]\; \langle y'\rangle \frown t_2 \triangleq \; \{\langle y\rangle \frown u \mid u \in t_1 \;[\![\,A\,]\!]\; \langle y'\rangle \frown t_2\}$$
$$\cup \{\langle y'\rangle \frown u \mid u \in \langle y\rangle \frown t_1 \;[\![\,A\,]\!]\; t_2\}$$

where $x, x' \in A \cup \{\checkmark\}$, $y, y' \notin A \cup \{\checkmark\}$, and $x \neq x'$.

- Given a trace $t \in \Sigma^{*\checkmark}$ and a set $A \subseteq \Sigma$, $(t \setminus A)$ is inductively defined by:

$$\langle\rangle \setminus A \triangleq \langle\rangle$$
$$(\langle x\rangle \frown t) \setminus A \triangleq \begin{cases} t \setminus A & \text{if } x \in A \\ \langle x\rangle \frown (t \setminus A) & \text{otherwise} \end{cases}$$

- Given a relation $R \subseteq \Sigma \times \Sigma$, $R^* \subseteq \Sigma^{*\checkmark} \times \Sigma^{*\checkmark}$ is defined to be the smallest set satisfying:

  - $(\langle\rangle, \langle\rangle) \in R^*$
  - $(\langle\checkmark\rangle, \langle\checkmark\rangle) \in R^*$
  - $(a,b) \in R \wedge (t,t') \in R^* \implies (a \frown t, b \frown t') \in R^*$

B.4.2 The Failures/Divergences Model $\mathcal{N}$

In the failures/divergences model $\mathcal{N}$, the denotation of a process P over an alphabet Σ is given by a pair $(F, D) = (\mathit{failures}_\perp(P), \mathit{divergences}(P))$ with

- $F \subseteq \Sigma^{*\checkmark} \times \mathcal{P}(\Sigma^\checkmark)$ and
- $D \subseteq \Sigma^{*\checkmark}$.

Some healthiness conditions over $\mathcal{N}$ recur to the function $\mathit{traces}_\perp(\_)$: $\Sigma^{*\checkmark} \times \mathcal{P}(\Sigma^\checkmark) \to \Sigma^{*\checkmark}$, defined as follows:

$$\mathit{traces}_\perp(F) \triangleq \{s \mid \exists X \subseteq \Sigma^{*\checkmark}. (s, X) \in F\}$$

The Domain $\mathcal{N}$ and Its Ordering

- Healthiness conditions on process denotations (F, D):

F1 $traces_\perp(F) \neq \emptyset$ and $traces_\perp(F)$ is prefix-closed, i.e.,
$\forall s \in \Sigma^{*\checkmark},\, t \in traces_\perp(F) \,.\, s \leq t \implies s \in traces_\perp(F)$

F2 $\forall (s, X) \in F,\, Y \subseteq \Sigma^\checkmark \,.\, Y \subseteq X \implies (s, Y) \in F$

F3 $\forall (s, X) \in F,\, Y \subseteq \Sigma^\checkmark \,.$
$(\forall b \in Y \,.\, s \frown \langle b \rangle \notin traces_\perp(F)) \implies (s, X \cup Y) \in F$

F4 $\forall s \in \Sigma^* \,.\, s \frown \langle \checkmark \rangle \in traces_\perp(F) \implies (s, \Sigma) \in F$

D1 D is extension closed, i.e., $\forall s \in D \cap \Sigma^*,\, t \in \Sigma^{*\checkmark} \,.\, s \frown t \in D$

D2 $\forall s \in D,\, X \subseteq \Sigma^\checkmark \,.\, (s, X) \in F$

D3 $\forall s \in \Sigma^* \,.\, s \frown \langle \checkmark \rangle \in D \implies s \in D$

- Domain: $\mathcal{N} \triangleq \{ (F, D) \subseteq (\Sigma^{*\checkmark} \times \mathcal{P}(\Sigma^\checkmark)) \times \Sigma^{*\checkmark} \mid$
  (F, D) fulfills F1, F2, F3, F4, D1, D2, D3$\}$
- For finite Σ it holds that $\mathcal{N}$ is a complete partial order with bottom element:
$$(\mathcal{N}, \supseteq \times \supseteq, (failures_\perp(Div), divergences(Div)).$$
- Failures/divergences refinement:

  - $P \sqsubseteq_\mathcal{N} Q \iff failures_\perp(Q) \subseteq failures_\perp(P) \wedge$
    $divergences(Q) \subseteq divergences(P)$
  - There is no most refined process over $\mathcal{N}$.
  - The process Div is the least refined process.

Semantic Clauses

Together, the functions $failures_\perp(\_\_)$ and $divergences(\_\_)$ map CSP processes to observations in the failures/divergences domain $\mathcal{N}$. These functions are defined relatively to a function $M = (M_{failures_\perp}, M_{divergence}) : PN \rightarrow \mathcal{N}$ that gives interpretations to process names. While we write $failures_{\perp,M}(\_\_)$ and $divergences_M(\_\_)$ in the main text, it is common practice to omit the index M when writing semantical clauses.

In the context of semantic clauses over $\mathcal{N}$ we overload the function $traces_\perp(\_\_)$ such that it takes CSP processes as its input:

$$traces_\perp(P) \triangleq \{s \mid \exists X.(s, X) \in failures_\perp(P)\}$$

In some cases, e.g., for the external choice operator, the expressions determining the function $failures_\perp(\_\_)$ utilise the function $divergences(\_\_)$. In spite of this, we always first present the function computing the failures followed by the presentation of the function computing the divergences.

The definition of the renaming operator applies the inverse of a binary relation R to a set X:

$$R^{-1}(X) \triangleq \{a \mid \exists a' \in X.a\,R\,a'\}.$$

As in $\mathcal{N}$ it is in general not possible to give the correct value for the hiding operator, we do not give a clause for it. We refrain from giving a semantic clause for the interrupt operator in $\mathcal{N}$.

$$
\begin{aligned}
failures_\perp(Stop) &\triangleq \{(\langle\rangle, X) \mid X \subseteq \Sigma^\checkmark\} \\
divergences(Stop) &\triangleq \{\} \\
failures_\perp(Skip) &\triangleq \{(\langle\rangle, X) \mid X \subseteq \Sigma\} \cup \{(\langle\checkmark\rangle, X) \mid X \subseteq \Sigma^\checkmark\} \\
divergences(Skip) &\triangleq \{\} \\
failures_\perp(Div) &\triangleq \Sigma^{*\checkmark} \times \mathcal{P}(\Sigma^\checkmark) \\
divergences(Div) &\triangleq \Sigma^{*\checkmark} \\
failures_\perp(N) &\triangleq M_{failures_\perp}(N) \\
divergences(N) &\triangleq M_{divergences}(N) \\
failures_\perp(a \rightarrow P) &\triangleq \{(\langle\rangle, X) \mid X \subseteq \Sigma^\checkmark \wedge a \notin X\} \\
&\cup \{(\langle a\rangle \frown s, X) \mid (s, X) \in failures_\perp(P)\} \\
divergences(a \rightarrow P) &\triangleq \{\langle a\rangle \frown t \mid t \in divergences(P)\} \\
failures_\perp(?x : A \rightarrow P) &\triangleq \{(\langle\rangle, X) \mid X \subseteq \Sigma^\checkmark \wedge X \cap A = \{\}\} \\
&\cup \{(\langle a\rangle \frown s, X) \mid a \in A \wedge \\
&\qquad\qquad (s, X) \in failures_\perp(P[a/x])\} \\
divergences(?x : A \rightarrow P) &\triangleq \{\langle a\rangle \frown t \mid a \in A \wedge t \in divergences(P[a/x])\}
\end{aligned}
$$

$$
\begin{aligned}
failures_\perp(P \mathbin{\square} Q) &\triangleq \{(\langle\rangle, X) \mid (\langle\rangle, X) \in \\
&\qquad failures_\perp(P) \cap failures_\perp(Q)\} \\
&\cup \{(t, X) \mid t \neq \langle\rangle \wedge \\
&\qquad ((t, X) \in failures_\perp(P) \vee \\
&\qquad (t, X) \in failures_\perp(Q))\} \\
&\cup \{(\langle\rangle, X) \mid X \subseteq \Sigma^\checkmark \wedge \\
&\qquad (\langle\rangle \in divergences(P) \vee \\
&\qquad \langle\rangle \in divergences(Q))\} \\
&\cup \{(\langle\rangle, X) \mid X \subseteq \Sigma \wedge \\
&\qquad (\langle\checkmark\rangle \in traces_\perp(P) \vee \\
&\qquad \langle\checkmark\rangle \in traces_\perp(Q))\} \\
divergences(P \mathbin{\square} Q) &\triangleq divergences(P) \cup divergences(Q) \\
failures_\perp(P \sqcap Q) &\triangleq failures_\perp(P) \cup failures_\perp(Q) \\
divergences(P \sqcap Q) &\triangleq divergences(P) \cup divergences(Q)
\end{aligned}
$$

$$
failures_\perp(\textbf{if } cond \textbf{ then } P \textbf{ else } Q) \triangleq
\begin{cases}
failures_\perp(P) \text{ ; } cond \\
\qquad\qquad \text{evaluates to true} \\
failures_\perp(Q) \text{ ; otherwise}
\end{cases}
$$

$$
divergences(\textbf{if } cond \textbf{ then } P \textbf{ else } Q) \triangleq
\begin{cases}
divergences(P) \text{ ; } cond \\
\qquad\qquad \text{evaluates to true} \\
divergences(Q) \text{ ; otherwise}
\end{cases}
$$

$$
\begin{aligned}
failures_\perp(P \mathbin{\raisebox{0.2ex}{\tiny 9}} Q) \;\triangleq\; & \{(s, X) \mid s \in \Sigma^* \wedge \\
& \quad (s, X \cup \{\checkmark\}) \in failures_\perp(P)\} \\
& \cup\, \{(s \frown t, X) \mid s \frown \langle\checkmark\rangle \in traces_\perp(P) \wedge \\
& \quad (t, X) \in failures_\perp(Q)\} \\
& \cup\, \{(s, X) \mid s \in divergences(P \mathbin{\raisebox{0.2ex}{\tiny 9}} Q) \wedge \\
& \quad X \subseteq \Sigma^{\checkmark}\} \\[4pt]
divergences(P \mathbin{\raisebox{0.2ex}{\tiny 9}} Q) \;\triangleq\; & divergences(P) \\
& \cup\, \{s \frown t \mid s \frown \langle\checkmark\rangle \in traces_\perp(P) \wedge \\
& \quad t \in divergences(Q)\}
\end{aligned}
$$

$$
\begin{aligned}
failures_\perp(P \,[\!|\, A \,|\!]\, Q) \;\triangleq\; & \{(u, Y \cup Z) \mid Y, Z \subseteq \Sigma^{\checkmark} \wedge \\
& \quad Y\backslash(A \cup \{\checkmark\}) = Z\backslash(A \cup \{\checkmark\}) \wedge \\
& \quad \exists s, t \in \Sigma^{*\checkmark}.\, u \in (s \,[\!|\, A \,|\!]\, t) \wedge \\
& \quad (s, Y) \in failures_\perp(P) \wedge \\
& \quad (t, Z) \in failures_\perp(Q)\} \\
& \cup\, \{(u, Y) \mid u \in divergences(P \,[\!|\, A \,|\!]\, Q) \wedge Y \subseteq \Sigma^{*\checkmark}\} \\
divergences(P \,[\!|\, A \,|\!]\, Q) \;\triangleq\; & \{u \frown v \mid v \in \Sigma^{*\checkmark} \wedge \\
& \quad \exists s \in traces_\perp(P), t \in traces_\perp(Q). \\
& \quad u \in (s \,[\!|\, A \,|\!]\, t) \cap \Sigma^* \wedge \\
& \quad (s \in divergences(P) \vee \\
& \quad t \in divergences(Q))\} \\
failures_\perp(P[\![R]\!]) \;\triangleq\; & \{(s', X) \mid \exists s.s R^* s' \wedge \\
& \quad (s, R^{-1}(X)) \in failures_\perp(P)\} \\
& \cup\, \{(s, X) \mid s \in divergences(P[\![R]\!]) \wedge \\
& \quad X \subseteq \Sigma^{\checkmark}\} \\
divergences(P[\![R]\!]) \;\triangleq\; & \{s' \frown t \mid \exists s \in divergences(P) \cap \Sigma^*. \\
& \quad s R^* s' \wedge t \in \Sigma^{*\checkmark}\} \\
failures_\perp(\sqcap_{j \in J} P_j) \;\triangleq\; & \bigcup_{j \in J} failures_\perp(P_j) \\
divergences(\sqcap_{j \in J} P_j) \;\triangleq\; & \bigcup_{j \in J} divergences(P_j)
\end{aligned}
$$

B.4.3 The Stable Failures Model $\mathcal{F}$

In the stable failures model $\mathcal{F}$ the denotation of a process P over an alphabet Σ is a pair $(T, F) = (traces(P), failures(P))$ with

- $T \subseteq \Sigma^{*\checkmark}$ and
- $F \subseteq \Sigma^{*\checkmark} \times \mathcal{P}(\Sigma^{\checkmark})$.

The Domain $\mathcal{F}$ and Its Ordering

- Healthiness conditions on process denotations (T, F):

T1 $T \neq \emptyset$ and T is prefix-closed, i.e., $\forall s \in \Sigma^{*\checkmark}, t \in T . s \leq t \implies s \in T$

T2 $\forall (t, X) \in \Sigma^{*\checkmark} \times \mathcal{P}(\Sigma^{\checkmark}) . (t, X) \in F \implies t \in T$

T3 $\forall t \in \Sigma^*, X \subseteq \Sigma^{\checkmark} . t ^\frown \langle \checkmark \rangle \in T \implies (t ^\frown \langle \checkmark \rangle, X) \in F$

F2 $\forall (s, X) \in F, Y \subseteq \Sigma^{\checkmark} . Y \subseteq X \implies (s, Y) \in F$

F3 $\forall (s, X) \in F, Y \subseteq \Sigma^{\checkmark} .$
$(\forall b \in Y . s ^\frown \langle b \rangle \notin T) \implies (s, X \cup Y) \in F$

F4 $\forall s \in \Sigma^* . s ^\frown \langle \checkmark \rangle \in T \implies (s, \Sigma) \in F$

- Domain:

$$\mathcal{F} \triangleq \{ (T, F) \subseteq \Sigma^{*\checkmark} \times (\Sigma^{*\checkmark} \times \mathcal{P}(\Sigma^{\checkmark})) \mid$$
$$(T, F) \text{ fulfills T1, T2, T3, F2, F3, F4}\}.$$

- $\mathcal{F}$ is a complete partial order with bottom element:

$$(\mathcal{F}, \subseteq \times \subseteq, (traces(Div), failures(Div)))$$

- Refinement:

  - $P \sqsubseteq_{\mathcal{F}} Q \iff traces(Q) \subseteq traces(P) \wedge failures(Q) \subseteq failures(P)$
  - The process Div refines all processes.
  - The process $RUN_{\Sigma}^{+} = (?x : \Sigma \to RUN_{\Sigma}^{+}) \sqcap Stop \sqcap Skip$ is the least refined process. Over $\mathcal{F}$ this process can equivalentely be represented by $CHAOS\_tick = (\sqcap x : \Sigma@x \to CHAOS\_tick) \sqcap Stop \sqcap Skip$. As $CHAOS\_tick$ solely is built using the internal choice operator, in proofs it behaves 'better' than its counterpart RUN_{Σ}^{+}.

Semantic Clauses

Together, the functions $traces(\_)$ and $failures(\_)$ map CSP processes to observations in the stable failures domain $\mathcal{F}$. These functions are defined relatively to a function $M = (M_{traces}, M_{failures}) : PN \to \mathcal{F}$ that gives interpretations to process names. While we write $traces_M(\_\_)$ and $failures_M(\_\_)$ in the main text, it is common practice to omit the index M when writing semantical clauses.

The definition of the renaming operator applies the inverse of a binary relation R to a set X:

$$R^{-1}(X) \triangleq \{a \mid \exists a' \in X . a \, R \, a'\}.$$

Again, we refrain from giving a semantic clause for the interrupt operator in $\mathcal{F}$.

$$
\begin{aligned}
traces(Stop) \quad &\triangleq \quad \{\langle\rangle\} \\
failures(Stop) \quad &\triangleq \quad \{(\langle\rangle, X) \mid X \subseteq \Sigma^{\checkmark}\} \\
traces(Skip) \quad &\triangleq \quad \{\langle\rangle, \langle\checkmark\rangle\} \\
failures(Skip) \quad &\triangleq \quad \{(\langle\rangle, X) \mid X \subseteq \Sigma\} \\
& \qquad \cup \{(\langle\checkmark\rangle, X) \mid X \subseteq \Sigma^{\checkmark}\} \\
traces(Div) \quad &\triangleq \quad \{\langle\rangle\} \\
failures(Div) \quad &\triangleq \quad \{\} \\
traces(N) \quad &\triangleq \quad M_{traces}(N) \\
failures(N) \quad &\triangleq \quad M_{failures}(N) \\
traces(a \to P) \quad &\triangleq \quad \{\langle\rangle\} \\
& \qquad \cup \{\langle a\rangle \frown s \mid s \in traces(P)\} \\
failures(a \to P) \quad &\triangleq \quad \{(\langle\rangle, X) \mid a \notin X\} \\
& \qquad \cup \{(\langle a\rangle \frown t', X) \mid (t', X) \in failures(P)\} \\
traces(?x : A \to P) \quad &\triangleq \quad \{\langle\rangle\} \\
& \qquad \cup \{\langle a\rangle \frown s \mid a \in A \wedge \\
& \qquad\qquad s \in traces(P[a/x])\} \\
failures(?x : A \to P) \quad &\triangleq \quad \{(\langle\rangle, X) \mid A \cap X = \emptyset\} \\
& \qquad \cup \{(\langle a\rangle \frown t', X) \mid a \in A \wedge \\
& \qquad\qquad (t', X) \in failures(P[a/x])\} \\
traces(P \,\square\, Q) \quad &\triangleq \quad traces(P) \cup traces(Q) \\
failures(P \,\square\, Q) \quad &\triangleq \quad \{(\langle\rangle, X) \mid (\langle\rangle, X) \in \\
& \qquad\qquad failures(P) \cap failures(Q)\} \\
& \qquad \cup \{(t, X) \mid t \neq \langle\rangle \wedge \\
& \qquad\qquad (t, X) \in failures(P) \cup failures(Q)\} \\
& \qquad \cup \{(\langle\rangle, X) \mid X \subseteq \Sigma \wedge \\
& \qquad\qquad \langle\checkmark\rangle \in traces(P) \cup traces(Q)\} \\
traces(P \,\sqcap\, Q) \quad &\triangleq \quad traces(P) \cup traces(Q) \\
failures(P \,\sqcap\, Q) \quad &\triangleq \quad failures(P) \cup failures(Q)
\end{aligned}
$$

$$traces(\textbf{if } cond \textbf{ then } P \textbf{ else } Q) \quad \triangleq \quad \begin{cases} traces(P) \text{ ; } cond \text{ evaluates to true} \\ traces(Q) \text{ ; otherwise} \end{cases}$$

$$failures(\textbf{if } cond \textbf{ then } P \textbf{ else } Q) \triangleq \begin{cases} failures(P) \text{ ; } cond \text{ evaluates to true} \\ failures(Q) \text{ ; otherwise} \end{cases}$$

$$traces(P \mathbin{\fatsemi} Q) \quad \triangleq \quad (traces(P) \cap \Sigma^*)$$
$$\cup \{s \mathbin{\frown} t \mid s \mathbin{\frown} \langle \checkmark \rangle \in traces(P) \wedge t \in traces(Q)\}$$
$$failures(P \mathbin{\fatsemi} Q) \triangleq \quad \{(t_1, X) \mid t_1 \in \Sigma^* \wedge (t_1, X \cup \{\checkmark\}) \in failures(P)\}$$
$$\cup \{(t_1 \mathbin{\frown} t_2, X) \mid t_1 \mathbin{\frown} \langle \checkmark \rangle \in traces(P) \wedge$$
$$(t_2, X) \in failures(Q)\}$$

$$traces(P \,[\hspace{-1pt}|\, A \,|\hspace{-1pt}]\, Q) \quad \triangleq \quad \bigcup \{s \,[\hspace{-1pt}|\, A \,|\hspace{-1pt}]\, t \mid s \in traces(P) \wedge t \in traces(Q)\}$$
$$failures(P \,[\hspace{-1pt}|\, A \,|\hspace{-1pt}]\, Q) \triangleq \quad \{(u, Y \cup Z) \mid Y \setminus (A \cup \{\checkmark\}) = Z \setminus (A \cup \{\checkmark\}) \wedge$$
$$\exists t_1, t_2. \ u \in t_1 \,[\hspace{-1pt}|\, A \,|\hspace{-1pt}]\, t_2 \wedge$$
$$(t_1, Y) \in failures(P) \wedge (t_2, Z) \in failures(Q)\}$$
$$traces(P \setminus A) \quad\quad \triangleq \quad \{s \setminus A \mid s \in traces(P)\}$$
$$failures(P \setminus A) \quad\quad \triangleq \quad \{(t \setminus A, Y) \mid (t, Y \cup A) \in failures(P)\}$$
$$traces(P[\![R]\!]) \quad\quad \triangleq \quad \{t \mid \exists s \in traces(P) : sR^*t\}$$
$$failures(P[\![R]\!]) \quad\quad \triangleq \quad \{(t, X) \mid \exists t'. (t', t) \in R^* \wedge$$
$$(t', R^{-1}(X)) \in failures(P)\}$$
$$traces(\sqcap_{j \in J} P_j) \quad\quad \triangleq \quad \bigcup_{j \in J} traces(P_j)$$
$$failures(\sqcap_{j \in J} P_j) \quad\quad \triangleq \quad \bigcup_{j \in J} failures(P_j)$$

Appendix C
Concrete CASL Syntax

In this appendix we provide a grammar for the concrete syntax of the discussed sublanguage of CASL. The grammar of the full language can be found in Peter D. Mosses, CASL Reference Manual, Springer, 2004. We take the freedom to resolve some chain rules, to add a start symbol, to simply the grammar of terms as well as the lexical syntax. However, overall our presentation follows closely the one of the CASL reference manual. We also use the following conventions established in the CASL context:

- Nonterminal symbols are written as uppercase words, possibly hyphenated, e.g., SORT, BASIC-SPEC.
- Terminal symbols are written as either:

  - lowercase words, e.g., free, op; or
  - special sequences of characters that are enclosed in double-quotes, e.g., ".", "::=".

- Optional symbols are followed by a / , e.g., end/. Options of terminal symbols are also separated by a / , e.g., sort/sorts.
- Alternative sequences are separated by vertical bars, e.g., true | false.
- Repetitions are indicated by ellipsis ... when between symbols; ellipses are also used to indicate omissions to the grammar when at the end of an alternative.
- Production rules are written with the nonterminal symbol followed by ::=, followed by one or more alternatives.

The lexical syntax of identifiers is given by the nonterminal symbol WORD. A WORD must start with a letter, and must not be a reserved key word.

Our grammar offers minimal structuring only—for the concrete syntax of structuring operations see the CASL reference manual.

© Springer Nature Switzerland AG 2022
M. Roggenbach et al., *Formal Methods for Software Engineering*,
Texts in Theoretical Computer Science. An EATCS Series,
https://doi.org/10.1007/978-3-030-38800-3

C.1 Specifications

```
NAMED-SPEC ::= spec SPEC-NAME "=" BASIC-SPEC end/
             | spec SPEC-NAME "="
                  SPEC-NAME then...then SPEC-NAME
                  then
                  BASIC-SPEC then...then BASIC-SPEC end/
SPEC-NAME ::= WORD
```

C.2 Signature Declarations

```
BASIC-SPEC      ::= BASIC-ITEMS...BASIC-ITEMS
BASIC-ITEMS     ::= SIG-ITEMS
                  | ...
SIG-ITEMS       ::= sort/sorts SORT-ITEM ";"..."; " SORT-ITEM ";"/
                  | op/ops OP-ITEM ";"..."; " OP-ITEM ";"/
                  | pred/preds PRED-ITEM ";"..."; " PRED-ITEM ";"/
SORT-ITEM       ::= SORT ","..."," SORT
OP-ITEM         ::= OP-NAME ","..."," OP-NAME ":" OP-TYPE
                  | ...
OP-TYPE         ::= SOME-SORTS "->"  SORT
                  | SOME-SORTS "->?" SORT
                  | SORT
                  | "?" SORT
SOME-SORTS      ::= SORT "*"..."*" SORT
PRED-ITEM       ::= PRED-NAME ","..."," PRED-NAME ":" PRED-TYPE
                  | ...
PRED-TYPE       ::= SORT "*"..."*" SORT | "()"
SORT            ::= WORD
OP-NAME         ::= WORD
PRED-NAME       ::= WORD
```

C.3 Formulae

We extend the notion of BASIC-ITEMS by constructs for formulae:

```
BASIC-ITEMS ::= ...
              | forall VAR-DECL ";"...";" VAR-DECL
                 "." FORMULA "."..."." FORMULA ";"/
              | axiom/axioms FORMULA ";"...";" FORMUA ";"/
              | ...
VAR-DECL    ::= VAR ","..."," VAR ":" SORT
FORMULA     ::= PRED-NAME "(" TERM ","..."," TERM ")"
              | TERM "=" TERM
              | TERM "=e=" TERM
              | def TERM
              | true
```

```
             | false
             | not FORMULA
             | FORMULA "/\" ... "/\" FORMULA
             | FORMULA "\/" ... "\/" FORMULA
             | FORMULA "=>" FORMULA
             | FORMULA if FORMULA
             | FORMULA "<=>" FORMULA
             | QUANTIFIER  VAR-DECL ";"..."; " VAR-DECL "." FORMULA
             | "(" FORMULA ")"
             | ...
QUANTIFIER  ::= forall | exists
             | ...
TERM        ::= OP-NAME
             | OP-NAME "(" TERM ","..."," TERM ")"
VAR         ::= WORD
```

C.4 Sort Generation Constraints

We extend the notion of BASIC-ITEMS by a construct for writing a sort generation constraint:

```
BASIC-ITEMS ::= ...
             | free type SORT "::=" ALTERNATIVE "|"..."|" ALTERNATIVE
             | ...
ALTERNATIVE ::= OP-NAME "(" COMPONENT ";"..."; " COMPONENT ")"
             | OP-NAME
             | ...
COMPONENT   ::= OP-NAME ":" SORT ","..."," OP-NAME ":" SORT
             | SORT
             | ...
```

Index

© Springer Nature Switzerland AG 2022
M. Roggenbach et al., *Formal Methods for Software Engineering*,
Texts in Theoretical Computer Science. An EATCS Series,
https://doi.org/10.1007/978-3-030-38800-3

Printed in the United States
by Baker & Taylor Publisher Services